# TRACE ELEMENTS IN HUMAN AND ANIMAL NUTRITION

THIRD EDITION

# TRACE ELEMENTS IN HUMAN AND ANIMAL NUTRITION

---

THIRD EDITION


**E. J. UNDERWOOD**

INSTITUTE OF AGRICULTURE
UNIVERSITY OF WESTERN AUSTRALIA
NEDLANDS, WESTERN AUSTRALIA



1971

ACADEMIC PRESS New York and London

A Subsidiary of Harcourt Brace Jovanovich, Publishers

ACADEMIC PRESS, INC.
111 Fifth Avenue, New York, New York 10003

*United Kingdom Edition published by*
ACADEMIC PRESS, INC. (LONDON) LTD.
24/28 Oval Road, London NW1

LIBRARY OF CONGRESS CATALOG CARD NUMBER: 75-137611

PRINTED IN THE UNITED STATES OF AMERICA

*To my wife, Erica, for her help*
*and for her continuing encouragement.*

# CONTENTS

## 8 ZINC



## 9 CHROMIUM



## 10 CADMIUM



## 11 IODINE



## 12 SELENIUM

## 13 FLUORINE



## 14 SILICON



## 15 VANADIUM



## 16 OTHER ELEMENTS

## 17 SOIL–PLANT–ANIMAL INTERRELATIONS

# PREFACE

The nine-year period since the second edition was written has witnessed a surge of interest and activity in almost every aspect of trace element research. Remarkable advances in analytical techniques have occurred enabling trace element concentrations to be determined with a speed, accuracy, and sensitivity previously impossible. Several trace elements, notably zinc and chromium, were shown to be more important to human health and nutrition than was earlier believed; the basic biochemical defects associated with deficiencies of other elements, particularly copper and manganese, were delineated; and increasing interest was aroused in the possible chronic deleterious effects of trace element burdens arising from the growing industrialization, motorization, and urbanization of large sections of the human race.

These developments have necessitated a considerable expansion of the chapters dealing with copper, zinc, manganese, and selenium; an enlargement of cadmium, chromium, nickel, silicon, and vanadium into separate chapters; and the introduction of entirely new sections on lead, mercury, tin, germanium, and zirconium. The general approach of the earlier editions has nevertheless been maintained, and the aim of the book remains, as before, namely, to enable specialists in nutrition to obtain a balanced appreciation of the nutritional roles of the trace elements and of their biochemical and pathological significance to animals and man.

This work could not have materialized without the help and encouragement of many individuals in Australia and overseas. In this connection I wish to mention especially Dr. W. H. Allaway who kept me informed of many aspects which I might otherwise have overlooked and Dr. E. H. Morgan who critically read the chapter on iron and made all his literature on this topic freely available. Finally, it is a particular pleasure to record the debt owed to Miss M. Keane who, with the help of Mrs. E. Maskell, again undertook the formidable task of typing the manuscript and checking the references.

E. J. UNDERWOOD

# 1

# INTRODUCTION

## I. The Nature of Trace Elements

Many mineral elements occur in living tissues in such small amounts that the early workers were unable to measure their precise concentrations with the analytical methods then available. For this reason they were often mentioned as occurring in "traces" and the term trace elements arose to describe them. This term has remained in popular usage despite the fact that virtually all the trace elements can now be estimated with great accuracy and precision. Other popular designations such as "minor elements" have found favor with plant nutritionists and "oligo elements" (from the Greek *oligos* meaning scanty) is a term frequently used by French workers. None of these terms is fully acceptable for obvious reasons. Nor is the more scientifically respectable "micronutrient elements" entirely satisfactory, since it implies that all the elements present in living organisms in low concentrations, and, therefore, of biological interest, function as nutrients. The term "trace elements" is retained here because it is brief, because it has historical associations, and because it has become hallowed by time and by custom.

A meaningful classification of the trace elements is difficult because the only characteristic that they have in common is their occurrence in the tissues of plants, animals, and microorganisms in low concentrations.

The magnitude of these concentrations varies greatly among different elements and, for particular elements, among different living organisms and parts of such organisms. Such differences apply equally to those elements known to be nutritionally essential and those for which no vital function has yet been found. Thus the requirements of mammals for zinc and copper are many times greater than those for iodine and selenium, and the concentrations of iron and zinc in animal tissues are normally much greater than those of manganese and cobalt. Furthermore, some apparently nonessential elements such as bromine, rubidium, and silicon frequently occur in the blood and soft tissues of the body at levels substantially higher than those of most of the known essential trace elements. Typical concentrations of the essential trace elements in two fluids of great biological importance, namely human blood plasma and cow's milk, are presented in Table 1.

The trace elements can be divided conveniently into three groups: the dietary essentials, the possibly essentials, and the nonessentials. At the present time 10 trace elements are known to be essential for the higher animals. These are iron, iodine, copper, zinc, manganese, cobalt, molybdenum, selenium, chromium, and tin. In addition, several elements, notably nickel, fluorine, bromine, arsenic, vanadium, cadmium, barium, and strontium can be considered as possibly essential on the basis of suggestive but not completely convincing evidence. Thus the fluoride ion, at appropriate intakes, assists in the prevention of human dental caries and in the maintenance of a normal skeleton in man, but a fluorine-deficient diet that limits growth, health, or well-being in animals in the conventional manner has not been produced. Bromine has been shown to induce a small but significant growth increase in chicks fed a semisynthetic diet and in mice fed a diet containing iodinated casein (11). The omission of either barium or strontium from a complete mineral supplement to a highly purified diet fed to rats and guinea pigs was reported to result in a growth depression, and the omission of strontium alone to result in impaired calcification of the bones and teeth (20). These claims need to be substantiated before bromine, barium, or strontium can be placed securely among the essential trace elements.

According to Cotzias (6), a trace element can be dignified with the formal title "essential" if it meets the following criteria: (*1*) it is present in all healthy tissues of all living things; (*2*) its concentration from one animal to the next is fairly constant, (*3*) its withdrawal from the body induces reproducibly the same structural and physiological abnormalities regardless of the species studied; (*4*) its addition either prevents or reverses these abnormalities, (*5*) the abnormalities induced by deficiency are always accompanied by pertinent, specific biochemical changes; and

TABLE 1

*Typical Essential Trace Element Levels in Normal Human Blood Plasma and in Cow's Milk*[a]

| Fluid | Iron | Zinc | Copper | Manganese | Molybdenum | Iodine | Cobalt | Selenium | Chromium |
|---|---|---|---|---|---|---|---|---|---|
| Plasma | 1.2 | 1.2 | 1.0 | – | 0.04 | 0.08[b] | – | 0.2 | 0.15 |
| Milk | 0.5 | 3.5 | 0.2 | 0.02 | 0.06 | 0.05 | 0.0005 | 0.2 | 0.01 |

[a] Measured in μg/ml.
[b] Protein-bound iodine.

(*6*) these biochemical changes can be prevented or cured when the deficiency is prevented or cured.

Some 20 to 30 trace elements which do not meet these exacting requirements occur more or less constantly in highly variable concentrations in living tissues. They include aluminum, antimony, mercury, cadmium, germanium, vanadium, silicon, rubidium, silver, gold, lead, bismuth, titanium, and others. Such elements are believed to be acquired by the animal body as environmental contaminants and to reflect the contact of the organism with its environment. Skewed (log-normal) distribution patterns have been reported for the concentrations of these elements in human organs, whereas the essential elements have a normal distribution (15, 19, 23). In fact, the shape of the distribution curve for a trace element in tissue has been proposed as a method of determining whether or not the element is essential (15). Liebscher and Smith (6) have provided evidence to support this claim, based upon a study of the levels of several essential and nonessential trace elements in tissues from healthy adults who died as a result of violence and who had no known industrial exposure to the elements in question. It was suggested that the difference in distribution patterns of the two types of elements might be explained by different ways of incorporation and retention. For the essential elements an internal control mechanism is postulated, leading to a normal or symmetrical distribution. For the nonessential elements, arising from contamination and having no significant function, external control of tissue concentration would occur, leading to a distribution pattern similar to the environmental level.

The trace elements are sometimes classified into a further group, known as toxic elements. This term is perhaps justified for a small number of elements, such as arsenic, lead, cadmium, and mercury, the biological significance of which is at present largely confined to their toxic properties at relatively low concentrations. However, this classification has little general application because all the trace elements are toxic if ingested at sufficiently high levels and for a long enough period. In fact every element has a whole spectrum of actions, depending upon the dose and the nutritional state of the animal or system in respect to the element in question. This was recognized many years ago by the great French investigator, Gabriel Bertrand, who formulated such dose dependence into Bertrand's Law (2). More recently Venchikov (25) expanded this concept and presented the dose response in the form of a curve with two maxima. The first part of the curve, showing an increasing effect with increasing concentrations until a plateau is reached, expresses the *biological* action of the element, and the plateau expresses optimal supplementation and normal function. The width of the plateau is determined by the homeostatic capacity of the animal or system. With further increasing doses the element enters a phase of irritation and stimulation of some function, expressing what may be called its *pharmacological* action. In this phase the element acts as a drug independent of a deficiency state. This is followed, at still higher doses, by the appearance of signs of toxicity and perhaps by death, expressing the *toxic* action of the element. These different phases of action differ widely among different trace elements. With some, such as fluorine in man and copper in sheep, the margin between doses or intakes expressing the biological and toxic actions of the elements can be exceedingly small.

## II. Discovery of Trace Elements

The first indications that trace elements are of importance in animal physiology were obtained over a century ago in Europe. At this time interest centered upon a number of special compounds which were shown to contain various metals not previously suspected of biological significance. These included turacin, a copper-containing porphyrin occurring in the feathers of certain birds (5); hemocyanin, another copper compound found in the blood of snails (9); sycotypin, a zinc-containing blood pigment in Mollusca (18); and a vanadium-containing respiratory compound present in the blood of sea squirts (10). At that time these substances were regarded largely as scientific curiosities and their discovery did little to stimulate studies of the possible wider significance of the component elements. This was to come from the investigations of Claude Bernard (1) and MacMunn (16) on cell respiration and iron and oxidative processes which pointed the way to later studies of metal-enzyme catalysis and of the existence of a range of trace element-containing metalloenzymes of profound importance to the structural and functional integrity of living cells. This early period also saw the publication of the remarkable observations of the French botanist, Chatin, on the iodine content of soils, waters, and foods and his conclusion that the occurrence of goiter in man was associated with a deficiency of environmental iodine (4). More than 50 years were to elapse before these pioneer researches were fully accepted and the existence of "area" problems involving naturally occurring trace element deficiences and toxicities in man and farm animals became firmly established.

During the first quarter of this century, progress was slow. Further studies were initiated on iron and iodine in human health and nutrition, and several investigations were undertaken of the mineral composition of plant and animal tissues, employing the then new technique of emission spectrography (7, 26). This technique proved particularly valuable because it permits the simultaneous estimation of some 20 to 30 elements in low concentrations, thus enabling the patterns of distribution in living organisms to be determined in health and disease. This type of investigation entered a new phase some 20 years later, after World War II. At this time it became apparent, as a result of the development of high-energy accelerators and nuclear reactors, that all elements could be made radioactive and thus potential hazards to living cells. Information on the biological distribution of the stable elements then became a necessary prelude to assessment of the maximum safe levels of the radionuclides and their movement in the food chain from the soil to man. Extensive investigations were then carried out of the trace element levels in human and animal tissues and in the foods, waters, and atmosphere from which they are derived (3, 12, 13, 19, 22, 23).

These and many other distribution studies referred to in later chapters dealing with particular elements, were responsible for (*1*) defining the wide limits of concentration of many of the trace elements in foods and animal tissues, (*2*) illuminating the significance of such factors as age, location, disease, and industrial contamination in influencing these concentrations, (*3*) stimulating studies of the possible physiological significance of several elements previously unsuspected of biological potentiality, and (*4*) discriminating between those elements most likely to have such potential and those which, on the basis of their distribution patterns, were more likely to be environmental contaminants.

The second quarter of this century was notable for spectacular advances in our knowledge of the nutritional importance of the trace elements. In the early 1920s, Bertrand of France and McHargue of the United States pioneered the purified diet approach to animal studies with these elements. Their findings were mostly inconclusive because the purified diets they used were so deficient in other respects that, even with the addition of the element under study, the animals usually made poor growth or soon died. When vitamin research had progressed to the point where many of these factors could be supplied in pure or semipure form and the ubiquitous occurrence of trace elements in water, utensils, and dietary supplements was realized, reduction of adventitious sources of these elements became possible and rapid progress was made with the purified diet approach. The Wisconsin school, led by E. B. Hart, was conspicuously successful with such techniques and, in 1928, initiated a new era with the demonstration that copper is a dietary essential for hemoglobin formation in the rat. Within a few years the same group showed that manganese and zinc are also dietary essentials for the higher animals, and workers elsewhere confirmed and extended this finding. Nearly 20 years elapsed before the purified diet technique was again successful in identifying further essential elements in animal nutrition. These were first molybdenum and then, as an outcome of brilliant researches by Schwarz and his collaborators, selenium and chromium. In recent years even more sophisticated techniques have been devised for the further exclusion of environmental, including atmospheric, contamination. The use of such methods, as earlier forecast by Smith and Schwarz (21), has recently led to the addition of tin and probably nickel and vanadium to the "essential" list.

While these basic studies were proceeding, several widely separated, naturally occurring, nutritional maladies of man and his domestic animals were found to be caused by deficient or excessive intakes of various trace elements. In 1931, mottled enamel in man was shown to result from the ingestion of excessive amounts of fluoride from the water; in the same year, copper deficiency in grazing cattle was demonstrated in parts of Holland and the United States; in 1933, "alkali disease" and "blind staggers" of stock in parts of the Great Plains region of the United States were established as manifestations of selenium toxicity; in 1935, cobalt deficiency was shown to be the cause of various wasting diseases of grazing ruminants in certain localized areas in Australia; in 1936, a dietary deficiency of manganese was found to be responsible for "nutritional chondrodystrophy" and "perosis" in poultry; in 1937, enzootic neonatal ataxia of lambs in Australia was shown to be a manifestation of copper deficiency in the ewe during pregnancy; and in 1938, excessive intakes of molybdenum from the herbage were found to cause the debilitating diarrhea affecting cattle confined to certain areas in England.

Some years after this remarkably productive decade, chronic copper poisoning was demonstrated in sheep and cattle in parts of Australia, as a consequence of either high gross intakes of copper from the herbage or of normal copper combined with abnormally low pasture molybdenum levels. Finally, in 1958, selenium deficiency affecting the growth, health, and fertility of livestock over considerable areas was revealed in New Zealand and U.S.A. In all these studies attention was initially focused upon discovering the cause and devising practical means of control of the acute disorders as they occurred in the field. It soon became evident that a series of milder nutritional maladies also existed, with no clearly discernible symptoms and affecting far more animals and much greater areas than the acute conditions which prompted the original investigations. It became evident, further, that the situation was usually far more complex than was at first supposed. Simple deficiency or toxicity states were found to be rare under natural conditions—they were often ameliorated or accentuated, i.e., "conditioned," by the extent to which other elements or nutrients were present or absent from the environment. The importance of dietary balance among the trace elements has been particularly stressed by Kovalsky (14) from his extensive biogeochemical studies in the Soviet Union and is strikingly illustrated by the discovery of Dick in Australia in the early 1950's of a metabolic interrelationship between copper, molybdenum, and inorganic sulfate. Many other interactions of this type, notably between manganese and iron, copper, zinc and iron, zinc and cadmium, and arsenic and selenium, have been shown to be of profound nutritional importance—so much so that studies with the individual elements can be seriously misleading unless their quantitative relationship to the other interacting elements is known and considered.

During the last 20 years great progress has been made in our understanding of the mode of action of the trace elements within the tissues and of their metabolic movements within the animal body. This progress has been made possible by concurrent developments in three major areas. The first of these involved the use of radioactive isotopes of suitable half-life to facilitate studies of absorption, retention, excretion, and localization within the cells and tissues which were tedious, difficult, or impossible with the stable elements. The second significant development was in the field of enzymology, especially the discovery of many metalloproteins with enzymic activity and of trace element-catalyzed reactions in living tissues. These notable advances have greatly illuminated the basic biochemical lesions associated with the diverse manifestations of trace element deficiencies, excesses, and imbalances in the animal body. The third important development during the period under consideration was in analytical techniques. Microcolorimetric, polarographic, catalytic, flame photometric, X-ray fluorescent, and atomic absorption techniques have been evolved of such sensitivity and precision that the term "trace" elements with its historical connotation of imprecision has been largely outmoded. The more recent advent of neutron activation analysis has raised the detection and determination of the trace elements to a new level of sensitivity and sophistication and has made reliable estimation of these elements possible in amounts and concentrations previously beyond the reach of even the most ambitious analyst.

During the last few years a growing concern with problems of human environmental health has given renewed vigor to studies with trace elements. The possibility of chronic deleterious effects upon human health from contamination of the air, water supply, and food with elements arising from modern agricultural and industrial practices, and from the increasing motorization and urbanization of large sections of the community, has stimulated long-term investigations with many trace elements, notably cadmium, lead, and mercury, previously neglected in these repects.

## III. Mode of Action of Trace Elements

Most of the trace elements serve a variety of functions, depending upon their chemical form or combination and their location in the body tissues and fluids. Two of these elements, iodine and cobalt, are unique because, on present evidence, their entire functional significance can be accounted for by their presence in single compounds, thyroxine (or triiodothyronine) and vitamin $B_{12}$ (cyanocobalamin), respectively. However, both thyroxine and vitamin $B_{12}$, and therefore iodine and cobalt, are involved in a variety of metabolic processes.

The functional forms of the trace elements and their characteristic concentrations must be maintained within narrow limits if the functional and structural integrity of the tissues is to be safeguarded and the growth, health, and fertility of the animal are to remain unimpaired. Continued ingestion of diets that are deficient, imbalanced, or excessively high in a particular trace element invariably induces changes in the functioning forms, activities, or concentrations of that element in the body tissues or fluids, so that they fall below, or rise above, the permissible limits. Under these circumstances biochemical defects develop, physiological functions are affected, and structural disorders may arise in ways which vary with different elements, with the degree and duration of the dietary deficiency or toxicity, and with the age, sex, and species of the animal involved. Protective mechanisms can be brought into play which may delay or minimize the onset of such diet-induced changes. With some elements these protective mechanisms can be extremely efficient and effective over prolonged periods, but to prevent deleterious changes completely the animal must be supplied with a diet that is palatable and nontoxic as well as containing the required elements in adequate amounts, in proper proportions, and in available forms.

The trace elements function primarily as catalysts in enzyme systems in the cells. As stated by Green (8) nearly 30 years ago, "enzymic catalysis is the only rational explanation of how a trace of some substance can produce profound biological effects." The roles that the trace elements play in enzymic reactions range from weak, ionic strength effects to highly specific associations known as metalloenzymes. In the metalloenzymes the metal is firmly associated with the protein and there is a fixed number of atoms per molecule of protein which cannot be removed from this association by dialysis. Removal of the metal by more drastic means results in a loss of enzymic activity which cannot readily be restored either by readdition of the metal or by any other metal. An increasing number of metalloenzymes has been isolated from living tissues in recent years, indicating the wide range of cellular activities in which the trace elements participate. The following examples of metalloenzymes are presented to give some idea of the wide range of such activities: metalloenzymes containing iron—the cytochromes, catalase, cytochrome c reductase, succinic dehydrogenase, and fumaric dehydrogenase; those containing zinc—alkaline phosphatase, carbonic anhydrase, carboxypeptidase, and alcohol dehydrogenase. In addition, pyruvate carboxylase contains manganese and xanthine oxidase, molybdenum.

The molecular mechanisms involved in the metalloenzymes and the nature of the metal-ion specificity in their reactions have been illuminated in recent years by sophisticated studies involving amino acid sequence determinations, chemical modification, and X-ray crystallography. With the large group of enzyme systems that have less rigid and less specific associations with a metal—the metal-ion-activated enzymes—it is more difficult to assign precise metabolic roles to trace metals, although there is ample evidence of their importance to the living organism. A "simple" metal-ion catalysis, in which the protein has an enhancing effect only, appears untenable. The main types of roles that can be visualized for a metal-ion coenzyme were put forward some years ago by Malmstrong and Rosenberg (17). These functions are (*1*) to induce or maintain the active confrontation state of the protein molecule, (*2*) to act as a bridge in the formation of a ternary complex, a purely coordinative role, (*3*) to change the electronic structure of the substrate molecule. These different effects are not mutually exclusive and may well act cooperatively.

During the last decade, more and more trace element-enzyme associations have been identified and related to the manifestations of deficiency or toxicity states in the animal. In fact the unraveling of many of the basic biochemical defects that lie at the root of the profound functional and structural disorders characteristic of such deficiencies and toxicities are among the most encouraging aspects of recent trace element research. However, many clinical and pathological disorders arise in the animal as a consequence of trace element deficiencies and excesses for which there are as yet no acceptable explanations in biochemical or enzymic terms. This suggests either that there are many trace element-dependent enzymes of great metabolic significance which have still to be discovered, and their precise loci identified, or that these elements participate in the activity and structure of other compounds in the tissues.

## IV. Trace Element Needs and Tolerances

The minimum requirements of animals and man for the essential trace elements are commonly expressed in proportions or concentrations of the total dry diet consumed daily. The maximum intakes of these and other elements that can be safely tolerated are usually expressed similarly. These requirements, or tolerances, are arrived at by relating the growth, health, fertility, or other relevant criteria in the animal to varying dietary mineral concentrations. The latter are found by the application of appropriate analytical techniques which measure the total amounts of the elements in the diets or their component foods and beverages. Such analyses are not normally affected by variations in the chemical forms of the element as they occur in foods of different types or from different environments. Since the availability of mineral elements to the animal is affected by the chemical form in which it is ingested, it is obvious that gross dietary intakes do not necessarily reflect minimum requirements or maximum tolerances of wide or universal applicability.

Trace element requirements and tolerances expressed as concentrations such as parts per million of the dry diet also carry the assumption that the whole diet is otherwise adequate and well-balanced for the purpose for which it is fed, and that it is effectively free from other toxic factors capable of adversely affecting the animal's health, appetite, or utilization of the element concerned. The question of appetite is especially important since the capacity of a particular dietary concentration of an element to supply the needs of an animal will clearly depend upon the amount of the diet consumed daily or over a given period. Equally important is the level of other minerals or other nutrients which influence the availability or utilization of the element in question. A "true" or basic minimum requirement can thus be conceived as one in which all the dietary conditions affecting the element in this way are at an optimum. It must, therefore, be emphasized that a series of minimum requirements exist depending on the extent to which such interacting factors are present or absent from the whole diet.

Similarly, a series of "safe" dietary levels of potentially toxic trace elements exist, depending on the extent to which other elements which affect their absorption and retention are present. These considerations apply to all the trace elements to varying degrees but with some elements such as copper they are so important that a particular level of intake of this element can lead to signs either of copper deficiency or of copper toxicity in the animal, depending on the relative intakes of molybdenum and sulfate, or of zinc and iron. Many other mineral interactions of this type exist and are discussed in later chapters, including those between zinc and calcium and between manganese and calcium. Indeed the original discoveries of manganese deficiency in poultry and of zinc deficiency in pigs arose from the use of diets so high in calcium that the availability of these elements was depressed. These facts indicate clearly that estimates of the minimum requirement or safe intake of an element are only valid for the particular total dietary conditions imposed.

Such estimates also vary with the criteria of adequacy or tolerance employed. As the amount of an essential trace element available to the animal becomes insufficient for all the metabolic processes in which it participates, as a result of inadequate intake and depletion of body reserves, certain of these processes fail in the competition for the inadequate supply. The sensitivity of particular metabolic processes to lack of an essential element, and the priority of demand exerted by them, vary in different species and, within species, with the age and sex of the animal and the rapidity with which the deficiency develops. In the sheep, for example, the processes of pigmentation and keratinization of wool appear to be the first to be affected by a low copper status, so that at certain levels of copper intake no other function involving copper is impaired. Thus if wool quality is taken as the criterion of adequacy, the copper requirements of sheep are higher than if growth rate and hemoglobin levels are taken as criteria. In this species also, Underwood and Somers (24) have shown that the zinc requirements for testicular growth and development and for normal spermatogenesis are significantly higher than they are for the support of normal live-weight growth and appetite. If body growth is taken as the criterion of adequacy, which would be the case with rams destined for slaughter for meat at an early age, the zinc requirements would be lower than for similar animals kept for reproductive purposes. The position is similar with manganese in the nutrition of pigs. Ample evidence is available that the manganese requirements for growth in this species are substantially lower than they are for satisfactory reproductive performance. Recent evidence relating zinc intakes to rate of wound healing, also raises important questions on the criteria of adequacy to be employed in assessing the zinc requirements of man.

Criteria of adequacy and of tolerance present particular problems with fluorine. Indisputable evidence is available that fluoride, additional to that consumed by man in most areas, is required to confer maximal resistance to dental caries, and there is growing evidence that additional fluoride is also necessary to assist in the maintenance of a normal skeleton and in the reduction of osteoporosis in the mature, adult population. If a reduced incidence of dental caries and of osteoporosis is taken as the criterion of adequacy then clearly the human fluoride requirements are higher than if it is not. With farm animals, certain levels of fluoride intake are tolerated for prolonged periods without any measurable decline in growth, appetite, well-being, fertility or productivity, despite elevated bone fluoride levels and mild dental and skeletal abnormalities. On the usual criterion applied to farm stock, namely performance, such levels should not therefore be ruled as excessive. On the other hand, similar fluoride intakes could be considered intolerable if applied to younger animals or if consumed for still longer periods.

## REFERENCES

1. C. Bernard, "Leçons sur les effets des substances toxiques et médicamenteuses". Baillière, Paris, 1857.
2. G. Bertrand, *8th Int. Congr. Appl. Chem.* Vol. 28, p. 30 (1912).
3. E. M. Butt, R. E. Nusbaum, T. C. Gilmour, and S. L. Di Dio, *in* "Metal-Binding in Medicine" (M. J. Seven and L. A. Johnson, eds.), p. 43. Lippincott, Philadelphia, Pennsylvania, 1960.
4. A. Chatin, *C. R. Acad. Sci.* **30-39** (1850-1854).
5. A. W. Church, *Phil. Trans. Roy. Soc. London* **159**, 627 (1869).
6. G. C. Cotzias, *Proc. 1st Annu. Conf. Trace Substances Environ. Health, 1967* (D. D. Hemphill, ed.), p. 5 (1967). Columbia. Missouri.
7. P. Dutoit and C. Zbinden, *C. R. Acad. Sci.* **188**, 1628 (1929).
8. D. E. Green, *Advan. Enzymol.* **1**, 177 (1941).
9. E. Harless, *Arch. Anat. Physiol.* p. 148 (1947).
10. M. Henze, Hoppe-Seyler's *Z. Physiol. Chem.* **72**, 494 (1911).
11. J. W. Huff, D. K. Prosshardt, O. P. Miller, and R. H. Barnes, *Proc. Soc. Exp. Biol. Med.* **92**, 216 and 219 (1956).
12. M. Kirchgessner, *Z. Tierphysiol., Tierernaehr. Futtermittelk.* **14**, 217, 270, and 278 (1959); M. Kirchgessner, G. Merz, and W. Oelschläger, *Arch. Tierernaehr.* **10**, 414 (1960).
13. H. J. Koch, E. R. Smith, N. F. Shimp, and J. Connor, *Cancer* **9**, 499 (1956).
14. V. V. Kovalsky, *Proc. 1st Int. Symp. Trace Element Metab. Anim., 1969* (C. F. Mills, ed.), p. 385. Livingstone, Edinburgh and London, 1970.
15. K. Liebscher and H. Smith, *Arch. Environ. Health* **17**, 881 (1968).
16. C. A. MacMunn, *Phil. Trans. Roy. Soc. London* **177**, 267 (1885).
17. B. G. Malmstrom and A. Rosenberg, *Advan. Enzymol.* **21**, 131 (1959).
18. L. B. Mendel and H. C. Bradley, *Amer. J. Physiol.* **14**, 313 (1905); **17**, 167 (1906).
19. H. M. Perry, Jr., I. H. Tipton, H. A. Schroeder, and M. J. Cook, *J. Lab. Clin. Med.* **60**, 245 (1962).
20. O. Rygh, *Bull. Soc. Chim. Biol.* **31**, 1052, 1403, and 1408 (1949); **33**, 133 (1951); *Research (London)* **2**, 340 (1949).
21. J. C. Smith and K. Schwarz, *J. Nutr.* **93**, 182 (1967).
22. S. R. Stitch, *Biochem. J.* **67**, 97 (1957).
23. I. H. Tipton and M. J. Cook, *Health Phys.* **9**, 103 (1963).
24. E. J. Underwood and M. Somers, *Aust. J. Agr. Res.* **20**, 889 (1969).
25. A. J. Venchikov, *Vop. Pitan.* **6**, 3 (1960).
26. N. C. Wright and J. Papish, *Science* **69**, 78 (1929).

# 2

# IRON

## I. Introduction

The relation of iron to blood formation did not become apparent until the seventeenth century when two English physicians, Sydenham and Willis, found simple salts of iron to be of value in the treatment of chlorosis in young women. This empirical observation was placed on a more rational basis by the discovery that iron is a characteristic constituent of blood (71) and by the demonstration by Frodisch in 1832 that the iron content of the blood of chlorotics is lower than that of healthy individuals (89, 90). Before this, Lecanu had shown that hemoglobin contains iron, and, by 1886, Zinoffsky had estimated the iron content of horse hemoglobin to be 0.335% (261).

Following these pioneer discoveries, attention was focused for many years upon iron in relation to hemoglobin formation. Extensive studies of iron absorption and retention in different species, and of the forms and distribution of iron in foods and tissues, were carried out in the late nineteenth and early twentieth centuries, particularly by Bunge, Abderhalden, and others in Germany. These studies did much to establish the importance of iron in nutrition. They also served to foster the erroneous belief that only "food iron" or organically bound iron is available to the animal. This false concept greatly influenced the subsequent course of nutritional research on iron and affected clinical practice with this element for several decades. These early investigations also revealed that little iron is lost in the urine, even when large doses are ingested or injected. Since such doses were found to result in accumulations of iron in the epithelial cells of the small intestine, it was believed that the intestine is an excretory as well as an absorptive organ for iron (142). In 1937 this belief was challenged by McCance and Widdowson (153) and the new concept established that "the amount of iron in the body must be regulated by controlled absorption."

A broader and more basic approach to biological studies with iron arose from the investigations of Keilin and others in the 1920s, following the much earlier observations of MacMunn (143) on iron and cell respiration. The outstanding contribution of Keilin was to establish that iron, through its presence in the hemoprotein enzymes, the cytochromes, is vitally concerned in the oxidative mechanisms of all living cells. Subsequently, iron-containing flavoprotein enzymes were discovered (144, 206), and it became clear that iron is intimately involved in oxygen utilization by the tissues as well as in oxygen transport as part of the hemoglobin molecule. Lack of dietary iron was shown to inhibit some of these iron-dependent enzymic processes in the animal, in addition to its effect in limiting hemoglobin formation (19, 53).

Following the important discovery of McCance and Widdowson (153), mentioned earlier, the whole field of iron metabolism in health and disease, and particularly the means by which the body achieves iron balance, was intensively studied. The "mucosal block" theory was advanced by Hahn and co-workers (91) and apparently confirmed by Granick (83). The advent of suitable radioactive isotopes of iron greatly facilitated these studies. Two, nonheme iron-protein compounds, transferrin and ferritin, were discovered and shown to play key roles in iron absorption, transport, and storage. The relation between the two iron storage compounds, ferritin and hemosiderin, the latter long known as the stainable iron of the tissues, was also greatly illuminated. In addition, a range of factors was identified which influence iron absorption and great progress was made in elucidating the mechanisms involved in the absorptive process and in iron utilization by the tissues.

Nutritional disorders involving iron proved to be of little practical importance in farm animals, with the exception of the baby pig. Uncomplicated iron deficiency has never been demonstrated satisfactorily in grazing stock. There has, therefore, been less incentive to undertake nutritional studies with iron in these species than in human populations in which iron deficiency and disease involving disturbances in iron metabolism are relatively common. However, significant interactions between iron and copper, zinc, and manganese, affecting the requirements of farm stock for iron, have recently been demonstrated.

## II. Iron in Body Tissues and Fluids

### 1. *Total Content and Distribution*

The total iron content of the animal body varies with species, age, sex, nutrition, and state of health. Normal adult man is estimated to contain 4-5 g of iron (82) or 60-70 ppm of the whole body of a 70-kg individual. This is approximately twice the level of total body zinc and more than 20 times that of copper. A similar proportion of total body iron, namely 50 ppm, has been recorded for the adult rat (226). Most of the body iron exists in complex forms bound to protein, either as porphyrin or heme compounds, particularly hemoglobin and myoglobin, or as nonheme protein-bound compounds such as ferritin and transferrin. In certain disease states very large amounts of iron may also be present as hemosiderin. The hemoprotein and flavoprotein enzymes together constitute less than 1% of the total body iron. Free, inorganic iron is present in negligible quantities. The quantitative distribution of these forms of iron in the healthy adult human body, as calculated by Granick (82), is given in Table 2.

Hemoglobin iron occupies a dominant position in all animals, although in myoglobin-rich species, such as the horse and the dog, the proportion is substantially lower than in man. Thus Hahn (90) estimates blood hemoglobin iron to be 57% and myoglobin iron to be 7% of total body iron in the adult dog, compared with 60 to 70% and 3%, respectively, in adult man. The high proportion present as hemoglobin indicates that any condition influencing the level of this compound in the blood greatly affects the iron status of the body. Furthermore, hemoglobin iron is continuously reutilized, thus constituting a buffer against dietary fluctuations in this element. The levels of tissue stores of iron, as ferritin and hemosiderin, also influence total body iron content and act similarly as a nutritional buffer.

Species differences in total body iron concentrations occur in the newborn but they become much less pronounced in the adult, as is seen from the data in Table 3. The differences at birth reflect differences in liver iron stores and in blood hemoglobin levels. For example, the pig has relatively little iron in its body at birth because it is normally born with low liver iron stores and has no polycythemia of the newborn, as in the human infant. The newborn rabbit, by contrast, has an exceptionally high body iron concentration due to its large liver iron stores (259).

Female rats have a higher total body iron than males and accumulate more iron in their livers on the same diet (190). Female mice and birds

TABLE 2

*Iron Compounds and Distribution in Adult Human Body*[a]

| Compounds | Prosthetic groups per molecule | Compound (g) | Iron (g) | Per cent of total iron |
|---|---|---|---|---|
| Iron porphyrin (heme) compounds | | | | |
| Blood hemoglobin | 4 hemes | 900 | 3.0 | 60-70 |
| Myoglobin | 1 heme | 40 | 0.13 | 3-5 |
| Heme enzymes | | | | |
| Mitochondrial cytochromes | | | | |
| c | 1 heme | 0.8 | 0.004 | 0.1 |
| $a_3$,a,$c_1$,b | – | – | – | – |
| Microsomal cytochrome $b_5$ | – | – | – | – |
| Catalase | – | 5.0 | 0.004 | 0.1 |
| Peroxidase | – | – | – | – |
| Nonheme compounds | | | | |
| Flavin-Fe enzymes | | | | |
| Succinic dehydrogenase | 1 FAD[b]:4 Fe | – | – | – |
| Xanthine oxidase of liver | 1 FAD:4 Fe:1 Mo | – | – | – |
| NADH[c]-cytochrome c reductase | 1 FAD:4 Fe | – | – | – |
| Iron chelate enzyme aconitase | – | – | – | – |
| Transferrin | 2 Fe | 10.0 | 0.004 | 0.1 |
| Ferritin | $4(FeOOH)_n$ | 2-4 | 0.4-0.8 | 7-15 |
| Total available iron stores | – | – | 1.2-1.5 | – |
| Total Iron | – | – | 4-5 | 100 |

[a]Taken from Granick (82).

[b]FAD = flavin adenine dinncleotide.

[c]NADH = reduced nicotinamide adenine dinncleotide.

TABLE 3
*Iron Content of Bodies of Different Species*[a]

| Age | Fe of fat-free tissue (ppm) | | | | | | |
|---|---|---|---|---|---|---|---|
| | Human | Pig | Cat | Rabbit | Guinea pig | Rat | Mouse |
| Adult | 74 | 90 | 60 | 60 | – | 60 | – |
| Newborn | 94 | 29 | 55 | 135 | 67 | 59 | 66 |

[a]Taken from Widdowson (259).

also carry greater concentrations of liver iron than males, but no such sex difference is apparent in rabbits or guinea pigs (258). Data on possible sex differences in total body iron in the larger farm animals or in man have not appeared. Lower values would be expected in women than in men because of their normally lower blood hemoglobin and muscle myoglobin levels.

Among the organs and tissues of the body, the liver and spleen usually carry the highest iron concentrations, followed by the kidney, heart, skeletal muscles, pancreas, and brain, which normally contain only one-half to one-tenth of the concentrations in liver and spleen. Variation among species is small but individual variation in iron content within species can be very high in the liver, kidney, and spleen. In some species, notably the rat, rabbit, sheep, and man, but not in the dog, the liver has a remarkably high storage capacity for iron. Large increases in the total iron content of the liver, up to a total of 10 g, occur in cases of human malignancy and chronic infection (107). In the final stages of hemochromatosis as much as 50 g of iron may accumulate in the human body (57). Conversely, the iron content of these tissues and of the bone marrow is reduced below normal in conditions such as iron deficiency and hemorrhagic anemia.

### 2. *Iron in Blood*

Iron occurs in blood as hemoglobin in the erythrocytes and as transferrin in the plasma, in a ratio of nearly 1000 to 1. Minute amounts of nonheme iron also occur in the erythrocytes of human blood (17), probably at least in part as ferritin (64).

a. *Hemoglobin.* Hemoglobin is a complex of the basic protein, globin, and four ferro-protoporphyrin or "heme" moieties. It was first synthesized by Fischer and Zeile in 1929 (69). A realistic three-dimensional picture of the molecule with its four attached hemes and the nature of the bond between iron and globin, were established later (110). This union stabilizes the iron in the ferrous state and allows it to be reversibly bonded to oxygen, thus permitting hemoglobin to function as an oxygen carrier. The molecule is similar in size (molecular weight 68,000) in all animal species and has an average iron content very close to 0.34%.

The synthesis of heme and its attachment to globin take place in the later stages of red cell development in the bone marrow, with production of the two parts of the molecule probably occurring together. The main steps in the biosynthesis of heme have been presented by Rimington (207). They include the condensation of two molecules of $\delta$ -aminolevulinic acid to form porphobilinogen, a process which involves the copper-containing enzyme delta aminolevulinic acid (ALA) dehydrase (111). The exact means by which iron is incorporated into protoporphyrin to form heme in the final biosynthetic step are still not clear but iron is known to be carried to the bone marrow in the ferric form as transferrin, reduced to the ferrous form, and detached from transferrin with the aid of reducing substances, thus facilitating its transfer to protoporphyrin (86, 207).

The normal ranges of blood hemoglobin levels in different species are as follows: man and rat, 13-15; dog, 13-14; cow and rabbit, 11-12; and pig, sheep, goat, and horse, 10-11 g Hb/100 ml. The total hemoglobin in mammals is directly proportional to body weight. Drabkin (57) found the mean of five species examined to be 12.7 g Hb/kg body weight. The levels in the blood vary with age, sex, nutrition, state of health, pregnancy, lactation, and environment (climatic and barometric). All species are not necessarily affected in the same way or to the same degree by all these factors. In man, the level falls rapidly from about 18 to 19 g/100 ml at birth to about 12 g/100 ml at 3 to 4 months, at which level it usually remains until the child is about a year old, when a slow rise to adult values normally begins. A striking rise occurs at puberty in males and the higher hemoglobin levels of the male continue throughout the life span. This is evidently a real sex difference, since a significant rise to the level of males does not take place after the menopause or hysterectomy, when menstrual blood losses no longer occur (246). Sex differences similar to those in the human species do not occur in rats (125) nor in cattle (31).

Blood hemoglobin concentrations decline in late pregnancy in normal rats and in healthy women, but not in ewes (156). In both species this decline arises partially from hydremia (24, 56, 77, 139), but iron deficiency is frequently a contributing factor. This problem in women has been critically assessed by de Leeuw *et al.* (139).

b. *Serum Iron.* The iron of the plasma was shown by Holmberg and Laurell (104) to be completely bound to a specific protein, designated transferrin or siderophilin (211). Transferrin is a $\beta_1$-globulin with a molecular weight of about 86,000 and has two, separate iron-binding sites, each capable of binding one atom of ferric iron (60). There are also four sialic acid side chains attached to the protein, the function of which is not yet known (117). The metal is very tightly bound to the protein. At physiological pH the equilibrium constants for the two iron-binding sites are of the order of $10^{30}$ (1), and it has been postulated that there is no redistribution *in vivo* of iron from one binding site to another (70, 176). Many different transferrins occur in human blood, which are genetically controlled [see Bearn and Parker (15)], and similar or identical iron-binding proteins occur widely in mammals and birds (118). These include conalbumin, the iron-binding protein found in eggs which functions in a manner similar to human transferrin (118), and lactoferrin, the iron-binding protein of human, bovine, and rabbit milk.

Transferrin serves as a true carrier for iron in the blood, in the way that hemoglobin acts as a carrier for oxygen. More transferrin iron-binding sites than atoms of iron are present in normal plasma, so that no free iron is present. Since even very small quantities of free iron are toxic (79), the prevention of the occurrence of free iron in the plasma can be regarded as a function of transferrin.

In normal individuals of most species, only 30–40% of the transferrin carries iron, the remainder being known as the latent iron-binding capacity. Serum iron and both total and latent iron-binding capacity vary greatly among species, among individuals of the same species, and in various disease states. In man a well-marked diurnal rhythm occurs. Vahlquist (245) found 15 normal men to have a mean serum iron of $135 \pm 10.6$ μg/100 ml at 8 AM and of $99 \pm 9.2$ μg/100 ml at 6 PM. High morning and low evening values have since been observed by several investigators, except in night workers where the diurnal rhythm is reversed and diminished (95, 192). Sleep deprivation in human subjects also results in a gradual decline in serum iron levels, the maximum drop being to half the original levels (136). Most of this decline occurs during the first 48 hours of sleep deprivation, with a slower decline during subsequent intervals. Return to normal values takes about 1 week.

Individual variability in serum iron and total iron-binding capacity (TIBC) is high in all species studied (196). Differences among species appear to be small, although there is evidence that the levels of these constituents are higher in the blood of pigs, sheep, and cattle (196, 244) than they are in man. In male birds and in chicks both serum iron and TIBC concentrations are similar to those of man. In hens and ducks during the laying season, serum iron is markedly elevated by a factor of almost 5. TIBC levels are also raised, although not to the same extent and the iron-binding capacity is fully saturated (195, 197). The transport of iron in the laying fowl presents special characteristics, involving the existence of mechanisms auxiliary to that of transferrin (195, 197). Panic (191) has established that the increase in plasma iron in laying hens is caused by the appearance of a specific protein, phosvitin, which binds about two-thirds of the iron present in plasma and transports iron to the ovocytes and egg yolk.

Representative values for serum iron, TIBC, and percentage saturation in several species are presented in Table 4. The differences in serum iron in man which appear in Table 4 can be understood if the iron of the serum is conceived as a pool into which iron enters, leaves, and is returned at varying rates for the synthesis and resynthesis of hemoglobin, ferritin, and other iron compounds. For instance, in iron deficiency, low serum iron levels result from low intake, depletion of body stores, and reduced hemoglobin destruction accompanying the anemia. In fact, there is evidence that the serum iron level begins to fall before the iron stores are completely mobilized (94, 152). The high serum and percentage saturation values characteristic of pernicious anemia and aplastic anemia (37) can be associated with increased iron absorption and a bone marrow block to hemoglobin synthesis in the presence of adequate iron

TABLE 4

*Serum Iron, Total Iron-Binding Capacity (TIBC), and Percentage Saturation in Various Species*

| Species and condition | No. of cases | Serum iron ($\mu$g/100 ml) | TIBC ($\mu$g/100 ml) | Mean saturation (%) | Ref. No. |
|---|---|---|---|---|---|
| Human adults | | | | | |
| Normal male | 35 | 127(67-191) | 333(253-416) | 33 | (36) |
| Normal female | 35 | 113(63-202) | 329(250-416) | 37 | (36) |
| Iron deficiency anemia | 35 | 32(0-78) | 482(304-705) | 7 | (36) |
| Late pregnancy | 106 | 94(22-185) | 532(373-712) | 18 | (36) |
| Hemochromatosis | 14 | 250(191-290) | 263(205-330) | 96 | (36) |
| Infections | 11 | 47(30-72) | 260(182-270) | 20 | (36) |
| Bovine adults | | | | | |
| Normal cows | 10 | 146(89-253) | 553(388-724) | 26 | (244) |
| Normal bulls | 10 | 145(92-270) | 432(332-521) | 33 | (244) |
| Ovine adults | | | | | |
| Normal ewes | 12 | 182(102-304) | 353(278-456) | 51 | (244) |
| Normal rams | 12 | 152(114-191) | 353(248-455) | 43 | (244) |
| Castrate males | 65 | 180(108-268) | 331(264-406) | 56 | (76) |
| Birds | | | | | |
| Chicks | 90 | 102±6 | 239±4 | 43 | (195) |
| Nonlaying hens | 40 | 158 | 258 | 61 | (195) |
| Laying hens | 38 | 516 | 333 | 100 | (195) |

stores. The high serum iron and percentage saturation of the iron-binding protein in hemochromatosis can be related to excessive absorption and deposition of iron in the body. The low serum iron and TIBC levels in the nephrotic syndrome (36) may be explained by losses of bound iron in the urine in the presence of considerable proteinuria (35).

Not all the changes in serum iron and TIBC levels can be so confidently explained. For instance, the low serum iron and TIBC levels of infections and malignancy present a problem, although it is known that the decrease in serum iron is not due to the reduction in iron-binding capacity (37). The reasons for the variations in TIBC levels in various disease states, and for the rise in this component in the third trimester of human pregnancy, are also far from clear. Such changes are not necessarily correlated with the absorption of iron, with the magnitude of the iron stores, or with the need of the body for greater or lesser transport of iron. However, Morgan (171) has demonstrated a good inverse relationship between TIBC and hemoglobin in the rat. He suggests that in this species "the main factor regulating plasma transferrin concentration is the balance between tissue supply and requirements for oxygen, i.e. relative oxygen supply."

### 3. *Storage Iron Compounds*

The reserve or storage iron of the body occurs predominantly as the two nonheme compounds, ferritin and hemosiderin, which occur widely in the tissues, with the highest concentrations normally present in the liver, spleen, and bone marrow. The two compounds are chemically dissimilar, although intimately related in function. Chemical methods for their estimation are available, based upon the fact that ferritin is soluble in water whereas hemosiderin is insoluble (75, 126). Results obtained by such means are supported by immunochemical and radiotracer techniques (75).

Ferritin is a brown compound, first isolated by Laufberger in 1937 (137). In the crystalline state ferritin contains up to 20% iron and consists of a central nucleus of iron, stored in six micelles arranged at the corners of a regular octahedron, surrounded by a shell of protein approximately spherical in shape (182). The colorless, iron-free protein, apoferritin, is a physicochemically homogeneous globulin with a molecular weight of 460,000 (84, 208). Hemosiderin is a relatively amorphous compound which may contain up to 35% iron, consisting mainly of ferric hydroxide condensed into an essentially protein-free aggregate (219, 220). It exists in the tissues as a brown, granular, readily stainable pigment.

Histochemical examination of aspirated samples of bone marrow for hemosiderin has been proposed as a reliable index of body iron stores (228) and as a valuable aid in diagnosing iron deficiency anemia (21). Shoden and Sturgeon (221) have emphasized the value of using both staining and chemical methods which estimate the soluble (ferritin) as well as the insoluble (hemosiderin) forms of storage iron. Ferritin takes the Prussian blue stain, but it is too finely dispersed in the cells to be demonstrable by histochemical techniques evaluated with the light microscope (217). Furthermore, in a study of 130 human necropsies involving histological and chemical estimates of hepatic and splenic storage iron, Morgan and Walters (178) found a general agreement between the two estimates, with considerable variation. Histological examination of these tissues gave only a very approximate idea of storage iron levels. A similar conclusion was reached by Kerr (131) with respect to histochemical examinations of bone marrow.

The factors affecting the amounts and proportions of ferritin and hemosiderin in the liver and spleen of rats, rabbits, and man have been studied by Shoden and Sturgeon (216, 218, 221) and by Morgan (169, 170, 178). Up to certain levels and rates of iron storage, iron is deposited in the liver and spleen readily and roughly equally as ferritin and as hemosiderin. Iron is utilized equally readily from these two compounds for the demands of erythropoiesis and for placental iron transfer to the fetus. A threefold increase in liver and spleen total iron storage was achieved in rats without change in the relative proportions of ferritin and hemosiderin, compared with those of normal rats not fed additional iron (169). Moreover, the ferritin iron-hemosiderin iron ratio changed little in these organs in induced chronic or acute hemolytic anemia. Davis *et al.* (53) similarly observed little change, after the first week, in the ferritin-hemosiderin ratio in the livers of chicks during a depletion period on an iron-deficient diet.

The main factor affecting the relative distribution of iron between ferritin and hemosiderin in mammals is the total storage iron concentration. When the iron level in the liver and spleen of rats and rabbits increase beyond about 2000 $\mu$g/g, hemosiderin begins to predominate. At levels beyond 3000-4000 $\mu$g/g, additional storage iron is deposited quantitatively as hemosiderin (169, 170, 216) (see Fig. 1). From their study of iron storage in human necropsies, Morgan and Walters (178) concluded that, with total storage iron in liver and spleen below 500 $\mu$g/g of tissue, more iron was stored as ferritin than as hemosiderin; with levels above 1000 $\mu$g/g, more was stored as hemosiderin. The situation may not be quite comparable in avian species. Larger amounts of hemosiderin than of ferritin were found in the livers of chicks at very much lower levels of

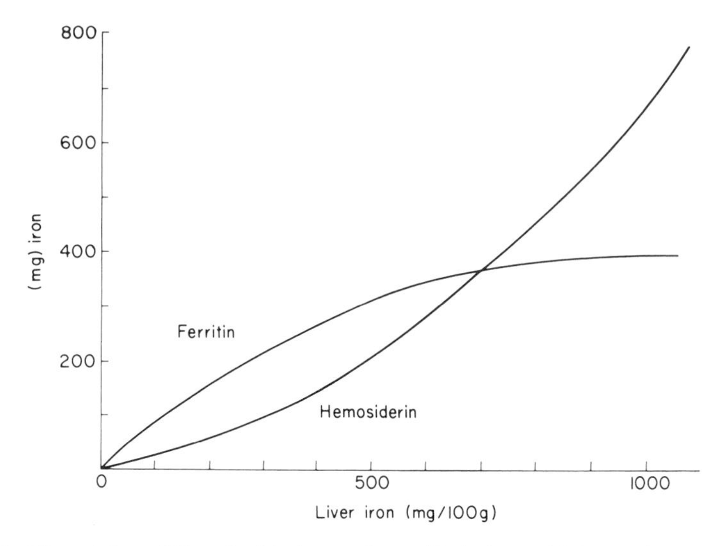


**Fig. 1.** The relation of liver ferritin iron to hemosiderin iron with increasing concentrations of total liver iron. [From Shoden and Sturgeon (219).]

total iron than those quoted for the rat and man (53). In human diseases such as hemochromatosis and transfusional siderosis, which are characterized by extremely high levels of iron in the tissues, most of the iron is present as hemosiderin (57,178). A similar situation probably exists in respect to the heavy iron deposits in liver and in spleen in nutritional siderosis in cattle (97) and in copper and cobalt deficiencies in sheep (167, 241).

The ratio of ferritin iron to hemosiderin iron is affected further by the rate of storage. When iron is injected at very high rates, or administered in a form such as saccharated iron which is rapidly cleared from the serum, hemosiderin is deposited rather than ferritin. With equivalent injections of iron dextran, which remains in the serum for a relatively long time, ferritin production is greater and hemosiderin smaller (219). This suggests that there is a limit to the amount of apoferritin which is present in, or can be produced by, the liver of the rabbit in a given interval of time. These findings apply only to very high rates of iron storage. Over quite a wide range of iron storage rates and levels and of iron depletion, the distribution of iron between ferritin and hemosiderin remains relatively constant, and the iron moves readily from one storage form to another.

The bone marrow and muscles contain considerable amounts of nonheme iron. The storage iron concentration of bone marrow in normal man is given by Hallgren (94) as about 100 $\mu$g/g. If the total active bone marrow weight is taken as 3000 g (256), it can be calculated that the bone marrow of the human body would contain approximately 300 mg of storage iron or about one-third to one-fifth of the total estimated storage iron (see Table 2). The concentration of nonheme iron in the muscles is low (94, 239) but, because of their large mass, the total amount is high. Torrance *et al.* (239) have shown that in human subjects with normal body stores, the total amount of nonheme iron in the muscles is at least equal to that in the liver. In subjects with iron overload the iron concentrations in muscle are raised but to a much smaller extent than in the liver. In rats this muscle iron represents a relatively nonmiscible pool which responds little to acute changes in the iron environment.

### 4. *Iron in Milk*

The iron content of milk varies with species and stage of lactation of the animal and is highly resistant to changes in the level of dietary iron. Individual variation within species is high but some of the reported variation probably reflects analytical inadequacies and insufficient care to avoid contamination. Contamination from metal receptacles can more than double the iron content of cow's milk (67, 112).

The average concentration of iron is very similar in human, cow's, and goat's milk. A high proportion of the most acceptable values falls between 0.3 and 0.6 μg/ml, with a mean close to 0.5 μg/ml (38, 67, 180, 198). The level of iron in colostrum is 3-5 times higher than that of the milk. After the initial fall there seems little evidence of a significant decline throughout lactation in the milk of women (65), cows (201), or sows (248). The milk of sows and of several other species is appreciably richer in iron than that of the species just cited. In an early study, values ranging from 1.4 to 2.4 μg/ml were reported for the milk of sows in mid-lactation (248). More recently, mean levels of 1.4 and 1.2 μg Fe/ml were obtained for the milk of sows iron-supplemented and unsupplemented, respectively, during the first 3 weeks of lactation (200). Rabbit's milk is still higher in iron, with levels mostly lying between 2 and 4 μg/ml (236). Rat's milk is exceptionally rich in this element and so is the milk of the Australian marsupials, *Setonyx brachyurus* (127) and *Trichosurus vulpecula* (106).

The high iron content of rat's milk, first shown by Cox and Mueller (45), has been further studied by Ezekiel and co-workers (63, 128). A marked effect of stage of lactation was found in this species. The iron concentration fell from the exceptionally high mean levels of 13.5 to 8.1 μg/ml in the first 4 days and then more slowly to a mean of 3.0 μg/ml at 24 days, with no change thereafter.

Administration of iron to lactating cows, sows, or women is largely ineffective in raising the iron content of the milk to levels above normal (129, 134, 198, 200, 248). This is of particular interest because of the naturally low iron content of the milk of these species, relative to the needs of the suckling. On the other hand, there is evidence that in the rat the level of iron in the milk can be raised by iron loading of the lactating animal (63). It is not entirely clear whether the reverse can occur, i.e., a reduction in the iron content of milk to subnormal levels in iron deficiency, although Ezekiel and Morgan (63) observed a decrease in the iron content of the milk of rats when their hemoglobin levels and iron stores were depleted by repeated bleeding.

Iron occurs in milk in combination with protein. An iron-protein compound named ferrilactin has been isolated from human (120) and bovine (87) milk, where it is present in low concentrations. This salmon-red compound, in which the iron is firmly bound, has two iron-binding sites and a molecular weight of 86,000. It is chemically and immunologically distinct from serum transferrin (55, 81, 87, 120). Rabbit milk whey is remarkable for its extremely high iron-binding capacity, 5-10 times that of serum (122), due to the presence of an iron-binding protein which has been called milk transferrin (10). Serum transferrin and rabbit milk transferrin are identical on the basis of molecular weight, amino acid and peptide composition, spectral absorption, iron-binding properties, and immunological characteristics. The only difference found was that most of the milk transferrin contained one sialic acid residue per molecule whereas serum transferrin contained two (10). The loss of this sialic acid does not affect its passage from serum to milk, absorption from the gut of suckling rabbits, or its ability to bind to and transfer iron to reticulocytes (8). In fact, the only metabolic difference found between rabbit serum transferrin and milk transferrin was that the latter protein was catabolized more quickly than the former (8). The physiological role of the high concentration of transferrin in rabbit milk is obscure. It does not appear to aid the transfer of iron from serum to milk in the lactating rabbit, or to aid iron absorption by the suckling young (122). The possibility that it possesses some small but important antibacterial function has been raised (8).

## III. Iron Metabolism

### 1. *Factors Affecting Iron Absorption*

The absorption of iron is affected by the age, iron status, and state of health of the animal, by conditions within the gastrointestinal tract, by the amount and chemical form of the iron ingested, and by the amounts and proportions of various other components of the diet, both organic and inorganic. In monogastric species this absorption takes place mainly in the duodenum (29, 74, 230) in the ferrous form (101, 189). Absorption does not depend on valency since some ferric compounds are more available than some ferrous compounds (73). There is some evidence that ferric salts are less effective for hemoglobin formation in man than ferrous forms (101, 189). This is not necessarily so with the rat (73, 230, 242), the dog (229), and the chick (73).

Iron occurs in foods in inorganic forms, in combination with protein, in heme compounds as constituents of hemoglobin and myoglobin, and in other organic complexes. Iron in heme compounds is absorbed directly into the mucosal cells of the intestine, without the necessity of release from its bound form. The inorganic forms of iron and the iron-protein compounds need to be reduced to the ferrous state and released from conjugation for effective absorption. These transformations are accomplished by the gastric juice and other digestive secretions. Normal gastric secretion is necessary for optimal absorption of iron by rats (184). Furthermore, iron absorption can be increased in achlorhydric subjects by the administration of HCl (116), and, in anemic patients, iron absorption is correlated with the amount of intrinsic acid secretion (115). However, achlorhydrics do not necessarily become iron-deficient (43), and there is no inhibition of iron absorption by gastric juice in patients with complete achlorhydria (114). In the rat, maximal iron absorption from the small intestine occurs in the pH range 2.0-3.5 (135). Jacobs and Miles (113) have recently shown that inorganic iron is able to form complexes with normal gastric juice constituents at a low pH. These remain soluble when the pH is raised to neutrality and enable the iron to be available in a suitable state for absorption in the small intestine.

Amino acids such as histidine and lysine assist in iron absorption (135, 247). Van Campen and Gross (247) found that histidine increased $^{59}$Fe uptake from isolated segments of rat duodenum only if the two were administered in the same solution. They suggested that a direct reaction between iron and histidine occurs and that an amino acid-iron chelate may be formed and subsequently absorbed. Since histidine is a product of protein hydrolysis in the gastrointestinal tract, it was further suggested that this amino acid may be involved in the normal absorption of iron. Organic acids and reducing substances in foods, such as ascorbic acid and cysteine, may also assist in reducing and releasing food iron (83). Ascorbic acid reduces ferric to ferrous iron *in vitro*, and the administration of large doses of this acid with iron salts and some foods can increase the efficiency of iron absorption (93, 165, 194). Ascorbic acid has no such effect on the absorption of hemoglobin iron (93).

In 1955, Walsh and co-workers (250) showed that humans absorbed significant quantities of iron from test doses of hemoglobin. Hemoglobin iron has since been found to be well absorbed in mice, rats, dogs, guinea pigs, and man (11, 32, 42, 93, 254). Hallberg and Solvell (93), using a double isotope technique permitting comparisons to be made on the same individuals, found that the iron of hemoglobin and of ferrous sulfate were equally well absorbed by normal human subjects when equivalent doses were given. The absorption of iron from ferrous sulfate was increased by ascorbic acid and reduced by sodium phytate, whereas the absorption of hemoglobin iron was unaffected by either. These studies are of considerable interest and importance, in view of the substantial amounts of heme iron, as hemoglobin and myoglobin, ingested by carnivorous animals and by omnivorous animals, including man, when consuming diets high in meat.

When phytate is added to diets as sodium phytate, iron absorption can be significantly reduced in man (93, 106, 214). Hussain and Patwardhan (106), in 8-day balance studies on healthy men, found that the addition of sodium phytate to the diet reduced the average retention of iron from 2.5 to 0.17 mg. There is some doubt if the normal phytates of foods are of much practical significance in respect to iron absorption, either in man (249) or in chicks (53). Furthermore, even added sodium phytate does not necessarily reduce iron absorption in rats (44).

Gross disturbances in dietary mineral balance can greatly affect the absorption of iron. Very high levels of phosphate reduce iron absorption (28), whereas its absorption tends to be increased with low phosphate diets (130). High intakes of zinc, cadmium, copper, and manganese also interfere with iron absorption, apparently through competition for protein-binding sites in the intestinal mucosa, and have been shown to raise iron requirements. These interactions with iron are considered in the chapters on copper, zinc, manganese, and cadmium. The findings of Settlemire and Matrone (213) with rats fed diets very high in zinc are particularly pertinent in the present context. These workers obtained evidence that zinc reduced iron absorption by interfering with the incorporation of iron into or release from ferritin. The increase in iron requirement brought about by the high zinc intakes was augmented by a shortening of the life-span of the red blood cells resulting in a faster turnover of iron.

Only 5-15% of food iron is normally absorbed by man from ordinary diets. This has been demonstrated with labeled foods (164) and with standard foods with small amounts of tracer iron added (194). Absorption of food iron may be increased to twice this level, or more, in children and in iron deficiency. Thus, Josephs (123) reported that 2-20% of an oral dose of radioiron was absorbed in normal subjects, compared with 20 to 60% in patients with iron deficiency anemia. In another study, higher values for iron absorption were obtained (225). These were 27±9.9% for subjects with no hematological abnormality and 62±11.8% for those with iron deficiency anemia. The absorption of hemoglobin iron is also increased in iron deficiency in rats (11) and in man (93), although the increase is less marked than it is with ferrous sulfate. In dogs rendered anemic by bleeding, radioactive iron absorption can increase to 5 to 15 times normal (91).

The efficiency of iron absorption changes in various disease states. Increased absorption occurs in aplastic anemia, hemolytic anemia, pernicious anemia, pyridoxine deficiency, hemochromatosis, and transfusional siderosis (37, 58, 193). Increased iron absorption is also related to increased erythropoiesis (26, 253), depletion of body iron stores (160, 251), increased iron turnover (27), and hypoxia (159). Decreased iron absorption, by contrast, occurs in transfusional polycythemia (26) and has been related to tissue iron overload (41).

### 2. *Mechanisms of Iron Absorption*

The mechanism of iron absorption from the gastrointestinal tract has been intensively studied since McCance and Widdowson (153) put forward the concept that iron absorption is quantitatively controlled by body needs. As Crosby (46) has said, "it is difficult to overstate the importance of this idea, for it placed the metabolic balance of iron in a unique category. If the amount of absorption is indeed a reflection of requirement, then it must follow: the intestine receives information concerning deficiency or surfeit and by means of sensory and effector mechanisms it acts in response to this information."

The "mucosal block" theory advanced by Hahn and co-workers (91) in 1943, and elaborated by Granick (83), constituted a major advance because it focused attention upon the mucosal cells. According to this theory the intestinal mucosa absorbs iron during periods of need and rejects it when stores are adequate. This was explained as follows: iron taken into mucosal cells is converted into ferritin and when the cells become physiologically saturated with ferritin, further absorption is impeded until the iron is released from ferritin and transferred to plasma. Later studies by Crosby and Conrad (40, 41, 255) provided evidence that, in the rat, the ultimate regulator of iron absorption is the iron concentration of the epithelial cells of the upper intestine. In normal rats with moderate intestinal iron concentrations, only a small part of the ingested iron taken up by mucosal cells is transferred to the bloodstream and is retained by the animal; the remainder stays in the mucosal cells and is lost into the gut lumen when the cells are sloughed from the tips of the intestinal villi. In iron-deficient rats with decreased intestinal iron concentration, most of the ingested iron is absorbed directly into the bloodstream with very little remaining in the mucosal cells. In rats given excessive iron stores parenterally the epithelial cells are "loaded from the rear" and are, therefore, unable to accept the ingested iron. Data recently reported by Allgood and Brown (5) indicate that this mechanism is not applicable to man. The nonheme iron concentrations of duodenal mucosa obtained from normal subjects did not differ from those of iron-deficient or iron-loaded subjects. Furthermore, these concentrations showed no significant correlations with simultaneous measurements of serum iron levels nor with radioiron absorption. The discrepancy between these findings and those cited earlier with rats suggests the existence of species differences in the regulation of gastrointestinal iron absorption.

Differences in the mechanism of iron absorption also occur within species, depending upon the chemical form in which the iron is ingested. The absorption of the iron of hemoglobin, unlike inorganic iron, is not affected by ascorbic acid, phytate, nor nonabsorbable chelating agents. This suggests that iron is not released from heme in the gut lumen but the complex is taken up direct by the intestinal epithelial cell, subsequently appearing in the plasma in nonheme form (42, 93, 254). Weintraub *et al.* (254) demonstrated, in duodenal mucosal homogenates from the dog, the presence of an enzymelike substance which is capable of releasing iron from hemoglobin *in vitro*. The rate at which the heme-splitting substance works *in vivo* appeared to be increased by the removal of the nonheme iron end product from the epithelial cell to the plasma. It seems, therefore, that the labile nonheme content of the intestinal cell determines its ability to accept heme from the lumen in dogs, as well as ionized iron from the lumen in rats.

Two other processes, both involving digestive secretions, have been invoked in attempts to throw further light on the mechanisms available to the body for regulating iron absorption. P. S. Davis and associates (54) found an iron-binding protein (gastroferrin) in normal gastric juice and observed that it was absent in hemochromatosis. In iron deficiency anemia caused by blood loss, the concentration of gastroferrin in gastric juice was reduced and returned to normal levels when hemoglobin values were restored (141) (see Table 5). The hypothesis was advanced that gastroferrin production is involved in the regulation of iron absorp-

TABLE 5
*Effect of Venesection on Iron-Binding Capacity (IBC) of Gastric Juice*[a]

| Day | Hemoglobin (g/100 ml) | Serum iron ($\mu$g/100 ml) | Iron saturation (%) | Gastric juice IBC | |
|---|---|---|---|---|---|
| | | | | Iron bound[b] (mg/ml) | Iron bound[c] (mg/ml) |
| 0[d] | 17.3 | 147 | 50 | 0.204 | 0.225 |
| 5 | 11.6 | 60 | 20 | 0.082 | 0.063 |
| 10 | 9.0 | 30 | 5 | 0.014 | 0.012 |
| 60 | 16.6 | 130 | 40 | 0.198 | 0.193 |

[a]Taken from Luke *et al.* (141).
[b]Whole gastric juice.
[c]Excluded fraction from gel filtration.
[d]2500 ml. of blood removed by venesection on days 2, 3, 4, 7 and 8.

tion—normal levels acting to inhibit the absorption of excessive intakes of iron and reduced levels permitting enhanced absorption in iron deficiency. It was further proposed that failure to produce gastroferrin through an inborn error of metabolism is a causal factor in hemochromatosis. No direct support for the proposal that this gastric inhibitor plays a role in the control of iron absorption has yet appeared. The excess absorption and deposition of iron characteristic of hemochromatosis has been further related to a primary pancreatic defect (23). A. E. Davis and Biggs (52) showed that iron absorption can be significantly reduced in hemochromatosis by the addition of a pancreatic extract with the oral dose of iron. A similar inhibitory effect of a pancreatic extract on iron absorption was demonstrated in the intact rat and with the isolated loop of the rat jejunum (23, 52). The active factor in pancreatic juice has not yet been characterized, and the physiological importance of the pancreatic secretion in the normal regulation of iron absorption remains to be evaluated.

### 3. *Excretion of Iron*

The limited ability of the body to excrete iron has been abundantly confirmed since the original observations of McCance and Widdowson (153). Even in hemolytic anemia and in the treatment of polycythemia with phenylhydrazine, when large amounts of iron are liberated in the body from the destruction of red cells, less than 0.5% of this iron appears in the urine and feces (154). Although absorbed iron is retained with great tenacity and, in the absence of bleeding, excretion is very small, the amounts lost cannot be neglected and are of nutritional importance, as Moore (164) has stressed.

The total iron in the feces of normal human adults usually lies between 6 and 16 mg/day, depending upon the amount ingested. Most of this consists of unabsorbed food iron. True excretory iron is estimated at about 0.2 mg/day by chemical balance studies (109) and at 0.3 to 0.5 mg/day by radioiron technique (59). This iron is derived from desquamated cells and from the bile. Iron occurs in the bile, mostly from hemoglobin breakdown, to the extent of about 1 mg/day. Most of this is reabsorbed and does not reach the feces. The amount of iron eliminated in the urine varies from as low as 0.02 mg/day (146) to as high as 2.0 mg/day (238). The mean urinary excretion of iron by normal adult men and women can be given as 0.2-0.3 mg/day (13, 98). This amount can be greatly increased, to as much as 10 mg/day, by the injection of certain chelating agents, and this technique can be used to increase iron loss in various excess iron storage diseases (68).

In addition to the iron excreted in the urine and feces, there is a continual dermal loss of iron in the sweat, hair, and nails. Most of this occurs in desquamated cells but cell-free sweat contains some iron. In one investigation an average iron content of 0.3 μg/ml was found in sweat low in cells and of 7.1 μg/ml for sweat high in cells (3). Hussain and Patwardhan (107) collected the sweat from the forearm of healthy Indian men and women and from iron-deficient anemic women. The "cell-rich" sweat of the healthy individuals averaged 1.15 and 1.61 μg Fe/ml compared with 0.34 and 0.44 μg/ml for the "cell free" sweat of men and women, respectively. In the anemic women the average iron content of the cell-rich sweat was 0.44 μg/ml and the cell-free fraction had no detectable iron. The total amount of iron lost daily in the sweat will depend upon the individual, the ambient temperature, and the cell content. Losses as high as 6.5 mg/day have been estimated in some circumstances (3, 72). Mitchell and Edman (161) in their 1962 review, assessed the average loss of iron through the skin of a healthy adult as about 0.5 mg/day. Iron lost by this route can be much greater in the tropics where the volume of sweat can be as much as 5 liters/day. Such losses have been proposed as a factor contributing to the high incidence of iron deficiency anemia in tropical areas (72).

The total quantity of iron lost in the urine, feces, and sweat amounts to 0.6 to 1.0 mg/day in the average individual. A loss of this magnitude is not inconsiderable when it is realized that the average amount of iron absorbed from ordinary diets is only 1.0-1.5 mg/day. In women the problem of iron balance is much more precarious because they are subject to regular additional iron losses in the menstrual blood, from the menarche to the menopause, apart from that lost from time to time with the newborn infant and its adnexa, and in the milk during lactation. Despite the low iron content of milk, lactation is of some significance in respect to iron loss. If the average concentration in human milk is taken as 0.5 μg Fe/ml and 800 ml is secreted daily, the loss of iron would amount to 0.4 mg/day. In some women the loss of iron in the milk would, of course, be much greater.

The results of numerous investigations of menstrual blood losses testify to the extremely large variation between women, with a very much smaller variation from period to period in the same women (92, 209). Barer and Fowler (12) reported an average blood loss of 50 ml, which would contain 23 mg of iron. In later studies, average losses of 35 to 70 ml, containing 16-32 mg iron were observed (121, 163). In a recent study of 12 young women over 12 months the mean individual menstrual losses ranged from 4 to 26 mg Fe per period (92). Higher losses have been reported in other investigations (80), with 60 to 80 ml of blood (27-37 mg Fe) regarded as the upper limit of normal blood loss (209). Assuming a normal 28-day cycle, it can be calculated that most women, between puberty and the menopause, lose 0.5-0.8 mg Fe/day as a consequence of the menstrual flow, while some lose considerably more.

### 4. *Intermediary Metabolism*

A high proportion of the absorbed iron is continuously redistributed throughout the body in several metabolic circuits, of which the cycle, plasma → erythroid marrow → red cell → senescent red cell → plasma, is quantitatively the most important. A series of subsidiary metabolic circuits exist, including the cycles, plasma → ferritin and hemosiderin → plasma, and plasma → myoglobin and iron-containing enzymes → plasma. It is apparent that the iron of the plasma provides the link between the cycles and facilitates ready exchange among them (Fig. 2).

The hemoglobin cycle dominates the intermediary metabolism of iron. The total mass of circulating hemoglobin, amounting to 800 to 900 g in a normal man, is synthesized and destroyed every 120-125 days, if the survival period of hemoglobin is taken as the same length of time as that of the erythrocyte. Some 21-24 mg of endogenous iron is, therefore, liberated daily, most of which is available for the replacement of the destroyed hemoglobin. This far transcends the amount of iron absorbed daily from ordinary diets. The removal of hemoglobin or nonviable erythrocytes, the breakdown of the heme moiety and release of iron, and the return of this iron to the plasma are the responsibility of the reticuloendothelial cells, particularly those of the liver, spleen, and bone marrow. Few red cells die intravascularly, except in certain hemolytic diseases, so that plasma hemoglobin is an unimportant pathway in normal hemoglobin metabolism.

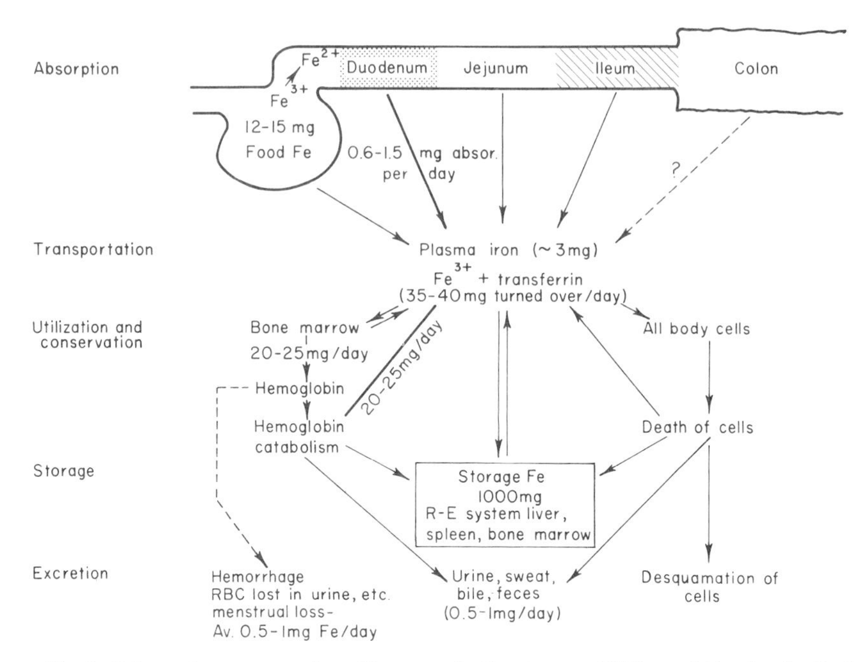


**Fig. 2.** Schematic representation of iron metabolism in man. RBC = red blood cells; R-E system = reticuloendothelial system. [From Moore (163).]

Kinetic studies, using radioiron, have shown that plasma iron has a very rapid turnover rate, with a normal half-time of only 90 to 100 min (198). In adult man, it has been calculated that 25-40 mg of total iron is transported in plasma every 24 hr, even though only 3-4 mg is present in the whole plasma volume at any one time. Pollycove (198) maintains that exchange of iron with the tissue stores, with myoglobin and the heme enzymes, and with the gastrointestinal tract and the sites of excretion is restricted to a few milligrams daily. He states further that more iron goes to the bone marrow than the 20-25 mg needed for hemoglobin synthesis. The remainder constitutes a labile marrow pool which is returned to the plasma. The large and rapid flow of plasma iron to the marrow is reflected in the completeness and promptness with which tracer doses of iron are used for hemoglobin synthesis. Tagged hemoglobin can be identified in the peripheral blood within 4 to 8 hr after administration of tracer iron; within 7 to 14 days from 70 to 100% of the isotope is found in circulating hemoglobin (163).

During the process of incorporation of iron into hemoglobin in the reticulocyte the first step is the binding of the transferrin in the cells (119, 177). The iron of transferrin is then released from transferrin and enters the pathway for heme synthesis. The release of iron from transferrin in reticulocytes is closely linked with the mitochondrial electron transport chain (175). The transferrin binding occurs in two stages, an initial, rapid, temperature-insensitive stage which is probably one of adsorption to the cell membrane, and a slower temperature-dependent stage which results in a firmer union of the transferrin to the cell (173). Recent experiments by Appleton and Morgan (6) suggest that in this second stage the transferrin molecules actually pass into the cells and are not exclusively localized to cell membrane receptors as previously thought (119). The mechanism of this process and its relation to intracellular organelles are as yet unknown, although it has been proposed that the process may be one of pinocytosis (6). Most of the iron used for hemoglobin synthesis comes from the transferrin of the plasma, in the manner described, but a portion of the iron released from red cell destruction in the phagocytic cells of the bone marrow can be used directly for this purpose, without going through the plasma cycle (18).

The mechanisms involved in the other major metabolic circuit of iron, namely the plasma→ferritin→plasma cycle have been greatly illuminated by the work of Mazur and associates (85, 149-151). The rapid process of incorporation of plasma iron into ferritin in the liver cells is dependent upon energy-yielding reactions in these cells for the continued synthesis of adenosine triphosphate (ATP) which, together with ascorbic acid, reduces the ferric iron of transferrin to the ferrous state, thus releasing it from its bond to protein and making it available for incorporation into ferritin. The reverse process, the release of iron from hepatic ferritin to the plasma, is mediated by the enzyme, xanthine oxidase, acting as a dehydrogenase. The reduced enzyme, formed as a result of oxidation of xanthine and hypoxanthine to uric acid, is reoxidized by some of the ferric iron of ferritin which is reduced to the ferrous state. In this reduced state the iron of ferritin dissociates easily from its bond to protein and is accepted by transferrin. Similar mechanisms are presumably involved in ferritin iron incorporation and its release in the spleen and bone marrow.

The iron of plasma is deposited in the liver and spleen just as readily in the form of hemosiderin as it is as ferritin, and iron is released from hemosiderin for utilization by the tissues just as readily as from ferritin. This occurs over quite a wide range of rates of iron deposition and depletion. The mechanism of release of iron from hemosiderin is unknown. Shoden and Sturgeon (222) have produced evidence that the iron of transferrin is not incorporated directly into hemosiderin as it is into ferritin, at least in rabbits. In this species ferritin has first to be formed, so that liver ferritin appears to be the immediate precursor of liver parenchymal cell hemosiderin.

In liver, spleen, and bone marrow, where the sinusoid walls are fenestrated, iron is transferred from the vascular to the extravascular space to reach the tissue cells, without prior dissociation from its firm complex with protein in transferrin. Morgan (172) has shown that in the rat and the rabbit, iron also remains bound to transferrin during passage through the capillary wall of relatively impermeable endothelial surfaces, as in the movement from the plasma to the lymph and to the peritoneal cavity. A different mechanism exists for the transfer of iron from the mother to the fetus. The route of iron transfer is from maternal plasma to placental tissue and from there to fetal plasma. Since the placental transfer of transferrin is many times less than that of iron (78,174), iron must dissociate from transferrin on the maternal side of the placenta and then be transferred across the placenta cells and be reassociated with new transferrin molecules on the fetal side. Using radioiron-labeled transferrin, Baker and Morgan (9) showed that, in the pregnant rabbit, transferrin is specifically bound to the fetal part of the placenta. They obtained no evidence of a specific placental binding of the transferrin present in the fetal circulation. The capacity of the fetal part of the placenta to bind transferrin increased markedly from the fifteenth to the twenty-eighth day of gestation, paralleling the sharp increase in the rate of iron transfer through the fetal part with advancing pregnancy. In studies with swine, Moustgaard and co-workers (181) showed that placental transfer of iron always occurs in protein-bound form but the mechanism is different from that of rodents. In the former species, injected radioiron is slower to appear in the fetal circulation and is transferred principally via the embryotrophic route and not via the hemototrophic route as in rodents and primates.

## IV. Iron Deficiency

### 1. *Manifestations of Iron Deficiency*

Iron deficiency in human adults is manifested clinically by listlessness and fatigue, palpitation on exertion, and sometimes by a sore tongue, angular stomatitis, dysphagia, and koilonychia (51). In children, anorexia, depressed growth, and decreased resistance to infection are commonly observed, as in other young, growing, iron-deficient animals, but the oral lesions and nail changes are rare. In all species iron deficiency results in the development of an anemia of the hypochromic, microcytic type, accompanied by a normoblastic, hyperplastic bone marrow which usually contains little or no hemosiderin. In addition, serum iron levels are subnormal, total iron-binding capacity is above normal, and there is a decreased saturation of transferrin. As Bainton and Finch (7) have pointed out, the presence of hypochromia and microcytosis of the circulating red cells is not essential for the diagnosis of iron deficiency anemia because an inadequate iron supply may retard erythropoiesis for some time before these characteristic abnormalities of the red cells are recognizable. Studies of a group of patients with iron deficiency anemia indicated that 16% saturation of plasma transferrin, or less, implies an inadequate supply of iron to the erythroid tissue and that this is associated in time with hypochromic, microcytic anemia. Marrow hemosiderin, together with the circulating hemoglobin level, are the best criteria for determination of total body iron. Marrow hemosiderin is not a reliable index of the adequacy of iron supply to the marrow. The best criterion of iron-deficient erythropoiesis is the per cent saturation of transferrin. Sideroblast count reflects both iron supply and hemoglobin synthesis by the red cells (7).

Abnormalities of the gastrointestinal tract, including gastric achlorhydria and associated histological lesions consisting of varying degrees of superficial gastritis and atrophy, have long been observed in iron deficiency anemia (99, 138). A high incidence of blood loss and loss of plasma proteins into the gastrointestinal lumen has also been reported in infants with iron deficiency anemia (260). More recently a diffuse and reversible enteropathy was observed in children as a result of iron deficiency (185). A high incidence of gastric achlorhydria, impaired absorption of xylose and vitamin A, and steatorrhea was evident in the small group studied, together with varying degrees of duodenitis and mucosal atrophy. Most of the abnormalities returned to normal following treatment with iron, and the abnormalities were generally absent from children with anemias not due to iron deficiency.

For many years it was believed that a lowering of hemoglobin levels was the primary and dominant manifestation of iron deficiency and that other heme proteins were little affected in this condition. Numerous investigations have now shown that various iron enzymes are depleted in the tissues under conditions of restricted iron supply. This raises the possibility that such changes underlie the absence of correlation, in some human patients, between the severity of the symptoms and the degree of anemia. Even muscle myoglobin, long considered to remain inviolate in iron deficiency, can be affected in some circumstances. A profound depression in myoglobin concentration has been observed in young iron-deficient pigs (88), rats (47,49), puppies (88), and chicks (53). Young animals are more susceptible to loss of myoglobin than mature animals (88), and skeletal muscles appear to be more susceptible to this loss than cardiac or diaphragmatic muscle (47).

Beutler (19,20) was the first to investigate systematically the heme enzymes in iron deficiency. In iron-deficient rats a marked decrease in cytochrome c was demonstrated in the liver and kidneys, with no reduction in catalase activity of red cells, a small reduction in cytochrome oxidase in the kidneys, but not in the heart, and a partial depletion of succinic dehydrogenase in the kidneys and heart but not in the liver. Beutler's findings have been confirmed and extended by many workers, indicating clearly that the body does not always accord priority to the heme enzymes under conditions of restricted iron supply. Cytochrome c can be reduced to as little as half of normal concentrations in the skeletal muscles, heart, liver, kidney, and intestinal mucosa of young iron-deficient rats (49,50,210), with a smaller reduction in the brain (49). Catalase activity appears to be very little affected in iron-deficient rats (47,227). Succinic dehydrogenase activity is reduced in the cardiac muscle of iron-deficient rats (20) and chicks (53). An actual increase has been reported in the brain, liver, and skeletal muscle of rats rendered anemic on a milk diet (227). The variations in degree of hemeprotein depletion in different tissues in response to iron deprivation are not understood. It has been suggested that they may be related to organ function, growth rate, and cell turnover (50).

### 2. *Iron Deficiency in Human Infants*

The fact that during the suckling period young mammals can become deficient in iron is now well documented, following the initial demonstration by Bunge as long ago as 1889 (30). The iron deficiency anemia of the suckling results from the provision of inadequate amounts of iron in the stores of the newborn and in the milk of the mother, relative to the rapid rate of growth of the young animal. It occurs even in the rat where the milk is many times richer in iron than is the milk of the human and most other species [see Ezekiel (62)]. In several studies of U.S. infants an incidence of iron deficiency anemia from as low as 8% (14) to as high as 64% (234) has been reported.

The anemia of human infants, which occurs most frequently between 4 and 24 months of age, is an iron deficiency anemia, since it normally responds rapidly to iron and is characterized by the typical abnormalities in erythrocyte morphology (234). A depletion of iron reserves occurs during the period of rapid growth of the infant, despite the considerable "store" of iron contained in the plethora of hemoglobin in the blood at birth. The blood of the newborn child normally contains 18-19 g Hb/100 ml which, if the blood volume is taken as 300 ml, represents 180-190 mg Fe. This is about 6 times the amount usually stored in the liver (155). At 6 months of age the average baby weighs 7 kg and has a blood hemoglobin level of about 12 g/100 ml. It, therefore, contains about 280 mg Fe in the form of hemoglobin at this time, of which 180-190 mg was present at birth. If this iron, plus the amount stored in the liver and absorbed from the milk, were all retained in the body, it is difficult to visualize iron deficiency arising at all. However, Cavell and Widdowson (38), in a study of babies at 1 week of age found significant negative iron balances, due to the excretion of some 10 times as much iron in the feces as was ingested in the milk daily. This is clearly an important cause of iron depletion, although how long such excess fecal iron excretion continues beyond the neonatal period is unknown.

A significant increase in hemoglobin levels in infants receiving oral iron in therapeutic doses has been demonstrated in a number of studies, with a greater rapidity of response from intramuscular injections of iron dextran than from oral doses (223). The regular need for supplementation with medicinal iron, or with iron-fortified foods, has been stressed by Sturgeon (234) and others (14,66), even with infants born of non-anemic mothers who have received ample iron during pregnancy. During normal pregnancy the level of maternal iron nutrition appears to exert little influence on the offspring's iron endowment at birth and, consequently, on his iron status during infancy. In premature babies and in babies born of frankly anemic mothers, the hemoglobin levels and the liver iron stores may be subnormal (155,233), indicating an even greater need for iron supplementation, especially during early infancy.

### 3. *Iron Deficiency in Human Adults*

Iron deficiency is the most frequently encountered, clinically manifest deficiency disease in man. In adult men and postmenopausal women the principal cause is chronic blood loss due to infections, malignancy, bleeding ulcers, and hookworm infestation. Iron deficiency anemia is much more common in women than in men because women of fertile age are subject to additional iron losses in menstruation, pregnancy, and lactation. The incidence of iron deficiency in women has been reported as 20-25% in different Swedish [see Rybo (209)] and English studies (137). In economically underprivileged groups, in both the developed and underdeveloped countries, a higher incidence has been observed, especially during the period of active child-bearing and during pregnancy

(25,183). This is related to a variety of factors, including infection, heavy reliance on foods of vegetable origin in which the iron is poorly available (108), and excessive sweating. Even where such adverse factors are not important, as in several U.S. studies with pregnant women, iron deficiency anemia ranging in incidence from 15 (4) to 58% (202) has been observed.

Normal adult females escape menstruation only by pregnancy. From the standpoint of evading iron loss this is unprofitable because some 350-450 mg Fe is lost in the fetus and its adnexa, compared with 200-300 mg yearly in the menstrual flow. Some compensation for the increased iron demands of pregnancy is achieved by increased absorptive efficiency, but this is not sufficiently effective to prevent signs of iron deficiency in many women in late pregnancy. In the third trimester of pregnancy, hemoglobin levels normally fall, due partly to an increase in plasma volume and partly, in most instances, to inadequate dietary intakes of iron. Supplemental iron has no effect upon the hydremia of the last trimester of pregnancy (139) but can significantly increase blood hemoglobin levels (139,168,235). The intramuscular injection of a single dose of 1000 mg of iron dextran, or the daily oral administration of 78 mg of ferrous iron for 24 weeks prepartum, was found equally effective for this purpose (139).

### 4. *Iron Deficiency in Baby Pigs*

Piglets denied access to sources of iron other than sow's milk develop anemia within 2 to 4 weeks of birth. The hemoglobin level of the blood may fall from a normal of about 10 g/100 ml to as low as 3-4 g/100 ml; breathing is labored and spasmodic; the skin becomes wrinkly and the coats rough; appetite declines; and growth is either poor or the piglets lose weight and a proportion die. Surviving members of litters usually begin a slow spontaneous recovery at 6 to 7 weeks of age, when they begin to eat significant amounts of the sow's food and to undertake such foraging as is permitted by their conditions of housing. The disease is an uncomplicated iron deficiency (96,157,248). Its hematology and pathology have been reviewed by Seamer (212). Piglet anemia is often associated with *Escherichia coli* infection or piglet edema disease. Decreased resistance to the endotoxin of *E. coli* has been demonstrated in anemic piglets (105).

The principal factors responsible for the particular sensitivity of the baby pig to iron deficiency anemia are its high growth rate and its relatively poor endowment with iron at birth. Piglets normally reach 4-5 times their birth weight at the end of 3 weeks and 8 times their birth weight at the end of 8 weeks. A growth rate of this magnitude imposes iron demands much greater than the amount of iron obtained from the milk, if there is to be no reduction in the concentration of iron in the body. At birth the pig has unusually low concentrations of total body iron and of liver iron stores (see Table 3). These are not significantly improved even when the sow's diet is supplemented with iron (199). Furthermore, there is no polycythemia of the newborn in the pig, as there is in the human and many other species, so that this source of endogenous iron is also denied to the baby pig. Anemia is, therefore, inevitable unless the piglet has access to or is supplied additional iron.

Many procedures have been adopted to raise the iron intake of piglets. The feeding of iron to the sow before or after farrowing is ineffective because such treatment neither increases the iron stores of the piglet at birth, nor raises the iron content of the sow's milk (200). Successful treatment involves direct increases in the iron intakes of the piglets. Maintaining a trough of soil or grass sods in the farrowing pen, or swabbing of the sow's udder with a concentrated iron solution, is quite effective, but the oral or parenteral administration of iron to the piglets within the first few days of life is now routine practice. Injection of compounds such as iron dextran or dextrin-ferric oxide possesses advantages in convenience and in response compared with treatment with oral iron, especially in promoting maximum hemoglobin levels (140,145,199). Oral administration of an iron tablet or solution containing 300-400 mg Fe within 4 days of birth is usually adequate to promote rapid growth and prevent mortality in the piglets. A second oral dose of this size is necessary for maximum hemoglobin levels. Even when an iron complex containing 100 mg Fe is injected at 2 to 4 days, a second injection within a further 2 weeks can increase hemoglobin levels at 3 to 4 weeks of age (140,145) (see Table 6).

TABLE 6

*Treatment of Piglet Anemia by Injection of Iron Complexes*[a]

| Treatment | No. of pigs | Hemoglobin (gm %) | | Body weight (lb) | |
|---|---|---|---|---|---|
| | | 0 | 3 weeks | 0 | 3 weeks |
| Untreated | 24 | 7.5 | 4.8 | 3.8 | 9.6 |
| Iron-dextran at 2 to 4 days (100 mg Fe) | 21 | 7.5 | 9.0 | 3.6 | 11.3 |
| Dextrin ferric oxide at 2 to 4 days (100 mg Fe) | 22 | 8.0 | 7.8 | 3.6 | 11.9 |
| Dextrin ferric oxide at 2 days and at 14 days (100 mg Fe each dose) | 16 | 8.3 | 10.5 | 3.8 | 11.5 |

[a]Taken from Linkenheimer *et al.* (140).

### 5. *Iron Deficiency in Newborn Lambs and Calves*

There is no convincing evidence that iron deficiency ever occurs in sheep or cattle of any age when grazing under natural conditions, except in circumstances involving blood loss or disturbance in iron metabolism as a consequence of parasitic infestation or disease. Heavy infestation with helminth intestinal parasites results in an iron deficiency type of anemia in lambs and calves (33,205). The effects of parasites such as the blood-sucking *Bunostomum* or *Haemonchus* can be accounted for largely by blood loss, through withdrawal of blood by the parasites. The effects of infestation with some trichostrongyle worms, such as *Chabertia* and *Ostertagia*, are less well understood. The anemia may occur through a rise in the rate of degradation of blood cells or through a depression of hematopoiesis from the action of toxic substances produced by the parasites. The relation of parasitic and infectious diseases to iron metabolism and anemia in farm animals has been reviewed by Kolb (132).

The postnatal development of anemia in lambs and calves, especially when fed an exclusive whole milk diet, and its alleviation by the administration of iron, is well documented [see Hibbs *et al.* (102)]. The anemia is usually mild, except under experimental conditions in which extraneous iron is carefully excluded, and occurs most frequently at 4 to 8 weeks of age. It is of limited practical significance under ordinary conditions of diet, housing, and management. Intramuscular injections of 150 mg Fe as iron dextran into newborn lambs or of 12 mg Fe per pound body weight into newborn calves produces significant improvements in hemoglobin levels and some improvement in body weight over several weeks (34). Bezau and Clark (22) observed no effect from such treatment on the growth of lambs or calves but some improvement in hemoglobin levels was noted in the calves. Rice *et al.* (204a) obtained increased blood hemoglobin levels from iron injections into beef calves at birth or at 2 months of age, together with increased weight gains during the period when the mild anemia was observed.

## V. Iron Requirements

### 1. *Man*

The demands of the body for iron are determined by the requirements for tissue growth and hemoglobin accretion and by the replacement needs imposed by iron losses in the urine, feces, and sweat and, in the female, in menstruation, gestation, and lactation. The former varies with the rapidity of growth at different periods of life and with changes in blood hemoglobin levels. The data of Heath and Patek (101) indicate that there are three periods when the demands for iron are greatest. These are during the first 2 years of life, during the period of rapid growth and hemoglobin increase of adolescence, and throughout the child-bearing period in women.

The iron requirement of the human infant during the first year of life may be as high as 1 mg /day but probably averages 0.6 mg/day (124). This rises to 1.5 to 2.5 mg in females from puberty onwards. At age 15-16 in males, the requirement also increases temporarily due to the rapid growth and hemoglobin accretion that occurs at this time. Estimates of total human requirements, which take all these factors into account, and which include the replacement needs for the various iron losses, have been made by Moore (163). Later studies by Monsen *et al.* (162) indicate that, to provide the iron needs of 95% of normal, menstruating women, enough iron must be consumed to permit absorption of approximately 2.0 mg/day. This average figure is equal to the maximum requirement for such women given by Moore. Estimates of the requirements for absorbed iron and for dietary iron recently made by the Council on Foods and Nutrition of the American Medical Association are given in Table 7 (39).

It is difficult to convert these physiological requirements into dietary requirements because of variations among individuals in absorptive capacity and among foods in the availability of their iron for absorption. The position is complicated further by the ability of the body to increase iron absorption during periods of iron deficiency. Normal subjects com-

TABLE 7
*Estimated Human Iron Requirements*[a]

| Subject | Absorbed iron requirement (mg/day) | Food iron requirement[b] (mg/day) |
|---|---|---|
| Normal men and nonmenstruating women | 0.5-1 | 5-10 |
| Menstruating women | 0.2-2 | 7-20 |
| Pregnant women | 2-4.8 | 20-48[c] |
| Adolescents | 1-2 | 10-20 |
| Children | 0.4-1 | 4-10 |
| Infants | 0.5-1.5 | 1.5(mg/kg)[d] |

[a]Taken from Committee on Iron Deficiency (39).
[b]Assuming 10% absorption.
[c]This amount cannot be derived from the diet and should be met by iron supplementation during the latter half of pregnancy.
[d]To a maximum of 15 mg.

monly absorb 5-10% and iron-deficient individuals 10-20% of dietary iron. Considerable divergence from these figures can occur with different individuals and different diets. Hussain and associates (108) have stressed that the absorption of iron by normal and by iron-deficient individuals from vegetable foods is significantly lower than it is from animal sources and iron salts (see Fig. 3). They found the iron in wheat to be 4% absorbed in normal and 7% in iron-deficient subjects, whereas the absorption of iron in hemoglobin in similar groups was 10 and 20%. In another study (194), ferrous salts of iron showed an average absorption for 5 mg of 7% in normal males and 44% in iron-deficient subjects. The latter value was reduced to 22% by the simple addition of food to the iron salts. The following dietary allowances for iron were made in 1964 by the U.S. National Research Council (204): children under 1 year, 6 mg/day; 1-3 years, 7 mg/day; 4-6 years, 8 mg/day; 7-9 years, 10 mg/day; 10-12 years, 12 mg/day; over 12 years, 15 mg/day; premeno-

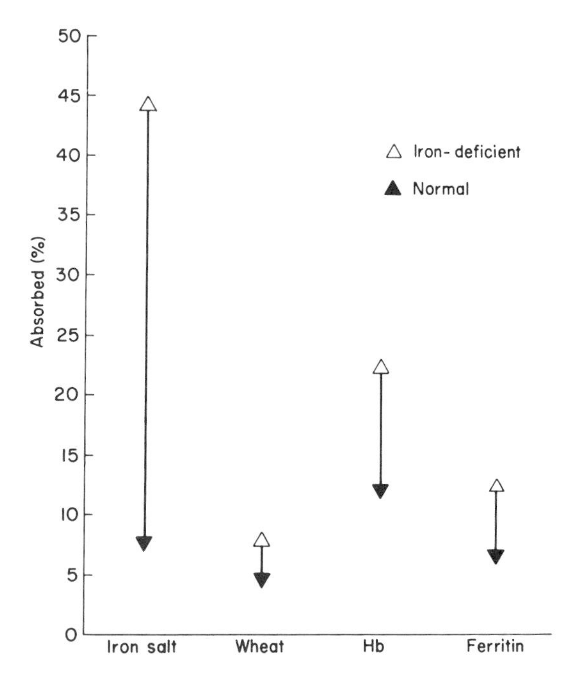


**Fig. 3.** Nutritive value of food iron. The average absorption of each form of iron in normal subjects is indicated by the solid arrow head and in iron-deficient subjects by the open arrow head. The distance between the two values indicates the ability of the absorptive mechanisms to modify absorption when iron deficiency is present. Hb = hemoglobin. [From Hussain *et al.* (108).]

pausal women, 15 mg/day; and men, 12 mg/day. More recent estimates, in which the absorption was taken as 10%, are given in Table 7.

2. *Pigs*

The iron requirements of pigs for growth cannot be given with any precision. Over 90% of the iron required in the first few weeks of life is for hemoglobin synthesis (27). Calculations have shown that the piglet must retain 7-11 mg Fe/day for this purpose, the actual amounts depending upon the rate of growth and the hemoglobin levels in the animals (133,248). Braude *et al.* (27) estimated that the piglet must retain 21 mg Fe/kg live weight increase in order to maintain a satisfactory level of iron in the body. This works out at less than the 7-11 mg Fe/day mentioned above, unless the rate of growth is exceptionally rapid.

The dietary concentrations of iron necessary to supply these quantities will vary with the percentage absorption of the iron in the diet. This will vary, in turn, with the iron intake and the proportion of other elements present in the diet, such as zinc, copper, and manganese. Ullrey and co-workers (240) found that 82, 61, and 50% of the iron intake was taken up from milk substitutes containing 25, 35, and 125 mg Fe/kg solids, respectively, whereas Matrone *et al.* (148) estimated that only 30% of the iron from fortified cow's milk containing 60 mg Fe/kg dry matter was utilized by pigs up to 60 days of age. A retention of about 30% of a 30-mg dose of labeled iron, given daily to baby pigs, is also apparent from the investigations of Braude *et al.* (27). Such variations in the availability of the iron in different diets and supplements and in the ability of animals to increase absorption in response to need, complicate the problem of defining the minimum iron requirements of pigs, as with other species, and no doubt explain some of the discrepant results reported. For instance, Ullrey *et al.* (240) found 125 ppm of iron to be necessary for full growth and hemoglobin production in baby pigs, whereas the results of Matrone *et al.* (148) indicate that 60 ppm is adequate. The U.S. National Research Council recommends an intake of 10 to 15 mg Fe/day for the first 6 weeks, which is comparable with the latter estimate.

With older pigs the problem is further complicated by the use of copper supplements as a means of stimulating growth. Such supplements significantly increase the pig's dietary requirements for iron (and zinc) to levels above "normal." This question is considered more fully in Chapter 3.

3. *Poultry*

From the results of experiments with diets almost devoid of iron, Hill and Matrone (103) suggested that a diet containing 50 ppm Fe and 5 ppm Cu would be close to the chick's requirements. Later experiments by Davis *et al.* (53) with soybean protein and with casein-gelatin diets, disclosed a higher iron requirement for chicks during the first 4 weeks of life. Their estimates, based on growth, blood data, myoglobin levels, liver iron stores, and succinic dehydrogenase activity, place the chick's requirements for iron at 75 to 80 ppm of the diet.

The demands for iron for egg production are extremely large compared with those of the nonlaying bird. An average hen's egg contains about 1 mg of iron, so that a heavy layer will lose 6 mg/week in the eggs produced or 150-200 mg a year. With the onset of laying, the efficiency of iron absorption is increased and serum iron levels are markedly raised. There is usually a small fall in blood hemoglobin levels, but no significant reduction in storage iron is apparent (203). It seems, therefore, that the increased iron requirement imposed by egg laying is met by these means and that this requirement is furnished by all normal laying rations without iron supplements.

### 4. *Cattle and Sheep*

Little is known of the iron requirements of sheep and cattle based upon definitive experiments, apart from some studies with calves at the preruminant stage. Matrone *et al.* (147) fed male calves on a milk diet containing 1 ppm Fe on the dry basis, alone or with supplemental iron at the rate of 30 or 60 mg/day. Normal growth and hemoglobin levels were maintained for 40 weeks from birth with either of the supplements. This suggests that the minimum iron requirements of calves for growth is not more than 30 mg/day—a level substantially lower than the requirement of nearly 100 mg/day indicated by the study carried out by Blaxter *et al.* (23a). More recently, Lawlor *et al.* (137a) fed graded increments of iron, as ferrous sulfate, to lambs consuming a purified diet under conditions designed to minimize extraneous sources of iron. In one experiment the lambs were 8-10 weeks of age at the commencement of a treatment period of 56 days. In a second experiment the lambs were 5-6 weeks of age at the commencement and treatment lasted for 98 days. The results of these experiments indicate that 10 ppm Fe is clearly inadequate and that the minimum dietary iron requirement of the growing-finishing lamb is greater than 25 ppm and not more than 40 ppm.

## VI. Sources of Iron

Compilations of the iron contents of the edible portions of a wide range of human foods have been made in several countries (16,-

100, 158, 187, 231, 263). Individual variability among samples of the same type of food is high in all these studies, as a reflection of varietal differences and variations in the soil and climatic conditions. Consistent differences in the iron content of various types of foods nevertheless occur. Rich sources of iron are the organ meats (liver, kidney, and heart), egg yolk, dried legumes, cocoa, cane molasses, shellfish, and parsley. Poor sources include milk and milk products, white sugar, white flour, and bread (unenriched), polished rice, sago, potatoes, and most fresh fruit. Foods of intermediate iron content are the muscle meats, fish, and poultry, nuts, green vegetables, and wholemeal flour, and bread. Boiling in liberal amounts of water can reduce the levels of iron in vegetables by as much as 20% (224). The iron content of cereals is also greatly reduced by removal of the germ and outer branny layers in the process of milling. In a study of North American wheats and the flours milled from these wheats, the mean iron content of the whole grain was found to be 43 ppm and that of the flour, 10.5 ppm (48).

The overall intake of iron from different diets varies greatly with the proportion of iron-rich and iron-poor foods that they contain, with the degree of contamination with iron to which they have been exposed and, to some extent, with the locality from which they are obtained. Average U.S. dietaries were reported over 30 years ago to supply 14-20 mg Fe per man value daily (215). Australian diets, which are typically high in meat, have been estimated to supply 20-22 mg Fe per adult male daily (188). A typical poor Indian diet was shown to provide only 9 mg of total iron daily, whereas an improved diet containing less milled rice and more pulses and green vegetables could provide as much as 60 mg/day (100). Such average figures can obscure important differences among individuals, families, and groups. Thus Stiebeling and associates (232) demonstrated significant differences in iron intakes in families classified according to their expenditure on food. At the three levels of expenditure included, the iron intake per "nutrition unit" averaged 12, 14, and 16 mg/day with increasing expenditure on food. In Australia, iron intake per nutrition unit fell, with increasing size of family, from an average of 15 to 11 mg/day (188). Iron intakes are, therefore, most likely to be inadequate where there is a heavy dependence on low-cost, high-energy foods, low in iron and with the iron that they do contain of low availability. Adequacy of the diet in other respects does not necessarily ensure adequacy in iron. Monsen *et al.* (162) found that the daily iron intakes of 13 young women, maintaining their normal eating habits, averaged only 9.2 mg when chemically analyzed and 9.9 mg when calculated. This is well below the 15 mg/day recommended for such individuals, although the diets were stated to be adequate in protein, calcium, vitamins A and C, and the B vitamins. However, the diet of these women provided only 1600 cal/day. In a recent review of the iron nutriture of girls and women in North America, consistent average iron intakes of the order of 10 to 12 mg/day were found (257). The importance of caloric intake in determining iron intake is illustrated by the study of Zook and Lehmann (262). The total diets (including the drinking water) were designed by these workers for 16 to 19-year-old boys and furnished no less than 4200 cal/day. An average of 35.6 mg Fe/day was supplied by these diets.

With normal men and postmenopausal women, there is little reason for general concern about iron intakes. Women during the fertile period are in a more precarious position, especially as food iron consumption may be declining, due to a progressive decrease in total calorie intake and to reduced opportunities for iron contamination as a result of improved cleanliness in handling of foods and a declining domestic use of iron cooking vessels. Some form of iron supplementation or food fortification is, therefore, probably necessary. The most popular vehicle for iron fortification is white flour, because of its universal consumption and the ease with which the iron can be incorporated. Several forms of iron are or have been used for this purpose, including reduced iron (ferrum redactum), ferric ammonium citrate, and ferric pyrophosphate. The results of a long-term clinical trial carried out with women consuming bread fortified with ferrum redactum gave no evidence of therapeutic value as judged by changes in hemoglobin levels (61). This contrasted with the significant mean increase shown by women receiving ferrous gluconate in tablet form and suggested that the iron in the fortified bread was poorly absorbed. Strong support for this conclusion comes from the work of Fritz (73) who compared the availability of various iron sources by measuring their capacity to promote hemoglobin regeneration in iron-deficient rats and chicks. Marked differences, similar in both species, were obtained. Good sources were ferric ammonium citrate, ferrous sulfate, gluconate, and fumarate, ferrous ammonium sulfate, and isolated soybean protein. Mediocre sources were reduced iron, ferric chloride and pyrophosphate, fish protein concentrate, flour, and corn meal enriched with reduced iron. Poor sources were ferric oxide, ferrous carbonate, sodium iron pyrophosphate, ferric orthophosphate and, smectile-vermiculite.

A possible influence on iron intakes of the source, as well as the choice of foods, was disclosed in a study of nutritional anemia in school children in rural areas in Florida (2). Typical diets consumed by these children provided only 4-6 mg Fe/day, and rapid restoration of hemoglobin levels was achieved by iron therapy. The low iron intakes were associated with poor food habits and with the consumption of a high proportion of locally grown foods, produced on poor sandy soils. Turnip greens, the most important source of iron to these people, contained only 50-60 ppm Fe when grown on the poor soils, compared with some 250 ppm on better soils. The anemia was increased in severity by hookworm infestation, in some instances, but it was concluded that the condition was due principally to poor food habits, accentuated by the dependence on locally grown materials subnormal in iron content.

The iron content of animal feeds, as with human foods, is highly variable. The concentration of iron in herbage plants is a reflection of the species and of the type of soil upon which the plants grow. Contamination with soil and dust can also greatly affect the iron content of plants as they are consumed by animals in the field. The normal range for leguminous pasture species is 200-300 ppm Fe on the dry basis and 100-250 ppm for grasses. Values as high as 700-800 ppm have been recorded for alfalfa and as low as 40 ppm for some grasses grown on poor, sandy soils [see Underwood (243)]. Most cereal grains contain 30-60 ppm Fe and species differences are small. The leguminous seeds and the oil seeds are appreciably richer in iron than the cereal grains. The oil seed meals commonly contain 100-200 ppm Fe, whether they are produced by expeller or by solvent processes.

Feeds of animal origin, with the exception of milk and milk products, such as skim milk, whey, and buttermilk powders, are rich sources of iron. Thus blood meal may contain over 3000 ppm Fe, and meat meals and fish meals commonly contain 400-600 ppm. Many of the inorganic compounds used as mineral supplements can contain very high concentrations of iron. For example, ground limestone, oyster shell, and many forms of calcium phosphate frequently contain 2000-5000 ppm or 0.2-0.5% iron.

All the figures just quoted refer to total iron, as determined chemically. Little is known of the availability of this iron to herbivorous species, although the iron in herbage plants is only about half as available for hemoglobin production in anemic rats as the iron in ferric chloride (237). A similar availability for this purpose has been found for the iron in wheat grain (186). These findings are of value in comparing different materials as sources of useful iron to the rat and no doubt to other monogastric species. They cannot be applied with any confidence to ruminants with their very different digestive and absorptive apparatus. However, differences in availability to ruminants of different chemical forms of iron have been demonstrated. Ammerman and co-workers (5a) showed that orally administered ferrous sulfate, ferrous carbonate, and ferric chloride ranked in decreasing order of availability to calves and sheep but were not significantly different on the basis of tissue $^{59}Fe$ deposition, whereas the iron in ferric oxide was significantly less available on this basis.

## REFERENCES


1. R. Aasa, B. G. Malmstrom, P. Saltman, and T. Vanngard, *Biochim. Biophys. Acta* **75**, 203 (1963).
2. O. D. Abbott, *Milbank Mem. Fund, Annu. Conf.*, New York (1940).
3. W. S. Adams, A. Leslie, and M. H. Levin, *Proc. Soc. Exp. Biol. Med.* **74**, 46 (1950).
4. B. I. Allaire and A. Campagna, *Obstet. Gynecol.* **17**, 605 (1961).
5. J. W. Allgood and E. B. Brown, *Scand. J. Haematol.* **4**, 217 (1967).

5a. C. B. Ammerman, J. M. Wing, B. G. Dunavant, W. K. Robertson, J. P. Feaster, and L. R. Arrington, *J. Anim. Sci.* **26**, 404 (1967).

6. T. C. Appleton and E. H. Morgan, *Nature (London)* **223**, 1371 (1969).
7. D. F. Bainton and C. A. Finch, *Amer. J. Med.* **37**, 62 (1964).
8. E. Baker, S. M. Jordan, A. A. Tuffery, and E. H. Morgan, *Life Sci.* **8**, 89 (1969).
9. E. Baker and E. H. Morgan, *Quart. J. Exp. Physiol.* **54**, 173 (1969).
10. E. Baker, D. C. Shaw, and E. H. Morgan, *Biochemistry* **7**, 1371 (1968).
11. R. M. Bannerman, *J. Lab. Clin. Med.* **65**, 944 (1965).
12. A. P. Barer and W. M. Fowler, *Amer. J. Obstet. Gynecol.* **31**, 979 (1936).
13. A. P. Barer and W. M. Fowler, *J. Lab. Clin. Med.* **23**, 148 (1937).
14. V. A. Beal, A. J. Meyers, and R. W. McCammon, *Pediatrics* **30**, 518 (1962).
15. A. G. Bearn and W. C. Parker, *in* "Transferrin in Glycoproteins, their Composition, Structure and Function" (A. Gottschalk, ed.), Elsevier, Amsterdam, 1966.
16. R. Belz, *Voeding* **21**, 236 (1960).
17. J. Bernard, M. Boiron, and C. Paoletti, *Rev. Fr. Etud. Clin. Biol.* **3**, 367 (1958).
18. M. C. Bessis and J. Breton-Gorius, *J. Biophys. Biochem. Cytol.* **3**, 503 (1957).
19. E. Beutler, *Amer. J. Med. Sci.* **234**, 517 (1957); *Acta Haematol.* **21**, 371 (1959); *J. Clin. Invest.* **38**, 1605 (1959).
20. E. Beutler and R. K. Blaisdell, *J. Lab. Clin. Med.* **52**, 694 (1958); *Blood* **15**, 30 (1960).
21. E. Beutler, M. J. Robson, and E. Buttenweiser, *Ann. Intern. Med.* **48**, 60 (1958).
22. L. M. Bezeau and R. D. Clark, *Can. J. Comp. Med.* **29**, 283 (1965).
23. J. C. Biggs and A. E. Davis, *Lancet* **1**, 814 (1963); *Aust. Ann. Med.* **15**, 36 (1966).

23a. K. L. Blaxter, G. A. M. Sharman, and A. M. MacDonald, *Brit. J. Nutr.* **11**, 234 (1957).

24. C. F. Bond, *Endocrinology* **43**, 180 (1954).
25. T. H. Bothwell and C. A. Finch, *in* "Iron Metabolism." Little, Brown, Boston, Massachusetts, 1962.
26. T. H. Bothwell, G. Pirzio-Biroli, and C. A. Finch, *J. Lab. Clin. Med.* **51**, 24 (1958).
27. R. Braude, A. G. Chamberlain, M. Kotarbinski, and K. G. Mitchell, *Brit. J. Nutr.* **16**, 427 (1962).
28. A. B. Brock and L. M. Diamond, *J. Pediat.* **4**, 445 (1934).
29. E. B. Brown and B. W. Justus, *Amer. J. Physiol.* **194**, 319 (1958).
30. G. von Bunge, *Hoppe-Seyler's Z. Physiol. Chem.* **13**, 399 (1889).
31. J. H. Byers, I. R. Jones, and J. R. Haag, *J. Dairy Sci.* **35**, 661 (1953).
32. S. T. Callender, B. J. Mallett, and M. D. Smith, *Brit. J. Haematol.* **3**, 186 (1957).

33. J. A. Campbell and A. C. Gardiner, *Vet. Rec.* **72**, 1006 (1960).
34. R. H. Carlson, M. J. Swenson, G. M. Ward, and N. H. Booth, *J. Amer. Vet. Med. Ass.* **139**, 457 (1961).
35. G. E. Cartwright, C. J. Gubler, and M. M. Wintrobe, *J. Clin. Invest.* **33**, 685 (1954).
36. G. E. Cartwright and M. M. Wintrobe, *J. Clin. Invest.* **28**, 86 (1949).
37. G. E. Cartwright and M. M. Wintrobe, *in* "Modern Trends in Blood Diseases" (J. F. Wilkinson, ed.), p. 183. Butterworth, London and Washington, D.C., 1954.
38. P. A. Cavell and E. M. Widdowson, *Arch. Dis. Childhood* **39**, 496 (1964).
39. Committee on Iron Deficiency, Council on Foods and Nutrition, *J. Amer. Med. Ass.* **203**, 407 (1968).
40. M. E. Conrad and W. H. Crosby, *Blood* **22**, 406 (1963).
41. M. E. Conrad, L. R. Weintraub, and W. H. Crosby, *J. Clin. Invest.* **43**, 963 (1964).
42. M. E. Conrad, L. R. Weintraub, D. A. Sears, and W. H. Crosby, *Amer. J. Physiol.* **211**, 1123 (1966).
43. J. D. Cook, G. M. Brown, and L. S. Valberg, *J. Clin. Invest.* **43**, 1185 (1964).
44. J. W. Cowan, M. Esfahani, J. P. Salji, and S. A. Azzam, *J. Nutr.* **90**, 423 (1966).
45. W. M. Cox and A. J. Mueller, *J. Nutr.* **13**, 249 (1937).
46. W. H. Crosby, *Blood* **22**, 441 (1963).
47. R. P. Cusack and W. D. Brown, *J. Nutr.* **86**, 383 (1965).
48. C. P. Czerniejewski, C. W. Shank, W. G. Bechtel, and W. B. Bradley, *Cereal Chem.* **41**, 65 (1964).
49. R. P. Dallman, *J. Nutr.* **97**, 475 (1969).
50. P. D. Dallman and H. C. Schwartz, *Pediatrics* **35**, 677 (1965); *J. Clin. Invest.* **44**, 1631 (1965).
51. W. J. Darby, *in* "Handbook of Nutrition," 2nd ed., McGraw-Hill (Blakiston), New York, 1951.
52. A. E. Davis and J. C. Biggs, *Aust. Ann. Med.* **13**, 201 (1964); *Gut* **6**, 140 (1965).
53. P. N. Davis, L. C. Norris, and F. H. Kratzer, *J. Nutr.* **94**, 407 (1968).
54. P. S. Davis, C. G. Luke, and D. J. Deller, *Lancet* **2**, 1431 (1966).
55. S. S. Derechin and P. Johnson, *Nature (London)* **194**, 473 (1962).
56. W. J. Dieckman and C. R. Wegner, *Arch. Intern. Med.* **53**, 71 (1934).
57. D. L. Drabkin, *Physiol. Rev.* **31**, 345 (1951).
58. R. Dubach, S. T. Callender, and C. V. Moore, *Blood* **3**, 526 (1948).
59. R. Dubach, C. V. Moore, and S. T. Callender, *J. Lab. Clin. Med.* **45**, 599 (1955).
60. A. Ehrenberg and C. B. Laurell, *Acta Chem. Scand.* **9**, 68 (1955).
61. P. C. Elwood, *Brit. Med. J.* **1**, 224 (1963).
62. E. Ezekiel, *J. Lab. Clin. Med.* **70**, 138 (1967).
63. E. Ezekiel and E. H. Morgan, *J. Physiol. (London)* **165**, 336 (1963).
64. M. Faber and I. Falbe-Hansen, *Nature (London)* **184**, 1043 (1959).
65. A. Fabiano, *Ann. Obstet. Gynecol.* **78**, 1043 (1956).
66. J. D. Farquhar, *Amer. J. Dis. Child.* **106**, 201 (1963).
67. Y. M. Fenillon and M. Plumier, *Acta Pediat. (Stockholm)* **41**, 138 (1952).
68. W. G. Figueroa, *in* "Metal-Binding in Medicine" (M. J. Seven and L. A. Johnson, eds.), p. 146. Lippincott, Philadelphia, Pennsylvania, 1960.
69. H. Fischer and K. Zeile, *Justus Liebigs Ann. Chem.* **468**, 98 (1929).
70. J. Fletcher and E. R. Huehns, *Nature (London)* **215**, 584 (1967).
71. W. M. Fowler, *Ann. Med. Hist.* **8**, 168 (1936).
72. H. Foy and A. Kondi, *J. Trop. Med. Hyg.* **60**, 105 (1957).
73. J. C. Fritz, G. W. Pla, T. Roberts, J. W. Boehne, and E. L. Hove, *Agr. Food Chem.* **18**, 647 (1970).
74. B. W. Gabrio and K. Salomon, *Proc. Soc. Exp. Biol. Med.* **75**, 124 (1950).

75. B. W. Gabrio, A. W. Shoden, and C. A. Finch, *J. Biol. Chem.* **204**, 815 (1953).
76. M. R. Gardiner, *J. Comp. Pathol.* **75**, 397 (1965).
77. C. A. Gemzell, H. Robbs, and T. Sjöstrand, *Acta Obstet. Gynecol. Scand.* **33**, 289 (1954).
78. D. Gitlin, J. Kumate, and C. Morales, *J. Clin. Invest.* **43**, 1938 (1964).
79. S. E. Gitlow and M. R. Beyers, *J. Lab. Clin. Med.* **39**, 337 (1952).
80. E. Goltner and H. J. Gailer, *Zentralbl. Gynaekol.* **34**, 1177 (1964).
81. W. G. Gordon, J. Ziegler, and J. J. Basch, *Biochim. Biophys. Acta* **60**, 410 (1962).
82. S. Granick, *in* "Trace Elements" (C. A. Lamb, O. G. Bentley, and J. M. Beattie, eds.), p. 365. Academic Press, New York, 1958.
83. S. Granick, *J. Biol. Chem.* **164**, 737 (1946); *Physiol. Rev.* **31**, 489 (1951).
84. S. Granick and L. Michaelis, *Science* **95**, 439 (1942).
85. S. Green and A. Mazur, *J. Biol. Chem.* **227**, 653 (1957).
86. S. Green, A. K. Saha, A. W. Carleton, and A. Mazur, *Fed. Proc., Fed. Amer. Soc. Exp. Biol.* **17**, 233 (1958).
87. M. L. Groves, *J. Amer. Chem. Soc.* **82**, 3345 (1960).
88. C. J. Gubler, G. E. Cartwright, and M. M. Wintrobe, *J. Biol. Chem.* **224**, 533 (1957).
89. R. L. Haden, *J. Amer. Med. Ass.* **111**, 1059 (1938).
90. P. F. Hahn, *Medicine (Baltimore)* **16**, 249 (1937).
91. P. F. Hahn, W. F. Bale, J. F. Ross, W. M. Balfour, and G. H. Whipple, *J. Exp. Med.* **78**, 169 (1943).
92. L. Hallberg and L. Nilsson, *Acta Obstet. Gynecol. Scand.* **43**, 352 (1964).
93. L. Hallberg and L. Sölvell, *Acta Med. Scand.* **181**, 335 (1967).
94. B. Hallgren, *Acta Soc. Med. Upsal.* **59**, 79 (1954).
95. L. D. Hamilton, C. J. Gubler, G. E. Cartwright, and M. M. Wintrobe, *Proc. Soc. Exp. Biol. Med.* **73**, 65 (1950).
96. E. B. Hart, C. A. Elvehjem, H. Steenbock, G. Bohstedt, and J. M. Fargo, *J. Nutr.* **2**, 227 (1930).
97. W. J. Hartley, J. Mullins, and B. M. Lawson, *N. Z. Vet. J.* **7**, 99 (1959).
98. W. W. Hawkins, *in* "Nutrition" (G. H. Beaton, ed.), Vol. 1, p. 309. Academic Press, New York, 1964.
99. J. C. Hawksley, R. Lightwood, and U. M. Bailey, *Arch. Dis. Childhood* **9**, 359 (1934).
100. Health Bull. No. 23. Govt. India Press, New Delhi, 1951.
101. C. W. Heath and A. J. Patek, *Medicine (Baltimore)* **16**, 267 (1937).
102. J. W. Hibbs, H. R. Conrad, J. H. Vandersall, and C. Gale, *J. Dairy Sci.* **46**, 1118 (1963).
103. C. H. Hill and G. Matrone, *J. Nutr.* **73**, 425 (1961).
104. C. G. Holmberg and C. B. Laurell, *Acta Chem. Scand.* **1**, 944 (1947).
105. Z. Horvath, *Proc. 1st Int. Symp. Trace Element Metab. Anim., 1969* p. 328 (1970).
106. R. Hussain and V. N. Patwardhan, *Indian J. Med. Res.* **47**, 676 (1959).
107. R. Hussain and V. N. Patwardhan, *Lancet* **1**, 1073 (1959); R. Hussain, V. N. Patwardhan, and S. Sriranachari, *Indian J. Med. Res.* **48**, 235 (1960).
108. R. Hussain, R. B. Walker, M. Layrisse, P. Clark, and C. A. Finch, *Amer. J. Clin. Nutr.* **20**, 842 (1967).
109. R. L. Ingalls and F. A. Johnston, *J. Nutr.* **53**, 351 (1954).
110. D. J. E. Ingram, J. F. Gibson, and M. F. Peratz, *Nature (London)* **178**, 906 (1956).
111. A. A. Iodice, D. A. Richert, and M. P. Schulman, *Fed. Proc., Fed. Amer. Soc. Exp. Biol.* **17**, 248 (1958).
112. A. G. Itzerott, *J. Aust. Inst. Agr. Sci.* **8**, 119 (1942).
113. A. Jacobs and P. M. Miles, *Gut* **10**, 226 (1969).

114. A. Jacobs and G. M. Owen, *Gut* **10**, 488 (1969).
115. A. Jacobs, J. Rhodes, D. K. Peters, H. Campbell, and J. D. Eakins, *Brit. J. Haematol.* **12**, 728 (1966).
116. P. Jacobs, T. H. Bothwell, and R. W. Charlton, *J. Appl. Physiol.* **19**, 187 (1964).
117. G. A. Jamieson, *J. Biol. Chem.* **240**, 2914 (1965).
118. J. H. Jandl, J. K. Inman, R. L. Simmons, and D. W. Allen, *J. Clin. Invest.* **38**, 161 (1959).
119. J. H. Jandl and J. Katz, *J. Clin. Invest.* **42**, 314 (1963).
120. B. Johanson, *Acta Chem. Scand.* **14**, 510 (1960).
121. F. A. Johnston and T. J. McMillan, *J. Amer. Diet. Ass.* **28**, 633 (1952).
122. S. M. Jordan, I, Kaldor, and E. H. Morgan, *Nature* (*London*) **215**, 76 (1967).
123. H. W. Josephs, *Blood* **13**, 1 (1958).
124. H. W. Josephs, *Acta Paediat.* (*Stockholm*) **48**, 403 (1959).
125. I. Kaldor, *Aust. J. Exp. Biol. Med. Sci.* **32**, 437 (1954).
126. I. Kaldor, *Aust. J. Exp. Biol. Med. Sci.* **36**, 173 (1958).
127. I. Kaldor, personal communication (1956).
128. I. Kaldor and E. Ezekiel, *Nature* (*London*) **196**, 175 (1961).
129. M. G. Kamarkar and C. V. Ramakrishnan, *Acta Paediat. Belg.* **49**, 599 (1960); quoted by Ezekiel and Morgan (63).
130. T. D. Kenney, D. M. Hegsted, and C. A. Finch, *J. Exp. Med.* **90**, 137 and 147 (1949).
131. L. M. H. Kerr, *Biochem. J.* **67**, 627 (1957).
132. E. Kolb, *Advan. Vet. Sci.* **8**, 49 (1963).
133. M. Kotarbinski, *Postepy Nauk Roln.* **22**, 23 (1960); quoted by Braude *et al.* (27).
134. W. E. Kraus and R. G. Washburn, *J. Biol. Chem.* **114**, 247 (1936).
135. D. J. Kroe, N. Kaufman, J. V. Klavins, and T. D. Kinney, *Amer. J. Physiol.* **211**, 414 (1966).
136. E. Kuhn, V. Brodan, M. Brodanova, and B. Friedmann, *Nature* (*London*) **213**, 1041 (1967).
137. V. Laufberger, *Bull. Soc. Chim. Biol.* **19**, 1575 (1937).
137a. M. J. Lawlor, W. H. Smith, and W. M. Beeson, *J. Anim. Sci.* **24**, 742 (1965).
138. F. Lees and F. D. Rosenthal, *Quart. J. Med.* **27**, 19 (1958).
139. N. K. M. de Leeuw, L. Lowenstein, and Y-S. Hsieh, *Medicine* (*Baltimore*) **45**, 219 (1966).
140. W. H. Linkenheimer, E. L. Patterson, R. Milstrey, J. A. Brockman, and D. D. Johnston, *J. Anim. Sci.* **19**, 763 (1960).
141. C. G. Luke, P. S. Davis, and D. J. Deller, *Lancet* **1**, 926 (1967).
142. A. B. Macallum, *J. Physiol.* (*London*) **16**, 268 (1894).
143. C. A. MacMunn, *Phil. Trans. Roy. Soc. London* **177**, 267 (1885).
144. H. R. Mahler and D. G. Elowe, *J. Amer. Chem. Soc.* **75**, 5769 (1953).
145. J. H. Maner, W. G. Pond, and R. S. Lowrey, *J. Anim. Sci.* **18**, 1373 (1959).
146. A. Marlow and F. H. L. Taylor, *Arch. Intern. Med.* **53**, 551 (1934).
147. G. Matrone, C. Conley, G. H. Wise, and R. K. Waugh, *J. Dairy Sci.* **40**, 1437 (1957).
148. G. Matrone, E. L. Thomason, and C. R. Bunn, *J. Nutr.* **72**, 459 (1960).
149. A. Mazur and A. Carleton, *Blood* **26**, 317 (1965).
150. A. Mazur, S. Green, and A. Carleton, *J. Biol. Chem.* **235**, 595 (1960).
151. A. Mazur, S. Green, A. Saha, and A. Carleton, *J. Clin. Invest.* **37**, 1809 (1958).
152. M. G. McCall, G. E. Newman, J. R. O'Brien, and L. J. Witts, *Brit. J. Nutr.* **16**, 305 (1962).
153. R. A. McCance and E. M. Widdowson, *Lancet* **2**, 680 (1937); *J. Physiol.* (*London*) **94**, 138 (1938).

154. R. A. McCance and E. M. Widdowson, *Nature* (*London*) **152**, 326 (1943).
155. R. A. McCance and E. M. Widdowson, *Brit. Med. Bull.* **7**, 297 (1951).
156. E. I. McDougal, *J. Agr. Sci.* **37**, 337 (1947).
157. J. P. McGowan and A. Crichton, *Biochem. J.* **17**, 204 (1923); **18**, 265 (1924).
158. *Med. Res. Council* (*Gt. Brit.*), *Spec. Rep. Ser. No. 235* (1942).
159. G. A. Mendel, *Blood* **18**, 727 (1961).
160. G. A. Mendel, R. J. Wiler, and A. Mangalik, *Blood* **22**, 450 (1963).
161. H. H. Mitchell and M. Edman, *Amer. J. Clin. Nutr.* **10**, 163 (1962).
162. E. R. Monsen, I. N. Kuhn, and C. A. Finch, *Amer. J. Clin. Nutr.* **20**, 842 (1967).
163. C. V. Moore, *Harvey Lect.* **55**, 67 (1951).
164. C. V. Moore, *Amer. J. Clin. Nutr.* **3**, 3 (1955).
165. C. V. Moore and R. Dubach, *Trans. Ass. Amer. Physicians* **64**, 245 (1951).
166. C. V. Moore, R. Dubach, V. Minnich, and H. K. Roberts, *J. Clin. Invest.* **23**, 755 (1944).
167. H. O. Moore, *Aust., Commonw. Sci. Ind. Res. Organ., Bull.* **113** (1938).
168. E. H. Morgan, *Lancet* **1**, 9 (1961).
169. E. H. Morgan, *Aust. J. Exp. Biol. Med. Sci.* **39**, 361 and 371 (1961).
170. E. H. Morgan, *J. Pathol. Bacteriol.* **84**, 65 (1962).
171. E. H. Morgan, *Quart. J. Exp. Physiol.* **47**, 57 (1962); **48**, 170 (1963).
172. E. H. Morgan, *J. Physiol.* (*London*) **169**, 339 (1963).
173. E. H. Morgan, *Brit. J. Haematol.* **10**, 442 (1964).
174. E. H. Morgan, *J. Physiol.* (*London*) **171**, 26 (1964).
175. E. H. Morgan and E. Baker, *Biochim. Biophys. Acta* **184**, 442 (1969).
176. E. H. Morgan, E. R. Huehns, and C. A. Finch, *Amer. J. Physiol.* **210**, 579 (1966).
177. E. H. Morgan and C. B. Laurell, *Brit. J. Haematol.* **9**, 471 (1963).
178. E. H. Morgan and M. N. I. Walters, *J. Clin. Pathol.* **16**, 101 (1963).
180. S. D. Morrison, *Tech. Commun., Bur. Anim. Nutr.* No. 18 (1952).
181. J. Moustgaard, B. Palludan, and I. Wegger, *in* "Trace Mineral Studies with Isotopes in Domestic Animals," Proc. Joint FAO/IAEA Panel. I.A.E.A., Vienna, 1968.
182. A. R. Muir, *Quart. J. Exp. Physiol.* **45**, 192 (1960).
183. C. Mukherjee and S. K. Mukherjee, *J. Indian Med. Ass.* **22**, 345 (1953).
184. M. J. Murray and N. Stein, *Proc. 1st Int. Symp. Trace Element Metab. Anim., 1969* p. 321 (1970).
185. J. L. Naiman, F. A. Oski, L. K. Diamond, G. F. Vawter, and H. Shwachman, *Pediatrics* **33**, 83 (1964).
186. F. I. Nakamura and H. H. Mitchell, *J. Nutr.* **25**, 39 (1943).
187. *Nat. Health Med. Res. Council* (*Aust.*), *Spec. Rep.* No. 2 (1946).
188. *Nat. Health Med. Res. Council* (*Aust.*), *Spec. Rep.* No. 1 (1945).
189. W. L. Niccum, R. L. Jackson, and G. Stearns, *Amer. J. Dis. Child.* **86**, 553 (1953).
190. L. Otis and M. C. Smith, *Science* **91**, 146 (1940).
191. B. Panic, *Proc. 1st Int. Symp. Trace Element Metab. Anim., 1969* p. 324 (1970).
192. J. C. S. Patterson, D. Marrack, and H. S. Wiggins, *J. Clin. Pathol.* **6**, 105 (1953).
193. R. E. Peterson and R. H. Ettinger, *Amer. J. Med.* **15**, 518 (1953).
194. G. Pirzio-Biroli, T. H. Bothwell, and C. A. Finch, *J. Lab. Clin. Med.* **51**, 37 (1958).
195. J. Planas, *Nature* (*London*) **215**, 289 (1967).
196. J. Planas and S. de Castro, *Nature* (*London*) **187**, 1126 (1960).
197. J. Planas, S. de Castro, and J. M. Recio, *Nature* (*London*) **189**, 668 (1961).
198. M. Pollycove, *in* "Iron in Clinical Medicine" (R. O. Wallerstein and S. R. Mettier, eds.), p. 43. Univ. of California Press, Berkeley, California, 1958.
199. W. G. Pond, R. L. Lowrey, J. H. Maner, and J. K. Loosli, *J. Anim. Sci.* **19**, 1286 (1960).

200. W. G. Pond, T. L. Veum, and V. A. Lazar, *J. Anim. Sci.* **24**, 668 (1965).
201. H. B. Prapulla and C. P. Ramakrishnan, *Acta Paediat. Belg.* **49**, 599 (1960); quoted by Ezekiel and Morgan (63).
202. J. A. Pritchard and C. F. Hunt, *Surg., Gynecol. Obstet.* **106**, 516 (1958).
203. W. N. M. Ramsay and E. A. Campbell, *Biochem. J.* **58**, 313 (1954).
204. Recommended Dietary Allowances, *Nat. Acad. Sci.—Nat. Res. Council, Publ.* **1146** (1964).
204a. R. W. Rice, G. E. Nelms, and C. O. Schoonover, *J. Anim. Sci.* **26**, 613 (1967).
205. R. M. Richard, R. F. Shumard, A. L. Pope, P. H. Phillips, C. A. Herrick, and G. Bohstedt, *J. Anim. Sci.* **13**, 274 and 674 (1954).
206. D. A. Richert and W. W. Westerfeld, *J. Biol. Chem.* **209**, 179 (1954).
207. C. Rimington, *Brit. Med. Bull.* **15**, 19 (1959).
208. A. Rothen, *J. Biol. Chem.* **152**, 679 (1944).
209. G. Rybo, *Acta Obstet. Gynecol. Scand.* **45**, Suppl. 7 (1966).
210. H. A. Salmon, *J. Physiol. (London)* **164**, 17 (1962).
211. A. L. Schade, R. W. Reinhart, and H. Levy, *Arch. Biochem.* **20**, 170 (1949).
212. J. Seamer, *Vet. Rev. Annot.* **2**, 79 (1956).
213. C. T. Settlemire and G. Matrone, *J. Nutr.* **92**, 153 and 159 (1967).
214. L. M. Sharpe, W. C. Peacock, R. Cook, and R. S. Harris, *J. Nutr.* **41**, 433 (1950).
215. H. C. Sherman, "Chemistry of Food and Nutrition." Macmillan, New York, 1935.
216. A. Shoden, B. W. Gabrio, and C. A. Finch, *J. Biol. Chem.* **204**, 823 (1953).
217. A. Shoden and G. W. Richter, *Folia Haematol. (Frankfurt am Main)* **4**, 180 (1960).
218. A. Shoden and P. Sturgeon, *Amer. J. Pathol.* **34**, 113 (1958).
219. A. Shoden and P. Sturgeon, *Acta Haematol.* **22**, 140 (1949); **23**, 376 (1960); **27**, 33 (1962).
220. A. Shoden and P. Sturgeon, *Nature (London)* **189**, 846 (1961).
221. A. Shoden and P. Sturgeon, *Proc. 1st Int. Congr. Histochem. Cytochem., 1960.*
222. A. Shoden and P. Sturgeon, *Brit. J. Haematol.* **9**, 471 (1963).
223. T. R. C. Sisson, *Fed. Proc., Fed. Amer. Soc. Exp. Biol.* **23**, 879 (1964).
224. O. Skeets, E. Frazier, and D. Dickins, *Miss., Agr. Exp. Sta., Bull.* **291** (1931).
225. M. D. Smith and B. J. Mallet, *Clin. Sci.* **16**, 23 (1957).
226. C. V. Smythe and R. C. Miller, *J. Nutr.* **1**, 209 (1929).
227. S. K. Srivastava, G. G. Sanwal, and K. K. Tewari, *Indian J. Biochem.* **2**, 257 (1965).
228. A. R. Stevens, D. H. Coleman, and C. A. Finch, *Ann. Intern. Med.* **38**, 199 (1953).
229. W. B. Stewart, *Bull. N. Y. Acad. Med.* **29**, 818 (1953).
230. W. B. Stewart, C. L. Yuile, H. A. Clairborne, R. T. Snowman, and G. H. Whipple, *J. Exp. Med.* **92**, 372 (1950).
231. H. K. Stiebeling, *U. S., Dep. Agr., Circ.* **205** (1932).
232. H. K. Stiebeling, D. Monroe, E. F. Phipard, S. F. Adelson, and F. Clark, *U.S., Dep. Agr., Misc. Publ.* **452** (1941).
233. M. B. Strauss, *J. Clin. Invest.* **12**, 345 (1933); K. V. Toverud, *Acta Paediat.* **17**, Suppl. 127 (1935).
234. P. Sturgeon, *Pediatrics* **17**, 341 (1956); **18**, 267 (1956).
235. P. Sturgeon, *Brit. J. Haematol.* **5**, 31 (1959).
236. H. Tarvydas, S. M. Jordan, and E. H. Morgan, *Brit. J. Nutr.* **22**, 565 (1968).
237. R. Thompson and A. M. Raven, *J. Agr. Sci.* **52**, 177 (1959); A. M. Raven and R. Thompson, *ibid.* p. 224.
238. I. H. Tipton, P. L. Stewart, and P. G. Martin, *Health Phys.* **12**, 1683 (1966).
239. J. D. Torrance, R. W. Charleton, A. Schmaman, S. R. Lynch, and T. H. Bothwell, *J. Clin. Pathol.* **21**, 495 (1968).

240. D. E. Ullrey, E. R. Miller, O. A. Thompson, I. M. Ackerman, D. A. Schmidt, J. A. Hoefer, and R. W. Luecke, *J. Nutr.* **72**, 459 (1960).
241. E. J. Underwood, *Aust. Vet. J.* **10**, 87 (1934).
242. E. J. Underwood, *J. Nutr.* **16**, 299 (1938).
243. E. J. Underwood, "The Mineral Nutrition of Livestock." FAO/CAB Publ. Central Press, Aberdeen, 1966.
244. E. J. Underwood and E. H. Morgan, *Aust. J. Exp. Biol. Med. Sci.* **41**, 247 (1963).
245. B. Vahlquist, *Acta Paediat.* **28**, Suppl. 5 (1941).
246. B. Vahlquist, *Blood* **5**, 874 (1950).
247. D. Van Campen and E. Gross, *J. Nutr.* **99**, 68 (1969).
248. J. A. J. Venn, R. A. McCance, and E. M. Widdowson, *J. Comp. Pathol. Ther.* **57**, 314 (1947).
249. A. R. P. Walker, F. W. Fox, and J. T. Irving, *Biochem. J.* **42**, 252 (1948).
250. R. J. Walsh, I. Kaldor, I. Brading, and E. P. George, *Aust. Ann. Med.* **4**, 272 (1955).
251. L. R. Weintraub, M. E. Conrad, and W. H. Crosby, *J. Clin. Invest.* **43**, 40 (1964).
252. L. R. Weintraub, M. E. Conrad, and W. H. Crosby, *Blood* **24**, 19 (1964).
253. L. R. Weintraub, M. E. Conrad, and W. H. Crosby, *Brit. J. Haematol.* **11**, 432 (1965).
254. L. R. Weintraub, M. E. Conrad, and W. H. Crosby, *J. Clin. Invest.* **47**, 531 (1968).
255. M. S. Wheby and W. H. Crosby, *Blood* **22**, 416 (1963).
256. L. Whitby, *Brit. Med. J.* **1**, 1279 (1964).
257. H. S. White, *J. Amer. Diet. Ass.* **53**, 563 (1968).
258. E. M. Widdowson and R. A. McCance, *Biochem. J.* **42**, 488 (1948).
259. E. M. Widdowson, *Nature (London)* **166**, 626 (1950); C. M. Spray and E. M. Widdowson, *Brit. J. Nutr.* **4**, 332 (1951); E. M. Widdowson and C. M. Spray, *Arch. Dis. Childhood* **26**, 205 (1951).
260. J. F. Wilson, D. C. Heiner, and M. E. Lahey, *J. Pediat.* **60**, 787 (1962).
261. O. Zinoffsky, *Hoppe-Seyler's Z. Physiol. Chem.* **10**, 16 (1886).
262. E. G. Zook and J. Lehmann, *J. Ass. Offic. Agr. Chem.* **48**, 850 (1965).
263. E. G. Zook and J. Lehmann, *J. Amer. Diet. Ass.* **52**, 225 (1968).

# 3

# COPPER

## I. Introduction

The presence of copper in plant and animal tissues was recognized more than 150 years ago (44). Its occurrence was assumed to be accidental until McHargue (215) produced suggestive evidence of its value in the diet of rats. Conclusive evidence of the essentiality of copper emerged from studies of hemoglobin regeneration in rats suffering from milk anemia. In 1928, Hart and associates (139) announced that copper, in addition to iron, was necessary for blood formation in this species. During the following decade this important discovery was confirmed and extended to other species. Hematopoiesis was considered to be the principal physiological activity involving copper until field studies with sheep and cattle in various parts of the world revealed wider functions for this element. Much later, in 1961, copper was shown to play an additional vital role in the formation of aortic elastin (235)—a role more critical for the survival of Cu-deficient chicks than blood formation (145). Cardiovascular lesions, resulting from a derangement of the elastic membranes of the major blood vessels, were also demonstrated in Cu-deficient pigs (63,80,272). A key role for copper in connective tissue metabolism thus became firmly established.

Long before copper was recognized as an essential element in the diets of mammals and birds, it was found to occur as a component of specific compounds in the blood of various gastropods, arthropods, and marine species. As early as 1847 copper was shown to exist in combination with the blood proteins of snails (138), and 30 years later the copper-containing pigment (hemocyanin) was found to behave as a respiratory compound (116). Hemocyanins from different marine organisms were subsequently shown to vary in composition and copper content and to unite with oxygen in a definite ratio to copper. In addition, the red pigment, turacin, found in the feathers of the South African bird, turaco, was found to be a copper compound, containing no less than 7% copper (73). Subsequently turacin was shown to be a derivative of the porphyrin pigments normally present in plants and animals (249), the porphyrin was identified as uroporphyrin III (231), and the copper was shown to occur in divalent form within the porphyrin moiety (31).

Shortly after the initial discovery that copper is required by rats, certain naturally occurring disorders of grazing sheep and cattle were found to be due to a dietary deficiency of copper, or to respond to copper therapy. The first indication that copper deficiency occurs naturally in livestock appeared in 1931 as an outcome of the studies of Neal and associates (230) of "salt-sick" cattle in Florida. Two years later, Sjollema (276) discovered that copper deficiency is a causal factor in a disease of sheep and cattle occurring in parts of Holland and known locally as *lechsucht*. In 1937, Bennetts and Chapman (27) found that a disease of lambs occurring in parts of Western Australia, designated enzootic neonatal ataxia, was a manifestation of inadequate intakes of copper from the grazing. These pioneer studies were followed by the demonstration of extensive areas of simple or conditioned copper deficiency in different parts of the world, affecting both crops and stock. Concurrent laboratory studies disclosed a remarkable variety of physiological functions involving copper. Defects in the processes of pigmentation, keratinization of wool, bone formation, reproduction, myelination of the spinal cord, cardiac function, and connective tissue formation, in addition to those of growth and hematopoiesis, were found to be manifestations of copper deficiency.

These investigations were accompanied by basic researches into the mode of action of copper at the cellular level. Many copper-protein compounds were isolated from living tissues, several of which were found to be enzymes with oxidative functions. Indeed, copper enzymes were shown to be uniquely important in catalyzing the reduction of molecular oxygen to water. As Frieden (117) has said "no metal ion surpasses copper salts in their versatility as catalysts for an impressive variety of reactions." This catalytic activity was found to be enhanced and made more specific when the copper is incorporated in a protein to form a copper enzyme. Tyrosinase, laccase, ascorbic acid oxidase, cytochrome oxidase, uricase, monoamine oxidase, δ-aminolevulinic acid dehydrase, and dopamine-β-hydroxylase were all identified as copper enzymes. Several of the manifold manifestations of copper deficiency in the animal were related to decreased tissue concentrations of certain of these enzymes. The basic biochemical defects were thus revealed.

Studies of the nutritional physiology of copper were given a further impetus by Australian investigations of chronic copper poisoning in sheep. Copper retention was shown to be dependent upon the molybdenum status of the diet, and the limiting effect of molybdenum was shown, in turn, to depend upon the inorganic sulfate status of the diet and of the animal (98). These impressive discoveries drew attention in a striking manner to interactions among the trace elements and to the importance of dietary mineral balance. The particular importance of such interactions with copper was exemplified further by the demonstration of a marked and reciprocal zinc-copper antagonism and by the discovery that the extent to which copper deficiency or copper toxicity actually occurs in animals depends not merely upon copper intakes but upon the relative proportions of these intakes to the dietary levels of zinc, iron, and calcium (146,220).

## II. Copper in Body Tissues and Fluids

### 1. *Distribution of Copper in the Body*

Over 30 years ago the healthy, adult human body was calculated to contain 110-120 mg of total copper (72). A lower estimate of 80 mg has been given by Cartwright and Wintrobe (68). Newborn and very young animals are normally much richer in copper per unit of body weight than adults of the same species (285, 319) (Table 8). The newborn levels are largely maintained throughout the suckling period, followed by a steady fall during growth from weaning when adult levels are reached.

TABLE 8
*Concentrations of Copper in the Whole Bodies of Various Species*[a]

| Age | Human | Pig | Cat | Rabbit | Rat | Guinea pig |
|---|---|---|---|---|---|---|
| Newborn[b] | 4.7 | 3.2 | 2.9 | 4.0 | 4.3 | 6.9 |
| Adult[c] | 1.7 | 2.5 | 1.9 | 1.5 | 2.0 | – |

[a]Concentrations in parts per million of fat-free tissue.
[b]From Widdowson (319).
[c]From Spray and Widdowson (285).

The distribution of the total body copper among the tissues varies with the species, age, and copper status of the animal. In a study of the distribution of copper in the tissues of 5 normal humans, Cartwright and Wintrobe (68) found a total of 23 mg in the liver, heart, spleen, kidneys, brain, and blood. Of this total 8 mg was present in the liver and, surprisingly, 8 mg in the brain. The data of Smith (277) for the human body point to a smaller proportion of total body copper in the brain, although the mean concentration of copper in the brain was very close to that of the liver. The data of Cunningham (84) for a range of species also indicate that a higher proportion of total body copper exists in the liver than in the brain. In ruminants, which have a high capacity for hepatic storage of copper, the proportion in the liver can be very high. Thus Dick (97) found the total body copper of 2 adult sheep with extremely high liver copper concentrations to be distributed as follows: liver, 72-79%; muscles, 8-12%; skin and wool, 9%; and skeleton, 2%.

Highly variable concentrations of copper occur in the tissues of all species. The glands (prostate, pituitary, thyroid, and thymus) are examples of tissues low in copper; the spleen, pancreas, muscles, skin, and bones represent organs of intermediate copper concentration; and the liver, brain, kidneys, heart, and hair are tissues of relatively high copper concentration (84,277). In a study of the variation of copper levels with age the brain was the only organ in which the concentration increased from birth to about double the level at maturity (261). In the liver, spleen, lung, and aorta, copper levels fell from birth to maturity, indicating that the high total body copper of the newborn is a reflection of the relatively high copper levels of several organs and not merely of the liver. The ability of the individual tissues to vary their copper content with variations in dietary copper intake differs within and among species. The muscles, heart, and endocrine glands are most resistant to change, and the liver, kidney, blood, spleen, and lungs are the least resistant (193). In addition, deprivation of dietary copper results in a significant reduction in the copper level in bone tissue and in the forms of copper present in the bones of chicks (256).

Exceptionally high concentrations of copper occur in the pigmented parts of the eye (34,296). Differences among species are large but in all of them the eye tissues can be placed in the following descending order of copper concentration: iris, choroid, vitreous humor, aqueous humor, retina (minus pigmented epithelium), optic nerve, cornea, sclera, and lens. Levels as high as 105 ppm (dry basis) for the iris and 88 ppm for the choroid of the eyes of freshwater trout and of 50 and 13.5 ppm for these tissues in sheep's eyes were reported. The copper is associated particularly with the melanins and is largely bound in ionic form to protein. The role of copper in these sites is not clear. Kikkawa *et al.* (168) have proposed that pigmentation is influenced by metal ions. This suggestion finds more support from studies with manganese (78) than with copper.

Black and white hair from ten different species was examined by Goss and Green (126). Their figures ranged from 10 to 47 ppm of washed, dry hair, with more than half the values lying between 14 and 20 ppm and with no relationship between color and copper concentration. Smith (277) examined 29 samples of human hair of unspecified color and found the copper concentration to range from 7.6 to 54.5 ppm with a mean of 23.1 ppm. Earlier claims that the copper content of pigmented hair is higher than that of unpigmented (168,292) could not be confirmed, at least in man (125,183). Lea and Luttrell (183), in contrast to the findings of others (125,211), found that kwashiorkor is not necessarily accompanied by a reduction in the copper content of the hair. The copper concentration in hair of humans does not rise significantly with age (247) as it does in rats (248).

Anke (8) reported a mean value of 7.8 ppm for 671 samples of black hair from dairy cattle. Van Koetsveld (314) obtained an average of 10 ppm, with the levels seldom rising above 15 ppm. He observed that values below 8 ppm were associated with signs of copper deficiency in the cattle. A significant decline in the copper content of the hair is not apparent in other studies with Cu-deficient cattle (88) nor with rats (101). O'Mary and co-workers (237) found red and white hair from Hereford cattle to range from 10 to 31 ppm Cu, with no consistent differences due to color but with significant effects of season or time of sampling.

A relatively high copper concentration of 15 to 30 ppm has been reported for the enamel of human teeth, with no change with age or the layer analyzed (43). In a later study employing activation analysis, no significant difference in the copper content of the inner and outer layers of enamel was found but the actual copper concentrations were substantially lower. The mean of the inner layer was 11.3 and that of the outer 9.5 ppm Cu (233).

### 2. *Copper in the Liver*

The copper content of the liver is determined by the species and the age of the animal, by the chemical composition of the diet, and by various disease conditions (see Table 9). There is no effect of sex, except on the Australian salmon (*Arripis trutta*), in which the female carried

TABLE 9

*Influence of Species, Age, and Copper Intake on the Concentration of Copper in the Liver*

| Species | Age and treatment | No. of animals | Copper concentration (ppm)[a] | Ref. |
|---|---|---|---|---|
| Man | Newborn (0-7 weeks); normal | – | 230 | Bruckmann and Zondek (42) |
| Man | Adult; normal diet | – | 35 | Bruckmann and Zondek (42) |
| Rat | Newborn; normal diet | 30 | 58 ± 4.0 | Lorenzen and Smith (194) |
| Rat | Mature; normal diet | 10 | 9 ± 0.4 | Lorenzen and Smith (194) |
| Rabbit | Newborn; normal diet | 30 | 37 ± 6.7 | Lorenzen and Smith (194) |
| Rabbit | Mature; normal diet | 10 | 23 ± 3.6 | Lorenzen and Smith (194) |
| Pig | Newborn; normal diet | – | 233 | Cunningham (85) |
| Pig | Mature; normal diet | 12 | 19 (12-48) | Cunningham (85) |
| Sheep | Newborn; normal diet | 27 | 168 (74-430) | Cunningham (85) |
| Sheep | Newborn; Cu-deficient diet | 29 | 13 (4-34) | Cunningham (85) |
| Sheep | Mature (aged); normal diet | 44 | 599 (186-1374) | Cunningham (85) |
| Sheep | Mature (aged); Cu-deficient diet | 35 | 27 (7-106) | Cunningham (85) |
| Cattle | Newborn; normal diet | 41 | 381 (143-655) | Cunningham (85) |
| Cattle | Newborn; Cu-deficient diet | 20 | 55 (8-109)[b] | Cunningham (85) |
| Cattle | Mature; normal diet | 23 | 200 (23-409) | Cunningham (85) |
| Cattle | Mature; Cu-deficient diet | 41 | 11.5 (3-32) | Cunningham (85) |
| Domestic fowl | Mature; normal diet | 51 | 14.8 (10-31) | Beck (21) |
| Domestic duck | Mature; normal diet | 34 | 153 (37-555) | Beck (21) |

[a] On the dry basis.
[b] Much lower levels have been obtained for Cu-deficient calves in Western Australia.

higher concentrations than the male (21). Individual variation is high in all species.

a. *Species.* The livers of normal adults of most species contain 10–50 ppm Cu on the dry basis, with a high proportion containing 15–30 ppm (21). These levels apply to species as diverse and unrelated in their environment as man, rats, rabbits, cats, dogs, foxes, pigs, kangaroos, whales, snakes, crocodiles, domestic fowls, turkeys, sharks, and herring. A few species, notably sheep, cattle, ducks, frogs, and certain fish exhibit consistently higher liver copper levels, with a normal range of 100 to 400 ppm.

These species differences cannot be explained on the basis of normal differences in copper intakes, since domestic fowls, turkeys, and ducks consume similar diets and yet the two former species carry much lower liver copper levels than do ducks. In fact, the characteristic species difference between fowls and ducks is maintained even when dietary copper intakes are raised two- to fivefold (22). The normally low liver copper concentrations of the rat are also maintained with elevated copper intakes, until very high intakes are reached (225). Beck (21) has suggested that the differences between high and low liver copper species arise from differences in the excretory mechanism. This is a plausible hypothesis but, as Milne and Weswig (225) have pointed out, cattle and sheep have probably also an enhanced capacity to bind copper in the liver, because blood copper levels do not rise correspondingly with increased copper intakes in these species, as they do in rats, except at very high intakes. A significant difference between two strains of chickens in ability to retain copper in the liver and kidney, following copper injections, has also been demonstrated (191).

b. *Influence of Age.* With most species, including man, liver copper concentrations are higher in the newborn than they are in adults. Sheep and cows provide exceptions. In the former they rise continuously from birth, and in the latter they change little throughout the whole growth period from birth to old age. The extent of intrauterine storage of copper and the time of maximum copper concentration in the fetal liver vary with the species. In the rat, rabbit, guinea pig, dog, and man the peak concentration occurs at or very shortly after birth (42,213). In the pig the peak occurs slightly earlier in embryonic life (322). In the bovine, copper storage keeps pace with the growth of the fetal liver because liver copper concentrations do not rise significantly during gestation (243). During the suckling period the liver copper concentrations generally fall, although the total amounts of liver copper continue to rise.

The decline in copper concentration of whole liver that occurs as the rat matures is accompanied by changes in the distribution of the copper among the subcellular fractions (129). At birth, over 80% of the total copper is present in the mitochondrial and nuclear fractions and less than 20% in the microsomes and soluble (supernatant) fractions. In the adult rat the supernatant contains about one-half the total copper of the liver, with the copper content of nuclei, mitochondria, and microsomes following in that order. Porter (241) has shown that the copper accumulating physiologically in newborn liver is chiefly accounted for by mitochondrocuprein, a protein compound extraordinarily high in copper (more than 3%), localized in the mitochondrial fraction and specific to the neonatal period.

c. *Influence of Diet.* In several species, subnormal intakes of copper during pregnancy are reflected in subnormal levels of copper in the liver of the newborn. Thus, Bennetts and Beck (26) found the liver copper of 5 ataxic lambs from Cu-deficient ewes to range from 4 to 8 ppm on the dry basis, compared with 120 to 350 ppm for normal lambs from ewes receiving adequate copper. When the mother is copper deficient, supplementary copper administered during pregnancy is clearly effective in raising fetal liver copper levels to normal (7).

Similar treatment of mothers already receiving adequate copper does not significantly increase liver copper storage in the newborn, presumably due to limited placental transfer.

Liver copper concentrations are sensitive to low copper intakes and are useful aids in the diagnosis of copper deficiency. Subnormal liver copper levels occur in rats and pigs suffering from milk anemia (267, 321a), in Cu-deficient rats (102, 225), chicks (256), and dogs (17), and in sheep and cattle grazing Cu-deficient pastures (15, 26, 85, 216). The response to high dietary copper intakes is different in ruminants and nonruminants. Dick (97) studied liver copper storage in groups of sheep ingesting graded increments of copper from 3.6 to 33.6 mg/day for 177 days. The liver copper levels increased steadily from 562 ppm (dry weight) at the lowest, to 2340 ppm at the highest intake. The proportion of the ingested copper stored in the liver was remarkably uniform at intakes up to 18.6 mg/day and the increase in liver storage was linear (see Fig. 4). Copper supplementation of the normal diets of rats has no comparable effect upon liver copper storage until high intakes are reached. At this threshold, which has been reported as 1 mg/day or 200 ppm of the ration (225), liver copper increases rapidly (see Fig. 5), apparently due to overloading of the excretory mechanism. However the actual liver copper concentrations do not approach those given above for sheep.

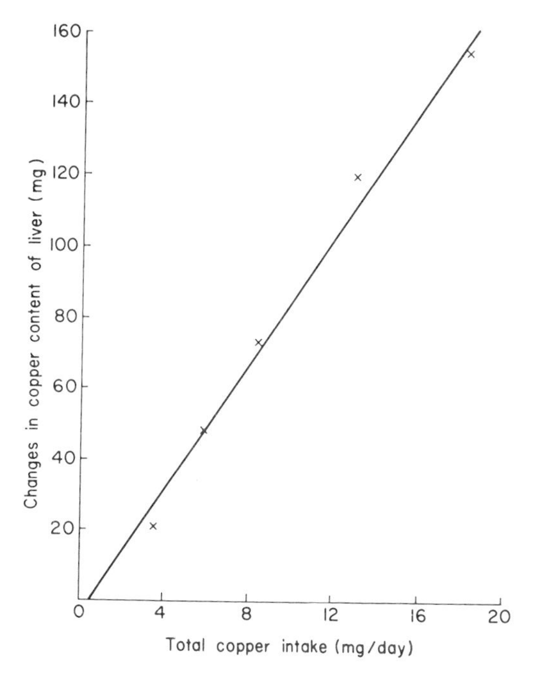


**Fig. 4.** Relationship between copper intake and increase in liver copper storage during a 177-day period in sheep. [From Dick (97).]

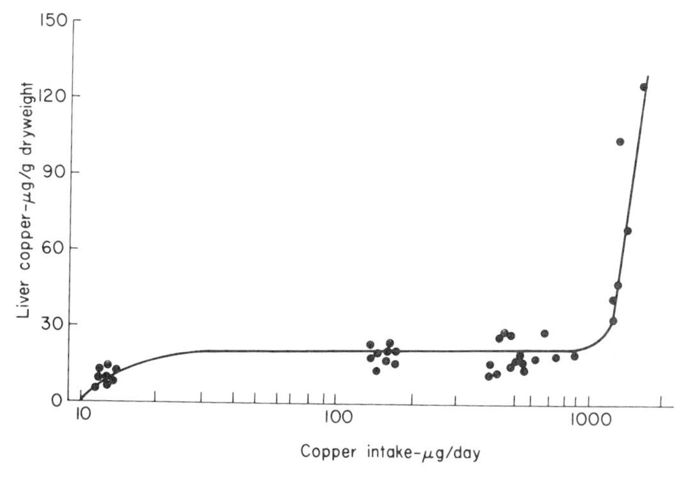


**Fig. 5.** Influence of copper intake on liver copper in the rat. [From Milne and Weswig (225).]

The intracellular distribution of liver copper does not appear to have been studied in ruminants with their characteristically high concentrations. In the rat, Gregoriadis and Sourkes (129) found that in copper loading, the mitochondria and nuclei hold most of the excess copper, with the microsomes and cytoplasm accumulating much less. Milne and Weswig (225), under different experimental conditions, obtained a linear increase in the amount of copper in each intracellular fraction with the total amount of copper in the liver. The relative amount in the mitochondria and the soluble fraction remained essentially constant, whereas the relative amount in the microsomes increased and in the debris decreased from the copper-depleted to the copper-supplemented groups (Table 10).

The concentration of copper in the liver is profoundly influenced by several dietary factors other than copper. Dick and Bull (99) first showed that the storage of copper in the livers of sheep and cattle could be reduced significantly by an increase in dietary molybdenum intake. Subsequently Dick (98) discovered that the inhibiting effect of molybdenum upon copper retention is dependent upon the inorganic sulfate content of the diet. These exciting discoveries were confirmed and extended to other species, and evidence was obtained that unidentified dietary factors, as well as molybdenum and sulfate, affect copper retention in the liver (98, 152).

Copper retention in the liver and other tissues is also influenced by the levels of zinc and iron and of calcium carbonate in the diet. High intakes of zinc depress both copper and iron absorption and retention (81, 128, 196, 311). A marked Zn-Cu antagonism is evident both when copper is limiting (146), and in copper toxicosis (295). A highly significant inverse correlation between hepatic iron and copper concentrations has also been demonstrated in rats (282). Rats consuming an Fe-deficient diet accumulated high concentrations of copper in the liver after 7 to 8 weeks, and rats fed a Cu-deficient diet accumulated excessive amounts of iron, beginning much earlier.

TABLE 10
*Distribution of Copper in Rat Liver Fractions*[a]

| Copper intake (ppm) | Whole liver copper (dry) (ppm) | Debris (%) | Mitochondria (%) | Microsomes (%) | Soluble (%) |
|---|---|---|---|---|---|
| 1 | 8.7 | 20.3 | 12.4 | 13.9 | 53.4 |
| 10 | 19.9 | 14.3 | 12.7 | 17.1 | 55.9 |
| 50 | 18.2 | 15.2 | 11.9 | 20.5 | 52.4 |
| 100 | 19.7 | 14.1 | 15.1 | 18.1 | 52.7 |
| 200 | 68.5 | 14.2 | 11.0 | 23.8 | 50.8 |

[a]From Milne and Weswig (225).

d. *Disease.* Abnormally high liver copper concentrations are characteristic of a number of diseases of man. These include Mediterranean anemia, hemochromatosis, cirrhosis, and yellow atrophy of the liver, tuberculosis, carcinoma, severe chronic diseases accompanied by anemia and Wilson's disease (hepatolenticular degeneration) (65). In Wilson's disease the excessive deposits of copper in the liver and in other body tissues arise in part from a breakdown of biliary excretion of the element. The relation of the high copper levels to the other diseases mentioned is obscure. Extremely high liver copper levels, as high as 4000 ppm on the dry fat-free basis, also occur in hemolytic jaundice of sheep.

3. *Copper in Blood*

a. *Forms and Distribution.* Bovine erythrocytes contain a blue copper protein, hemocuprein, with a molecular weight of 35,000, 0.34% Cu in the cupric form, and 2 atoms of copper per molecule (197a). A similar but not identical, nearly colorless copper protein, called erythrocuprein, has been isolated from human erythrocytes (169, 198, 273). This compound is distinct from ceruloplasmin and from the human hepatocuprein isolated by Shapiro *et al.* (271). Erythrocuprein has a molecular weight of 31,000, contains 3.4 μg Cu/mg protein, and comprises 60% or more of the total red cell copper (273). It remains extremely constant in quantity under a wide range of conditions in man (68). The remainder of the erythrocyte copper is more loosely bound to unidentified proteins and is much more labile than erythrocuprein (273).

The copper in plasma also occurs in two main forms—one firmly and one loosely bound. The former consists of the blue copper protein, ceruloplasmin, which is an $\alpha_2$-globulin with a molecular weight of 151,000, containing 8 atoms of copper per molecule (153). In normal rats, dogs, pigs, sheep, and man (57, 65, 225, 287, 324), about 80% of the plasma copper exists as ceruloplasmin, so that highly significant correlations have been found between ceruloplasmin and plasma, serum, and whole blood copper (209, 305). In chicks (287) and turkeys (320), a very much smaller proportion of the normally low level of plasma copper is present as ceruloplasmin. In fact the amount is barely detectable. Ceruloplasmin is a true oxidase which can catalyze the oxidation of a variety of substrates including various polyphenols (153) and biological compounds such as serotonin and epinephrine (205). Ceruloplasmin cannot play a major role in copper absorption and transport, because the amount of ceruloplasmin copper exchanged daily is extremely small compared with the amounts of copper absorbed from the intestinal tract.

The plasma copper not present as ceruloplasmin is known as "direct-reacting" copper, because it reacts directly with dithizone, is nondialyzable and is loosely bound to protein, probably serum albumin (33, 55, 228, 324). This albumin-bound plasma copper is believed to constitute true transport copper. In addition, the plasma contains copper enzymes, such as cytochrome oxidase and monoamine oxidase, in concentrations that vary with the copper status of the animal.

In humans (65), rats (225), pigs (180), sheep (57, 210), and cattle (2, 29) the concentration of copper is higher in the plasma than it is in the erythrocytes. The amount of copper in the individual human erythrocyte has been calculated to be 65±10.8 $\mu\mu\mu g$, and the amount in the individual leukocyte and platelet to be close to one-quarter of this figure (181). Since the number of the latter cells is much smaller than the number of erythrocytes, they must contain an insignificant portion of the total blood copper. Failure to separate them therefore results in a negligible error. However, plasma copper is more labile than corpuscular copper and is a more sensitive and reliable indicator of the copper status of an animal than whole blood copper (102).

b. *Normal Blood Copper Levels.* The normal range of concentration of copper in the blood of healthy animals is wide but of similar magnitude in all the higher mammals. For humans, rats, pigs, dogs, cats, rabbits, horses, sheep, and cattle this range can be given as 0.5-1.5 $\mu$g/ml, with a high proportion of values lying between 0.8 and 1.2 (22). Levels consistently below 0.6 $\mu$g Cu/ml in the whole blood or plasma of sheep and cattle are indicative of copper deficiency, simple or conditioned. A narrower normal range, with lower mean values approximating one-half of those just quoted, have been found by Beck (22) for poultry, fish, frogs, and marsupials (see Table 11).

Variations within species can be accounted for more by individual differences than by diurnal or day-to-day variations. This has been established for man (65, 142a, 310) and for sheep (57, 212). Significant breed differences have also been demonstrated in sheep (321, 321a). There appears to be no cyclic pattern of variation in man, and the small diurnal variation in plasma copper observed in some studies is of doubtful statistical significance (310). Plasma copper does not increase following meals nor decrease during fasting (65). Physical exertion in man produces no change, but violent exercise induces a significant increase in whole blood copper in sheep (96). There are no significant sex differences in whole blood or plasma copper in most species, but plasma copper is higher in human females than in males. Normal values reported by Cartwright (65) are 105.5 ± 5.03 $\mu g$ Cu/100 ml for men and 114.0 ± 4.67 $\mu$g Cu/100 ml for women. No explanation of this sex dif-

TABLE 11
*Copper Concentration in the Blood of Different Species*

| Species | Age and condition | Mean copper concentration ($\mu$g/ml) | Ref. |
|---|---|---|---|
| Human | Healthy, adult male | 1.10 ± 0.12[a] | 232 |
| Human | Healthy, adult female | 1.23 ± 0.16[a] | 232 |
| Human | Pregnant female, at delivery | 2.69 ± 0.49[a] | 232 |
| Human | Female, late pregnancy | 2.80[a] | 255 |
| Human | Healthy, adult female | 1.00[a] | 255 |
| Ovine | Healthy, mature | 1.01 ± 0.96[b] | 21 |
| Ovine | Healthy, mature | 0.91[b] | 85 |
| Bovine | Healthy, mature | 0.93[b] | 85 |
| Guinea pig | Healthy, mature | 0.50 ± 0.006[b] | 21 |
| Domestic fowl | Healthy, mature | 0.23 ± 0.008[b] | 21 |
| Domestic duck | Healthy, mature | 0.35 ± 0.007[b] | 21 |

[a]Serum.
[b]Whole blood.

ference has been advanced, but it is of interest that plasma copper levels can be significantly increased by the administration of estradiol in humans (165), by stilbestrol in swine (177) and in rats (307), and by thyroxine in sheep (253). Estrogenic preparations have no such effect upon sheep, even when administered at dose rates and for a sufficient period to induce significant increases in teat length in both ewes and wethers (226). In chicks, a five- to sevenfold increase in serum ceruloplasmin activity results from the administration of adrenocorticotropic hormone (ACTH) and hydrocortisone (287). The taking of oral contraceptives by women also increases serum copper levels. Halsted *et al.* (135a) reported a mean of 3.0 ± 0.7 $\mu$g/ml in 10 such women, compared with 1.18 ± 0.2 $\mu$g/ml in 20 controls. Clemetson (74) found the mean plasma copper of women taking oral contraceptives containing an estrogen to be 2.16 $\mu$g/ml, compared with 1.24 $\mu$g/ml in normal women and 1.07 $\mu$g/ml in normal men.

c. *Influence of Pregnancy and Parturition.* Marked changes in plasma copper occur in some species during pregnancy. Thus Nielsen (232), in a study of 31 pregnant women, found serum copper to increase from the third month to an average of 2.7 $\mu$g/ml, compared with a normal nonpregnant level of 1.2 $\mu$g/ml. Similar results were reported by Krebs (177), Rottger (255), Fay *et al.* (112), and Halsted *et al.* (135a). In these studies the red cell copper remained at normal levels throughout pregnancy. A much smaller but significant rise in serum copper during the last 30 days of pregnancy has been observed in mares

(290). The position is quite different in pregnant ewes. In several early studies with sheep no characteristic changes in blood copper were observed during pregnancy (20, 105, 212). In a later study, Butler (57) found that housed ewes on a constant diet exhibited a decline in whole blood, plasma, and ceruloplasmin copper and in erythrocyte copper levels during pregnancy. The decline was more evident when the copper intake was deficient and was still significant when the copper intake was adequate. Studies with grazing sheep also indicate that whole blood and plasma copper levels (5, 58) and also ceruloplasmin levels (157) fall during pregnancy (see Fig. 6). Unfortunately, erythrocyte copper levels were not measured in these studies, so that Butler's finding of a fall in this component could not be confirmed.

In women, the high serum copper levels of pregnancy return to normal in the first few weeks postpartum; the serum copper of the newborn infant is much lower than that of the mother, and by the second week of life the values for infants rise to adult levels and remain at these levels throughout childhood (112). In the ewe, blood copper levels continue to fall until about 1 month after parturition and then rise to premating levels (57). Howell and co-workers (157) found that blood copper and ceruloplasmin levels rose at the time of parturition to reach the highest levels recorded 1 week after lambing (Fig. 6). In lambs, blood copper

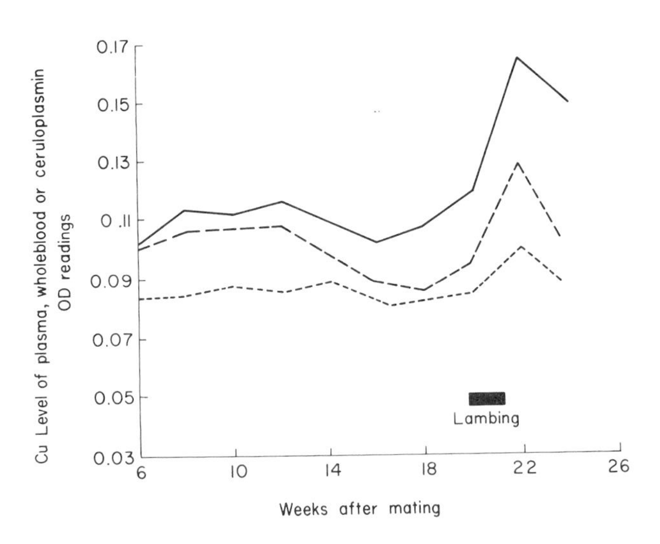


**Fig. 6.** Mean values of the levels of copper in blood and plasma and of ceruloplasmin in the serum of 7 pregnant ewes. (_____) Ceruloplasmin; (_ _ _) plasma; (. . .) whole blood copper. [From McC. Howell *et al.* (157).]

and ceruloplasmin levels were low at birth and 24 hrs. later. At 1 week after birth, values were well within the range of normal adults. These results are very similar to those obtained much earlier by McDougall (212). Bingley and Dufty (29) have shown that whole blood and plasma copper levels are significantly lower and erythrocyte copper levels significantly higher in newborn calves than in their mothers.

d. *Influence of Diet.* Blood copper concentration is influenced by the level of dietary copper and by the ratio of this copper to other components of the diet, notably molybdenum and inorganic sulfate, zinc, and iron. Hypocupremia occurs in rats, pigs, and dogs suffering from milk anemia (17, 181, 267), in the anemia of human infancy (291, 327), and in sheep and cattle grazing Cu-deficient pastures (15, 26, 104). Levels as low as 0.1 μg/ml whole blood have been reported in sheep and cattle under such conditions and of 0.2 μg/ml in Cu-deficient pigs. In the latter species the level of erythrocyte copper was reduced, although to a lesser extent than of the plasma (180). A fall in erythrocyte copper, as well as in the ceruloplasmin and direct-reacting copper of the plasma, has also been observed in Cu-deficient rats (225). No such decline in erythrocyte copper in this species was apparent in the studies of Dreosti and Quicke (102), even in severe copper deficiency. Hypocupremia can occur in rats on normal copper intakes if the Zn:Cu ratio is excessively high (128). Ceruloplasmin estimations have been shown by Todd (305) to provide a rapid screening method for detecting copper deficiency in cattle. Such estimates possess advantages over blood or plasma copper determinations because of the relative stability of the enzyme, the technical convenience of the assay, and the small size of the serum sample required (0.1 ml, compared with 1 ml or more for copper estimation). The specificity of the enzyme reaction also makes extreme cleanliness to avoid contamination unnecessary.

Moderate additions of copper to normal diets have little effect on blood copper levels in most species. Thus, Milne and Weswig (225) observed no increase in plasma copper in rats when the copper content of the diet was raised from 10 to 50 ppm. Increasing the copper intake tenfold, from 10 to 100 ppm, doubled the plasma copper concentration from 1.13 to 2.34 μg/ml. Copper intakes of 100 ppm or more are also necessary to induce significant increases in plasma copper in pigs. At toxic levels of dietary copper, such as 750 ppm, a severe hypercupremia develops in pigs which can largely be prevented by the concurrent administration of 500 ppm zinc (295). Hypercupremia occurs in other species as a consequence of extremely high dietary copper intakes or transiently by single, massive injections of copper salts (35, 85, 104). During the terminal stages of copper poisoning, i.e., within 24 to 48 hrs. of the hemolytic crisis in sheep, blood copper levels as high as 10–14 $\mu$g/ml have been reported (11, 304, 306).

Hypercupremia can be produced by very high intakes of molybdenum and sulfate (95). The effect of molybdenum depends upon the level of this element and of sulfate in the diet, and the status of the animal in respect to these nutrients and copper (128a). Under some circumstances, molybdenum can reduce blood copper levels or it can maintain them in the face of copper deficiency (86, 201).

e. *Influence of Disease.* In man, hypocupremia is associated with nephrosis and Wilson's disease, both of which are accompanied by increased urinary excretion of copper. Hypocupremia occurs also in kwashiorkor and cystic fibrosis associated with a low dietary intake of protein. In one study the mean plasma copper of nephrotics was found to be 0.6±0.2 $\mu$g/ml compared with 1.2 $\mu$g in normal subjects (66). The mean total serum copper of 36 patients with Wilson's disease was 0.61±0.21 $\mu$g/ml and that of 205 normal subjects, 1.14±0.17 $\mu$g/ml. Five normal relatives of Wilson's disease patients were found to have values below 0.79 $\mu$g Cu/ml. In this disease, serum copper levels are highly positively correlated with ceruloplasmin concentrations (67). Almost all patients with Wilson's disease have less than 23 mg ceruloplasmin per 100 ml of serum, which can be taken as the lower limit of normality. Some patients have even too little ceruloplasmin to be detectable. However, all cases of Wilson's disease do not exhibit such low levels, and the correlation between ceruloplasmin activity and the duration and severity of the clinical manifestations is poor. Furthermore, some individuals without Wilson's disease have abnormally low levels of ceruloplasmin (67, 260). It seems that a decreased concentration of this copper compound is not, as Cartwright and associates (67) have pointed out, "the single uncomplicated determinant of the disease."

Hypercupremia is evident in most chronic and acute infections in man, and in leukemia, Hodgkin's disease, various anemias, "collagen" disorders, hemochromatosis, and myocardial infarction (310, 324) (Table 12). In hyperthyroidism an increase in plasma copper is accompanied by a decrease in erythrocyte copper. Significant variations from normal adult levels of whole blood and plasma copper, largely attributable to changes in ceruloplasmin levels, have been demonstrated by McCosker (210) in several disease conditions in sheep. Elevations of plasma copper in infection have also been reported by others (21, 252). Significant depressions of whole blood and plasma copper occur in calves infested with the intestinal parasite *Bunostomum phlebotomum*

TABLE 12
*Blood Copper in Various Clinical Conditions in Man*[a]

| Condition | No. of subjects | Whole blood Cu (μg/%) | Plasma Cu (μg/%) | Cell Cu (μg/%) | Volume of packed RBC[b] (m/100 ml) | Plasma Fe (μg/%) |
|---|---|---|---|---|---|---|
| Normal | 63 | 98±13 | 109±17 | 115±22 | 47 | 115±42 |
| Pregnancy | 30 | 169 | 222 | 130 | 37 | 91 |
| Infection | 37 | 141 | 167 | 116 | 41 | 57 |
| Acute leukemia | 19 | 195 | 236 | 98 | 27 | 171 |
| Chronic leukemia | 21 | 119 | 148 | 101 | 39 | 113 |
| Hodgkin's disease | 14 | 142 | 171 | 109 | 40 | 78 |
| Pernicious anemia | 10 | 111 | 121 | 98 | 27 | 173 |
| Aplastic anemia | 8 | 130 | 152 | 86 | 28 | 203 |
| Iron deficiency | | | | | | |
| anemia: adults | 9 | 114 | 132 | 109 | 30 | 26 |
| infants | 24 | 155 | 168 | 152 | 28 | 31 |
| Hemochromatosis | 14 | 103 | 134 | – | – | 234 |
| Wilson's disease | 3 | 79 | 55 | 110 | 42 | 64 |
| Nephrosis | 3 | 70 | 80 | 119 | 44 | 62 |

[a]From Wintrobe *et al.* (325).
[b]RBC = red blood cells.

(39) and in sheep with chronic hemonchosis (210). Infection of chicks with *Salmonella gallinarum* produced a sixfold increase in ceruloplasmin activity (287). A similar increase was observed with other stressors, including ACTH and hydrocortisone. These findings led Starcher and Hill (287) to conclude that any stress or any situation resulting in increased corticosteroid levels could increase ceruloplasmin concentrations.

### 4. *Copper in Milk*

The level of copper in milk varies with the species, the stage of lactation, and the copper status of the diet. Individual variability is high and treatment after milking, e.g., pasteurizing, drying, or holding in metal containers, can result in a variable contamination with copper, sometimes more than doubling the original copper concentration (91). With increasing use of glass and stainless steel containers such contamination is less common than it used to be.

Rat's milk appears to be uniquely rich in copper, according to Cox and Mueller (82) who reported 7 μg/ml, compared with 0.6 μg/ml for cow's milk. More recently, Dreosti and Quicke (101) report values of 2.8 to 3.8 μg Cu/ml for milk from normal rats and of 0.7 to 1.5 for milk from Cu-deficient rats. In all species, colostrum is substantially richer in copper than later milk, and in most cases there is a progressive fall throughout lactation. Thus Beck (20) found the milk of normal ewes to decline from 0.2-0.6 μg Cu/ml in early lactation to 0.04-0.16 μg Cu/ml several months later. Similar falls were observed in the milk of some cows but not in others. Copper levels of the order just quoted, with a tendency for lower values in later lactation, have been reported by others (170, 174). Beck (19) also examined the milk of 8 mares and several women with similar results. The copper content of the mare's milk fell from a mean of 0.36 μg/ml in the first week of lactation to a mean of 0.17 μg/ml several weeks later. With women the fall was from 0.62-0.89 μg/ml in the first few weeks to 0.15-0.17 μg/ml several months later. Cavell and Widdowson (69) also obtained a mean of 0.62 μg Cu/ml (range 0.51-0.77) for the milk of 10 women taken at the end of the first week of lactation. Higher levels, with a similar decline during lactation, were observed by Lesne *et al* (187) in 4 of the 5 women studied.

Deficient intakes of copper are reflected in subnormal levels of copper in the milk. This has been demonstrated in rats (101) and in sheep and cattle. Levels as low as 0.01-0.02 μg Cu/ml have been reported in the milk of ewes and cows grazing Cu-deficient pastures (20). The addition of copper to diets already adequate in this element has little influence upon the copper content of the milk of cows and goats (108), ewes (300), and women (229). On the other hand, Dunkley *et al.* (102a) obtained a substantial elevation of milk copper for at least 4 weeks following subcutaneous injections of the cows with 300 mg Cu as copper glycinate. This was accomplished without any increase in the incidence of spontaneous oxidized flavor in the milk, despite an earlier finding that the incidence of oxidized flavor is positively correlated with the natural copper content of the milk, except during very early lactation (170). This change in susceptibility to oxidized flavor is probably related to a change in the distribution of copper in the milk. After 2 to 4 weeks of lactation, only about 15% of the milk copper is associated with fat, whereas after 15 weeks the proportion so associated rises to 35% (171).

## III. Copper Metabolism

Copper is absorbed mainly from the upper jejunum in dogs (257), the small intestine and the colon in pigs (33), and from the duodenum in man (302) and in chicks (286). In the rat, copper is absorbed to about the same extent from the stomach as from the small intestine (313). Studies with stable and radioactive copper indicate that ingested copper is poorly absorbed in most species (33, 75). In normal human subjects, according to the calculations of Cartwright and Wintrobe (68), orally administered copper is better absorbed, to the extent of about 32%. Copper absorption and retention is affected by the chemical forms in which the metal is ingested, by the dietary levels of several other minerals and organic substances, and by the acidity of the intestinal contents in the absorptive area. Relatively little is known of the mechanism of copper absorption. A copper-binding protein has been demonstrated by Starcher (286) in the mucosal cells of the duodenum of the chick which may play a role in copper absorption in this species.

A marked depression in copper absorption in sheep can be brought about by high dietary intakes of calcium carbonate and of ferrous sulfide (97). The former presumably reduces absorption by raising the intestinal pH and the latter by the formation of insoluble copper sulfide. Many years earlier, anemic rats were shown to be unable to use the copper of copper sulfide or copper porphyrin, whereas the oxide, hydroxide, iodide, glutamate, glycerophosphate, aspartate, citrate, nucleinate, and pyrophosphate were readily utilized (267). Pigs also absorb the copper of cupric sulfide much less efficiently than that of cupric sulfate (33). Lassiter and Bell (182) studied the availability of copper to sheep by administering, orally and intravenously, cupric sulfate, nitrate, chloride, carbonate, and oxide and cuprous oxide and copper wire, all labeled with $^{64}Cu$. The copper in the copper wire was largely unavailable, and the copper in the oxides less available than that in the water-soluble forms or the carbonate.

Little is known of the chemical forms in which copper exists in foods. Changes in these forms affecting availability must occur because fresh herbage is significantly less effective in promoting body copper stores than hay or dried herbage of equivalent total copper content (140). In pasture herbage, much of the copper exists in bound form as water-soluble organic complexes. Aqueous extracts of fresh herbage have been shown by Mills (218) to contain a small proportion of their copper as free ions or as positively charged complexes. The greater part exists in the form of neutral or negatively charged complexes. When fed to Cu-deficient rats these complexes induce a more rapid response, and greater liver copper storage, than does the feeding of equivalent amounts of copper as copper sulfate. As an outcome of these studies, Mills suggested that copper may be transported through the intestinal mucosa both as ionic copper and in the form of complexes such as those encountered in herbage. This suggestion receives support from an earlier observation that the sodium salt of a copper-allylthioureabenzoic acid complex is more effective in stimulating growth and hematopoiesis in the rabbit than is inorganic copper (269) and from the findings of Kirchgessner and co-workers (172, 173). These workers have shown that the affiinity of copper ions for inorganic and organic ligands in the food can reduce the rate of absorption, depending upon the size and stability of the resulting complexes. Small stable complexes may be superior to copper sulfate in absorbability. With single amino acids as ligands the rate of copper absorption depends on the type of amino acid, its configuration, and the degree of polymerization.

The importance of organic complexes to copper absorption is further exemplified by the fact that phytate can reduce the assimilation of this element (93) and that high dietary ascorbic acid increased the severity of copper deficiency in chicks (61, 148) and in rabbits (160). High dietary levels of ascorbic acid were shown by Van Campen and Gross (312) to depress significantly $^{64}Cu$ absorption when the acid was placed into a ligated segment of rat intestine along with the radiocopper. Whole-body retention studies provided further evidence that the depressing effect of ascorbic acid upon copper retention is achieved primarily by reduced intestinal absorption rather than by increased excretion.

Several inorganic dietary factors markedly affect copper absorption, retention, and distribution within the body. The depressing effects of zinc and of molybdenum were mentioned earlier and are considered in more detail later. High dietary levels of cadmium also severely depress copper uptake (147, 311) and change the distribution of copper in the tissues. The relation of silver and mercury to copper utilization is less clear. Hill and co-workers (150) found that silver, but not mercury, tended to accentuate the effects of copper deficiency in chicks, whereas mercury had an adverse effect on copper-adequate chicks. Van Campen (311), on the other hand, found that mercury produced a moderate, but not statistically significant, lowering of $^{64}Cu$ uptake from the intestine of the rat, whereas silver had very little effect. Each of these elements induced significant and quite different changes in the distribution of the retained copper among the tissues and organs of the body. The diversity of the physiological responses and interactions with copper produced by the four elements, zinc, cadmium, silver and mercury, which are chemically very similar, is surprising and, at present, inexplicable. Starcher (286) has recently obtained evidence that the depressing effect of Zn and Cd on copper absorption in the chick arises from the fact that these elements bind to and displace copper from a duodenal mucosal protein with molecular weight of about 10,000. Matrone (207) suggests that molybdenum interferes with copper absorption by the formation of a Cu:Mo complex, probably $CuMo0_4$, of low availability.

Dick (97) contends that molybdenum and inorganic sulfate reduce copper retention in the sheep by reducing the absorption of ingested copper and by increasing the urinary excretion of absorbed and stored copper, due in each case to interference with membrane transport of copper. Mills (219) found that high intakes of molybdate and sulfate do not increase the rate of radioactive copper excretion in sheep and suggests that these anions restrict copper retention by depressing copper solubility within the digestive tract. Neither molybdenum nor sulfate alone, nor salts of zinc, nickel, iron, tungsten, vanadium, chromium, rhenium, uranium, or tantalum, within the limits fed, have any such limiting effect upon copper retention (86, 97). However, there is evidence that other dietary factors can exert a modifying influence upon the copper-molybdenum interaction and that there are unrecognized factors capable of affecting copper absorption and utilization. This is apparent from the wide variations in the response of individual animals kept under identical conditions and from the fact that copper deficiency can occur in sheep and cattle on herbage of apparently normal copper, molybdenum, and sulfate contents (189). In fact, excessive copper storage has been observed in stalled sheep maintained on diets containing concentrations of these three constituents in similar proportions to those of herbage from which neonatal ataxia can occur in lambs (189).

Copper entering the blood plasma from the intestine becomes loosely bound to serum albumin to form the small, direct-reacting pool of plasma copper, in which form it is distributed widely to the tissues and can pass readily into the erythrocytes (55). The Cu-albumin serum pool also receives copper from the tissues. The copper in ceruloplasmin does not appear to be so readily available for exchange or for transfer.

The copper reaching the liver, the key organ in the metabolism of this element, is incorporated into the mitochondria, microsomes, nuclei, and soluble fraction of the parenchymal cells in proportions which vary with the age (129, 241, 242), the strain (144, 299), and the copper status (129, 225) of the animal. The copper is either stored in these sites or is released for incorporation into erythrocuprein and ceruloplasmin and the various copper-containing enzymes of the cells. Ceruloplasmin is synthesized in the liver (199, 289) and secreted into the serum. Erythrocuprein is probably synthesized in normoblasts in the bone marrow (55). The hepatic copper is also secreted into the bile and excreted via this route back to the intestinal contents. Much smaller amounts of copper also pass directly from the plasma into the urine or through the intestinal wall (see Fig. 7).

A high proportion of ingested copper appears in the feces. Most of this normally consists of unabsorbed copper but active excretion via the bile occurs in all species. The biliary system is the major pathway of excretion in humans (315), pigs and dogs (33, 197), mice (123), and ducks and fowls (22). It has been estimated that of the 2-5 mg Cu in-

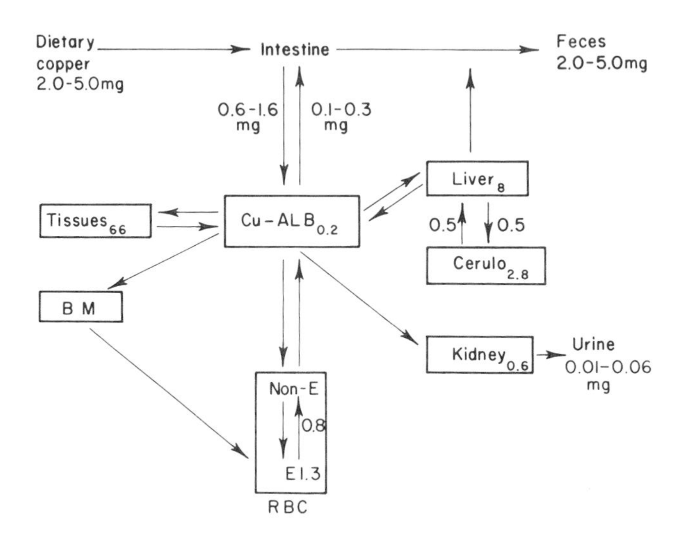


**Fig. 7.** Schematic representation of some metabolic pathways of copper in man. The numbers in the boxes refer to milligrams of copper in the pool. The numbers next to the arrows refer to milligrams of copper transversing the pathway each day. Cu-ALB = direct-reacting fraction; cerulo = ceruloplasmin; NON-E = nonerythrocuprein; BM = bone marrow; RBC = red blood cell. [From Cartwright and Wintrobe (68).]

gested daily by adult man, 0.6-1.6 mg (32%) is absorbed, 0.5-1.3 mg is excreted in the bile, 0.1-0.3 mg passes directly into the bowel, and 0.01-0.06 mg appears in the urine (68). Intravenous injection of copper, resulting in elevated blood and tissue copper levels, is followed by a greater excretion of copper in the bile and, hence, in the feces but does not normally raise urinary copper output. Thus Mahoney *et al.* (197) found that dogs excreted 0.6% of a dose of injected radiocopper in the urine, 1.5% passed directly through the intestinal wall, and 7-10% was excreted in the bile. When the biliary route was obstructed the excretion of copper through the kidney and intestinal wall increased. Increased urinary excretion of copper occurs similarly in patients with liver cirrhosis accompanied by biliary obstruction, but not in patients with Laennec's cirrhosis without significant biliary obstruction (18.). In Wilson's disease the presence of a markedly elevated hepatic copper pool does not raise bile copper concentrations (56), pointing to biliary obstruction in this condition.

Reported values for the small amounts of copper excreted daily in the urine, and for human urinary copper concentrations, are extremely variable. Some of this variation probably reflects analytical difficulties in estimating the very low concentrations of copper normally present.

Giorgio *et al.* (122) developed an accurate colorimetric method, with which they estimated the mean 24-hr excretion of copper in the urine of 20 normal adult persons to be $21 \pm 5.2$ μg. This is substantially higher than the 9 μg/24 hr reported earlier by the same group for 10 normal subjects in Utah (66) but compares favorably with the mean values of 18 (71) and 30 μg (315) reported by others. Schroeder *et al* (261), using direct atomic absorption spectrometry, obtained very variable urinary copper concentrations with some unusually high values up to 350 μg Cu/liter. These workers give 60 μg Cu/day as the average urinary output by adult man. When radioactive copper is administered orally or intravenously, radioactivity can be detected in the urine (56), so that the traces of copper normally found in urine cannot be attributed to contamination. It is probable that this copper represents the copper which has been dissociated from the copper-albumin complex of the serum during its passage through the kidneys (56).

In Wilson's disease, levels of urinary copper of 1500 μg/day may be reached. Excessive excretion of copper in the urine is a constant feature of the condition. Amino acid excretion is also high in this disease, with no evidence that the copper is eliminated as copper-amino acid complexes. In nephrosis the excess urinary copper occurs as a nondialyzeable copper-protein complex, partly as ceruloplasmin. The amount of urinary copper in nephrotics can be correlated with the high urinary excretion of zinc and iron. This, in turn, can be correlated with high urinary protein, resulting from the firm binding of these metals to proteins in the plasma.

Negligible amounts of copper are lost in the sweat (136) and comparatively small amounts in the normal menstrual flow. An average of 0.5 mg Cu per period has been estimated which is slightly less than 0.02 mg ($0.5 \times 13/365$) per day (188). This is only one-twenty-fifth or less of the average loss of iron per day through menstruation, whereas the average intake of dietary copper is close to one-fifth of that of iron. The loss of copper in the milk at the height of human lactation is much higher (approximately 0.4 mg/day). There is no evidence that this loss imposes any nutritional hazard.

## IV. Copper Deficiency and Functions

A wide variety of disorders has been associated with a dietary deficiency of copper or with responses to copper therapy. They include anemia, depressed growth, bone disorders, depigmentation of hair and wool, abnormal wool growth, neonatal ataxia, impaired reproductive performance, heart failure, cardiovascular defects, and gastrointestinal disturbances. The extent to which one or more of these dysfunctions is actually revealed depends upon the species and its age and sex, upon the environment, and upon the severity and duration of the copper deficiency. It appears that as the copper available to the animal becomes insufficient for all the metabolic processes involving copper, as a result of inadequate intake or depletion of body reserves, certain of the processes fail in the competition for the inadequate supply. The degree of sensitivity of particular processes to lack of copper varies with the species and, within a species, with the stage of maturity of the animal and the rapidity with which the copper deficiency develops. For instance, in the sheep the processes of pigmentation and keratinization of wool are among the first to be affected by a lowered copper status. At certain levels of copper intake these defects can arise without any other signs of copper deficiency becoming apparent. Neonatal ataxia occurs readily in lambs from Cu-deficient ewes in some areas but occurs only rarely in calves in the same areas. Further, the cardiovascular lesions affecting the major blood vessels in Cu-deficient pigs, dogs, and chicks have not been observed in Cu-deficient sheep or cattle.

There are also certain environmental or "area" differences in the incidence of various naturally occurring nutritional disorders which respond to copper. For instance, specific lesions in the wool of sheep do not occur in all areas where other signs of copper deficiency or responses to copper supplements are manifest; fibrosis of the myocardium, with sudden deaths, has only been observed in association with copper deficiency in bovines in parts of Australia; and neonatal ataxia occurs in parts of England where copper intakes from the grazing are adequate by standards found adequate elsewhere. Copper utilization in the animal is affected by many other elements so that the discrepancies just cited may ultimately be explained on the basis of environmental differences in dietary factors other than copper. It is also possible that dietary factors exist in some environments which are themselves responsible for some of the apparent manifestations of copper deficiency, which only arise when insufficient copper is present to afford protection or to counteract their effects. Genetic variations in copper metabolism, such as those recently disclosed by Wiener and Field (321a) for sheep in England, may constitute a further factor contributing to certain "area" differences.

### 1. *Copper and Anemia*

Anemia is a common but not inevitable expression of copper deficiency in all species studied. The extent to which it occurs depends on the severity of the deficiency and the length of time over which the deficiency develops. Copper deficiency is first expressed in a slow depletion of body copper stores, including a fall in the copper concentration of the blood plasma until the level is below that necessary to maintain a normal rate of hematopoiesis. A blood copper level of 0.10 to 0.12 μg/ml has been found to limit blood formation in the sheep (20, 203) and 0.2 μg/ml is suggested as the minimum level at which normal hematopoiesis can take place in the pig (180). If such low levels are maintained for any length of time, anemia develops and may progress to a fatal termination. Young rats, pigs, and dogs placed on a milk diet plus iron quickly become anemic because of the rapid fall in blood copper to limiting levels.

The morphological characteristics of the anemia of copper deficiency vary in different species. In rats, rabbits, and pigs, it is hypochromic and mycrocytic (180, 280, 281). In lambs the anemia is also hypochromic and microcytic (26); in cattle and in ewes, it is hypochromic and macrocytic (26, 85, 212); and in chicks (206) and dogs (17), normocytic and normochromic. In the last-named species the anemia is characterized by a reduction in the number of erythrocytes and the maintenance of relatively normal red cell indices. The bone marrow does not undergo a normoblastic hyperplasia as in the Cu-deficient pig, and there is evidence of a defective maturation of the erythrocytes (17). Further evidence that copper is concerned in the erythrocyte maturation process comes from the observation that copper anemic pigs and dogs contain a lower percentage of reticulocytes in their blood than those with iron anemia (17, 180) and the fact that the addition of copper to the diet of rats, rabbits, and pigs suffering from a combined iron and copper deficiency anemia elicits a marked and persistent reticulocyte response, whereas iron has little effect (180, 279, 280). In addition, the survival time of the erythrocytes is shorter than normal in the Cu-deficient pig (54). It seems, therefore, that copper is an essential component of adult red cells and that a certain minimum of copper must be available both for their production and for the maintenance of their integrity in the circulation.

The defect in hemoglobin synthesis of copper deficiency could arise either from abnormalities in the biosynthesis of protoporphyrin and heme, or of globin, or from abnormalities in iron metabolism. The synthesis of globin in copper deficiency does not appear to have been specifically investigated but protein synthesis is generally normal in this deficiency (120). Lee *et al.* (184) were unable to demonstrate any abnormality in the heme biosynthetic pathway in Cu-deficient pigs. In fact, as the anemia developed the activity of heme biosynthetic enzymes increased. Copper could not, therefore, be a cofactor in any of the reactions studied.

For many years it was accepted that iron absorption is unaffected in copper deficiency (107) and that copper is necessary only for "the utilization of iron by the blood-forming organs and for the mobilization of iron from the tissues" (265). In 1952, Chase *et al.* (70) reported that copper increased radioiron absorption in rats and produced evidence of an impaired ability of Cu-deficient pigs to absorb iron, to mobilize iron from the tissues, and to utilize iron in hemoglobin synthesis (131). Lee and associates (185) carried out further studies of iron metabolism in copper-deficient pigs. They confirmed that these animals fail to absorb iron at a normal rate and observed increased amounts of stainable iron in fixed sections of the duodenal mucosa. When $^{59}$Fe was administered orally the mucosa extracted iron from the duodenal lumen at a normal rate but transfer to the plasma was impaired. When the iron was given intramuscularly, increased amounts were found in the reticuloendothelial system, the hepatic parenchymal cells, and in the normoblasts. Hypoferremia was observed in the early stages of copper deficiency, even though the iron stores were normal or increased. When red cells damaged by long storage were administered the reticuloendothelial system failed to extract and transfer the erythrocyte iron to the plasma at a normal rate. Administration of copper to the Cu-deficient pigs resulted in a prompt increase in plasma iron.

The abnormalities in iron metabolism just described were explained in terms of an impaired ability of duodenal mucosa, the reticuloendothelial system, and the hepatic parenchymal cells of the Cu-deficient animal to release iron to the plasma. This hypothesis is compatible with the suggestion made by others that the transfer of iron from tissues to plasma required the enzymic oxidation of ferrous iron and that ceruloplasmin is the enzyme (ferroxidase) that catalyzes the reaction. The authors proposed an additional defect in iron metabolism in copper deficiency, residing within the normoblast itself. The excessive amounts of iron found in the normoblasts suggested that a defect in these cells plays a major role in the development of the anemia. As a result of this defect, iron cannot be incorporated into hemoglobin and, instead, accumulates as nonhemoglobin iron.

### 2. *Copper and Bone Formation*

Bone abnormalities have been reported in Cu-deficient rabbits (161), chicks (119, 235, 256), pigs (115, 180, 298), dogs (17), and foals (25). Poor mineralization of the bones, responsive to copper supplementation, also occurs in mice fed on a meat diet (133). Bone defects are not a conspicuous feature of copper deficiency in ruminants, but a low incidence of spontaneous bone fractures has been observed in sheep and cattle grazing Cu-deficient pastures (26, 86, 92). The fractured bones are almost normal in appearance and shaft thickness, exhibit a mild degree of osteoporosis, and do not reveal any macroscopic defect that could account for their brittleness (86).

In young dogs rendered severely copper deficient, a gross bone disorder has been demonstrated, with fractures and severe deformities in many of the animals (17). The skeletal changes are a specific effect of the copper deficiency unrelated to the concurrent anemia. Studies of the bones of these dogs and of Cu-deficient pigs with essentially similar histological changes (115) disclosed a marked failure of deposition of bone in the cartilage matrix, accompanied by normal growth of cartilage. The bones were characterized by abnormally thin cortices, deficient trabeculae, and wide epiphyses, together with normal ash, $Ca_1P$ and $CO_2$ contents. These changes are distinct from those seen in rickets. show similarities to those that occur in scurvy, and contrast with most deficiency states in growing animals, in which osteoblastic and chondroblastic activities usually fail together.

The ash content of the bones and the Ca, P, and Mg contents of this ash remain within normal levels in Cu-deficient chicks (256). Bone catalase is also unaffected by the deficiency, but there is a marked reduction in both cytochrome oxidase and amine oxidase activity in the Cu-deficient bones. In addition, collagen extracted from the bones of the Cu-deficient chicks contains less aldehyde and is more easily solubilized than collagen from control bones (256). Since collagen, like elastin, contains intramolecular cross-links and its solubility is inversely related to the degree of cross-linking (32, 137), it seems that copper is involved in promoting the structural integrity of bone collagen, in a manner analogous to its function in the biosynthesis of aortic elastin. Although the exact nature of the cross-links in collagen is unknown, it is probable that the biosynthetic sequence is similar to that of elastin and that the primary biochemical lesion in the bones of Cu-deficient chicks is a reduction in the copper enzyme, amine oxidase.

### 3. *Copper and Neonatal Ataxia*

A nervous disorder of lambs, characterized by incoordination of movement, has long been recognized in various parts of the world and given local names such as "swayback," "lamkruis," "renguera," and "Gingin rickets." All these conditions appear to be pathologically identical and the term "enzootic neonatal ataxia" can properly be applied to them. In 1937, Bennetts and Chapman (27) showed that the ataxia of lambs occurring in Western Australia was associated with subnormal levels of copper in the pastures and in the blood and tissues of both ewes and affected lambs, and could be prevented by copper supplementation of the ewe during pregnancy (26). Subsequent investigations in several countries confirmed the efficacy of copper supplements to the ewe in preventing the ataxia. In some areas the incidence of ataxia could not be explained in terms of a simple dietary deficiency of copper (4, 164). Thus in the swayback areas of England, the copper content of the pastures is in the normal range, the copper of the blood and liver is subnormal, but not severely so, and no strict correlation is observed between the blood copper of the ewe during pregnancy and the appearance of ataxia in her lamb. High dietary intakes of molybdenum and inorganic sulfate by ewes during pregnancy can result in the birth of ataxic lambs under particular experimental conditions (222, 224), but these substances are in the normal range in swayback pastures (189). The incidence of swayback in those parts of England must, therefore, be attributed to a "conditioned" copper deficiency brought about by the existence in the environment of other factors affecting copper utilization or distribution in the animal. Barlow and associates (15), investigating swayback as it occurs in southeast Scotland, also found no correlation between the copper status of the animal and the severity of the ataxia. They showed that a low copper status can occur in flocks completely free from the disease. However, many of their values for the copper contents of affected pastures and of the blood and tissues of ataxic lambs and their mothers are close to the copper deficiency levels reported from Australia. It should be noted also that Lewis *et al.* (190) have induced swayback in lambs from stall-fed ewes fed a semipurified, copper-deficient diet, apparently adequate and normal in other respects.

For some years it was accepted that the sheep was unique in its particular susceptibility to nervous system abnormalities and ataxia in copper deficiency. This belief is no longer tenable. Neonatal ataxia has been reported in Cu-deficient goats (142, 238, 263) and has been demonstrated experimentally in the newborn guinea pig (110). Before this, ataxia was observed in young pigs, associated with low liver copper levels and with "demyelination of all areas of the spinal cord except the dorsal areas" (166, 323). Subsequently, McGavin and co-workers (214) described demyelination of the spinal cord extending into the medulla oblongata and cerebellum in 3 to 4-month-old pigs with ataxia and subnormal liver copper concentrations, and Carlton and Kelly (62) reported neural lesions in the offspring of female rats fed a Cu-deficient diet, accompanied in some cases by behavioral disturbances. Cerebral edema and cortical necrosis were prominent in the affected rats but nerve fiber degeneration in the brain stem and spinal cord was not a feature of the disease, as it is in the ataxia of Cu-deficient lambs. The lesions in the rats were consistent with alterations produced by severe tissue hypoxia. They most resembled the lesions described in sheep by Roberts *et al.* (251) and considered to represent an acute, delayed form of swayback.

In lambs two types of ataxia have been recognized under field conditions—the common form in which the lambs are affected when born and a delayed type in which clinical signs may not appear for some weeks. Incoordinated movements of the hind limbs, a stiff and staggering gait, and swaying of the hind quarters are evident as the disease develops. Some lambs are completely paralyzed or are ataxic at birth and soon die; others appear normal at birth but the condition develops progressively until locomotion becomes impossible. Very mild cases also occur in which the ataxia is only revealed when the lambs are startled or driven. Appetite usually remains unimpaired.

The histopathology and biochemistry of the neonatal ataxia of copper deficiency, and especially of swayback in lambs, have been intensively studied. Innes and Shearer (164) initially characterized swayback as a demyelinating encephalopathy, with massive collapsed areas in the cerebral hemispheres, cavitation, and cerebral convolutions. Barlow and co-workers (15) found cerebral lesions absent in approximately one-half of the affected lambs examined, whereas cell necrosis and fiber degeneration occurred in the brainstem and spinal cord of all the affected animals. The pathological criteria of swayback were given as "cavitation or gelatenous lesions of the cerebral white matter and/or a characteristic picture of chromatolysis, neurone necrosis and myelin degeneration in the brain stem and spinal cord." Later work showed that degenerative changes were particularly prominent in the large nerve cells of the brainstem and spinal cord and that these changes accompany the late stages of chromatolysis, with a gradual loss of lysosomes and the formation of Golgi agglomerates (12-14). Mills and co-workers (222, 224) found the ataxia to be associated with demyelination of the nuclei of the large motor neurones of the red nucleus in the brainstem and with hypertrophy of the oligodendroglia in the caudate nucleus. Howell *et al.* (156) also observed changes in the neurones throughout the brain and spinal cord of swayback lambs but products of degenerating myelin were not detected by histochemical or biochemical methods. It was suggested that the lesion in the white matter of the spinal cord may be one of myelin aplasia.

It seems that the primary lesion in swayback is a low content of copper in the brain, leading to a deficiency of cytochrome oxidase in the motor neurones (113). An early and marked loss of cytochrome oxidase in heart, liver, and bone marrow of the Cu-deficient rat was demonstrated by Schultze (264) and in the heart and liver of Cu-deficient pigs by Gubler *et al.* (130). Subsequently, Gallagher and co-workers (120) demonstrated a significant fall in cytochrome oxidase activity of the brain, as well as of the heart and liver, of rats in copper deficiency, together with a depression in phospholipid synthesis. These workers maintained that the loss of activity of cytochrome oxidase, a copper enzyme that contains a heme prosthetic group and possibly a lipid (316), results from a failure of synthesis of the prosthetic group, heme $\alpha$, and that this must be regarded as one of the basic functions of copper. Gallagher (119) then put forward an hypothesis to account for the particular sensitivity of the lamb to ataxia. He pointed out that (*a*) myelin is composed largely of phospholipid and phospholipid synthesis is depressed in copper deficiency, (*b*) inhibition of cytochrome oxidase activity can lead to demyelination (163), so that a severe deficiency of this enzyme should produce the same effect, (*c*) myelination in the lamb proceeds most rapidly during late gestation, and (*d*) lambs can be severely Cu-deficient *in utero* and at birth. Demyelination of the central nervous system (CNS) could, therefore, occur as a consequence first of a depletion of cytochrome oxidase activity leading to inhibition of aerobic metabolism and, second, as a consequence of a decrease in phospholipid and, hence, in myelin synthesis. Both conditions are present at the critical period when myelin is being laid down most rapidly in the lamb. These arguments apply similarly to the guinea pig which, like the sheep, undergoes considerable myelination during gestation and can be severely Cu-deficient *in utero* and at birth. (110).

Howell and Davison (155) first demonstrated a significant lowering of copper content and of cytochrome oxidase in the brain of swayback lambs. The reduction in cytochrome oxidase was much more pronounced in the brain than it was in the liver. Mills and Williams (224) also found significantly lower cytochrome oxidase activity in the brainstem of swayback than of normal lambs and provided evidence that brain copper levels are of greater significance to the integrity and function of the CNS than are liver copper levels. None of the 18 ataxic lambs examined by them had a brain copper content as high as 3 $\mu$g/g dry weight of tissue, whereas of the 37 animals in the normal group only 2 had a brain copper content of less than 3 $\mu$g/g. Subsequent histochemical studies indicated that the most severe reduction in cytochrome oxidase activity occurs in those groups of nerve cells that show the morphological lesions of the disease (12). Finally, Fell *et al.* (113) examined histochemically the brains of clinically normal, copper-deficient lambs and found that the large motor neurones of the red nucleus and of the ventral horns of gray matter in the spinal cord were most severely lacking in cytochrome oxidase. These findings led to the suggestion, quoted earlier, "that a low copper content in the brain leads to a deficiency of cytochrome oxidase in the motor neurones and that this is probably the primary lesion in swayback."

### 4. *Copper in Relation to Pigmentation of Hair and Wool*

Achromotrichia occurs in the Cu-deficient rat, rabbit, guinea pig, cat, dog, goat, sheep, and cattle but has not been observed in the pig. Achromotrichia and alopecia are a more sensitive index of copper deficiency in the rabbit than is anemia (279). The situation is similar in the sheep. Lack of pigment production in the wool of black-wooled sheep occurs at copper intakes sufficient to prevent anemia or other signs of copper deficiency. The pigmentation process is so sensitive to changes in copper status of the animal that, once a condition of copper deficiency has been established, alternating bands of pigmented and unpigmented fibers can be produced by adding or withholding copper from the diet (Fig. 8). Even on fairly high copper intakes, Dick (95) showed that it is possible to block the functioning of copper in the pigmentation process in sheep within 2 days by raising the molybdenum and inorganic sulfate intakes sufficiently. He suggests that this effect may take place within hours of giving the first dose of molybdenum and sulfate. A marked loss of pigmentation in the hair of cattle as a result of a "conditioned" copper deficiency has also been reported in Japan (308).

The precise mechanism involving copper in the pigmentation process is unknown, although a breakdown in the conversion of tyrosine to melanin is the most likely explanation since this conversion is catalyzed by copper-containing polyphenyl oxidases (244).

### 5. *Copper and Keratinization*

Changes in the growth and physical appearance of hair, fur, or wool have been noted in Cu-deficient rats, rabbits, guinea pigs, dogs, cattle, and sheep. A reduction in the quantity and quality of wool produced by sheep in Cu-deficient areas was recognized early in the Australian investigations (25, 204). Copper supplementation of such sheep increases the weight of wool produced, the rate of wool growth being related to the degree of copper deficiency imposed (202). The lowered wool weights are probably mainly an expression of an inadequate supply of substrate to the wool follicles, consequent upon a reduced food intake by the Cu-deficient animals. The deterioration in the process of keratinization, signified by the failure to impart crimp, appears to be a specific effect of copper deficiency.

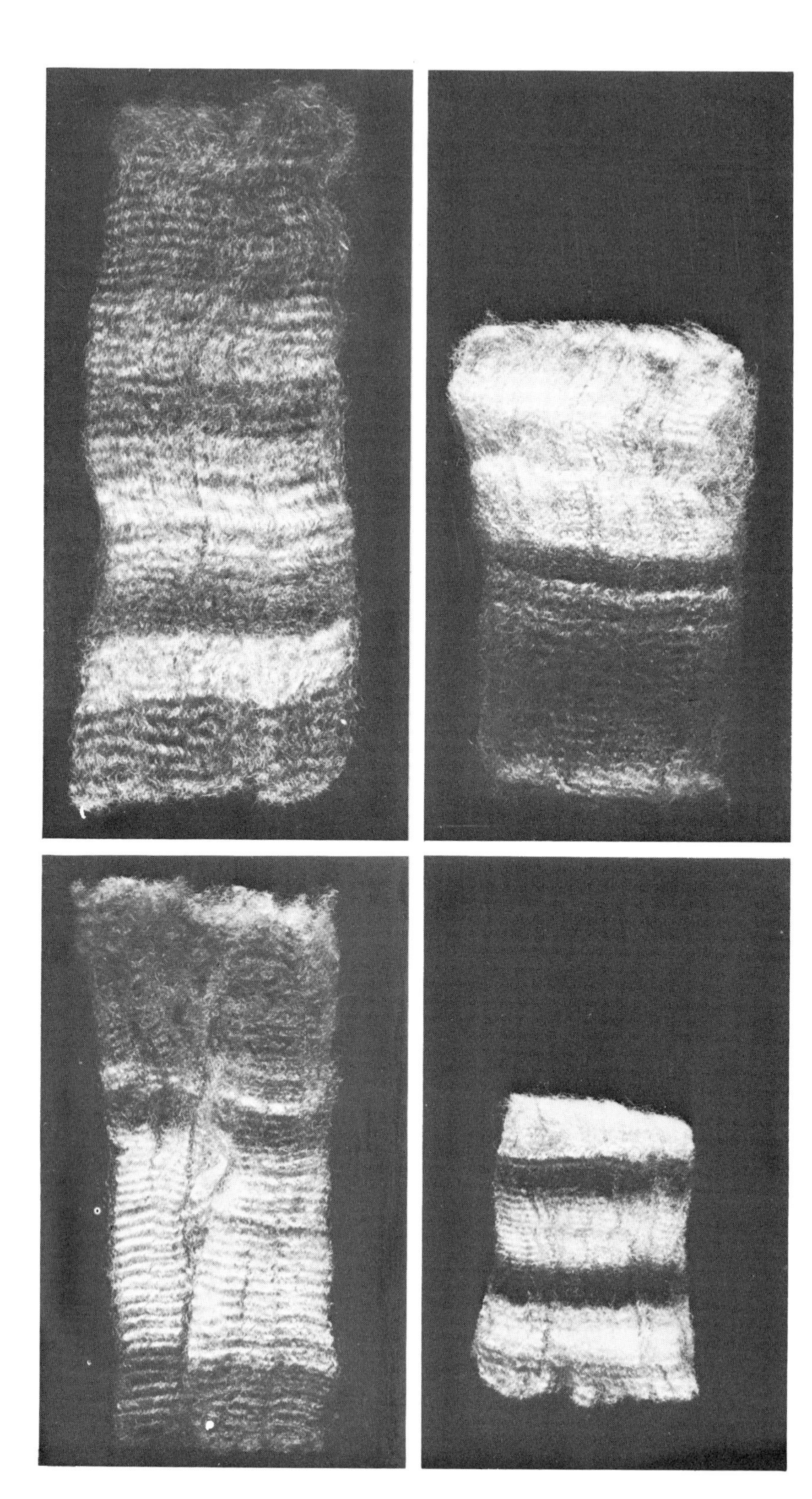

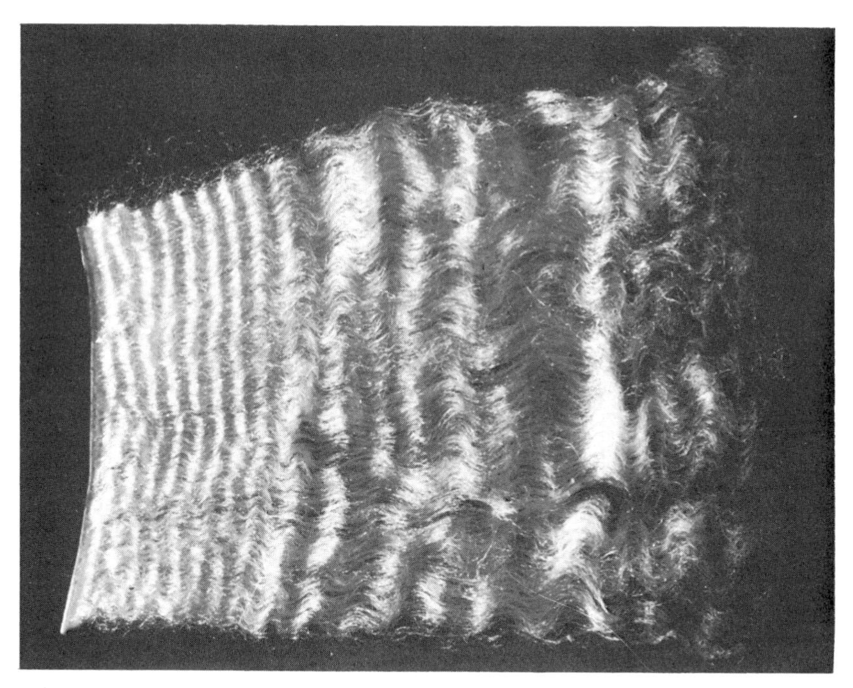

**Fig. 9.** Wool from Merino sheep showing loss of crimp when diet was lacking in copper and restoration of crimp with the provision of supplemental copper. (Photo by courtesy of H. J. Lee.)

As the sheep's reserves of copper are depleted and the blood copper concentration falls, the crimp in the wool becomes progressively less distinct in the newly grown staple until the fibers emerge as almost straight hairlike growths, to which the terms "stringy" or "steely" wool have been given (Fig. 9). A spectacular restoration of the crimp occurs when copper supplements are given. The tensile strength and affinity for dyes of steely wool are reduced, the elastic properties are abnormal, and it tends to set permanently when stretched (200).

These abnormalities are more obvious in Merino wool than in British breed wool because the former is normally more heavily crimped. However, Lee (186) has demonstrated fleece abnormalities in four British breeds comparable with those encountered in Cu-deficient Merinos in Australia. This particular manifestation of copper deficiency has not been

---

**Fig. 8.** Wool from black-wooled sheep, showing unpigmented bands when the animals were suffering from copper deficiency. (Photos by courtesy of H. J. Lee.)

observed in British breeds in other countries. This raises the question as to whether a further dietary factor, additional to copper and peculiar to the environment where the wool lesions arise, may not be involved.

The characteristic physical properties of wool, including crimp, are dependent upon the presence of disulfide groups which provide the cross-linkages or bonding of keratin and upon the alignment or orientation of the long-chain keratin fibrillae in the fiber. Both of these are adversely affected in copper deficiency (200). Straight, stringy wool has more sulfhydryl groups and fewer disulfide groups than normal wool (52,200), suggesting that copper is required for the formation or incorporation of disulfide groups in keratin synthesis. Furthermore, wool from Cu-deficient sheep contains more N-terminal glycine and alanine and sometimes more N-terminal serine and glutamic acid than normal wool, indicating that a lack of copper can interfere with the arrangement of the polypeptide chains in keratin synthesis (53).

### 6. *Copper and Fertility*

In female rats, copper deficiency results in reproductive failure, due to fetal death and resorption (103,135,158). The estrus cycles remain unaffected and conception appears to be uninhibited. Howell and Hall (158) showed that normal fetal development in their Cu-deficient rats ceased on the thirteenth day of pregnancy when the fetal tissues were seen to be disintegrating. Necrosis of the placenta became apparent on the fifteenth day of pregnancy. Savage (258) reports the results of an experiment in which hens were fed a severely Cu-deficient (0.7-0.9 ppm Cu) diet for 20 weeks. Egg production and the copper content of eggs, plasma, and liver were decreased and hatchability showed a rapid initial drop and approached zero in 14 weeks. Embryos from hens receiving the low-copper diet showed anemia, retarded development, and a high incidence of hemorrhage after 72 to 96 hr of incubation. Preliminary results indicated a reduction in monoamine oxidase activity in the Cu-deficient embryos, as well as in the Cu-deficient hens. Of great interest, also, is the report that the deficient embryos revealed abnormalities of the spinal cord. It appears that the hemorrhages in the chick embryos are similar to those observed in Cu-deficient rat embryos. The mortality is probably due to a defect in red blood cell and in connective tissue formation (275) during early embryonic development.

Low fertility in cattle has been associated with copper deficiency in several areas (6,28,314). The nature of the reproductive disturbance has not been thoroughly studied but depressed or delayed estrus is common in dairy cows in such areas.

### 7. *Copper and Cardiovascular Disorders*

The first evidence of cardiac lesions in copper deficiency emerged from studies carried out by Bennetts and co-workers (28) of a disease of cattle occurring in Western Australia, known locally as "falling disease." The essential lesion of falling disease is an atrophy of the myocardium with replacement fibrosis. The morbid process is a progressive one, extending even over a period of years, commencing with the presence of occasional areas of small-celled infiltration and proceeding to the replacement of large areas of atrophied myocardium with dense collagenous tissue. The sudden deaths characteristic of the disease are believed to be due to acute heart failure, usually after mild exercise or excitement. The disease can be completely prevented by copper therapy of the animals or by treatment of the pastures with copper compounds to raise their naturally very low copper levels (1-3 ppm) to normal. Falling disease has never been observed in sheep or horses grazing the same areas and occurs only rarely in Cu-deficient areas elsewhere (92). Sudden cardiac failure, associated with cardiac hypertrophy greater than could be accounted for by the occurrence of anemia, has been reported in Cu-deficient pigs (130), but cardiac hypertrophy is not a prominent feature of falling disease. Lesions in the aorta and other large vessels were not described in this condition.

In 1961, O'Dell and associates (235) presented histological evidence of a derangement in the elastic tissue of the aortas of Cu-deficient chicks and found the mortality in these chicks to be caused by a rupture of the major blood vessels. At about the same time the Utah group described cardiovascular changes in Cu-deficient pigs which resulted in ruptures of the major blood vessels and consequent death of the animals (63, 80, 272). The tensile strength of the aorta was markedly reduced and the myocardium was abnormally friable. In these species the histological evidence indicated a derangement of the elastic membranes of the aorta and of the coronary and pulmonary arteries. Subsequently, aortic rupture, with degeneration of the elastic membrane, was demonstrated by others (60,274) in Cu-deficient chicks. Extensive internal hemorrhage with a high incidence of aortic aneurisms were also observed in young Cu-deficient guinea pigs (110). In swayback lambs and their dams no such vascular lesions could be demonstrated. Whether this constitutes a real species difference or is a reflection of differences in the intensity or duration of the copper deficiency or of the existence of other complicating factors is not yet known (79).

During the next few years studies in several laboratories combined to elucidate the role of copper in elastin biosynthesis. The most significant findings may be given as follows. The elastin content of the aortas of Cu-deficient pigs (318) and chicks (288) is decreased. The elastin from such animals contains a greatly elevated content of lysine and significantly less desmosine and isodesmosine than that from normal animals (217,234). Desmosine, a tetracarboxylic, tetraamino acid, and its isomer, isodesmosine, are the key cross-linkage groups in elastin (239), and at least two and possibly four lysine residues condense to form these substances (301). For this reaction to take place the ε-amino group of the lysine residues needs to be removed and the carbon oxidized, possibly to an aldehyde – a reaction catalyzed by amine oxidases, which are copper-containing enzymes (45,151,326). The amine oxidase of the plasma of Cu-deficient pigs is lower than that of pigs given a copper supplement (30) and that of ewes and lambs with low plasma copper levels is also subnormal (221). The chick is hatched without detectable amine oxidase activity in the aorta or liver. When the diet contains copper the activity appears in both tissues by the third day at levels which are essentially maintained thereafter. When the diet is deficient in copper the amine oxidase activity remains undetectable in the aorta for 2 to 3 weeks and that in the liver remains substantially below that of copper-supplemented chicks (149) (see Table 13). Finally, the aortas from copper-supplemented chicks are much more capable of incorporating lysine into desmosine than are those from Cu-deficient chicks, and addition of amine oxidase to the culture medium of Cu-deficient aortas decreases the lysine-to-desmosine ratio to approximately that of the aortas from copper-supplemented chicks (149).

The role of copper in the formation of aortic elastin can best be stated in the words of Hill *et al.* (149) as follows: "The primary biochemical lesion is a reduction in amine-oxidase activity of the aorta. This reduction in enzymatic activity results, in turn, in a reduced capacity for oxidatively deaminating the epsilon-amino group of the lysine residues in elastin. The reduction in oxidative deamination results, in turn, in less

TABLE 13

*Effect of Copper Deficiency on Amine Oxidase Activity of Chick Aorta and Liver Mitochondria*[a]

| | Specific activity[b] | | | | | |
|---|---|---|---|---|---|---|
| | Days of age: | | | | | |
| Mitochondria | 1 | 3 | 5 | 10 | 17 | 26 |
| Aorta + Cu | 0 | 15.3 | 20.7 | 19.4 | 14.8 | 22.2 |
| − Cu | 0 | 0 | 0 | 0 | 0 | 4.0 |
| Liver + Cu | 0 | 9.6 | 8.8 | 7.3 | 8.4 | 6.8 |
| − Cu | 0 | 0 | 1.5 | 5.6 | 0 | 1.9 |

[a]From Hill *et al.* (149).
[b]Expressed as ΔOD/min/g protein.

lysine being converted to desmosine. The reduction in desmosine, which is the cross-linkage group of elastin, results in fewer cross-linkages in this protein, which, in turn, results in less elasticity of the aorta." It is probable, as mentioned previously, that the impaired collagen synthesis which occurs in the bones of Cu-deficient chicks, is similarly a manifestation of the low amine oxidase activity of the bones.

### 8. *Copper and Scouring (Diarrhea) of Cattle*

Intermittent diarrhea in cattle occurs in the severely copper-deficient areas of Australia (28). Scouring or diarrhea was such a prominent feature of a copper-responsive disease of cattle occurring in parts of Holland that the name "scouring disease" was applied to it (276). In New Zealand a disease of cattle restricted to pastures on peat land has been named "peat scours" (86) and a clinically similar condition in cattle, characterized by severe, chronic diarrhea, occurs in England (6). In all these cases the scouring is associated with a low copper status of the blood and tissues of affected animals and is prevented or cured by copper therapy. Diarrhea is not a common manifestation of copper deficiency in other species and does not always occur in cattle in copper-deficient areas (92). Molybdenum has been incriminated as a factor in peat scours (86), but it is unlikely that a high Mo:Cu ratio can account for the occurrence of scouring in cattle in other copper-deficient areas.

## V. Copper Requirements

It was emphasized earlier that copper absorption and utilization in the animal can be markedly affected by several other mineral elements and dietary components. A series of minimum copper requirements must, therefore, exist, depending upon the extent to which these influencing factors are present or absent from the diet. A basic minimum requirement can be conceived as one in which all the dietary conditions affecting copper are at an optimum. Since all the factors affecting copper utilization are not yet known and the quantitative interactions of copper with other elements are not completely defined, the basic minimum requirements for copper cannot be given with great accuracy. The minimum requirements for copper also depend upon the criteria of sufficiency employed. For instance, the copper requirements of rats necessary to maintain pigmentation in the hair are substantially higher than those required for growth (223), whereas lesions appear in the wool of sheep in some areas at levels of copper intake sufficient to prevent anemia or other signs of copper deficiency.

1. *Laboratory Species*

Copper supplements as low as 0.005 mg Cu/day were early found to give a growth and hemoglobin response in young rats fed a milk diet plus iron and manganese; supplements of 0.01 to 0.05 mg Cu/day were stated to be optimum (266). Since then numerous workers have developed solid diets containing 1 ppm Cu or less, upon which copper deficiency develops rapidly in young rats and more slowly in adult rats. Mills and Murray (223), using such a diet containing only 0.3-0.4 ppm Cu, established the following tentative minimum requirements for rats of about 70 g live weight fed a 10-g diet daily: for hemoglobin production, 1 ppm; for growth, 3 ppm; and for melanin production in hair, 10 ppm. The minimum requirements for reproduction and lactation were not investigated. It was stated that at a level of 50 ppm Cu these requirements were fully met through at least one generation.

The importance of the length of time over which the copper deficiency is imposed is evident from the work of Everson *et al.* (110) with guinea pigs. Young female guinea pigs fed a diet providing 0.5-0.7 ppm Cu grew well and reproduction was equal to that of females fed the control (6 ppm Cu) diet. Eventually they developed a mild anemia and their hair coats became wiry and depigmented. Furthermore, the growth of the young from these animals began to slow down at about day 12. Surviving animals maintained on the copper-deficient diet were markedly stunted by 50 or 60 days of age.

2. *Pigs*

The minimum copper requirements of pigs for growth and for reproduction and lactation cannot be stated with any precision. Ullrey and coworkers (309) compared dietary copper levels of 6, 16, and 106 ppm and observed no significant treatment differences in baby pigs in growth rate, efficiency of feed use, or level of formed elements in the blood. It was, therefore, concluded that 6 ppm Cu is adequate for the growth of these animals. The Agricultural Research Council of Great Britain (3) came to the conclusion that 4 ppm Cu was probably adequate for growing pigs up to 90 kg live weight. All ordinary swine rations would contain appreciably higher levels of copper than 4 or 6 ppm. The copper concentrations of cereal grains mostly lie between 4 and 8 ppm, and the leguminous seeds and oil seed meals provided as protein supplements generally contain 15-30 ppm. On these grounds, copper supplementation of normal pig rations would appear unnecessary.

As long ago as 1928, Evvard and associates (111) demonstrated an improvement in the growth of pigs when their normal ration was supplemented with copper sulfate. Many years later, Carpenter (64) published a similar observation and Braude (36) observed that young pigs fed a normal diet exhibit a craving for copper. Subsequently, Braude and associates (9) showed that the addition of copper to a normal ration at the rate of 250 ppm resulted in increased rate of weight gains in growing pigs. Numerous experiments in Great Britain and elsewhere have confirmed this finding. In 1967, Braude (37) assessed the results of trials in that country up to mid-1965 and presented evidence that the daily live weight gain can be accelerated by about 8% and the efficiency of feed use increased by about 5.5% by the inclusion of 250 ppm Cu in pig rations. In a coordinated trial carried out at twenty-one centers and involving 245 pigs per treatment, such supplementation increased the mean daily live weight gain by 9.7% and the efficiency of feed use by 7.9% (38). The response in weight gain is related to the amount of copper in the gut, and copper sulfate is more effective than copper sulfide or oxide. However, it is the copper and not the sulfate radicle that is effective (10,33,141). In some experiments the feeding of antibiotics has given similar growth responses to those obtained with copper (10,106), suggesting that the mode of action is similar; in others, the effects have been additive (195); and in still others the antibiotic has been effective when the copper has not (317). Copper sulfate apparently does not stimulate the growth of pigs by suppressing or increasing bacterial multiplication in the gastrointestinal tract because differences in the bacterial flora of the feces (118) or throughout the alimentary canal (278) between copper-supplemented and unsupplemented pigs are not apparent. The mode of action of copper in stimulating growth in this species remains obscure.

Beneficial effects from copper supplementation of normal rations have not been obtained under all conditions. In fact, reports from U.S.A. have indicated that copper supplementation at a level of 250 ppm can significantly reduce live weight gain and efficiency of feed use (16) and can lead to anemia and copper toxicosis in some animals (250,317). In Australia, also, deaths were reported in pigs receiving a diet containing 250 ppm of copper (236). In this experiment, O'Hara and co-workers (236) made the important observation that the feeding of the high copper ration led to skin lesions similar to those of parakeratosis. Zinc supplementation of the ration rectified this condition and brought about a marked improvement in live weight gain. Subsequently, Suttle and Mills (295), in a series of studies of copper poisoning in pigs, established clearly that high intakes of copper raise the animal's requirements for zinc and for iron. Modest increases in the zinc and iron contents of the rations were shown to prevent adverse effects from copper supplementation at the rate of 250 ppm or even higher. These increases also facilitated the improvement in growth rate and feed efficiency observed by Braude and others from this rate of copper treatment on diets naturally higher in zinc and iron. Evidence that the zinc and iron requirements of pigs fed rations high in copper are greater than normal was almost simultaneously obtained by workers in Iowa and in Michigan. Thus, Bunch *et al.* (51) found that increasing the iron content of the ration prevented a fall in hemoglobin levels in copper-fed pigs. Ritchie *et al.* (250), from a consideration of the results of their experiments, suggested that zinc supplements "appeared to furnish complete protection against copper poisoning." The quantitative relationships among zinc, iron, and copper cannot yet be fully defined.

The results of the experiments just described indicate that the discrepancies in the reported responses to copper can be largely explained by differences in the zinc and iron contents of the rations used. Suttle and Mills (295), in their studies of copper toxicity with different protein sources, showed that differences in the calcium content of these rations constitute a further factor of importance. Diets high in calcium have a lower zinc availability, so that the copper toxicity is enhanced. Evidence for a three-way interaction between Cu, Zn, and Ca was also shown by Guggenheim (132) in mice suffering from the induced copper deficiency anemia arising on meat diets. This was accentuated by adding zinc, an effect which was nullified by the simultaneous addition of calcium. From a practical standpoint it is clear that the high requirements of growing pigs for copper to achieve maximal growth and efficiency of feed use can only be safely and satisfactorily exploited if the zinc and iron contents of their diets are higher than those found adequate at lower copper intakes, especially if these diets are also high in calcium.

### 3. *Poultry*

There appear to be no definitive data, based on critical experiments with graded increments of copper, on the minimum copper requirements of chicks for growth or hens for egg production. Nor is there any evidence that poultry have high copper requirements relative to mammals, as they have for manganese. Diets containing 4–5 ppm Cu can be presumed to be adequate. All rations composed of normal feeds are likely to contain more than 5 ppm of this element.

### 4. *Sheep and Cattle*

The basic minimum copper requirement of sheep is much lower than is generally believed. Dick (97) found that under appropriate conditions cross-bred sheep can be maintained in copper balance at intakes of 1 mg Cu/day or less. At normal feed intakes this amount of copper would be supplied by a diet containing only about 1 ppm Cu on the dry basis. The normal wastage of copper by Dick's sheep was calculated to be less than 3 mg/day. This suggests that where more than this amount is required factors are operating which either impose a limitation on copper assimilation and retention or increase the animal's requirements for this element. Such factors apparently operate in many environments. Thus, Marston and co-workers (202,203) found that Merino sheep grazing on the calcareous soils of South Australia, where the herbage contains about 3 ppm Cu, rapidly develop signs of copper deficiency when supplied with additional cobalt. A supplement of 5 mg Cu/day, equivalent to a total pasture content of some 8 ppm, was insufficient to ensure normal blood copper levels and wool keratinization in all animals, although it was adequate to prevent any other evidence of copper deficiency. Under these conditions, where there is a high consumption of calcium carbonate from the environment and of moderate intakes of molybdenum and sulfate from the herbage, the minimum copper requirements of wool sheep are close to 10 mg/day or about 10 ppm of the diet.

In Western Australia, sheep and cattle grazing pastures containing 3-4 ppm Cu, or less, and with molybdenum concentrations usually below 1.5 ppm, exhibit a wide range of copper deficiency symptoms and carry subnormal concentrations of copper in the blood and liver (23). Pastures containing 4-6 ppm Cu on the dry basis, and similarly relatively low in molybdenum, provide sufficient copper for the full requirements of cattle and of British breed and cross-bred sheep. If Merino sheep, which are apparently less efficient than British breeds in securing copper from herbage, are to avoid all signs of copper deficiency, including defective keratinization of the wool, a minimum level of 6 ppm Cu in such pastures is necessary. The minimum copper requirements of cattle and of British breeds and cross-bred sheep can, therefore, be placed at about 4 ppm of the moisture-free diet and those of Merino sheep nearer to 6 ppm. However, even the relatively low levels of molybdenum in these pastures may be having some effect on the requirements. Dick (97) showed that molybdenum intakes as low as 0.5 mg/day can adversely affect copper retention in the sheep, provided the sulfate intake is high. Excess sulfate in the herbage in certain parts of Greece, in the presence of "normal" levels of copper and molybdenum, has been associated by Spais (283,284) with copper deficiency and ataxia in lambs. Furthermore, pastures in parts of England containing 7-14 ppm Cu, or more, and normal levels of molybdenum and sulfate, can result in a subnormal copper status in cattle (6,189) and in sheep, together with ataxia in lambs (4,

162,164,189). These findings illustrate the difficulty of assigning precise minimum copper requirements to sheep or cattle and the impossibility of basing adequacy solely on copper intakes.

Three main methods of preventing copper deficiency in grazing sheep and cattle are employed, the choice of method depending upon the environment and the conditions of husbandry. In many areas the application of copper-containing fertilizers is an effective means of raising the copper content of the herbage to adequate levels. Australian experience indicates that 5-7 lb/acre (or kg/hectare) of copper sulfate or its equivalent in copper ores is sufficient for this purpose for several years and may also increase herbage yields. Under range conditions where fertilizers are not normally applied or on calcareous soils where copper absorption by plants is poor, the copper may be provided in the form of salt licks containing 0.5-1.0% copper sulfate. Such licks are usually consumed by sheep and cattle in Cu-deficient areas in sufficient quantities and with sufficient regularity to maintain adequate total copper intakes. Dosing or drenching the animals at intervals with soluble copper salts is also practised effectively in some areas. Frequent dosing is less important with copper than it is with cobalt, because of the considerable capacity of ruminants to store copper in their livers during periods of excess intake and to draw upon these stores during periods of inadequate intake.

The third method of preventing copper deficiency in grazing ruminants is to give subcutaneous or intramuscular injections of some safe and slowly absorbed organic copper complex. Injections of copper glycinate or copper calcium edetate [CuCa ethylenediaminetetraacetate (EDTA)] at intervals of 3 months, in doses of 30 to 40 mg Cu for sheep, or 120-240 mg for cattle, have been found satisfactory (59,87,227,293). A single injection in midpregnancy of 40 to 50 mg Cu either as copper glycinate, as CuCa EDTA, or as Cu methionate, has been found effective in preventing swayback in lambs and in maintaining satisfactory blood and liver copper concentrations in ewes and lambs (143).

## IV. Copper in Human Health and Nutrition

Hypocupremia occurs in a number of disease conditions in man. The cause in these cases is not a dietary deficiency of copper. It can be related to a defect in ceruloplasmin synthesis, poor absorption, excessive excretion, or some other disturbance associated with the disease. In kwashiorkor, the hypocupremia appears to arise from the low protein intakes, although the diets are frequently also low in copper (211, 246).

Hypocupremia occurs occasionally in infants, accompanied by hypoproteinemia, hypoferremia, and anemia (179,262,291,327). The anemia and the hypoproteinemia usually respond to iron and not to copper. The hypocupremia responds to copper and not to iron. It appears that the primary deficiency in these cases is iron. The hypocupremia results from an inability of the infants to obtain sufficient copper from their copper-low milk diets to prevent copper depletion in the face of increased loss of the plasma copper protein into the bowel (68). Even normal breast-fed infants are often in negative copper balance in early postnatal life and are unable to obtain sufficient copper from the milk to prevent some body copper depletion (69). However, hypocupremia does not necessarily arise in infants on exclusive milk diets (323a,324). Thus Wilson and Lahey (323a) could detect no significant differences in weight gains, hemoglobin values, serum protein or plasma copper levels in two groups of premature infants, one of which received a milk diet supplying only 14 $\mu$g Cu/kg body weight daily for 7 to 10 weeks and one supplying 5-6 times this amount of copper.

More convincing evidence of copper deficiency in infants has been obtained in Peru. Cordano and co-workers (76) studied four severely malnourished infants, aged 8-15 months, who were being rehabilitated on a high-calorie, low-copper milk diet. They exhibited anemia, marked neutropenia, "scurvylike" bone changes, and hypocupremia. In two cases, prompt and dramatic response to copper supplementation was obtained, and in the other two there was a response to vitamin $B_{12}$, with copper producing additional responses. In a later, wider study, Cordano *et al.* (77) reported that copper deficiency is often found in malnourished infants being rehabilitated on a milk diet. The earliest and most constant manifestation was a marked and persistent neutropenia, responsive to copper therapy. This is of particular interest in view of the evidence of neutropenia in Cu-deficient piglets (324). Furthermore, the hypocupremia of these infants was not accompanied by hypoproteinemia. The copper requirements of rapidly growing infants with poor body stores was estimated at between 42 and 135 $\mu$g/kg body weight/day. This is substantially higher than the 14 $\mu$g/kg/day deduced from the results of Wilson and Lahey (323a), quoted above. It conforms well with the daily allowance of 50 $\mu$g Cu/kg/day recommended by the U.S. National Research Council (245).

Estimates of the copper requirements of humans of different ages have been made from balance studies. Scoular (268) indicated that the average 3- to 6-year-old child weighing up to 20 kg requires 1.0-1.6 mg Cu daily. More recently, Engel *et al.* (109) estimated, from balance studies carried out on 36 preadolescent girls, aged 6-10 years, that 1.3 mg

Cu/day is necessary for equilibrium. Their suggested daily allowance was 2.5 mg. Copper balance is maintained in adults on an intake of about 2 mg/day (65,188) or less (303).

Estimates of the actual daily copper intakes mostly lie between 2 and 4 mg (188, 261, 302), with lower estimates for certain Dutch (24) and poor English and Scottish diets (90). Indian adults consuming rice and wheat diets were shown to ingest 4.5 and 5.8 mg Cu/day (94). In a study of total diets designed for 16- to 19-year old North American boys and supplying 4200 cal/day, the overall average copper intake was found to be 3.8 mg, with relatively little variation in different localities or in composite samples taken at different times (328). Copper intakes of this magnitude are believed to be fully adequate for all normal individuals.

The amount of copper ingested daily from the diet is determined mainly by the choice or selection of foods making up the total diet. It can also be influenced by the locality in which the foods are grown and by the extent of contamination with adventitious copper during processing, storage, and treatment with copper fungicides and anthelmintics. The important of the drinking water, particularly soft water, as a further source of copper has been stressed by Schroeder *et al.* (261). These workers observed a progressive increase in the copper in water from brook, to reservoir, to hospital tap, as well as a considerable increment in soft water compared with hard water from private homes. It was claimed that some soft waters, with their capacity to corrode metallic copper, could raise copper intakes by as much as 1.4 mg/day, whereas hard waters would reduce this source of copper to 0.05 mg or less.

The different classes of human foods differ greatly in copper content, with considerable individual variation within particular food items. The richest sources of copper are crustaceans and shellfish, especially oysters, and the organ meats (liver, kidney, and brain), followed by nuts, dried legumes, dried vine, and stone fruits and cocoa (192, 208, 261). These foods range in copper content from 20–30 to 300–400 ppm. The poorest sources of copper are the dairy products, milk, butter and cheese, and white sugar and honey, which rarely contain more than 0.5 ppm Cu. The nonleafy vegetables, most fresh fruits, and refined cereals, including white flour and bread, generally contain up to 2 ppm Cu. This classification of foods as sources of copper is similar to that made earlier for iron, except for Crustacea and shellfish which are not particularly rich in iron as they are in copper. The refining of cereals for human consumption results in a significant loss of copper, as with most other minerals. Thus the mean copper content of the whole grain of North American hard wheats was reported to be 5.3 ppm, whereas the copper content of the white (72% extraction) flour made from these wheats averaged only 1.7 ppm (89).

The copper content of plants, especially leafy vegetables, reflects in some degree the copper status of the soils upon which they have grown. Communities living in areas of low soil-copper status would, therefore, be expected to have lower daily intakes of copper than communities with the same food habits in areas of high soil-copper status. However, evidence of malnutrition attributable to lack of dietary copper has never been presented for any area, even where copper deficiency in grazing livestock is severe. In this respect it should be noted that studies of human foods, including fruits (329), vegetables (154), and infant foods (159) have disclosed great variability in copper content but little evidence of an effect of locality.

## VII. Copper Toxicity

### 1. *General*

Chronic copper poisoning may occur in animals under natural grazing conditions or as a consequence of excessive consumption of copper-containing salt licks or mixtures, or of the unwise use of copper-containing drenches of various kinds. Contamination of feeds with copper compounds from horticultural or industrial sources is a further cause. In pigs the use of copper supplements as growth stimulants can also induce copper toxicosis, if the basal diet is not suitably balanced. Such supplements can also affect the composition of the body fat by significantly raising its iodine number and lowering its physical consistency (297).

In all animals the continued ingestion of copper in excess of requirements leads to some accumulation in the tissues, especially in the liver. The capacity for hepatic copper storage varies greatly among species, and the liver copper levels that can be tolerated without signs of copper toxicosis also vary greatly among species. In sheep and cattle the existence of liver copper values above very high critical levels, is usually followed by a catastrophic liberation of copper into the blood, with resulting hemolysis and jaundice. Hemolytic jaundice has not been reported in rats or rabbits suffering from chronic copper poisoning (35, 134) or in birds. Nor is jaundice a marked feature of this condition in pigs. Adult birds exposed to excessive intakes of copper exhibit a marked loss of body weight and anemia, with no evidence of intravascular hemolysis (124). Turkeys subjected to copper loading through copper injections developed diarrhea and "pale, ruffled feathers" but no hemolysis was noted (320). Rats appear to be extremely tolerant of high copper intakes. Normal growth and health was maintained on diets containing 500 ppm Cu, or about 100 times normal, despite a fourteen-fold increase in liver copper (35). However, the actual liver copper levels attained in the experiment just cited were very much lower than those characteristic of copper poisoning and hemolytic jaundice in sheep.

## 2. *Pigs*

When considering the copper requirements of pigs, mention was made of the fact that signs of copper poisoning can appear when certain diets are supplemented with copper sulfate at the rate of 250 ppm Cu (236, 250, 317). It was stated further that pigs consuming such Cu-supplemented diets could be completely protected against the deleterious effects of copper at this level by raising the animal's zinc and iron intakes (51, 250, 295). A similar protection against toxicosis in pigs, at copper intakes much higher than 250 ppm, has been demonstrated by Suttle and Mills (295), provided the Zn and Fe intakes are sufficiently large. Dietary copper levels of 425, 450, 600, and 750 ppm caused severe toxicosis, manifested by a marked depression in feed intake and growth rate, hypochromic microcytic anemia and jaundice, and marked increases in liver copper and serum copper levels and in serum aspartate aminotransferase (AAT) activities. The latter can be taken as an indication of general tissue damage (1). A limited study of ornithine carbamoyltransferase (OCT) activity in the serum, which more specifically indicates liver damage (325), revealed elevated levels of this enzyme in pigs suffering from copper toxicosis (294). All these signs of copper toxicosis were eliminated by simultaneously providing an additional 150 ppm Zn + 150 ppm Fe to the diets containing 425 or 450 ppm Cu (see Table 14). Addition of 500 ppm Zn or 750 ppm Fe to diets containing 750 ppm Cu eliminated the jaundice and produced normal serum copper and AAT values, but only iron afforded protection against the anemia.

These important findings, together with those of Hill and co-workers (50, 147, 150), previously discussed, provide a secure base for further

TABLE 14
*Mean Serum Copper and Serum Aspartate Aminotransferase (AAT) Activities in Pigs Fed a Cereal-Fishmeal Ration with and without Copper Supplementation*[a,b]

| Trace element supplement (ppm) | Serum Cu (μg/ml) | | Serum AAT (μg/ml) | | |
|---|---|---|---|---|---|
| | 24 days | 46 days | 24 days | 34 days | 46 days |
| 0 | 1.5 | 1.6 | 15 | 25 | 25 |
| 250 Cu | 2.1 | 3.0 | 40 | 65 | 85 |
| 450 Cu | 4.1 | 3.3 | 140 | 240 | 160 |
| 450 Cu + 150 Zn + 150 Fe | 2.5 | 2.6 | 30 | 25 | 25 |

[a]From Suttle and Mills (295).
[b]Four animals were used per treatment.

studies designed to elucidate the mechanisms involved and to identify the sites at which the interactions among Zn, Fe, and Cu occur. It seems likely that these interrelationships involve competition between elements with similar affinities for binding sites on protein in the digesta, or in the tissues, as already demonstrated by Starcher (286) for a duodenal mucosal protein in the chick. The ability of an excess of one element to displace other elements from binding sites on protein molecules has already been demonstrated *in vitro* (40, 240). However, it is already evident that the toxicity of a particular dietary intake of copper is determined not only by the level of copper, but by the ratio of this level to the dietary levels of zinc, iron, calcium, molybdenum, inorganic sulfate, and of other dietary components, some of which have yet to be identified. Copper intakes alone can be as misleading in determining potential copper toxicity as they are in determining potential copper deficiency.

### 3. *Sheep and Cattle*

Sheep are the most susceptible of all domestic livestock to copper toxicosis. Cattle appear to be less so but the experimental evidence is conflicting, no doubt as a result of differences in the ratio of copper to other components in the diets employed. Adult cows have been fed 1.2–2.0 g copper sulfate daily for 5 to 18 weeks without harm (114), and 0.8–5.0 g of copper sulfate fed daily for 9 months to calves and cows produced no observable ill-effects (85). On the other hand, Kidder (167) induced generalized icterus, hemoglobinuria, and death in a steer fed 5 g copper sulfate for 122 days, and Shand and Lewis (270) produced typical signs of chronic copper poisoning in calves fed a milk substitute powder averaging 115 ppm Cu. Hemoglobinemia and hemoglobinuria associated with marked jaundice and widespread necrosis were observed, together with very high levels of copper in the livers of fatal cases. Marked spongy lesions, particularly of the white matter in the midbrain, pons, and cerebellum, but not in the telencephalon, have also been reported in young sheep fed a diet containing 80 ppm Cu for 6 weeks (100).

Losses of sheep from chronic copper poisoning, with hemolytic icterus and hemoglobinuria as characteristic signs, have been reported from many parts of the world. In some cases the excess copper has come from the ingestion of herbage in orchards and vineyards previously sprayed with copper compounds (178, 259) or from pastures sprayed with copper sulfate as a molluscicide (127). Chronic copper poisoning may also occur on dry feed (175), and has been reported at pasture (176) from continued free-choice consumption of a trace mineral salt mixture containing a recommended amount of copper. In these cases the condition was associated with very low intakes of molybdenum.

In parts of Australia, chronic copper poisoning occurs in sheep under natural grazing conditions (46). The disease has been characterized in the following terms: "a more or less sudden hemoglobinemia and hemoglobinuria, usually associated with the occurrence of icterus which increases rapidly. The hemolytic crisis is associated with necrosis of the liver, kidney dysfunction and so-called uremia. The copper concentration in the liver is usually higher than 1000 ppm on the dry, fat-free basis . . . . The presence of hemoglobinuria in the sheep is highly presumptive of chronic copper poisoning. If this is associated with a central necrosis of the liver nodules, with a premonitory rise in blood copper which may be tenfold in magnitude and high copper content of the liver, then a diagnosis of chronic copper poisoning can be made."

The critical factor in the precipitation of the hemolytic crisis has been postulated by Bull (46) as a high concentration of mobile or active copper. Copper becomes toxic to the cell when it is liberated, in sufficient concentration, from organic combination within the cell. The toxic concentration may be reached either by such liberation through calls upon the food reserves of the liver or directly by the cell taking up sufficient copper by assimilation from a favorable diet.

A significant rise in serum transaminases and in lactic dehydrogenase precedes the hemolytic crisis by several weeks (254, 306). Determination of the level of glutamic oxalacetic transaminase (GOT) has, therefore, been proposed as an aid in the early detection of chronic copper poisoning in sheep and in the assessment of the rate of recovery. The magnitude of the changes in the activities of these enzymes in chronic copper poisoning in sheep is illustrated in Table 15. The blood copper levels remain normal until 1 to 2 days before clinical signs appear and then increase to 5 to 20 times normal as mentioned earlier. Todd (305)

TABLE 15
*Serum Enzyme Activities in Copper Poisoning in Sheep*[a]

| | LD[b] (IU/liter) | GOT[b] (IU/liter) | GPT[b] (IU/liter) |
|---|---|---|---|
| Normal controls | 345-840 | 80-135 | 5-9 |
| 4 weeks before crisis | 365-2575 | 90-1600 | 4-11 |
| 3 weeks before crisis | 870-5900 | 305-1600 | 6-10 |
| 2 weeks before crisis | 465-3825 | 75-1600 | 6-12 |
| 1 week before crisis | 1500-3260 | 100-700 | 7-9 |
| Onset of crisis | 2600-9500 | 340-1600 | 175-1200 |

[a]From Todd and Thompson (306).
[b]LD = Lactic dehydrogenase; GOT = glutamic oxalacetic transaminase; GPT = glutamic pyruvic transaminase.

has shown further that the hemolytic crisis is accompanied by marked falls in hemoglobin and glutathione concentrations, whereas methemoglobin levels rise.

In Australia, chronic copper poisoning occurs in sheep under three different sets of conditions: (*a*) when the copper content of the soils and pastures are high, (*b*) when these are "normal" but the molybdenum levels are very low, and (*c*) in association with liver damage due to poisoning by the plant *Heliotropium europaeum*. The copper poisoning which occurs under the first of these conditions appears to be due to a straightforward high gross copper intake from the abnormally high copper content of the pastures growing on the cupriferous soils of the area and to the copper taken in by the animal from soil and dust. Some plant species growing on these soils contain as much as 50-60 ppm Cu. The disease which occurs under the second set of conditions is usually seasonal in occurrence, appears less in Merinos than in British breeds or crosses, occurs only on the more acid soils of the region, and is favored by dominance of the pastures with the clover, *Trifolium subterraneum*. This plant usually contains 10-15 ppm Cu and extremely low levels of molybdenum which rarely exceed 0.1-0.2 ppm. Such a level of copper in the grazing, coupled with the very low molybdenum, favors the development of a high copper status in the sheep and leads to copper poisoning. Providing molybdate-containing salt licks (99) or dosing of the animals with ammonium molybdate and sodium sulfate (254) is highly effective in reducing liver copper levels and reducing mortality from the disease.

The disease which occurs in association with the plant *Heliotropium europaeum* has been named "yellows," enzootic jaundice, and hemolytic jaundice. This plant contains hepatotoxic alkaloids, including heliotrine and lasiocarpine (47, 83), which initiate an irreversible change in liver parenchyma cells reflected in an increase in size of cell and nucleus. These enlarged cells (megatocytes) have a shortened life and are not replaced by new liver cells. As these cells disappear the liver becomes atrophic and the attempted regeneration results in extensive new bile duct formation (47, 48). Copper retention in heliotrope-damaged livers is extremely high, usually over 1000 ppm on the dry, fat-free basis, due to an increased avidity of the cells for this element, thereby increasing the susceptibility of the sheep to death from chronic copper poisoning. The accumulation of copper in the liver is due directly to the liver damage by the heliotrope alkaloids and not just to the high copper content of the plant (49).

Sheep are much more subject to chronic copper poisoning from this cause than are cattle, although cattle are extremely susceptible to the toxic effects of *H. europaeum* (50). It seems that the toxic effects of the plant are so rapid and so severe in cattle that a progressive disease of the same chronicity as in sheep is unlikely. Furthermore, cattle find the plant more unpalatable than do sheep and, therefore, generally consume less.

Chronic copper poisoning in sheep, associated with high liver copper levels, has also been reported in Western Australia from the consumption of toxic, alkaloid-containing lupins (121). The copper intakes of the sheep from the lupins and other components of the diet were low to normal, but marked increases in the copper concentrations in the blood and liver and in 24-hr urinary copper excretions were observed in sheep consuming the lupin diets. It was suggested that the disease, lupinosis, is, in part, a conditioned form of copper poisoning and that hemolytic factors in toxic lupins and hemolysis resulting from release of liver copper may both operate to cause the jaundice of lupinosis.

### 4. *Copper Poisoning in Man*

Copper poisoning can occur as an industrial hazard in workers engaged in copper mining or processing. Continuous ingestion of copper from the food or water supply at intakes sufficient to induce chronic copper poisoning in man is extremely unlikely, judging by the amounts required for this purpose in other monogastric species. Public health measures in most countries effectively control the use of copper compounds in the coloring and preserving of foods and copper compounds are not now so prominant among available fungicides and insecticides as they were several decades ago. Contamination of foods from these sources is, therefore, less important than it was. There are no chronic degenerative disorders in man known to result from modern exposures to copper, other than the industrial exposures mentioned above. Wilson's disease, which is characterized by excessive concentrations of copper in the tissues, arises from metabolic defects involving absorbed copper, and not from the ingestion of excessive amounts of copper.

## REFERENCES

1. R. Abderhalden, "Clinical Enzymology," Van Nostrand, Princeton, New Jersey, 1961.
2. F. W. Adams and J. R. Haag, *J. Nutr.* **63**, 585 (1957).
3. Agricultural Research Council, *Nutr. Requir. Farm Livestock* No. 3 (1967).
4. R. Allcroft, *Vet. Rec.* **64**, 17 (1952); R. Allcroft and G. Lewis, *J. Sci. Food Agr.* **8**, Suppl., 96 (1957).
5. R. Allcroft, F. G. Clegg, and O. Uvarov, *Vet. Rec.* **71**, 884 (1959).

6. R. Allcroft and W. H. Parker, *Brit. J. Nutr.* **3**, 205 (1949).
7. R. Allcroft and O. Uvarov, *Vet. Rec.* **71**, 797 (1959).
8. M. Anke, *Chem. Abstr.* **66**, 92830j (1967).
9. R. S. Barber, R. Braude, K. G. Mitchell, and J. Cassidy, *Chem. Ind. (London)* **21**, 601 (1955); R. S. Barber, R. Braude, and K. G. Mitchell, *Brit. J. Nutr.* **9**, 378 (1955).
10. R. S. Barber, R. Braude, K. G. Mitchell, and J. F. Rook, *Brit. J. Nutr.* **11**, 70 (1957).
11. P. J. Barden and A. Robertson, *Vet. Rec.* **74**, 252 (1962).
12. R. M. Barlow, *J. Comp. Pathol.* **73**, 51 and 61 (1963).
13. R. M. Barlow and P. Cancilla, *Acta Neuropathol.* **6**, 175 (1960).
14. R. M. Barlow, A. C. Field, and N. C. Ganson, *J. Comp. Pathol.* **74**, 530 (1964).
15. R. M. Barlow, D. Purves, E. J. Butler, and I. J. McIntyre, *J. Comp. Pathol.* **70**, 396 and 411 (1960).
16. B. Bass, J. T. McCall, H. D. Wallace, G. E. Combs, A. Z. Palmer, and J. E. Carpenter, *J. Anim. Sci.* **15**, 1230 (1956).
17. J. H. Baxter and J. J. Van Wyk, *Bull. Johns Hopkins Hosp.* **93**, 1 (1953); J. H. Baxter, J. J. Van Wyk, and R. H. Follis, Jr., *ibid.* p. 25; J. J. Van Wyk, J. H. Baxter, J. H. Akeroyd, and A. G. Motulsky, *ibid.* p. 51.
18. A. G. Bearn and H. G. Kunkel, *J. Clin. Invest.* **33**, 400 (1954).
19. A. B. Beck, personal communication (1960).
20. A. B. Beck, *Aust. J. Exp. Biol. Med. Sci.* **19**, 145 and 249 (1941).
21. A. B. Beck, *Aust. J. Zool.* **4**, 1 (1956).
22. A. B. Beck, *Aust. J. Agr. Res.* **12**, 743 (1961).
23. A. B. Beck and R. Harley, *J. Dep. Agr. West. Aust.* [2] **18**, 285 (1941); *Dep. Agr. West. Aust., Leaf.* No. 678 (1951).
24. R. Belz, *Voeding* **6**, 236 (1960).
25. H. W. Bennetts, *Aust. Vet. J.* **8**, 137 and 183 (1932).
26. H. W. Bennetts and A. B. Beck, *Aust., Commonw., Counc. Sci. Ind. Res., Bull.* **147** (1942).
27. H. W. Bennetts and F. E. Chapman, *Aust. Vet. J.* **13**, 138 (1937).
28. H. W. Bennetts and H. T. B. Hall, *Aust. Vet. J.* **15**, 52 (1939); H. W. Bennetts, R. Harley, and S. T. Evans, *ibid.* **18**, 50 (1942); H. W. Bennetts, A. B. Beck, and R. Harley, *ibid* **24**, 237 (1948).
29. J. B. Bingley and J. H. Dufty, *Clin. Chim. Acta* **24**, 316 (1969).
30. H. Blaschko, F. Buffoni, N. Weissmann, W. H. Carnes, and W. F. Coulson, *Biochem. J.* **96**, 4c (1965).
31. W. E. Blumberg and J. Peisach, *J. Biol. Chem.* **240**, 870 (1965).
32. P. Bornstein, A. H. Kang, and K. A. Piez, *Proc. Nat. Acad. Sci. U.S.* **55**, 417 (1966).
33. J. P. Bowland, R. Braude, A. G. Chamberlain, R. F. Glascock, and K. G. Mitchell, *Brit. J. Nutr.* **15**, 59 (1961).
34. J. M. Bowness, R. A. Morton, M. H. Shakir, and A. L. Stubbs, *Biochem. J.* **51**, 521 (1952); J. M. Bowness and R. A. Morton, *ibid.* p. 530.
35. R. Boyden, V. R. Potter, and C. A. Elvehjem, *J. Nutr.* **15**, 397 (1938).
36. R. Braude, *J. Agr. Sci.* **35**, 163 (1945).
37. R. Braude, *World Rev. Anim. Prod.* **3**, 69 (1967).
38. R. Braude, M. J. Townsend, G. Harrington, and J. G. Rowell, *J. Agr. Sci.* **58**, 251 (1962).
39. K. C. Bremner, *Aust. J. Agr. Res.* **10**, 471 (1959); *Aust. Vet. J.* **35**, 389 (1959).
40. E. Breslow and E. R. N. Gurd, *J. Biol. Chem.* **238**, 1332 (1963).

41. W. M. Britton and C. H. Hill, *Fed. Proc., Fed. Amer. Soc. Exp. Biol.* **23**, 133 (1964).
42. S. Bruckmann and S. G. Zondek, *Nature* (*London*) **146**, 3 (1940).
43. F. Brudevold and L. T. Steadman, *J. Dent. Res.* **34**, 209 (1955).
44. C. F. Bucholz, *Rep. Pharm.* **2**, 253 (1816); E. J. Boutigny, *Chim. Med.* **9**, 147 (1833).
45. F. Buffoni and H. Blaschko, *Proc. Roy. Soc., London, Ser. B* **161**, 153 (1964).
46. L. B. Bull, *Brit. Commonw. Spec. Conf. Agr. Australian, 1949* p. 300 (1951).
47. L. B. Bull, C. C. J. Culvenor, and A. T. Dick, "The Pyrrolizidine Alkaloids," Vol. 9. North-Holland Publ., Amsterdam, 1968.
48. L. B. Bull, A. T. Dick, J. C. Keast, and G. Edgar, *Aust. J. Agr. Res.* **7**, 281 (1956).
49. L. B. Bull and A. T. Dick, *J. Pathol. Bacteriol.* **78**, 483 (1959).
50. L. B. Bull, E. S. Rogers, J. C. Keast, and A. T. Dick, *Aust. Vet. J.* **37**, 37 (1961).
51. R. J. Bunch, V. C. Speers, V. M. Hays, and J. T. McCall, *J. Anim. Sci.* **22**, 56 (1963).
52. R. W. Burley, *Nature* (*London*) **174**, 1019 (1954).
53. R. W. Burley and W. T. de Koch, *Arch. Biochem. Biophys.* **68**, 21 (1957).
54. J. A. Bush, W. N. Jensen, J. W. Athens, H. Ashenbrucker, G. E. Cartwright, and M. M. Wintrobe, *J. Exp. Med.* **103**, 701 (1956).
55. J. A. Bush, J. P. Mahoney, C. J. Gubler, G. E. Cartwright, and M. M. Wintrobe, *J. Lab. Clin. Med.* **47**, 898 (1956).
56. J. A. Bush, J. P. Mahoney, M. Markowitz, C. J. Gubler, G. E. Cartwright, and M. M. Wintrobe, *J. Clin. Invest.* **34**, 1766 (1955).
57. E. J. Butler, *Comp. Biochem. Physiol.* **9**, 1 (1963).
58. E. J. Butler and R. M. Barlow, *J. Comp. Pathol. Ther.* **73**, 107 (1963).
59. W. V. Camargo, H. J. Lee, and D. W. Dewey, *Proc. Aust. Soc. Anim. Prod.* **4**, 12 (1962).
60. W. W. Carlton and W. Henderson, *J. Nutr.* **81**, 200 (1963).
61. W. W. Carlton and W. Henderson, *J. Nutr.* **85**, 67 (1965).
62. W. W. Carlton and W. A. Kelly, *J. Nutr.* **97**, 42 (1969).
63. W. H. Carnes, G. S. Shields, G. E. Cartwright, and M. M. Wintrobe, *Fed. Proc., Fed Amer. Soc. Exp. Biol.* **20**, 118 (1961).
64. L. E. Carpenter, *Hormel Inst. Univ. Minn., Ann. Rep.* p. [illegible] (19[illegible]); cited by Mills (220).
65. G. E. Cartwright, *in* "Symposium on Copper Metabolism" (W. D. McElroy and B. Glass, eds.), p. 274, Johns Hopkins Press, Baltimore, Maryland, 1950.
66. G. E. Cartwright, C. J. Gubler, and M. M. Wintrobe, *J. Clin. Invest.* **33**, 685 (1954).
67. G. E. Cartwright, H. Markowitz, G. S. Shields, and M. M. Wintrobe, *Amer. J. Med.* **28**, 555 (1960).
68. G. E. Cartwright and M. M. Wintrobe, *Amer. J. Clin. Nutr.* **14**, 224 (1964); **15**, 94 (1964).
69. P. A. Cavell and E. M. Widdowson, *Arch. Dis. Childhood* **39**, 496 (1964).
70. M. S. Chase, C. J. Gubler, G. E. Cartwright, and M. M. Wintrobe, *J. Biol. Chem.* **199**, 757 (1952); *Proc. Soc. Exp. Biol. Med.* **80**, 749 (1952).
71. P. E. Chen, *Chin. Med. J.* **75**, 917 (1957); cited by Georgio *et al.* (122).
72. T. P. Chou and W. H. Adolph, *Biochem. J.* **29**, 476 (1935).
73. A. W. Church, *Phil. Trans. Roy. Soc. London* **159**, 627 (1869); *Proc. Roy. Soc., London, Ser. B* **51**, 399 (1892).
74. A. Clemetson, *Aust. Vet. J.* **42**, 34 (1966).
75. C. L. Comar, *in* "Symposium on Copper Metabolism" (W. D. McElroy and B. Glass, eds.), p. 191, Johns Hopkins Press, Baltimore, Maryland, 1950.
76. A. Cordano, J. M. Baertle, and G. G. Graham, *Pediatrics* **34**, 324 (1964).
77. A. Cordano, R. P. Placko, and G. G. Graham, *Blood* **28**, 280 (1966).

78. G. C. Cotzias, P. S. Papavasiliou, and S. T. Miller, *Nature* (*London*) **201**, 1228 (1964).
79. W. F. Coulson, R. M. Barlow, P. Cancilla, N. Weissmann, A. Linker, J. Waisman, and W. H. Carnes, *Amer. J. Vet. Res.* **28**, 815 (1967).
80. W. F. Coulson and W. H. Carnes, *Amer. J. Pathol.* **43**, 945 (1963).
81. D. H. Cox and L. H. Harris, *J. Nutr.* **70**, 514 (1960).
82. W. J. Cox and A. J. Mueller, *J. Nutr.* **13**, 249 (1937).
83. C. C. J. Culvenor, L. J. Drummon, and J. R. Price, *Aust. J. Chem.* **7**, 277 (1954).
84. I. J. Cunningham, *Biochem. J.* **25**, 1267 (1931).
85. I. J. Cunningham, *N. Z. J. Sci. Technol., Sect. A* **27**, 372 and 381 (1946).
86. I. J. Cunningham, *in* "Symposium on Copper Metabolism" (W. D. McElroy and B. Glass, eds.), p. 246. Johns Hopkins Press, Baltimore, Maryland, 1950.
87. I. J. Cunningham, *N. Z. Vet. J.* **7**, 15 (1959).
88. I. J. Cunningham and K. G. Hogan, *N. Z. J. Agr. Res.* **1**, 841 (1958).
89. C. P. Czerniejewski, C. W. Shank, W. G. Bechtel, and W. B. Bradley, *Cereal Chem.* **41**, 65 (1964).
90. L. S. P. Davidson, H. W. Fullerton, J. W. Howie, J. M. Croll, J. B. Orr, and W. M. Godden, *Brit. Med. J.* **2**, 685 (1933).
91. W. L. Davies, "The Chemistry of Milk," Chapman & Hall, London, 1936.
92. G. K. Davis, *in* "Symposium on Copper Metabolism" (W. D. McElroy and B. Glass, eds.), p. 216. Johns Hopkins Press, Baltimore, Maryland, 1950.
93. P. N. Davis, L. C. Norris, and F. H. Kratzer, *J. Nutr.* **77**, 217 (1962).
94. H. N. De, *Indian J. Med. Res.* **37**, 301 (1949).
95. A. T. Dick, *Aust. Vet. J.* **28**, 30 (1952); **29**, 18 (1953); **30**, 196 (1954); *Nature* (*London*) **172**, 637 (1953).
96. A. T. Dick, Doctoral Thesis, University of Melbourne, Australia, 1954.
97. A. T. Dick, *Aust. J. Agr. Res.* **5**, 511 (1954).
98. A. T. Dick, *Soil Sci.* **81**, 229 (1956).
99. A. T. Dick and L. B. Bull, *Aust. Vet. J.* **21**, 70 (1945).
100. P. C. Doherty, R. M. Barlow, and K. W. Angus, *Res. Vet. Sci.* **10**, 303 (1969).
101. I. E. Dreosti and G. V. Quicke, *S. Afr. J. Agr. Sci.* **9**, 365 (1966).
102. I. E. Dreosti and G. V. Quicke, *Brit. J. Nutr.* **22**, 1 (1968).
102a. W. L. Dunkley, M. Ronning, and J. Voth, *J. Dairy Sci.* **46**, 1059 (1963).
103. B. Dutt and C. F. Mills, *J. Comp. Pathol.* **70**, 120 (1960).
104. A. Eden, *J. Comp. Pathol. Ther.* **52**, 429 (1939); **53**, 90 (1940).
105. A. Eden, *Biochem. J.* **35**, 813 (1941).
106. D. B. Ellis, *Anim. Prod.* **3**, 89 (1961).
107. C. A. Elvehjem, *Physiol. Rev.* **15**, 471 (1935).
108. C. A. Elvehjem, H. Steenbock, and E. B. Hart, *J. Biol. Chem.* **83**, 27 (1929).
109. R. W. Engel, N. O. Price, and R. F. Miller, *J. Nutr.* **92**, 197 (1967).
110. G. J. Everson, M. C. Tsai, and T. Wang, *J. Nutr.* **93**, 533 (1967).
111. J. M. Evvard, V. E. Nelson, and W. E. Sewell, *Proc. Iowa Acad. Sci.* **35**, 211 (1928).
112. J. Fay, G. E. Cartwright, and M. M. Wintrobe, *J. Clin. Invest.* **28**, 487 (1949).
113. B. F. Fell, C. F. Mills, and R. Boyne, *Res. Vet. Sci.* **6**, 10 (1965).
114. W. S. Ferguson, *J. Agr. Sci.* **33**, 116 (1943).
115. R. H. Follis, Jr., J. A. Bush, G. E. Cartwright, and M. M. Wintrobe, *Bull. Johns Hopkins Hosp.* **97**, 405 (1955).
116. L. Fredericq, *Arch. Zool. Exp. Gen.* **7**, 535 (1878).
117. E. Frieden, *in* "Horizons in Biochemistry" (M. Kasha and B. Pullman, ed.), p. [illegible]. Academic Press, New York, 1962.

118. R. Fuller, L. G. M. Newland, C. A. E. Briggs, R. Braude, and K. G. Mitchell, *J. Appl. Bacteriol.* **23**, 195 (1960).
119. C. H. Gallagher, *Aust. Vet. J.* **33**, 311 (1957).
120. C. H. Gallagher, J. D. Judah, and K. R. Rees, *Proc. Roy. Soc., London, Ser. B* **145**, 134 and 195 (1956).
121. M. R. Gardiner, *J. Comp. Pathol.* **76**, 107 (1966).
122. A. P. Giorgio, G. E. Cartwright, and M. M. Wintrobe, *Amer. J. Clin. Path.* **41**, 22 (1964).
123. D. Gitlin, and C. A. Janeway, *Nature* (*London*) **185**, 693 (1960); D. Gitlin, W. A. Hughes, and C. A. Janeway, *ibid.* p. 151.
124. A. Goldberg, C. B. Williams, R. S. Jones, M. Yanagita, G. E. Cartwright, and M. M. Wintrobe, *J. Lab. Clin. Med.* **48**, 442 (1956).
125. C. Gopalan, *J. Pediat.* **63**, 646 (1963).
126. H. Goss and M. M. Green, *Science* **122**, 330 (1955).
127. J. Gracey and J. R. Todd, *Brit. Vet. J.* **116**, 405 (1960).
128. D. R. Grant-Frost and E. J. Underwood, *Aust. J. Exp. Biol. Med. Sci.* **36**, 339 (1958).
128a. L. F. Gray and L. J. Daniel, *J. Nutr.* **84**, 31 (1964).
129. G. Gregoriadis and T. L. Sourkes, *Can. J. Biochem.* **45**, 1841 (1967).
130. C. J. Gubler, G. E. Cartwright, and M. M. Wintrobe, *J. Biol. Chem.* **224**, 533 (1957).
131. C. J. Gubler, M. E. Lahey, M. S. Chase, G. E. Cartwright, and M. M. Wintrobe, *Blood* **7**, 1075 (1952).
132. K. Guggenheim, *Blood* **23**, 786 (1964).
133. K. Guggenheim, E. Tal, and U. Zor, *Brit. J. Nutr.* **18**, 529 (1964).
134. E. M. Hall and E. M. Mackey, *Amer. J. Pathol.* **7**, 327 (1931).
135. G. A. Hall and J. McC. Howell, *Brit. J. Nutr.* **23**, 41 (1969).
135a. J. A. Halsted, B. Hackley, and J. C. Smith, *Lancet* **2**, 278 (1968).
136. T. S. Hamilton and H. H. Mitchell, *J. Biol. Chem.* **178**, 345 (1949).
137. J. J. Harding, *Advan. Protein Chem.* **20**, 109 (1965).
138. E. Harless, *Arch. Anat. Physiol.* p. 148 (1847).
139. E. B. Hart, H. Steenbock, J. Waddell, and C. A. Elvehjem, *J. Biol. Chem.* **77**, 797 (1928).
140. J. Hartmans and M. S. M. Bosman, *Proc. 1st Int. Symp. Trace Element Metab. Anim., 1969,* p. 362 (1970).
141. J. A. Hawbaker, V. C. Speers, J. D. Jones, V. M. Hays, and D. V. Catron, *J. Anim. Sci.* **18**, 1505 (1959).
142. R. S. Hedger, D. A. Howard, and M. L. Burdin, *Vet. Rec.* **76**, 493 (1964).
142a. L. Heilmeyer, W. Keiderling, and G. Stuwe, "Kupfer u. Eisen als Korpereigne Wirkstoffe u. ihre Bedeutung Reim Krankheitgeschaben," Fischer, Jena, 1941.
143. R. G. Hemingway, A. Macpherson, and N. S. Ritchie, *Proc. 1st Int. Symp. Trace Element Metab. Anim., 1969* p. 264 (1970).
144. G. E. Herman and E. Kun, *Exp. Cell Res.* **22**, 257 (1961).
145. C. H. Hill and G. Matrone, *J. Nutr.* **73**, 425 (1961).
146. C. H. Hill and G. Matrone, *Proc. 12th World Poultry Congr., 1962* p. 219 (1962).
147. C. H. Hill, G. Matrone, W. L. Payne, and C. W. Barber, *J. Nutr.* **80**, 227 (1963).
148. C. H. Hill and B. Starcher, *J. Nutr.* **85**, 67 (1965).
149. C. H. Hill, B. Starcher, and C. Kim, *Fed. Proc., Fed. Amer. Soc. Exp. Biol.* **26**, 129 (1968).
150. C. H. Hill, B. Starcher, and G. Matrone, *J. Nutr.* **83**, 107 (1964).
151. J. M. Hill and P. G. Mann, *Biochem. J.* **85**, 198 (1962).

152. K. G. Hogan, D. F. L. Money, and A. Blayney, *N. Z. J. Agr. Res.* **11**, 435 (1968).
153. C. G. Holmberg and C. B. Laurell, *Acta Chem. Scand.* **1**, 944 (1947); **2**, 250 (1948); **5**, 476 (1951).
154. H. Hopkins and J. Eisen, *J. Agr. Food Chem.* **7**, 633 (1959).
155. J. McC. Howell and A. N. Davison, *Biochem. J.* **72**, 365 (1959).
156. J. McC. Howell, A. N. Davison, and J. Oxberry, *Res. Vet. Sci.* **5**, 376 (1964).
157. J. McC. Howell, N. Edington, and R. Ewbank, *Res. Vet. Sci.* **9**, 160 (1968).
158. J. McC. Howell and G. A. Hall, *Brit. J. Nutr.* **23**, 47 (1969).
159. G. Hughes, V. J. Kelley, and R. A. Stewart, *Pediatrics* **25**, 477 (1960).
160. C. E. Hunt and W. W. Carlton, *J. Nutr.* **87**, 385 (1965).
161. C. E. Hunt, W. W. Carlton, and P. M. Newberne, *Fed. Proc., Fed. Amer. Soc. Exp. Biol.* **25**, 432 (1966).
162. A. H. Hunter, A. Eden, and H. H. Green, *J. Comp. Pathol.* **55**, 19 (1945).
163. E. W. Hurst, *Aust. J. Exp. Biol. Med. Sci.* **20**, 297 (1942).
164. J. R. M. Innes and G. D. Shearer, *J. Comp. Pathol.* **53**, 1 (1940); G. D. Shearer, J. R. M. Innes, and E. I. McDougal, *Vet. J.* **96**, 309 (1940); G. D. Shearer and E. I. McDougal, *J. Agr. Sci.* **34**, 207 (1944).
165. N. C. Johnson, T. Kheim, and W. B. Kountz, *Proc. Soc. Exp. Biol. Med.* **102**, 98 (1959); E. M. Russ and J. Raymunt, *ibid.* **92**, 465 (1956).
166. J. M. Joyce, *N. Z. Vet. J.* **3**, 157 (1955).
167. R. W. Kidder, *J. Anim. Sci.* **8**, 623 (1949).
168. H. Kikkawa, Z. Ogita, and S. Fujito, *Science* **121**, 43 (1955).
169. J. R. Kimmel, H. Markowitz, and D. M. Brown, *J. Biol. Chem.* **234**, 46 (1959).
170. R. L. King and W. L. Dunkley, *J. Dairy Sci.* **42**, 420 (1959).
171. R. L. King and W. F. Williams, *J. Dairy Sci.* **46**, 11 (1963).
172. M. Kirchgessner and E. Grassman, *Proc. 1st Int. Symp. Trace Element Metab. Anim., 1969* p. 277 (1970).
173. M. Kirchgessner and U. Weser, *Z. Tierphysiol. Tierernaehr. Futtermittelk.* **20**, 44 (1965).
174. C. A. Koppejan and H. Mulder, *Proc. 13th Int. Dairy Congr., 1953* Vol. 3, p. 1400 (1953).
175. T. Kowalczyk, A. L. Pope, and D. K. Sorensen, *J. Amer. Vet. Ass.* **141**, 362 (1962).
176. T. Kowalczyk, A. L. Pope, K. C. Berger, and B. A. Muggenberg, *J. Amer. Vet. Ass.* **145**, 352 (1964).
177. H. A. Krebs, *Klin. Wochenschr.* **7**, 584 (1928).
178. H. Lafenetre, L. Monteil, and F. Galter, *Vet. Bull.* **5**, 855 (1935).
179. M. E. Lahey, *Amer. J. Clin. Nutr.* **5**, 516 (1957).
180. M. E. Lahey, C. J. Gubler, M. S. Chase, G. E. Cartwright, and M. M. Wintrobe, *Blood* **7**, 1053 (1952).
181. M. E. Lahey, C. J. Gubler, G. E. Cartwright, and M. M. Wintrobe, *J. Clin. Invest.* **32**, 322 and 329 (1953).
182. J. W. Lassiter and M. C. Bell, *J. Anim. Sci.* **19**, 754 (1960).
183. C. M. Lea and V. A. S. Luttrell, *Nature (London)* **206**, 413 (1965).
184. G. R. Lee, G. E. Cartwright, and M. M. Wintrobe, *Proc. Soc. Exp. Biol. Med.* **127**, 977 (1968).
185. G. R. Lee, S. Nacht, J. N. Lukens, and G. E. Cartwright, *J. Clin. Invest.* **47**, 2058 (1968).
186. H. J. Lee, *J. Agr. Sci.* **47**, 218 (1956).
187. E. Lesne, P. Zizine, and S. B. Briskas, *Rev. Pathol. Comp. Hyg. Gen.* **36**, 1369 (1936).

188. R. M. Leverton and E. S. Binkley, *J. Nutr.* **27**, 43 (1944).
189. G. Lewis, and R. Allcroft, *Proc. 5th Int. Congr. Nutr.*, Washington D.C. (1960).
190. G. Lewis, S. Terlecki, and R. Allcroft, *Vet. Rec.* **81**, 415 (1967).
191. D. V. Lillis, V. L. Miller, G. E. Bearse, and C. M. Hamilton, *Toxicol. Appl. Pharmacol.* **5**, 12 (1963).
192. C. W. Lindow, C. A. Elvehjem, and W. H. Peterson, *J. Biol. Chem.* **82**, 465 (1929).
193. C. W. Lindow, W. H. Peterson, and H. Steenbock, *J. Biol. Chem.* **84**, 419 (1929).
194. E. J. Lorenzen and S. E. Smith, *J. Nutr.* **33**, 143 (1947).
195. I. A. M. Lucas and A. F. C. Calder, *J. Agr. Sci.* **49**, 184 (1957).
196. A. C. Magee and G. Matrone, *J. Nutr.* **72**, 233 (1960).
197. J. P. Mahoney, J. A. Bush, C. J. Gubler, W. H. Moretz, G. E. Cartwright, and M. M. Wintrobe, *J. Lab. Clin. Med.* **46**, 702 (1955).
197a. T. Mann and D. Keilin, *Proc. Roy. Soc., London Ser. B* **126**, 303 (1938).
198. H. Markowitz, G. E. Cartwright, and M. M. Wintrobe, *J. Biol. Chem.* **234**, 40 (1959).
199. H. Markowitz, C. J. Gubler, J. P. Mahoney, G. E. Cartwright, and M. M. Wintrobe, *J. Clin. Invest.* **34**, 1498 (1955).
200. H. R. Marston, *Proc. Symp. on Fibrous Proteins,* Soc. Dyers and Colorists, Leeds, 1946.
201. H. R. Marston, *Physiol. Rev.* **32**, 66 (1952).
202. H. R. Marston and H. J. Lee, *Aust. J. Sci. Res., Ser. B* **1**, 376 (1948); *J. Agr. Sci.* **38**, 229 (1948).
203. H. R. Marston, H. J. Lee, and I. W. McDonald, *J. Agr. Sci.* **38**, 216 and 222 (1948).
204. H. R. Marston, R. G. Thomas, D. Murnane, E. W. Lines, I. W. McDonald, H. O. Moore, and L. B. Bull, *Aust. Commonw. Counc. Sci. Ind. Res., Bull.* **113** (1938).
205. G. M. Martin, M. A. Derr, and E. P. Benditt, *Lab. Invest.* **13**, 282 (1964).
206. G. Matrone, *Fed. Proc., Fed. Amer. Soc. Exp. Biol.* **19**, 659 (1960).
207. G. Matrone, *Proc. 1st Int. Symp. Trace Element Metab. Anim., 1969* p. 354 (1970).
208. R. A. McCance and E. M. Widdowson, "The Chemical Composition of Foods." Chem. Publ. Co., New York, 1947.
209. P. J. McCosker, *Nature (London)* **190**, 887 (1961).
210. P. J. McCosker, *Res. Vet. Sci.* **9**, 91 and 103 (1968).
211. I. McDonald and P. J. Warren, *Brit. J. Nutr.* **15**, 593 (1961).
212. E. I. McDougall, *J. Agr. Sci.* **37**, 329 (1947).
213. W. D. McFarlane and H. I. Milne, *J. Biol. Chem.* **107**, 309 (1934).
214. M. D. McGavin, P. D. Ranby, and L. Tammemagi, *Aust. Vet. J.* **38**, 8 (1962).
215. J. S. McHargue, *Amer. J. Physiol.* **72**, 583 (1925); **77**, 245 (1926).
216. K. J. McNaught, *N. Z. J. Sci. Technol., Sect. A* **30**, 26 (1948).
217. E. J. Miller, E. R. Martin, C. E. Mecca, and K. A. Piez, *J. Biol. Chem.* **240**, 3623 (1965).
218. C. F. Mills, *Biochem. J.* **57**, 603 (1954); **63**, 187 and 190 (1956); *Brit. J. Nutr.* **9**, 398 (1955).
219. C. F. Mills, *Rowett Res. Inst. Coll. Pap.* **17**, 57 (1961).
220. C. F. Mills, *Feed Forum* **3**, 21 (1968).
221. C. F. Mills, A. C. Dalgarno, and R. B. Williams, *Biochem. Biophys. Res. Commun.* **24**, 537 (1966).
222. C. F. Mills and B. F. Fell, *Nature* **185**, 20 (1960); B. F. Fell, R. B. Williams, and C. F. Mills, *Proc. Nutr. Soc.* **20**, XXVII (1961).
223. C. F. Mills and G. Murray, *J. Sci. Food Agr.* **9**, 547 (1960).
224. C. F. Mills and R. B. Williams, *Biochem. J.* **85**, 629 (1962).
225. D. B. Milne and P. H. Weswig, *J. Nutr.* **95**, 429 (1968).

226. D. F. L. Money, K. G. Hogan, and S. M. Hare, *N. Z. J. Agr. Res.* **10**, 345 (1967).
227. G. R. Moule, A. K. Sutherland, and J. M. Harvey, *Queensl. J. Agr. Sci.* **18**, 93 (1959).
228. J. Moustgaard and N. J. Hojgaard-Olsen, *Nord. Veterinaermed.* **3**, 763 (1951).
229. S. Munch-Petersen, *Acta Paediat. (Stockholm)* **39**, 378 (1951).
230. W. M. Neal, R. B. Becker, and A. L. Shealy, *Science* **74**, 418 (1931).
231. R. E. H. Nicholas and C. Rimington, *Biochem. J.* **50**, 194 (1951).
232. A. L. Nielsen, *Acta Med. Scand.* **118**, 87 and 92 (1944).
233. G. S. Nixon, H. Smith, and H. D. Livingston, *Int. At. Energy Ag., Symp. Nucl. Activation Tech. Life Sci. 1967.*
234. B. L. O'Dell, D. W. Bird, D. F. Ruggles, and J. E. Savage, *J. Nutr.* **88**, 9 (1966).
235. B. L. O'Dell, B. C. Hardwick, G. Reynolds, and J. E. Savage, *Proc. Soc. Exp. Biol. Med.* **108**, 402 (1961).
236. P. J. O'Hara, A. P. Newman, and R. Jackson, *Aust. Vet. J.* **36**, 225 (1960).
237. C. C. O'Mary, W. T. Butts, R. A. Reynolds, and M. C. Bell, *J. Anim. Sci.* **28**, 268 (1969).
238. E. C. Owen, R. Proudfoot, J. M. Robertson, R. M. Barlow, E. J. Butler, and B. S. W. Smith, *J. Comp. Pathol.* **75**, 241 (1965).
239. S. Partridge, D. F. Elsden, J. Thomas, A. Dorfman, A. Telser, and P. L. Ho, *Biochem. J.* **93**, 30c (1964).
240. D. F. Plocke and B. L. Vallee, *Biochemistry* **1**, 1039, (1962).
241. H. Porter, *Proc. 1st. Symp. Trace Element Metab. Anim., 1969* p 237 (1970).
242. H. Porter, W. Wiener, and M. Barker, *Biochim. Biophys. Acta* **52**, 419 (1961).
243. W. J. Pryor, *Res. Vet. Sci.* **5**, 123 (1964).
244. H. S. Raper, *Physiol. Rev.* **8**, 245 (1928).
245. Recommended Daily Allowances, *Nat. Acad. Sci.—Nat. Res. Counc., Publ.* **598** (1958).
246. B. Reiff and H. Schneiden, *Blood* **14**, 967 (1959).
247. J. G. Reinhold, G. A. Kfoury, N. A. Galamber, and J. C. Bennett, *Amer. J. Clin. Nutr.* **18**, 294 (1966).
248. J. G. Reinhold, G. A. Kfoury, and T. A. Thomas, *J. Nutr.* **92**, 173 (1967).
249. C. Rimington, *Proc. Roy. Soc., London, Ser. B* **127**, 126 (1939).
250. H. D. Ritchie, R. W. Luecke, B. V. Baltzer, E. R. Miller, D. E. Ullrey, and J. A. Hoefer, *J. Nutr.* **79**, 117 (1963).
251. H. E. Roberts, B. M. Williams, and A. Harvard, *J. Comp. Pathol.* **76**, 279 and 285 (1966).
252. H. A. Robertson, Doctoral Thesis, University of Edinburgh, 1950; cited by McCosker (210).
253. H. A. Robertson and A. V. X. Broome, *J. Sci. Food Agr.* **8**, S82 (1957).
254. D. B. Ross, *Vet. Rec.* **76**, 875 (1964).
255. H. Röttger, *Arch. Gynäekol.* **177**, 650 (1950).
256. R. B. Rucker, H. E. Parker, and J. C. Rogler, *J. Nutr.* **98**, 57 (1969).
257. A. Sacks, V. E. Levine, F. C. Hill, and R. C. Hughes, *Arch. Intern. Med.* **71**, 489 (1943).
258. J. E. Savage, *Fed. Proc., Fed. Amer. Soc. Exp. Biol.* **27**, 927 (1968).
259. Schaper and Luitje, *Vet. Bull.* **1**, 172 (1931).
260. H. Scheinberg and I. Sternlieb, *Lancet* **2**, 1420 (1963).
261. H. A. Schroeder, A. P. Nason, I. H. Tipton, and J. J. Ballassa, *J. Chronic Dis.* **19**, 1007 (1966).
262. W. K. Schubert and M. E. Lahey, *Pediatrics* **24**, 710 (1959).

263. K. C. A. Schultz, P. K. Van der Merwe, P. J. Van Rensburg and J. S. Swart, *Onderstepoort J. Vet. Res.* **25**, 35 (1961).
264. M. O. Schultze, *J. Biol. Chem.* **129**, 729 (1939); **138**, 219 (1941).
265. M. O. Schultze, *Physiol. Rev.* **20**, 37 (1940).
266. M. O. Schultze, C. A. Elvehjem, and E. B. Hart, *J. Biol. Chem.* **106**, 735 (1934).
267. M. O. Schultze, C. A. Elvehjem, and E. B. Hart, *J. Biol. Chem.* **115**, 453 (1936); **116**, 93 and 107 (1936).
268. F. I. Scoular, *J. Nutr.* **16**, 437 (1938).
269. M. Seeleman and F. Baudissin, *Zentralbl. Veterinaermed.* **1**, 354 (1954); cited by Mills (218).
270. A. Shand and G. Lewis, *Vet. Rec.* **69**, 618 (1957).
271. J. Shapiro, A. G. Morell, and I. H. Scheinberg, *J. Clin. Invest.* **40**, 1081 (1961).
272. G. S. Shields, W. F. Coulson, D. A. Kimball, W. H. Carnes, G. E. Cartwright, and M. M. Wintrobe, *Fed. Proc., Fed. Amer. Soc. Exp. Biol.* **20**, 118 (1961).
273. G. S. Shields, H. Markowitz, W. H. Klassen, G. E. Cartwright, and M. M. Wintrobe, *J. Clin. Invest.* **40**, 2007 (1961).
274. C. F. Simpson and R. H. Harms, *Exp. Mol. Pathol.* **3**, 390 (1964).
275. C. F. Simpson, J. E. Jones, and R. H. Harms, *J. Nutr.* **91**, 283 (1967).
276. B. Sjollema, *Biochem. Z.* **267**, 151 (1933); **295**, 372 (1938).
277. H. Smith, *J. Forensic Sci. Soc.* **7**, 97 (1967).
278. H. W. Smith and J. E. T. Jones, *J. Appl. Bacteriol.* **26**, 262 (1963).
279. S. E. Smith and G. H. Ellis, *Arch. Biochem.* **15**, 81 (1947).
280. S. E. Smith and M. Medlicott, *Amer. J. Physiol.* **141**, 354 (1944).
281. S. E. Smith, M. Medlicott, and G. H. Ellis, *Amer. J. Physiol.* **142**, 179 (1944).
282. T. L. Sourkes, K. Lloyd, and H. Birnbaum, *Can. J. Biochem.* **46**, 267 (1968).
283. A. G. Spais, *Rec. Med. Vet.* **135**, 161 (1959).
284. A. G. Spais, T. K. Lazaridis, and A. G. Agiannidis, *Res. Vet. Sci.* **9**, 337 (1968).
285. C. M. Spray and E. M. Widdowson, *Brit. J. Nutr.* **4**, 361 (1951).
286. B. Starcher, *J. Nutr.* **97**, 321 (1969).
287. B. Starcher and C. H. Hill, *Comp. Biochem. Physiol.* **15**, 429 (1965).
288. B. Starcher, C. H. Hill, and G. Matrone, *J. Nutr.* **82**, 318 (1964).
289. I. Sternlieb, A. G. Morell, and I. H. Scheinberg, *Trans. Ass. Amer. Physicians* **75**, 228 (1962).
290. H. D. Stowe, *J. Nutr.* **95**, 179 (1968).
291. P. Sturgeon and C. Brubaker, *Amer. J. Dis. Child.* **92**, 254 (1956).
292. U. Surata, *Jap. J. Med. Sci.* **3**, 79 (1935); H. Yosikawa *ibid.* p. 195.
293. A. K. Sutherland, G. R. Moule, and J. M. Harvey, *Aust. Vet. J.* **31**, 141 (1955).
294. N. F. Suttle, Doctoral Thesis, University of Aberdeen, 1964; cited by Mills (220).
295. N. F. Suttle and C. F. Mills, *Brit. J. Nutr.* **20**, 135 and 149 (1966).
296. F. W. Tauber and A. C. Krause, *Amer. J. Opthalmol.* **26**, 260 (1943).
297. M. Taylor and S. Thomke, *Nature* (*London*) **201**, 1246 (1964).
298. H. S. Teague and L. E. Carpenter, *J. Nutr.* **43**, 389 (1951).
299. R. E. Thiers and B. L. Vallee, *J. Biol. Chem.* **226**, 911 (1957).
300. B. H. Thomas, *Iowa Agr. Exp. Sta. Rep.* p. 87 (1937).
301. J. Thomas, D. F. Elsden, and S. Partridge, *Nature* (*London*) **200**, 661 (1963).
302. S. L. Thompsett, *Biochem. J.* **34**, 961 (1940).
303. I. H. Tipton, P. L. Stewart, and P. G. Martin, *Health Phys.* **12**, 1683 (1966).
304. J. R. Todd, *Vet. Bull.* (*London*) **32**, 573 (1962).
305. J. R. Todd, *Proc. 1st Int. Symp. Trace Element Metab. Anim., 1969* p. 448 (1970).
306. J. R. Todd and R. H. Thompson, *Brit. Vet. J.* **119**, 161 (1963).

307. R. Turpin, H. Jerome, and H. Schmidt-Jubeau, *C. R. Soc. Biol.* **146**, 1703 (1952).
308. S. Uesaka, R. Kawashima, A. Iritani, K. Namikawa, N. Nakanishi, and T. Nishimo, *Bull. Res. Inst. Food Sci., Kyoto Univ.* **24** (1960).
309. D. E. Ullrey, E. R. Miller, O. A. Thompson, C. L. Zutaut, D. A. Schmidt, H. D. Ritchie, J. A. Hoefer, and R. W. Luecke, *J. Anim. Sci.* **19**, 1298 (1960).
310. B. L. Vallee, *Metabolism* **1**, 420 (1954).
311. D. R. Van Campen, *J. Nutr.* **88**, 125 (1966).
312. D. R. Van Campen and E. Gross, *J. Nutr.* **95**, 617 (1968).
313. D. R. Van Campen and E. A. Mitchell, *J. Nutr.* **86**, 120 (1965).
314. E. E. Van Koetsveld, *Diergeneesk. Jaarb.* **83**, 229 (1958).
315. A. H. Van Ravesteyn, *Acta Med. Scand.* **118**, 163 (1944).
316. W. W. Wainio, C. Van de Wender, and N. F. Shimp, *J. Biol. Chem.* **234**, 2433 (1959).
317. H. D. Wallace, J. T. McCall, B. Bass, and G. E. Combs, *J. Anim. Sci.* **19**, 1155 (1960).
318. N. Weismann, G. S. Shields, and W. H. Carnes, *J. Biol. Chem.* **238**, 3115 (1963).
319. E. M. Widdowson, *Nature (London)* **166**, 626 (1960).
320. R. E. Wiederanders, *Proc. Soc. Exp. Biol. Med.* **128**, 627 (1968).
321. G. Wiener and A. C. Field, *Nature (London)* **209**, 835 (1966); *J. Comp. Pathol.* **79**, 7 (1969).
321a. G. Wiener and A. C. Field, *Proc. 1st Int. Symp. Trace Element Metab. Anim., 1969* p. 92 (1970).
322. V. A. Wilkerson, *J. Biol. Chem.* **104**, 541 (1934).
323. W. J. Willkie, *Aust. Vet. J.* **35**, 203 (1959).
323a. J. F. Wilson and M. E. Lahey, *Pediatrics* **25**, 40 (1960).
324. M. M. Wintrobe, G. E. Cartwright, and C. J. Gubler, *J. Nutr.* **50**, 395 (1953).
325. B. Wretlind, K. Orstadius, and P. Lindberg, *Zentralbl. Veterinaer med.* **6**, 693 (1959).
326. H. Yamada and K. T. Yasunobu, *J. Biol. Chem.* **237**, 1511 (1962).
327. A. Zipursky, H. Dempsey, H. Markowitz, G. E. Cartwright, and M. M. Wintrobe, *Amer. J. Dis. Child.* **95**, 148 (1958).
328. E. G. Zook and J. Lehmann, *J. Ass. Offic. Agr. Chem.* **48**, 850 (1965).
329. E. G. Zook and J. Lehmann, *J. Amer. Diet. Ass.* **52**, 225 (1968).

# 4

# MOLYBDENUM

## I. Introduction

The first evidence of a biological role for molybdenum was obtained in 1930 when Bortels (22) showed that this element is an essential nutrient for the growth of *Azotobacter*. Molybdenum was later found to be necessary for all nitrogen-fixing organisms, and for *Aspergillus niger* (108), and to occur regularly in low concentrations in all plant and animal tissues (109). By 1939, Arnon and Stout (8) had shown that molybdenum is required by the higher plants, independent of its role in symbiotic nitrogen fixation. Molybdenum-deficient soils were then discovered in many parts of the world and striking yield responses were obtained from treatment of these soils with small amounts of molybdenum salts or ores, especially with pasture legumes [see Anderson (5)]. The indirect effects of such treatment upon livestock productivity were very great, but there was no evidence in the Mo-deficient areas of molybdenum deficiency directly affecting the grazing animal, in the way that copper or cobalt deficiency affects animals grazing on soils deficient in those elements.

Attention was first directed to the metabolic significance of molybdenum in animals in England in 1938, when the drastic scouring disease of cattle, known as "teart," was shown to be a manifestation of molybdenum poisoning, arising from the ingestion of excessive amounts of this element from the herbage of affected areas, and controllable by treatment of the cattle with large amounts of copper (52). Further evidence of a Cu:Mo interaction in animals came from Australia in 1945 when molybdenum was found to be effective in the treatment of chronic copper poisoning in sheep (42). This important discovery arose independently as a result of earlier observations of unusually high concentrations of molybdenum in the livers of cattle suffering from hematuria vesicalis or red water (42). From these investigations came a realization of the profound effect of molybdenum upon copper metabolism and of the dependence of this effect upon inorganic sulfate. A copper-molybdenum-sulfate interrelationship was subsequently demonstrated with other animal species, and investigations were undertaken in many laboratories to define the quantitative nature of this interaction and the physiological and biochemical mechanisms involved. Certain aspects of these investigations were considered in Chapter 3.

The first indication of an essential role for molybdenum in animal nutrition came in 1953 when two groups of workers independently discovered that the flavoprotein enzyme, xanthine oxidase, is a molybdenum-containing metalloenzyme which is dependent for its activity upon the presence of this metal (99, 100). More direct evidence that molybdenum is essential in the diet of lambs, chicks, and turkey poults was subsequently obtained, using highly purified diets (48, 65, 98). Under naturally occurring conditions, uncomplicated molybdenum deficiency has never been reported either in man or in farm animals. Claims that dietary molybdenum intakes influence the incidence of human dental caries are inconclusive. In fact, there is little evidence that molybdenum plays a significant role in any aspect of human health or disease.

## II. Molybdenum in Animal Tissues and Fluids

### 1. *General Distribution*

The molybdenum concentrations in animal tissues of all species are normally very low, comparable with those of manganese (see Table 16).

TABLE 16

*Normal Molybdenum Concentrations in Animal Organs*[a]

| Species | Liver | Kidney | Spleen | Lung | Brain | Muscle | Ref. |
|---|---|---|---|---|---|---|---|
| Adult man | 3.2 | 1.6 | 0.20 | 0.15 | 0.14 | 0.14 | (111) |
| Adult rats | 1.8 | 1.0 | 0.52 | 0.37 | 0.24 | 0.06 | (65) |
| Chickens | 3.6 | 4.4 | – | – | – | 0.14 | (65) |

[a]Concentration expressed in parts per million on dry basis.

The liver and kidneys carry consistently higher levels of molybdenum than do the other body organs but molybdenum does not accumulate in the liver in excessively high concentrations.

The levels of molybdenum in the tissues do not change significantly with age. Changes in dietary molybdenum intakes are reflected in changes in the levels of molybdenum in the tissues, especially in the liver, kidneys, bones, and skin and in wool (30). Thus Davis (36) increased the concentration in the bones of rats from 0.2 to 9-12 ppm on the dry basis and in the livers from 1-2 to an average of 11-12 ppm by raising the molybdenum content of the diet from below 1 to 30 ppm. When a basal diet low in molybdenum was supplemented with 2000 ppm Mo, as molybdate, the concentration of molybdenum in the tibia of chicks was increased 100-fold (35). Even higher concentrations than these were observed in the bones and soft tissues of guinea pigs and rabbits fed highly toxic or lethal doses of calcium molybdate or of molybdenum trioxide (50).

The response of the tissues to changes in dietary molybdenum intakes is greatly influenced by the level of inorganic sulfate in the diet. A reduction in molybdenum retention in the animal and, hence, lower tissue molybdenum concentrations were first demonstrated by Dick (38) as a consequence of high sulfate intakes in the sheep. A similar effect was subsequently reported in cattle (31, 32) and in rats (84). High dietary intakes of tungstate also reduce the levels of molybdenum in the tissues of rats and chicks (34, 35, 65). The magnitude of the sulfate effect on molybdenum retention in the tissues of the sheep is illustrated in Table 17. The figures in this table indicate that the amounts of molybdenum in all the tissues examined are much smaller at high than at low sulfate intakes, both when dietary molybdenum intakes are high and when they are low. Calculations from these figures indicate, further, that from one-half to three-quarters of the total body molybdenum of the sheep is situated in the skeleton, with the next largest proportions in the skin, wool, and muscles and only about 2% of the total in the liver. This contrasts markedly with the distribution of total body copper, in which a high proportion can occur in the liver in sheep and very little in the skeleton.

The molybdenum content of cattle hair and sheep's wool can be markedly raised by increasing dietary molybdenum intakes, except where sulfate intakes are also high (30). Healy *et al.* (62) reported a mean concentration of 0.16 ppm Mo (range 0.03-0.58) for twenty samples of wool. This may be compared with a mean concentration of 0.06 ppm (range 0.02-0.13) for human hair obtained from the same area in New Zealand (15).

TABLE 17
*Influence of Dietary Molybdenum and Inorganic Sulfate Intake on the Molybdenum Content of Tissues in Sheep*

| Tissue | Molybdenum content (mg) | | | |
|---|---|---|---|---|
| | 0.3 mg/day Mo intake | | 20.8 mg/day Mo intake | |
| | 0.9 g/day sulfate intake | 6.3 g/day sulfate intake | 0.9 g/day sulfate intake | 6.3 g/day sulfate intake |
| Liver | 1.58 | 0.48 | 5.79 | 1.93 |
| Kidney | 0.17 | 0.02 | 1.17 | 1.32 |
| Spleen | 0.52 | 0.02 | 0.57 | 0.14 |
| Heart | 0.18 | 0.01 | 0.94 | 0.04 |
| Lung | 0.65 | 0.09 | 3.96 | 0.42 |
| Muscle | 5.84 | 0.08 | 28.6 | 1.92 |
| Brain | 0.01 | 0.01 | 0.09 | 0.01 |
| Skin | 6.62 | 1.50 | 58.9 | 3.44 |
| Wool | 15.2 | 0.99 | 26.9 | 1.14 |
| Small intestine | 0.26 | 0.03 | 0.88 | 0.79 |
| Cecum | 0.24 | 0.01 | 1.64 | 0.58 |
| Colon | 0.63 | 0.04 | 4.21 | 0.62 |
| Skeleton | 61.0 | 13.0 | 164.0 | 16.0 |
| Total body Mo (mg) | 92.9 | 16.8 | 297.7 | 28.4 |

[a]From Dick (41).

## 2. *Molybdenum in the Liver*

On normal diets the level of molybdenum in the liver is of the same order, namely 2-4 ppm, in several species of widely differing dietary habits (29, 42, 65, 111). Similar concentrations occur in the livers of newborn lambs, indicating that this element is not normally stored in the fetal liver during pregnancy. However, molybdenum concentrations 3-10 times normal were observed in the livers of newborn lambs from ewes receiving a high molybdenum diet (29). This suggests that molybdenum readily passes the placental barrier in this species.

Adult sheep and cows retain molybdenum concentrations in their livers of 25 to 30 ppm, so long as they are ingesting large or moderately large amounts of molybdenum. The levels rapidly return to normal when the administration of the extra molybdenum ceases. The extent of retention in the liver and other tissues and the rate at which the levels fall depend upon the amount and proportion, relative to molybdenum, of the inorganic sulfate intake. The molybdenum level in the liver of an animal, therefore, gives little indication of its dietary molybdenum status and is of limited diagnostic value for this purpose, unless the sulfate and protein status of the diet of the animal are also known. An indication of the variation which can occur is given by the data of Moore (90) who found the molybdenum concentration of the livers of 125 abattoir horses of unknown origin to range from 3 to the unusually high value of 85 ppm on dry basis.

### 3. *Molybdenum in Blood*

An average of 1 $\mu$g Mo/100 ml of whole blood was found by Beck (16) in sheep grazing pastures normal in copper and low in molybdenum. The same sheep consuming a proportion of their diet as low-sulfate cereal hay averaged 6 $\mu$g/100 ml. A level of 6 $\mu$g Mo/100 ml was also found by Cunningham (29) to be normal for sheep and cattle grazing herbage normal in copper and low in molybdenum. When the animals were dosed with molybdate, equivalent to a dietary intake of 30 ppm Mo, the levels rose to 60 to 80 $\mu$g/100 ml in young cattle and to 240 to 340 $\mu$g/100 ml in breeding ewes. Comparable increases in the level of molybdenum in the blood of sheep and cattle have been observed in England as a consequence of raising the dietary molybdenum intakes from normal to 40 ppm (58). When the intake of molybdenum is increased in this way the blood molybdenum level immediately rises and continues to rise until a steady value is reached, the level of which depends upon the amount of molybdenum ingested daily. Thus Dick (40) found the steady values of sheep to increase from 2 $\mu$g Mo/100 ml of whole blood to 495 $\mu$g/100 ml when the molybdenum intake was raised from 0.4 mg Mo/day to 96 $\mu$g Mo/day. This large increase was accomplished without change in the ratio of blood Mo : Mo intake.

The concentration of molybdenum in the blood of sheep is not only proportional to the molybdenum intake, but within certain limits, it is markedly and inversely dependent upon the inorganic sulfate intake (40). Values for the magnitude of these effects upon whole blood molybdenum and the changes in the distribution of this molybdenum between red cells and plasma which occur at various dietary levels of molybdenum and sulfate are given in Table 18. It is apparent from the data that, at low intakes of molybdenum and sulfate, over 70% of the small amount of blood molybdenum is present in the red cells; whereas at high intakes of both, a very small proportion of the much larger amount of blood molybdenum is present in the red cells, with most of the increase occurring in the plasma. A depressing effect of high sulfate intakes upon whole blood molybdenum levels has also been demonstrated in rats consuming high-Mo diets (84). Little is known of the chemical form or combination in which Mo exists in blood, although Scaife (102) has presented evidence that in the sheep, both red cell and plasma molybdenum

TABLE 18
*Blood Molybdenum Concentration and Distribution between Red Cells and Plasma When Molybdenum and When Molybdenum and Sulfate Intakes Are Increased*[a]

| Sheep No. | Stage of experiment | Intake per day: Mo (mg) | Intake per day: $SO_4$ (g) | Molybdenum (μg/100 ml): Whole Blood | Plasma | Red cells | Percentage in red cells |
|---|---|---|---|---|---|---|---|
| D804 | 1 | 0.5 | 1.7 | 9.5 | 4.3 | 17.8 | 71.6 |
| | 2 | 50.0 | 1.7 | 968 | 1208 | 561 | 21.5 |
| | 3 | 50.0 | 7.2 | 68.4 | 91.8 | 16.6 | 4.5 |
| D862 | 1 | 0.5 | 1.7 | 6.8 | 2.5 | 13.8 | 70.4 |
| | 2 | 50.0 | 7.2 | 72.8 | 100 | 12.3 | 5.2 |

[a]From Dick (40).

is readily dialyzable and is wholly present as an anion, probably as molybdate.

In a study of the blood of male adults resident in nineteen cities in different parts of the United States, Allaway and co-workers (3) found that more than 80% of the samples contained less than 0.5 μg Mo/100 ml and only about 3% contained more than 10 μg/100 ml. Considerable variation from site to site was apparent, with little evidence of a broad geographical pattern.

Bala and Liftshits (13) studied the levels of molybdenum (and chromium) in the blood of healthy human subjects and of patients with leukemia, iron deficiency anemia, post-hemorrhagic anemia and anemia secondary to malignant neoplasms. The mean molybdenum content of the whole blood of the healthy subjects was 1.47±0.12 μg/100 ml, evenly distributed between the erythrocytes and the plasma. It was stated that the molybdenum is "firmly bound only to the erythrocyte and plasma proteins." In the leukemia patients the molybdenum level was significantly increased in the whole blood and in the erythrocytes but not in the plasma. In all types of anemia studied the molybdenum content of the whole blood and erythrocytes was reduced and there was also a substantial decrease in plasma molybdenum in iron-deficiency and cancer anemia. In post-hemorrhagic anemia the molybdenum content of the blood decreased solely as a result of a decrease in the number of erythrocytes. With the other anemias there appears to be a disturbance in the metabolism of molybdenum because erythrocyte saturation with molybdenum is greatly reduced. This reduction in erythrocyte molybdenum is the result of a decrease in the molybdenum content of the nonhemoglobin protein fraction of the cells. The authors contend that in iron deficiency anemia there is a concurrent molybdenum deficiency arising from a metabolic disturbance linked with iron. They, therefore, propose that the therapeutic treatment of iron-deficiency anemia and perhaps cancer anemia should include preparations containing molybdenum (and chromium) as well as iron.

### 4. *Molybdenum in Milk*

Published values for the molybdenum content of milk are so variable that it is impossible to state whether real species differences exist or whether there are significant changes related to stage of lactation. Individual variability in the molybdenum content of milk is high (7,72), even among animals consuming the same diets, but dietary differences in molybdenum and sulfate contents constitute the main source of variation. The molybdenum content of the milk is extremely susceptible to changes in dietary molybdenum intakes in cows (7, 61, 71), goats (61), and ewes (67). Thus Archibald (7) found "normal" cow's milk to average 73 μg Mo/liter (range 18-120). This concentration was raised to a mean of 371 μg/liter by feeding the same cows 500 mg Mo as ammonium molybdate daily. Hart *et al.* (61) raised the molybdenum content of milk from cows consuming a low-Mo diet supplying only 5 mg/cow/day, from about 20 to 30 μg/liter to over 40 μg/liter, by increasing the molybdenum intake as sodium molybdate to 50 mg/cow/day, and then raised the content in the milk to as high as 60 μg Mo/liter by increasing the molybdenum intake to 100 mg/cow/day. Similar molybdenum concentrations in the milk and similar increases were obtained with goats when the low-Mo diet of these animals was supplemented with sodium molybdate to provide approximately one-fifth of the daily molybdenum intakes of the cows. Hogan and Hutchinson (67) examined the milk of two groups of ewes, one grazing on a low-Mo pasture (less than 1 ppm) and the other grazing a high-Mo (13 ppm) pasture. The milk of the former group averaged less than 10 μg Mo/liter, whereas that of the latter averaged no less than 980 μg/liter. These workers also studied the effect of dosing ewes on a high Mo pasture (25 ppm Mo) with sodium sulfate at the rate of 23 g sulfate ion per day. The effect of the sulfate in reducing the molybdenum content of the milk was remarkable. At the end of 3 days of dosing, the milk of the dosed ewes had fallen to 137 μg Mo/liter, compared with 1043 μg/liter in comparable ewes receiving no additional sulfate.

The forms of molybdenum in milk have been studied mainly in relation to xanthine oxidase. In ewe's milk a high proportion of the molybdenum is associated with the aqueous phase, i.e., it remains after separation of the fat by centrifugation and the casein and albumin by clotting (67). All the molybdenum in the milk of cows on normal rations is bound to enzymically active xanthine oxidase, so that the xanthine oxidase activity of such milk is proportional to its molybdenum content (61, 71). The oral administration of molybdate produces a rapid rise in the molybdenum content of the milk, as stated above, but does not increase its xanthine oxidase activity (61, 71). Molybdate administration to goats also increases the molybdenum content of the milk without affecting xanthine oxidase activity. However, the xanthine oxidase activity of goat's milk is lower than that of cow's milk, and there is no correlation between molybdenum content and xanthine oxidase activity of the milk of this species, as there is with cows (61). The ingestion of doses of up to 6 g sodium tungstate (56 mg W/kg body weight) by goats diminished the amount of xanthine oxidase secreted in their milk, so that in some samples the enzyme became undetectable (97). A reduction in milk xanthine oxidase activity also occurred in cows dosed with up to 20 g sodium tungstate in early lactation (97).

## III. Molybdenum Metabolism

Studies with both stable and radioactive molybdenum ($^{99}$Mo) indicate that molybdenum is readily and rapidly absorbed from most diets. The hexavalent water-soluble forms, such as sodium and ammonium molybdate, and the molybdenum of high-Mo herbage, most of which is water-soluble, are particularly well absorbed by cattle (52). Even such insoluble compounds as $MoO_3$ and $CaMoO_4$, but not $MoS_2$, are well absorbed by rabbits and guinea pigs when fed in large doses (50). In a comparison of yearling cattle and growing swine on similar molybdenum, copper, and sulfate intakes (18), the former species absorbed oral doses of $^{99}$Mo much less rapidly than the latter. The administered $^{99}$Mo reached a peak in the blood of the swine at 2-4 hr, whereas the average time in the cattle was 96 hr.

Species differences also exist in the main routes of excretion of molybdenum. Thus, in the experiment just quoted, over 75% of both orally and intravenously administered $^{99}$Mo was excreted in the urine of swine in 120 hr. Fecal excretion was the main route in the cattle and only 15% was excreted in the urine in 168 hr by the intravenously dosed cattle (18). The importance of the urine as the major route of excretion of molybdenum is evident from other studies with pigs (103), rats (94), and man (101), employing $^{99}$Mo, and from the balance studies of Tipton *et al.* (112) with two adult human subjects.

Studies of molybdenum metabolism are of limited value unless the inorganic sulfate status of the diet is known. The potent influence of sulfate upon molybdenum absorption and retention and upon the route of excretion of absorbed molybdenum (38) can be illustrated as follows: sheep fed a diet of oaten chaff (< 0.1% sulfate) plus 10 mg Mo/day, excreted 63% of this molybdenum in the total excreta during a period of 4 weeks, of which 3-4.6% appeared in the urine. When fed a diet of alfalfa chaff (0.3% sulfate) plus 10 mg/ Mo/day the recovery in the total excreta was 96%, of which 50-54% appeared in the urine. The actual outputs in the feces were similar on each diet. By fractionation of the alfalfa chaff diet the factor responsible for this marked effect on molybdenum excretion was identified as inorganic sulfate and the administration of a single oral dose of potassium sulfate to the sheep on the cereal hay (low-sulfate) diet induced a rapid rise in urinary molybdenum excretion (Fig. 10). Sodium sulfate produced the same effect without the diuresis that accompanies the use of potassium sulfate, whereas potassium chloride induced diuresis but had no effect on molybdenum excretion (38). Similar results were later obtained by Scaife (102), who fed sheep a low-sulfate and a high-sulfate diet plus 50 mg Mo/day in each case. On the low-sulfate diet, only 5% of the molybdenum appeared in the urine, compared with 30 to 40% on the high-sulfate diet.

Inorganic sulfate increases urinary excretion of molybdenum in the marsupial, *Setonix brachyurus*, without significantly increasing urine volume (14) and alleviates molybdenum toxicity in all species studied.

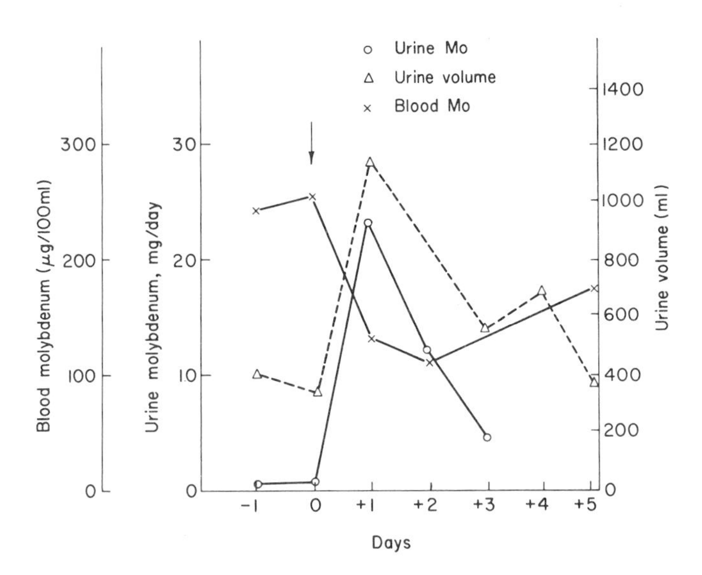


**Fig. 10.** Effect on blood and urine molybdenum of 11 g potassium sulfate given b' mouth, at the time indicated by the arrow, to a sheep on a diet of chaffed hay and a ' intake of 10 mg/day. [From Dick (38).]

Whether the effect of sulfate on toxicity is mediated entirely through an influence on the pattern of molybdenum excretion is doubtful. In rats (84) and in cattle (32), as in sheep, sulfate reduces molybdenum retention in the tissues, presumably through increased urinary excretion. No such reduction in tissue retention of molybdenum by sulfate occurs in the chick, despite a marked beneficial effect on the manifestations of molybdenum toxicity (34,35).

In the sheep, sulfate limits molybdenum retention both by reducing intestinal absorption and increasing urinary excretion, the extent of each depending upon the previous history of the animal with respect to molybdenum and sulfate intakes (40,102). The increased urinary excretion is not a passive result of the greater urinary volume that occurs on high-sulfate diets (38). Several diuretics increase urine volume without increasing molybdenum excretion (102). The sulfate effect is highly specific, and, in sheep, is not shared by such anions as tungstate, selenate, silicate, permanganate, phosphate, malonate, and citrate (40,102). The capacity of sulfate to alleviate molybdenum toxicity in the rat is also not shared by citrate, tartrate, acetate, bromide, chloride, or nitrate (114). However, sulfate of endogenous origin can be just as effective as dietary inorganic sulfate. This is indicated by the effects of high protein diets, by the catabolic breakdown of body tissue, and by the administration of thiosulfate and methionine to sheep (40,102). The protective action of methionine, cystine, and thiosulfate administration to rats fed high-Mo diets (55,115) may also result from the production of endogenous sulfate.

The influence of sulfate on molybdenum absorption and excretion is explained by Dick (40) on the hypothesis that inorganic sulfate interferes with, and if its concentration is high enough, prevents the transport of molybdenum across membranes. Such an effect would increase urinary molybdenum excretion through the rise in the sulfate concentration in the ultrafiltrate of the kidney glomerulus which follows high-sulfate intakes, impeding or blocking reabsorption of molybdenum through the kidney tubule. The mechanism of this postulated interference with membrane transport is unknown.

## IV. Molybdenum As an Essential Element

Xanthine oxidase or xanthine dehydrogenase, which is widely distributed in animals, and the nitrate reductase of plants, are the only known molybdenum-containing metalloproteins. The only known biochemical function of molybdenum in animals, other than the interaction with copper, is related to the formation and activity of xanthine oxidase. The purified enzyme isolated from cow's milk is a reddish-brown protein with a molecular weight of 275,000 (6), containing 8 atoms of iron, 2 molecules of flavin adenine dinucleotide (FAD), and variable amounts, usually 1.3-1.5 atoms, of Mo per molecule of protein (12,23). The molybdenum is firmly bound to the protein and cannot be removed without destroying the oxidase activity (19).

The need for dietary molybdenum to maintain normal xanthine oxidase levels in rat tissues, especially the intestine, was clearly established, but definite evidence that molybdenum is an essential element was not obtained in those early experiments (99,100), because neither the growth nor the purine metabolism of the depleted rats on the low-Mo diets was affected and no improvement from molybdenum supplements was observed. The rat has an extremely low requirement for molybdenum and for tissue xanthine oxidase, so much so that rats have been shown to grow normally, reproduce, accumulate xanthine oxidase in the tissues other than the intestine, and to oxidize xanthine normally on diets containing only 0.2 ppm Mo (65). Even when tungstate, a Mo antagonist, was added to such diets, at levels equivalent to a W:Mo ratio of 100:1 or 1000:1, growth and xanthine oxidation were normal. Similar low-Mo diets were equally well tolerated by chicks. In this species the addition of tungstate to give W:Mo ratios 1000:1 and 2000:1 reduced growth and tissue molybdenum and xanthine oxidase concentrations, as well as the capacity of the birds to oxidize xanthine to uric acid (65). There was also a 25% mortality in the tungstate-fed birds. All of these effects could be prevented by additional dietary molybdenum. It seems, therefore, that the chicks were suffering from a tungsten-induced molybdenum deficiency and that molybdenum can be considered an essential element for this species. This has been confirmed by Leach and Norris (78) with chicks fed a purified casein diet containing 0.5-0.8 ppm Mo plus added tungsten, and with highly depleted chicks from hens fed a special low-mineral diet.

A significant growth response from molybdenum has been reported in lambs fed a semipurified diet containing 0.36 ppm Mo, to which molybdate was added to raise its Mo content to 2.36 ppm (48). Digestibility trials revealed a significant improvement in cellulose digestibility on the Mo-supplemented diets, and the authors suggest that molybdenum stimulates growth in lambs by increasing cellulose degradation by rumen organisms. A growth response from supplementary molybdenum in lambs fed a low-Mo purified diet was subsequently confirmed by Ellis and Pfander (47). These interesting findings deserve further intensive study with a view to defining the minimum molybdenum requirements of lambs and other animals and the dietary factors that affect this requirement. Many pastures grazed regularly by sheep and cattle contain lower concentrations of molybdenum than the 0.36 ppm present in the deficient diets just cited. Such Mo-low pastures favor the accumulation of copper in the tissues of sheep and, under certain conditions, can lead to chronic copper poisoning. There is no evidence that the consumption of these pastures adversely affects growth or feed utilization.

The only other disability affecting animals which has been even tenuously related to molybdenum insufficiency is the occurrence of renal xanthine calculi in sheep. A high incidence of such calculi, especially in castrate male sheep, was recognized in the Moutere Hills of New Zealand (46). The affected pastures contained only 0.03 ppm Mo or less, compared with concentrations up to 0.4 ppm in "noncalculi" areas nearby (11). There were also indications of subnormal levels of molybdenum in the livers of sheep from the affected areas and of a correlation between these levels and the incidence of calculi. Resowing of the pastures and treatment with lime, which raises pasture molybdenum levels, prevents the development of the calculi. Since such treatment would also raise the levels of protein and other constituents of the herbage and since pastures as low in molybdenum as those of the Moutere Hills exist in many parts of the world, without xanthine calculus formation in sheep, the condition described is unlikely to be an expression of molybdenum deficiency alone.

## V. Molybdenum Toxicity

### 1. *General*

Tolerance to high dietary intakes of molybdenum varies with the species and age of the animal; the amount and chemical form of the ingested molybdenum; the copper status and copper intake of the animal; the inorganic sulfate content of the diet and its content of substances such as methionine, cystine, and protein, capable of oxidation to sulfate in the body; and with the level of intake of other metals, including zinc and lead (57).

Some of the comparisons made of the relative susceptibility of different species to elevated intakes of molybdenum are of dubious validity because of a failure to recognize the profound importance of differences in the other dietary components, as just indicated. Nevertheless, real species differences in tolerance to molybdenum undoubtedly exist. Cattle are by far the least tolerant, followed by sheep, whereas horses and pigs are the most tolerant of farm livestock. The high tolerance of horses is illustrated by their failure to show any signs of toxicity on "teart" pastures, which severely affect cattle, and the high tolerance of pigs, by the report of Davis (36) that levels of molybdenum of 1000 ppm, fed to these animals for 3 months, induced no ill-effects. This level of molybdenum is 10-20 times greater than the levels which result in drastic scouring in cattle. The high tolerance of pigs cannot be due to poor absorption, as previously suggested, judging by the results of studies with $^{99}Mo$ which indicate rapid absorption and rapid excretion of this element in this species (18,103). Rats, rabbits, guinea pigs, and poultry have a lower tolerance to high-molybdenum intakes than pigs but are much more tolerant than cattle. Molybdenum intakes of 2000 ppm induce severe growth depression in chicks, accompanied by anemia when the dietary level is raised to 4000 ppm Mo (10,35,73). At 200 ppm Mo some inhibition of chick growth results and, at 300 ppm of dietary Mo, turkey poult growth can be depressed (73).

Growth retardation or loss of body weight is an invariable manifestation of high-molybdenum intakes in all species. Diarrhea is a conspicuous feature of molybdenosis in cattle, with anemia and loss of coat color developing as the high-molybdenum intakes are prolonged. Diarrhea occurs sometimes in rats but has not been reported in rabbits or guinea pigs ingesting sufficient molybdenum to produce marked loss of weight and death (50). In young rabbits the molybdenosis syndrome is characterized further by alopecia, dermatosis, and severe anemia, with a deformity in the front legs of some animals, but no achromotrichia or diarrhea (9). In rats, high-molybdenum intakes result in the usual loss of body weight often with marked anorexia. Anemia has been reported in some studies (94,96) and not in others (57,70). Deficient lactation and male sterility associated with some testicular degeneration (70), and connective tissue changes resulting in mandibular or maxillary exostoses, have also been observed (96,114).

The extent to which these various manifestations of molybdenum toxicity actually arises in the animal depends upon the level of intake of molybdenum relative to those of copper and inorganic sulfate or of sulfate-producing substances. The complex interactions of molybdenum and copper may be illustrated by two experiments with rats. The addition of as little as 80 ppm Mo as molybdate inhibited growth and induced mortality in young rats fed a low-Cu diet, whereas no such disabilities arose when the copper content of the diet was raised to 35 ppm with copper sulfate (28). Young rats fed a diet containing 77 ppm Cu grew poorly when Mo was added as sodium molybdate at levels of 500 or 1000 ppm but grew normally when an additional 200 ppm Cu was supplied (94). Comparable additions of iron, cobalt, or zinc were ineffective but 5% dried whole liver gave considerable protection against 400 ppm Mo. Supplementation with copper has similarly been shown to alleviate a Mo-induced growth depression in rabbits (9) and in chicks (10,73).

Supplementary methionine and cystine can be as effective as copper in alleviating molybdenum toxicity in rats (55,115) and in sheep (102) and thiosulfate administration is equally effective in sheep (102). As these substances are all capable of oxidation to sulfate in the tissues, it is probable that they act primarily as inorganic sulfate acts by reducing Mo retention in the tissues through promoting its excretion in the urine. This mode of action has been demonstrated for these substances in the rat (84) and the sheep (38,102) and in a marsupial (14). An ameliorating effect of high-protein diets, which would also be expected to yield endogenous sulfate, has been similarly demonstrated in Mo toxicity in rats (83) (see Table 19). The effect was associated with a significant depression in blood and liver Mo levels in the rats receiving the high-casein diets. In the chick, neither methionine nor sulfate reduces Mo retention in the tissues, although the latter is highly effective in preventing the growth depression induced by high-Mo diets (35).

The activities of two enzymes, alkaline phosphatase and sulfide oxidase, are affected in the molybdenotic animal. Alkaline phosphatase activity is decreased in the liver and increased in the kidney and intestine of rats suffering from molybdenum toxicity (113,117). Liver sulfide oxidase activity is substantially depressed in these animals but liver cysteine desulfhydrase and kidney aryl sulfatase activities remain unaffected. The oxidation of L-cysteine sulfinate by liver homogenates is also unaffected. The depression in liver sulfide oxidase activity suggests that the syndrome of molybdenum toxicity is contributed to by an accumulation of sulfide in the tissues. This is supported by the fact that feeding calcium sulfide (86), or excessive quantities of cystine, which could lead

TABLE 19

*Effect of Dietary Casein Level upon Molybdenum Toxicity*[a]

| | Treatment[b]: | | | |
|---|---|---|---|---|
| Casein (%) | $-SO_4$ $-$Mo | $+SO_4$[c] $-$Mo | $-SO_4$ +Mo[d] | $+SO_4$[c] +Mo[d] |
| 8 | 60 | 66 | 26 | 52 |
| 12 | 109 | 115 | 74 | 104 |
| 18 | 126 | 140 | 133 | 138 |

[a]From Miller and Engel (83).

[b]Values given are 6-week weight gains (in grams); 4-6 rats per treatment; six replicates.

[c]Sulfate, 2000 ppm.

[d]Molybdenum, 100 ppm as $Na_2MoO_4$.

to excessive generation of sulfide in the liver through the activities of the cysteine desulfhydrase system, cause anemia, diarrhea, and death in rats consuming high-Mo diets (60). In both cases the disorders were prevented by supplying additional copper. The endogenous production of sulfide may, therefore, be a significant factor in molybdenosis in the rat and in the appearance of signs of copper deficiency.

Of further interest is the observation that the anorexia of molybdenosis results from a voluntary rejection of the diets, consequent upon the development by the rats of an ability to recognize the presence of molybdenum in the diet (89). This phenomenon implies sensory, probably olfactory, recognition of the presence of molybdate following an interaction of the molybdate with other dietary constituents to form compounds with a characteristic odor detectable by rats. The ability to reject high-Mo diets in this way requires a learning or conditioning period. It is lacking or weak with freshly prepared diets and extends to a discrimination between a toxic (high-molybdate) and a nontoxic (high-molybdate + sulfate) diet. Since sulfate alleviates the diarrhea which develops in rats fed the high-Mo diet, Monty and Click (89) suggest that the rats learn to associate a gastrointestinal disturbance with a sensory attribute of diets containing toxic levels of Mo.

2. *Molybdenosis in Cattle*

Severe molybdenosis occurring under natural conditions in cattle confined to certain grazings has been reported from England (52), from California (24), Nevada (45), Oregon (95), Ireland (93), and New Zealand, where the condition is known as "peat scours" (29). All cattle are susceptible to molybdenosis, with milking cows and young stock suffering most. Sheep are much less affected, and horses are not affected at all in teart areas. The characteristic scouring varies from a mild form to a debilitating condition so severe that cattle can suffer permanent injury or death. Within a few days of being turned on to teart pastures, cattle begin to scour profusely and develop harsh, staring, discolored coats. They usually recover rapidly when transferred to normal pastures lower in molybdenum (52). The molybdenum levels in typical teart pastures range from 20 ppm to as high as 100 ppm on dry basis, compared with 3 to 5 ppm, or less, in nearby healthy herbage (79). Confirmation that teart is a molybdenosis consequent upon the ingestion of excessive amounts of Mo from the "affected" herbage was obtained by feeding sodium or ammonium molybdate directly to cattle or by applying these compounds to healthy pastures. The effects on cattle were comparable to those of teart pastures (52,53,79).

Following a report from Holland of a scouring condition in cattle responsive to copper therapy (25), treatment of teart with copper sulfate was tried by Ferguson and co-workers (52). Such treatment at the very high rate of 2 g/day to cows and 1 g/day to young stock, or the intravenous injection of smaller quantities of the order of 200 to 300 mg/day, effectively controlled the Mo induced scouring and provided a practical means of field control. Although this discovery was made over 30 years ago, the mechanism of the protective action of copper is still obscure. If the high-molybdenum intakes are prolonged, a depletion of tissue copper to deficiency levels occurs with associated hypocupremia, but the scouring and loss of condition of severe molybdenosis can occur without a concomitant hypocuprosis. In peat scours, on the other hand, the molybdenum intakes from the herbage are lower than those from typical teart pastures and the molybdenosis is, therefore, less acute. Under these conditions the onset of scouring is delayed until the normal tissue copper stores are depleted. It can be overcome by supplying smaller amounts of copper, administered parenterally or by mouth, or by copper treatment of the pastures (29). In teart there is no inflammation or local damage to the intestines and the scouring can be produced by intravenous injections of molybdate over 2 to 3 weeks (4). The molybdenum injected at these levels probably reaches the lumen of the intestine. It can, therefore, be argued that high concentrations of molybdenum in the gastrointestinal tract, relative to those of copper, disturb bacterial activity in this site, with resulting diarrhea.

The possibility that the sulfate radicle may also be of some significance in the treatment of Mo toxicity has been suggested by Dick (37). This worker showed that Mo-induced scouring in a calf can occur within 4 days by daily drenching with 2 g copper sulfate and within 6 days by drenching with an equivalent amount of sulfate as $K_2SO_4$.

At very high Mo intakes, scouring and weight losses in cattle are so obvious that other disorders tend to be obscured. In some areas, moderately high levels of molybdenum in the herbage are accompanied by a disturbance of P metabolism, giving rise to lameness, joint abnormalities, osteoporosis, and high serum phosphatase levels (36). Under such conditions, cows may conceive with difficulty and young male bovines exhibit a complete lack of libido. The testes reveal marked damage to interstitial cells and germinal epithelium with little spermatogenesis (110).

Very high levels of molybdenum in herbage plants, resulting in the signs of molybdenosis in cattle just described, are usually associated with alkaline soils, as in the teart areas of England (52,79). Molybdenum is not readily absorbed by plants from most acid soils so that the liming of such soils, as well as treatment with Mo-containing fertilizers, can be effective means of raising the Mo content of herbage (5,43,66). High and potentially toxic levels of molybdenum also occur in legumes in poorly drained acid soils in some areas (75,93), and an effect of soil wetness upon the Mo content of herbage plants has been observed under both field (76) and greenhouse conditions (77). Very wide variations occur in the molybdenum levels in pasture species, ranging from less than 0.1 to over 200 ppm on dry basis. Most of this variation is an expression of the soil differences mentioned above but species differences also exist. Grasses are usually higher in molybdenum than clovers growing with them (17,43) but this is not always so, judging from the data of Kubota *et al.* (75) for grasses and legumes growing on high-Mo soils in Oregon.

## VI. Molybdenum-Copper Interrelations

The quantitative nature of the relationship between molybdenum and copper can be illustrated by comparing the conditions under which the two diseases of cattle, teart and peat scours, exist and are controlled. The very high-Mo levels of teart pastures cause rapid scouring, despite a normal copper content, and massive oral dosing with copper is required to counteract this amount of Mo. The moderate excess of Mo in peat scours pastures causes scouring only when the pasture copper is below normal and the animal is depleted of copper. At such intakes of Mo, control of the scouring can be achieved merely by raising the copper content of the pastures to normal levels. At such levels severe scouring occurs in cattle on teart herbage with its very high Mo levels.

These field findings are supported by considerable experimental evidence with cattle and other species which show clearly that the toxicity of any particular level of dietary Mo is affected by the ratio of the Mo to dietary Cu. The toxicity of a particular level of dietary Cu is also greatly influenced by the ratio of this copper to dietary Mo and sulfate. In the original work of Dick (38,39), neither Mo nor sulfate alone appeared to interfere with copper retention. However, some of his data and of those of Wynne and McClymont (118) indicated that when the Mo intakes were less than 0.5 mg/day or 8 ppm, respectively, inorganic sulfate additions resulted in an appreciable reduction in Cu storage. Spais (105,106) reported conditioned Cu deficiency in lambs grazing herbage very high in sulfate and in a mining area rich in sulfur minerals. In addition, Spais *et al.* (107) demonstrated high rumen sulfide production on high-sulfate diets. They suggested that this reduces copper absorption through the formation of insoluble copper sulfide. Finally, Bird (21) demonstrated increased ruminal sulfide concentrations and a decreased flow of soluble copper to the omasum when the intake of cystine-S or sulfate-S by sheep was increased under steady-state feeding conditions.

Molybdenum and sulfate can either increase or decrease the copper status of an animal, depending on their intakes relative to that of copper.

Thus, as shown in Chapter 3, chronic copper poisoning can occur in sheep with moderate copper intakes and very low levels of Mo and sulfate, and, conversely, the depletion of the animal's copper reserves, to the extent of clinical signs of copper deficiency, can arise on normal copper and high-Mo and -sulfate intakes (118). Furthermore, sulfate can either aggravate or ameliorate the toxic effects of Mo, depending upon the copper status of the animal. Gray and Daniel (56) showed that when the copper stores of rats were low and a Cu-deficient diet was fed, small amounts of Mo produced toxic symptoms which were intensified by the simultaneous addition of sulfate. By contrast, when the copper stores and dietary copper were adequate, larger amounts of Mo were required to produce molybdenosis, and sulfate completely prevented the harmful effects of molybdenum. The copper status of the rat, therefore, not only influences, it can actually reverse the direction of the sulfate effect on the Mo-Cu interaction.

The complexity of the Cu-Mo-$SO_4$ interrelationship is exemplified further by the existence of other dietary factors that have a modifying influence. Dick (41) showed that high-manganese intakes can block or antagonize the limiting effect of Mo on copper retention in the sheep, even in the presence of adequate sulfate. No such effect was observed by Mylrea (91) in steers. Dick (41) showed further that when sheep are on a high-protein diet, Mn and Mo together exert a severely limiting effect on copper retention. Dietary differences involving the presence or absence of these or other factors capable of affecting copper and molybdenum metabolism probably explain some of the apparently conflicting results obtained with these elements by different workers. Thus, Mills and co-workers (51,88) produced Cu-deficient lambs by feeding high amounts of molybdate and sulfate to penned sheep fed grass cubes, whereas Butler and Barlow (26) found no such effect when the basic diet consisted of hay and oats. Hogan and co-workers (69) were also unable to reduce the liver and brain copper of lambs to deficiency levels by dosing their mothers with molybdate and sulfate for 9 months while grazing pastures normal to low in copper content. Subsequently, Hogan *et al.* (68) obtained evidence with penned sheep of the presence of some dietary factor in the concentrate fed which facilitated excessive copper storage in the liver. The effect of this unidentified factor upon liver copper storage was significantly reduced but was not eliminated by feeding additional molybdate and sulfate.

Several theories have been advanced to explain the limiting effect of Mo and sulfate upon copper retention but no fully adequate explanation of the Cu-Mo-$SO_4$ interaction is yet available. Dick (40) first put forward the hypothesis that a membrane the permeability of which to molybdenum is impeded or blocked by sulfate impedes or blocks copper transport. He states the position as follows: ". . . under the influence of Mo and sulfate in the diet, the absorption of copper will be reduced and consequently the rate of accumulation of copper in the liver will be reduced. As the Mo and sulfate intakes are increased, the amount of copper will be insufficient to make good the normal wastage of copper by excretion, the animal's copper reserves will be depleted and a state of Cu-depletion deficiency will be established if the animal is exposed to these conditions for a sufficiently long period. There will, however, be no effect on blood copper until the Mo intake is increased sufficiently that, in spite of a high sulfate intake, sufficient Mo is absorbed to establish a high concentration at the membranes through which copper is normally excreted. Thus not only will copper absorption but also copper excretion be impeded."

Several alternative explanations of the Cu-Mo-$SO_4$ interaction have been proposed. Mills (87) found that high intakes of molybdate and sulfate did not increase the rate of excretion of intravenously injected $^{64}Cu$ in sheep. He suggests that these anions restrict copper utilization in this species by depressing copper solubility in the digestive tract through the precipitation of insoluble cupric sulfide. Higher sulfide levels were observed in the rumen of sheep fed molybdate and sulfate than in those fed molybdate alone. In addition, the concentration of soluble copper in this site was inversely related to the sulfide level. The lowest concentrations of soluble copper in the abomasum were also found when sulfate as well as molybdate were fed, but in this case there was no correlation with abomasal sulfide level which remained very low (86). Spais (105,107) has also suggested that the rumen sulfide formed from high-sulfate intakes acts by binding feed copper to insoluble and unavailable copper sulfide. The recent data of Bird (21), mentioned earlier, give support to this suggestion.

Accumulation of sulfide in the tissues, as a consequence of the depression in sulfide oxidase activity known to occur in the molybdenotic rat (113, 117), has been proposed as a further causal factor in the effect of molybdenum on copper utilization. It is argued that such tissue sulfide accumulation leads to a precipitation of insoluble cupric sulfide in which the copper is unavailable for physiological purposes (60,104,107). Sulfide oxidase activity appears to be dependent upon the *in vivo* supply of copper (104), so that high dietary levels of copper will help to maintain sulfide oxidase activity in the face of the inhibiting effect of molybdenum on this enzyme. In this way an endogenous supply of sulfate will emerge which, in turn, will tend to prevent Mo accumulation in the tissues. The protective action of copper against Mo toxicity could be explained, to some degree, by such a mechanism the effectiveness of which will clearly depend upon the intake of Cu relative to that of Mo.

A further hypothesis has been put forward by Dowdy and associates (44). These workers observed that $CuSO_4$ and $Na_2MoO_4$ form a complex, with a Cu:Mo ratio of 4:3, which precipitates in a near-neutral solution. This Cu-Mo complex can exist *in vivo*, and the copper bound in this complex is biologically unavailable to pigs and sheep. Pigs receiving the Cu-Mo complex showed an increase in serum ceruloplasmin activity which was no greater than that in pigs fed a diet without supplemental copper, and their tissue copper levels were generally lower than those in pigs fed either copper sulfate or copper citrate with equivalent amounts of sodium molybdate. In sheep given an intravenous injection of the Cu-Mo complex made from $^{64}Cu$ and $^{99}Mo$, the rates of removal from the blood of the copper and molybdenum were equal, and this rate was more rapid than that of $^{99}Mo$ injected alone. Conversely, the rate of urinary excretion of molybdenum from the $^{64}Cu$-$^{99}Mo$-injected sheep was slower than from the $^{99}Mo$-injected animal. Copper in the form of the Cu-Mo compound, synthetic lindgrenite, was also shown to be less available to weanling rats than copper from the sulfate salt.

Marcilese and co-workers (82) compared the effects of 0.4% sulfate alone and of sulfate (0.4%) plus molybdenum (50 ppm) upon plasma clearance of injected $^{64}Cu$ and its uptake by the liver and incorporation into ceruloplasmin in sheep. The sulfate alone had no effect but when both Mo and sulfate were present in the diet, radiocopper uptake by the liver was reduced, and ceruloplasmin synthesis was impaired, resulting in a slower removal of radiocopper from the plasma. In addition, stable copper in the liver and in the ceruloplasmin fraction of plasma was significantly reduced. These data clearly indicate a metabolic interference with copper by sulfate and molybdenum in the liver but the mechanism of this interference remains to be established.

## VII. Molybdenum in Human Health and Nutrition

### 1. *Molybdenum in Foods and Dietaries*

Molybdenum is apparently of so little practical significance in human nutrition, either in health or disease, that few people have been stimulated to undertake studies with this element. Relevant data are, therefore, exceedingly meager. Tipton *et al.* (112) found the mean dietary intake of 2 adults to be close to 100 $\mu$g Mo/day. These individuals were in slight negative Mo balance and more Mo was lost in the urine than in the feces. This figure compares well with the results of an earlier balance study carried out on 24 girls, 7-9 years old (85). The average intake in this case was 75 μg Mo/day, most of which appeared in the urine. An interesting feature of this study was that at high-protein intakes of 2.5 g/kg body weight, Mo retention was very low, whereas this retention increased fivefold when the protein intake of the girls was reduced to 0.7 g/kg body weight. It seems that an adequate protein intake is necessary for the elimination of dietary molybdenum.

The molybdenum content of foods varies greatly, both within and among the different classes of foodstuffs comprising normal human dietaries. Legumes, cereal grains, leafy vegetables, and liver and kidney are among the richest sources of molybdenum, with fruits, root and stem vegetables, muscle meats, and dairy products among the poorest (20, 109,116). Westerfeld and Richert (116) found the leguminous seeds to vary from 0.2 to 4.7 ppm Mo and the cereal grains from 0.12 to 1.14 ppm. In a later study (33) of the mineral content of North American wheat, flour, and bread the wheats ranged from 0.30 to 0.66 ppm Mo, with a mean of 0.48 ppm. The white flours made from these wheats averaged 0.25 ppm Mo. Molybdenum is thus concentrated to some extent in the outer layers of the grain removed in flour-milling, but the degree of such concentration is much less than with iron, manganese, zinc, or copper. Overall dietary intakes of molybdenum would, therefore, not be markedly influenced by the proportions of whole grain, as compared with refined cereals, consumed. Studies of the availability of the molybdenum in human foods do not appear to have been carried out, but determinations of the response in the intestinal xanthine oxidase levels of rats fed low-Mo diets indicate that 50-100% of the total Mo in various foods is available for this purpose (116).

2. *Molybdenum and Dental Caries*

Several claims have been made that molybdenum exerts a beneficial effect upon the incidence and severity of dental caries and that it can enhance the well-established effect of fluoride. These claims are based upon epidemiological studies carried out in Hungary (2,92) and in New Zealand (64,80) and upon a number of experiments with rats. These have been critically assessed by Hadjimarkos (59), who rightly points out that in neither the Hungarian nor the New Zealand study was it established that total dietary intakes of molybdenum by children in the towns with the low caries incidence were significantly higher than those of comparable children in the towns with a higher incidence. Furthermore, no significant differences were found in the levels of molybdenum in the permanent teeth (63) nor the hair (15) of the two groups of New Zealand children.

TABLE 20

*Effect of Prenatal and Postnatal Administration of Molybdenum and Fluoride on Dental Caries in Rats*[a]

| Diet (ppm) | Mean No. of carious teeth | Mean No. of carious lesions | *t* Values |
|---|---|---|---|
| Control | 4.5 ± 0.2 | 7.2 ± 0.6 | – |
| 25 Mo | 4.5 ± 0.6 | 7.7 ± 1.1 | – |
| 50 Mo | 4.1 ± 0.3 | 6.9 ± 0.6 | – |
| 50 F | 2.2 ± 0.5 | 3.0 ± 0.8 | 4.4 |
| 25 Mo + 50 F | 1.1 ± 0.3 | 1.2 ± 0.3 | 1.8 |
| 50 Mo + 50 F | 0.5 ± 0.2 | 0.6 ± 0.3 | 2.2 |

[a]From Buttner (27).

The results of experiments with rats are inconclusive and confusing. In two of the earliest experiments (1,74), a reduction in caries incidence was observed when the molybdenum was supplied during the formative period of the teeth. Subsequent experiments, in which the rats were fed a cariogenic diet and given supplemental molybdenum in the drinking water at various levels up to 50 ppm, failed to demonstrate any significant effect during this period or thereafter (27,81,116). The results obtained by Buttner (27) are of particular interest because of his finding that the combined administration of 25 or 50 ppm Mo and 50 ppm F in the water was more effective in reducing caries incidence in rats than the ingestion of water containing only 50 ppm F (Table 20). However, Malthus *et al.* (81) found no significant difference between the caries scores of animals fed a combination of Mo and F and those given fluoride alone. Moreover, the uptake of fluoride by intact human enamel is not increased by the addition of Mo salts to solutions of NaF (54), and neither fluoride retention in the bones of rats (27,49) nor the fluoride content of the saliva (49) is increased when $^{99}$Mo and $^{18}$F are given simultaneously by stomach tube, compared with animals similarly receiving $^{18}$F alone. It is obvious that further critical studies are necessary before molybdenum can unequivocally be assigned a preventive, or even an ameliorative, role in dental caries.

## REFERENCES


1. P. Adler, *Odontol. Revy* **8**, 202 (1957).
2. P. Adler and J. Straub, *Acta Med. Acad. Sci. Hung.* **4**, 221 (1953).
3. W. H. Allaway, J. Kubota, F. L. Losee, and M. Roth, *Arch. Environ. Health* **16**, 342 (1967).
4. R. Allcroft and G. Lewis, *Landbouwk. Tijdschr.* **68**, 711 (1956).
5. A. J. Anderson, *Advan. Agron.* **8**, 163 (1956).

6. P. Andrews, R. C. Bray, and K. V. Shooter, *Biochem. J.* **93**, 627 (1964).
7. J. G. Archibald, *J. Dairy Sci.* **34**, 1026 (1951).
8. D. I. Arnon and P. R. Stout, *Plant Physiol.* **14**, 599 (1939).
9. L. R. Arrington and G. K. Davis, *J. Nutr.* **51**, 295 (1953).
10. D. Arthur, I. Motzok, and H. D. Branion, *Poultry Sci.* **37**, 1181 (1958).
11. H. O. Askew, *N. Z. J. Agr. Res.* **1**, 447 (1958).
12. P. G. Avis, F. Bergel, and R. C. Bray, *J. Chem. Soc.* p. 1219 (1956).
13. Y. M. Bala and V. M. Liftshits, *Probl. Gematol. Perelv. Krovi* **10**, 23 (1965); *Fed. Proc., Fed. Amer. Soc. Exp. Biol.* **25**, Trans. Suppl., 370 (1966).
14. S. Barker, *Nature* (*London*) **185**, 41 (1960).
15. L. C. Bate and F. F. Dyer, *Nucleonics* **23**, 74 (1965).
16. A. B. Beck, personal communication (1960).
17. A. B. Beck, *Aust. J. Exp. Agr. Anim. Husb.* **2**, 40 (1962).
18. M. C. Bell, B. G. Higgs, R. S. Lowrey, and P. L. Wright, *Fed. Proc., Fed. Amer. Soc. Exp. Biol.* **23**, 873 (1964) (abstr.).
19. F. Bergel and R. C. Bray, *Biochem. Soc. Symp.* **15**, 64 (1958).
20. D. Bertrand, *C. R. Acad. Sci.* **208**, 2024 (1939).
21. P. R. Bird, *Proc. Aust. Soc. Anim. Prod.* **8**, 212 (1970).
22. H. Bortels, *Arch. Mikrobiol.* **1**, 333 (1930).
23. R. C. Bray, R. Petersson, and A. Ehrenberg, *Biochem. J.* **81**, 178 (1956).
24. J. W. Britton and H. Goss, *J. Amer. Vet. Med. Ass.* **108**, 176 (1946).
25. F. Brouwer, A. M. Frens, P. Reitsma, and C. Kalesvaart, *Versl. Landbouwk. Onderzoek.* **44C**, 267 (1938).
26. E. J. Butler and R. M. Barlow, *J. Comp. Pathol. Ther.* **73**, 208 (1963).
27. W. Buttner, *Arch. Oral Biol.* Suppl. 6, 40 (1961); *J. Dent. Res.* **42**, 453 (1963).
28. C. L. Comar, L. Singer, and G. K. Davis, *J. Biol. Chem.* **180**, 913 (1949).
29. I. J. Cunningham, *in* "Symposium on Copper Metabolism" (W. D. McElroy and B. Glass, eds.), p. 246 Johns Hopkins Press, Baltimore, Maryland, 1950.
30. I. J. Cunningham and K. G. Hogan, *N. Z. J. Agr. Res.* **1**, 841 (1958).
31. I. J. Cunningham and K. G. Hogan, *N. Z. J. Agr. Res.* **2**, 134 (1959).
32. I. J. Cunningham, K. G. Hogan, and B. M. Lawson, *N. Z. J. Agr. Res.* **2**, 145 (1959).
33. C. P. Czerniejewski, C. W. Shank, W. G. Bechtel, and W. B. Bradley, *Cereal Chem.* **41**, 65 (1964).
34. R. E. Davies, B. L. Reid, and J. R. Couch, *Fed. Proc., Fed. Amer. Soc. Exp. Biol.* **18**, 522 (1959).
35. R. E. Davies, B. L. Reid, A. A. Kurnich, and J. R. Couch, *J. Nutr.* **70**, 193 (1960).
36. G. K. Davis, *in* "Symposium on Copper Metabolism" (W. D. McElroy and B. Glass, eds.), p. 216. Johns Hopkins Press, Baltimore, 1950.
37. A. T. Dick, Doctoral Thesis, University of Melbourne, Australia, 1954.
38. A. T. Dick, *Aust. Vet. J.* **28**, 30 (1952); **29**, 18 and 233 (1953); **30**, 196 (1954).
39. A. T. Dick, *Aust. J. Agr. Res.* **5**, 511 (1954).
40. A. T. Dick, *in* "Inorganic Nitrogen Metabolism" (W. D. McElroy and B. Glass, eds.), p. 445. Johns Hopkins Press, Baltimore, Maryland, 1956.
41. A. T. Dick, *Soil Sci.* **81**, 229 (1956).
42. A. T. Dick and L. B. Bull, *Aust. Vet. J.* **21**, 70 (1945).
43. A. T. Dick, C. W. E. Moore, and J. B. Bingley, *Aust. J. Agr. Res.* **4**, 44 (1953).
44. R. P. Dowdy and G. Matrone, *J. Nutr.* **95**, 191 and 197 (1968); R. P. Dowdy, G. A. Kunz and H. E. Sauberlich, *ibid.* **99**, 491 (1969).
45. W. B. Dye and J. L. O'Hara, *Nev. Agr. Exp. Sta., Bull.* **208** (1959).
46. T. H. Easterfield, T. Rigg, H. O. Askew, and J. A. Bruce, *J. Agr. Sci.* **19**, 573 (1929).

47. W. C. Ellis and W. H. Pfander, *J. Anim. Sci.* **19**, 1260 (1960).
48. W. C. Ellis, W. H. Pfander, M. E. Muhrer, and E. E. Pickett, *J. Anim. Sci.* **17**, 180 (1958).
49. Y. Ericsson, *Acta Odontol. Scand.* **24**, 405 (1966).
50. L. T. Fairhall, R. D. Dunn, N. E. Sharpless, and E. A. Pritchard, *U.S. Pub. Health Serv., Bull.* **293** (1945).
51. B. F. Fell, C. F. Mills, and R. Boyne, *Res. Vet. Sci.* **6**, 170 (1965).
52. W. S. Ferguson, A. H. Lewis, and S. J. Watson, *Nature (London)* **141**, 553 (1938); *Jealott's Hill Bull. No. 1* (1940).
53. W. S. Ferguson, A. H. Lewis, and S. J. Watson, *J. Agr. Sci.* **33**, 44 (1943).
54. F. Goodman, *J. Dent. Res.* **44**, 565 (1965).
55. L. F. Gray and L. J. Daniel, *J. Nutr.* **53**, 43 (1954).
56. L. F. Gray and L. J. Daniel, *J. Nutr.* **84**, 31 (1964).
57. L. F. Gray and G. H. Ellis, *J. Nutr.* **40**, 441 (1950).
58. H. H. Green, *Proc. Spec. Conf. Agr. Aust., 1949* p. 293 (1951).
59. D. M. Hadjimarkos, *Advan. Oral Biol.* **3**, 253 (1968).
60. A. W. Halversen, J. H. Phifer, and K. J. Monty, *J. Nutr.* **71**, 95 (1960).
61. L. I. Hart, E. C. Owen, and R. Proudfoot, *Brit. J. Nutr.* **21**, 617 (1967).
62. W. B. Healy, L. C. Bate, and T. G. Ludwig, *N. Z. J. Agr. Res.* **7**, 603 (1964).
63. W. B. Healy and T. G. Ludwig, *41st Gen. Meet. Int. Ass. Dent. Res.* Abstr. No. 379 (1963).
64. W. B. Healy, T. G. Ludwig, and F. L. Losee, *Soil Sci.* **92**, 359 (1961).
65. E. S. Higgins, D. A. Richert, and W. W. Westerfeld, *J. Nutr.* **59**, 539 (1956).
66. K. G. Hogan and I. J. Cunningham, *N. Z. J. Agr. Res.* **7**, 174 (1964).
67. K. G. Hogan and A. J. Hutchinson, *N. Z. J. Agr. Res.* **8**, 625 (1965).
68. K. G. Hogan, D. F. L. Money, and A. Blayney, *N. Z. J. Agr. Res.* **11**, 435 (1968).
69. K. G. Hogan, D. R. Ris, and A. J. Hutchinson, *N. Z. J. Agr. Res.* **9**, 691 (1966).
70. M. A. Jeter and G. K. Davis, *J. Nutr.* **54**, 215 (1954).
71. F. Kiermeier and K. Capellari, *Biochem. Z.* **330**, 160 (1958).
72. M. Kirchgessner, *Z. Tierphysiol., Tierernaehr. Futtermittelk.* **14**, 270 and 278 (1959).
73. F. H. Kratzer, *Proc. Soc. Exp. Biol. Med.* **80**, 483 (1952).
74. B. J. Kruger, "The Effect of Trace Elements on Experimental Caries in the Rat." Univ. of Queensland Press, Brisbane, Australia, 1959.
75. J. Kubota, V. A. Lazar, G. H. Simonson, and W. W. Hill, *Soil Sci. Soc. Amer., Proc.* **31**, 667 (1967).
76. J. Kubota, V. A. Lazar, L. N. Langam, and K. C. Beeson, *Soil Sci. Soc. Amer., Proc.* **25**, 227 (1961).
77. J. Kubota, E. R. Lemon, and W. H. Allaway, *Soil Sci. Soc. Amer., Proc.* **27**, 679 (1963).
78. R. M. Leach, Jr. and L. C. Norris, *Poultry Sci.* **36**, 1136 (1957).
79. A. H. Lewis, *J. Agr. Sci.* **33**, 52 and 58 (1943).
80. T. G. Ludwig, W. B. Healy, and F. L. Losee, *Nature (London)* **186**, 695 (1960).
81. R. S. Malthus, T. G. Ludwig, and W. B. Healy, *N. Z. Dent. J.* **60**, 291 (1964).
82. N. A. Marcilese, C. B. Ammerman, R. M. Valsecchi, B. G. Dunavant, and G. K. Davis, *J. Nutr.* **99**, 177 (1969).
83. R. F. Miller and R. W. Engel, *Fed. Proc., Fed. Amer. Soc. Exp. Biol.* **19**, 666 (1960).
84. R. F. Miller, N. O. Price, and R. W. Engel, *J. Nutr.* **60**, 539 (1956).
85. R. F. Miller, N. O. Price, and R. W. Engel, *Fed. Proc., Fed. Amer. Soc. Exp. Biol.* **18**, 538 (1959).
86. C. F. Mills, *Proc. Nutr. Soc.* **19**, 162 (1960).

87. C. F. Mills, *Rowett Res. Inst. Collect. Pap.* **17**, 57 (1961).
88. C. F. Mills and B. F. Fell, *Nature (London)* **185**, 20 (1960).
89. K. J. Monty and E. M. Click, *J. Nutr.* **75**, 303 (1961).
90. A. T. Moore, *Nature (London)* **182**, 1175 (1958).
91. P. J. Mylrea, *Aust. J. Agr. Res.* **9**, 373 (1958).
92. Z. Nagy and E. Polyik, *Forgovosi Szemle* **48**, 154 (1954); quoted by Hadjimarkos (59).
93. M. Neenan, T. Walsh, and L. B. Moore, *Proc. 7th Int. Grassland Congr., 1956* Pap. 31A (1957).
94. J. B. Nielands, F. M. Strong, and C. A. Elvehjem, *J. Biol. Chem.* **172**, 431 (1948).
95. *Oreg., Agr. Exp. Sta., Misc. Pap.* No. 64 (1958).
96. C. A. Ostrom, C. W. Miller, and R. Van Reen, *J. Dent. Res.* **40**, 520 (1961).
97. E. C. Owen and R. Proudfoot, *Brit. J. Nutr.* **22**, 331 (1968).
98. B. L. Reid, A. A. Kurnich, R. L. Svacha, and J. R. Couch, *Proc. Soc. Exp. Biol. Med.* **93**, 245 (1956).
99. E. C. de Renzo, E. Kaleita, P. Heytler, J. J. Oleson, B. L. Hutchings, and J. H. Williams, *J. Amer. Chem. Soc.* **75**, 753 (1953); *Arch. Biochem. Biophys.* **45**, 247 (1953).
100. D. A. Richert and W. W. Westerfeld, *J. Biol. Chem.* **203**, 915 (1953).
101. B. Rosoff and H. Spencer, *Nature (London)* **20**, 410 (1964).
102. J. F. Scaife, *N. Z. J. Sci. Technol., Sect. A.* **38**, 285 and 293 (1963).
103. R. L. Shirley, M. A. Jeter, J. P. Feaster, J. T. McCall, J. C. Cutler, and G. K. Davis, *J. Nutr.* **54**, 59 (1954).
104. L. M. Siegel and K. J. Monty, *J. Nutr.* **74**, 167 (1961).
105. A. G. Spais, *Rec. Med. Vet.* **135**, 161 (1959).
106. A. G. Spais, *Ann. Med. Vet.* **3**, 131 (1962).
107. A. G. Spais, T. K. Lazaridis, and A. G. Agiannidis, *Res. Vet. Sci.* **9**, 337 (1968).
108. R. A. Steinberg, *J. Agr. Res.* **52**, 439 (1936).
109. H. Ter Meulen, *Nature (London)* **130**, 966 (1932).
110. J. W. Thomas and S. Moss, *J. Dairy Sci.* **34**, 929 (1951).
111. I. H. Tipton and M. J. Cook, *Health Phys.* **9**, 103 (1963).
112. I. H. Tipton, P. L. Stewart, and P. G. Martin, *Health Phys.* **12**, 1683 (1966).
113. R. Van Reen, *Arch. Biochem. Biophys.* **53**, 77 (1954).
114. R. Van Reen, *J. Nutr.* **68**, 243 (1959).
115. R. Van Reen and M. A. Williams, *Arch. Biochem. Biophys.* **63**, 1 (1956).
116. W. W. Westerfeld and D. A. Richert, *J. Nutr.* **51**, 85 (1953).
117. M. A. Williams and R. Van Reen, *Proc. Soc. Exp. Biol. Med.* **91**, 638 (1956).
118. K. N. Wynne and G. L. McClymont, *Aust. J. Agr. Res.* **7**, 45 (1956).

# 5

# COBALT

## I. Introduction

Cobalt was shown to occur regularly in plant and animal tissues more than 40 years ago (22a,45). Unusually high concentrations of cobalt were not observed in any organisms or parts of organisms, other than in *Archidoris tuberculata* (51) and in *Pleurobranchus*, a tectibranch mollusk (171). Much later, the leaves of the swamp black gum (*Nyssa sylvatica*) were shown to contain 60–250 times the level of cobalt found in native grasses growing under the same conditions (20).

None of the early investigations provided satisfactory evidence of a physiological role for cobalt in plants or animals. The pioneer studies of Bertrand and Nakamura (23) using purified diets with mice, and those of Stare and Elvehjem (156) with rats, were unsuccessful in demonstrating that cobalt is a dietary essential for those species. The first clear evidence that cobalt possesses biological activity came from the experiments of Waltner and Waltner (170) who reported, in 1929, that cobalt in large doses stimulates erythropoiesis in rats and induces polycythemia. This discovery, subsequently confirmed and extensively studied, is of historic significance because it generated interest in a possible physiological role for cobalt and was responsible for this element being given an early trial in "coast disease," the first of the naturally occurring diseases of sheep shown to respond to cobalt therapy. However, there is no evidence that the polycythemic action of cobalt is related to its normal functioning in any species.

The first conclusive evidence that cobalt is a dietary essential came in 1935 as an outcome of Australian researches into the cause of certain debilitating diseases of sheep and cattle, known locally as "coast disease" and "wasting disease" (104,167). This important discovery led to the delineation of cobalt deficiency areas in many parts of the world and to the development of practical means of prevention and control in the field. Studies of the distribution of cobalt in soils, plants, and animal tissues were undertaken, and the minimum dietary requirements of sheep and cattle for this element were determined.

The discovery, in 1948 (135,144), that the antipernicious anemia factor in liver is a compound containing 4% cobalt, later designated vitamin $B_{12}$, gave a further impetus to studies with cobalt in ruminant nutrition, as well as to investigations of the structure of the vitamin (77) and its functional activities in the animal body [see Smith (145)]. The presence of cobalt in vitamin $B_{12}$ suggested that this vitamin might be the functional form of cobalt in ruminant metabolism. Within 3 years of its original discovery as a cobalt compound, Smith and co-workers (150) demonstrated that injections of vitamin $B_{12}$ effected complete remission of all signs of cobalt deficiency in lambs. This was soon confirmed in other laboratories, and it became apparent that cobalt deficiency in ruminants is, in effect, a vitamin $B_{12}$ deficiency brought about by the inability of the rumen microorganisms to synthesize sufficient amounts of the vitamin to meet the needs of the host animal's tissues in the presence of inadequate dietary cobalt. A unique nutritional situation was disclosed—a situation in which animals appear to utilize a trace element solely as an integral part of a vitamin and to be completely dependent upon the symbiotic activities of their own microorganisms for their supply of this vitamin.

Attention was then focused upon identifying the basic biochemical defects arising in the vitamin $B_{12}$-deficient ruminant and relating these to the clinical and pathological manifestations of the deficiency. An explanation for the much higher cobalt requirements of ruminants than of nonruminants was sought, and improved procedures for the diagnosis and control of cobalt-vitamin $B_{12}$ deficiency under practical conditions were developed.

Efforts were also made during this period to produce cobalt deficiency per se in animals other than ruminants. So far these efforts have been unsuccessful, although some responses to cobalt supplements in monogastric species have been reported. Evidence was also obtained that cobalt plays an essential role in nitrogen fixation by the root nodules of legumes (68,133), and significant growth responses to cobalt were demonstrated in pasture legumes in certain areas (123,130).

## II. Cobalt in Animal Tissues and Fluids

Cobalt is widely distributed throughout the body, without excessive accumulation in any particular organ or tissue. The highest concentrations generally occur in the liver, kidneys, and bones. Distribution of retained cobalt in this manner has been demonstrated in mice, rats, rabbits, pigs, sheep, cattle, dogs and chicks, with relatively small differences among species (24,31,92,139). The cobalt concentrations reported for normal human tissues conform well with those of other species (163,174), with the exception of those of Butt *et al.* (26) and of Leddicotte (91), which are appreciably higher. There is no evidence that cobalt accumulates in human tissues with age (140,163). The total content in the body of a normal 70-kg man has been reported to average 1.1 mg Co (91). Representative values for human and ovine tissues are given in Table 21.

It is apparent from the figures given for sheep in Table 21 that the concentrations of cobalt in the tissues are reduced below normal in cobalt deficiency (15). They can also be increased above normal by cobalt injections or dietary supplements. In one study of normal newborn calves (162) the cobalt content of the tissues ranged from 0.10 to 0.22 ppm on dry basis, except for the skin and hair which contained the exceptionally high level of 1.0 ppm. Newborn calves, from cows that had received a dietary supplement of cobalt for 21 to 120 days before calving, carried substantially higher cobalt concentrations in their tissues, especially in the liver and kidneys, where the increases were of the order of 50%. Little is yet known of the forms in which cobalt occurs in the tissues, other than as vitamin $B_{12}$. The existence of other bound forms of

TABLE 21

*Cobalt Concentrations in Tissues*[a]

| Species | Liver | Spleen | Kidney | Heart | Pancreas |
|---|---|---|---|---|---|
| Normal human[b] | 0.18 | 0.09 | 0.23 | 0.10 | 0.06 |
| Healthy sheep[c] | 0.15 | 0.09 | 0.25 | 0.06 | 0.11 |
| Co-deficient sheep[c] | 0.02 | 0.03 | 0.05 | 0.01 | 0.02 |

[a]Concentration expressed in parts per million cobalt on dry basis.
[b]From Tipton and Cook (163).
[c]From Askew and Watson (15).

this metal has been demonstrated in the tissues of sheep (116), in the blood plasma of the dog, and in the intestinal wall of the chicken (92).

The concentration of cobalt in the liver has received special attention because of its possible value in diagnosing cobalt deficiency in the field. Thus in two separate Australian studies the mean liver cobalt levels of groups of Co-deficient sheep were 0.06 and 0.09 ppm on dry basis, compared with 0.28 and 0.34 ppm for healthy sheep (109,168). Similar values for the cobalt content of the livers of cobalt-deficient and healthy cattle have been reported from three different countries (34,60,113). From an extensive New Zealand study, McNaught (113) suggests that 0.04-0.06 ppm Co, or less, in the livers of sheep and cattle indicate cobalt deficiency and that 0.08-0.12 ppm, or more, indicate a satisfactory cobalt status. This worker showed further that cobalt, unlike copper and iron, does not normally accumulate in the fetal liver. On the other hand, the cobalt (and vitamin $B_{12}$) content of the liver of the newborn lamb and calf is reduced below normal when the mother has been on a Co-deficient diet and can be raised to normal levels by prepartum cobalt administration (120).

Liver cobalt concentrations can be increased to 10 or more times normal levels by cobalt injections or massive oral doses, without this store being available to the animal for vitamin $B_{12}$ synthesis (6,-19,107,127,132). Furthermore, ruminants on Co-deficient rations can be maintained in health by vitamin $B_{12}$ injections without raising their liver cobalt levels to normal. Freedom from signs of cobalt deficiency is, therefore, compatible with a relatively low liver cobalt level, just as a deficiency can occur in the presence of a high liver cobalt level. For these reasons, liver cobalt concentration is not a reliable criterion of the cobalt-vitamin $B_{12}$ status of ruminants. The proportion of the liver cobalt that is present in the form of vitamin $B_{12}$ varies with the cobalt status of the animal. In cobalt sufficiency, virtually all the liver cobalt exists as vitamin $B_{12}$, whereas in cobalt deficiency, only about one-fifth of the liver cobalt exists in this form (9). This indicates that there is a greater reduction of vitamin $B_{12}$-cobalt than of nonvitamin $B_{12}$-cobalt. Conversely, oral dosing of Co-deficient lambs can increase liver vitamin $B_{12}$ levels to normal but the increases in total liver cobalt tend to be even greater.

Reported values for the cobalt content of blood are so variable that they almost certainly reflect methodological inaccuracies. Thus for normal human blood the following ranges have appeared: 0.007-0.036 (mean 0.018) (161); 0.17-1.5 (173); 0.35-6.3 (mean 4.3) $\mu$g Co/100 ml (88). The concentration of cobalt is appreciably higher in the red cells than in the plasma (75,88,161). The proportion of the blood cobalt that exists in the form of vitamin $B_{12}$ cannot be estimated with any confidence until more reliable total blood cobalt values are available. Calculations based on present data suggest that it is quite low.

Normal cow's milk is very low in cobalt, with concentrations ranging from 0.4 to 1.1 $\mu$g/liter and with a mean close to 0.5 $\mu$g Co/liter (13,-46,85,124). Cow's colostrum is 4-10 times higher in cobalt than later milk (85,93) and can be further increased by prepartum cobalt supplementation of the cow's ration (162). Supplementing normal rations with liberal amounts of cobalt salts can also increase the level of cobalt in the milk. Such treatments are effective means of increasing the cobalt intakes of suckling lambs and calves in deficient areas, but this cobalt cannot be used until the young animal has developed a rumen population able to incorporate it into vitamin $B_{12}$. Cobalt supplementation of Co-deficient rations can also significantly increase the vitamin $B_{12}$ content of milk and particularly of colostrum (71,72,115,120). Supplementation of rations already adequate in cobalt is ineffective in raising the vitamin $B_{12}$ activity of the milk above the normal range (73).

## III. Cobalt Metabolism

Several studies with farm and laboratory animals have shown that dietary cobalt, or cobalt administered orally as soluble salts, is poorly absorbed and is excreted mainly in the feces. Comar and co-workers (32) gave radioactive cobalt in a single physiological dose, orally and by injection, to several species of animals with very similar results. After oral administration to rats, 80% of the radioactivity was found in the feces, 10% in the urine, with little in tissues other than the liver. With cattle, 80% was also found in the feces but only 0.5% in the urine, indicating a greater initial retention and slower loss from the tissues. A proportion of the fecal cobalt represents absorbed cobalt that has been re-excreted via the bile and to a small extent through the intestinal wall. The pancreatic juice is not a significant route of excretion of cobalt (117).

Injected cobalt is excreted principally in the urine, with a small fraction present in the bile to become part of the fecal cobalt. Thus in the experiments of Comar (32), about 65% of injected radiocobalt appeared rapidly in the urine, with 7 and 30% in the feces in different experiments. In dogs and chicks (92), injected radiocobalt rapidly reaches an equilibrium, with a standard and uniform partition among the body fluids, tissues, and intestinal contents and with little accumulation in any part of the body. In ruminants the time required to reach "uniform labeling" is longer following injection (31). Following oral administration, $^{60}Co$ is more slowly and less completely absorbed, indicating some immobilization in the digestive tract, presumably due to "binding" of cobalt as vitamin $B_{12}$ and related compounds by the rumen microorganisms (32,139). A similar binding of cobalt by the bacteria of the cecum apparently occurs in chicks (92). In ruminants, cobalt can accumulate in the liver to give concentrations many times higher than normal, as a consequence of injections (19,107) or of massive oral doses (6). Preferential retention in the liver, although not to the extent that occurs in sheep, has also been demonstrated in pigs given radioactive cobalt administered with the food over a period of 6 weeks (24).

In man, the pattern of absorption and excretion of cobalt appears to differ markedly from that of other mammals, and is characterized by high absorption and predominant excretion in the urine. This unusual pattern for a transitional cation such as cobalt is apparent from three separate balance studies with human subjects employing different methods of analysis and at very different estimated levels of dietary cobalt intake and excretion (70,140,164). Harp and Scoular (70), who detected only 6.6-7.9 μg cobalt in the daily diets of women, calculated that 73-97% was absorbed, with a mean of 67% appearing in the urine and less than 15% in the feces. Tipton and co-workers (164), who measured mean daily intakes of 160 and 170 μg Co, found mean daily urinary excretions of 140 and 190 μg Co and fecal losses of 40 and 60 μg. On the other hand, Engel *et al.* (46a), in balance studies with preadolescent girls, found 90% of the excreted cobalt in the feces and only 10% in the urine.

## IV. Cobalt in Ruminant Nutrition

### 1. *The Discovery of the Need for Cobalt*

Restricted areas in several parts of the world were long known to be unsatisfactory for the raising of sheep and cattle, in spite of apparently satisfactory pastures. Horses and other nonruminants thrived in these areas, whereas sheep and cattle became weak and emaciated, progressively anemic, and usually died. Various local names were given to these maladies, such as "pining," "vinquish," "salt-sick," "nakuruitis," "bush-sickness," "wasting disease," and "coast disease." The condition was essentially similar in each locality and could only be controlled by periodic removal of the animals to healthy areas for varying periods.

The first serious scientific work on any of these diseases began in New Zealand at the end of the 19th century. A series of investigations led by Aston (16) culminated in the claim that bush-sickness in cattle was due to a deficiency of iron. This was based upon the occurrence of anemia in affected animals, the low iron content of bush-sick pastures and soils compared with those of healthy areas, and upon the effectiveness of crude iron salts and ores in preventing or curing the disease. Within a few years Aston's iron deficiency theory received support from several parts of the world where iron compounds were also found effective in preventing or curing similar maladies (42,64,121,136). Subsequent work in New Zealand revealing insignificant differences in the iron content of healthy and bush-sick pastures (136) and little correlation between the iron content of different compounds and their curative effects (66) cast some doubt on the iron deficiency hypothesis. Indisputable evidence against this hypothesis was obtained by Filmer and Underwood (47,49,165,167) in their investigation of a wasting disease of sheep and cattle (enzootic marasmus) occurring in the south coastal areas of Western Australia. These workers became suspicious of the very large doses of iron compounds required to cure the disease. They could find little relation between the size of an effective dose and the amount of iron it supplied. Furthermore, they discovered that the liver and spleen of affected animals contained large stores of iron and that whole liver was curative in doses that supplied insignificant amounts of iron. Even more convincing was their finding that an *iron-free* extract of one of the curative compounds (limonite, $Fe_2O_3 \cdot H_2O$) was just as potent as whole limonite.

The hypothesis was advanced that enzootic marasmus was due to a deficiency in the soils and herbage of some trace element which occurred as a contaminant of the iron compounds successfully used. Underwood and Filmer (167) then chemically fractionated limonite and, after some misleading tests with nickel suggested by the large amounts of that element present, found that the potency of the limonite resided in the cobalt that it contained. Normal growth and health of sheep and cattle on the deficient pastures were then secured by the administration of small oral doses of a cobalt salt, and the soils and pastures and the livers of affected animals were shown to contain subnormal cobalt concentrations (168).

While the above investigations were proceeding, studies of coast disease of sheep occurring on the calcareous sandy dunes of South Australia were under way. The possibility that this disease was due to a deficiency of a mineral element was early recognized. Supplements of phosphorus and copper proved ineffective and the relatively small doses of the iron compounds used produced only a transitory improvement in the condition of "coasty" sheep (110). A mineral mixture supplying small amounts of iron, copper, boron, manganese, cobalt, nickel, zinc, arsenic, bromine, fluorine, and aluminum, on the other hand, was highly effective (110). The fact that coast disease is accompanied by anemia, coupled with Waltner and Waltner's earlier finding (170) that cobalt stimulates hemopoiesis in rats, led to the suggestion that cobalt might be the element responsible for the beneficial effects of the mineral mixture. Experiments by Marston and by Lines (104) revealed a dramatic improvement in the condition of coasty sheep from the oral administration of 1 mg cobalt/Day. Subsequent experiments disclosed that supplementation with copper, as well as with cobalt, was necessary for completely successful treatment of coast disease.

Within a few years of these discoveries, cobalt supplements were found to be equally effective in the cure and prevention of all the diseases previously shown to respond to massive doses of iron compounds, and the soils and herbage of the affected areas were shown to contain subnormal levels of cobalt [see Russell and Duncan (139a)]. Subsequently it became apparent that larger areas existed in many countries in which the deficiency was less severe.

### 2. *Manifestations of Cobalt Deficiency*

The appearance of a severely cobalt-deficient animal is one of extreme emaciation and listlessness, indistinguishable from that of a starved animal, except that the visible mucous membranes are blanched and the skin is pale and fragile. The emaciation or wasting of the musculature (marasmus) results from the failure of appetite which is an early and conspicuous feature of the disease, and the paleness of the skin and mucous membranes from the anemia which usually develops progressively with the severity of the deficiency.

The cobalt deficiency syndrome varies from this acute and fatal condition, through a series of less acute stages, to a mild, ill-defined and often transient state of unthriftiness that is difficult to diagnose. When ruminants are confined to Co-deficient pastures or are fed Co-deficient rations, there is a characteristic response. At first they thrive and grow normally, for a period of several weeks or months, depending upon their age, previous history, and the degree of deficiency of the diet. During this time they are drawing upon the vitamin $B_{12}$ reserves in the liver and other tissues. This period is followed by a gradual loss of appetite and failure of growth or loss of body weight, succeeded by extreme inappetence, rapid wasting, and anemia, culminating in death (47,110). In some areas the cobalt deficiency is less severe so that the acute disease conditions do not occur. Milder manifestations, characterized by unthriftiness in young stock and in mature stock also by suppression or diminution of lactation and birth of weak lambs and calves that do not long survive, are the only evidence of cobalt deficiency in such areas. In these mild or marginal areas the unthriftiness can be apparent in some years and absent in others (95).

At autopsy, the body of severely affected animals presents a picture of extreme emaciation, often with a total absence of body fat. The liver is fatty, the spleen hemosiderized, and in some animals there is hypoplasia of the erythrogenic tissue in the bone marrow (47). The red cell numbers and blood hemoglobin levels are always subnormal and sometimes very low. The anemia was reported by Filmer (47) to be normocytic and hypochromic in lambs and microcytic and hypochromic in calves. The nature of the anemia in Co-deficient calves does not appear to have been reinvestigated. More recent studies have characterized the anemia in lambs as normocytic and normochromic (59,149). However, it is not the anemia that is responsible for the main signs of cobalt deficiency. Inappetence and marasmus invariably precede any considerable degree of anemia. The first discernible response to cobalt feeding, or vitamin $B_{12}$ injections, is a rapid improvement in appetite and body weight. Improvement in the blood picture may be equally dramatic but is sometimes delayed. On occasions there is a small temporary *fall* in hemoglobin and red cell numbers accompanying the initial improvement in appetite following cobalt treatment.

These clinical and pathological manifestations of cobalt deficiency in ruminants are accompanied by biochemical changes in the tissues, many of which have only recently been demonstrated. The decline in cobalt concentrations in the tissues, particularly of the liver and kidney, has already been mentioned (Table 21). The level of cobalt in the rumen fluid also falls, as would be expected from the subnormal dietary intake of the element. When the level of cobalt in the rumen liquor has fallen below a critical level, tentatively set at about 20 $\mu\mu g/m$ (106), vitamin $B_{12}$ synthesis by the rumen microorganisms is inhibited and the levels of the vitamin decline in the rumen, the blood, the liver, and other tissues (9,38,78,84) (see Table 22). The tissue vitamin $B_{12}$ depletion is accompanied by a marked depression of appetite and by a metabolic inefficiency (146). The latter was reflected in a 30% faster loss of body weight, a higher excretion of fecal nitrogen, and a higher fasting energy expenditure than was observed in pair-fed sheep injected with vitamin $B_{12}$. On the other hand, retention of combustible energy from the diet by vitamin $B_{12}$-deficient animals was not significantly different from that of pair-fed animals treated with cobalt or vitamin $B_{12}$ (146).

TABLE 22
*Vitamin $B_{12}$ Activity of Blood, Liver and Rumen Ingesta of Sheep*[a]

| Sample | Co-deficient | | Co-sufficient (full-fed) | | Co-sufficient (limited fed) | |
|---|---|---|---|---|---|---|
| | No. | Mean±SD | No. | Mean±SD | No. | Mean±SD |
| Whole blood (mg/ml) | 16 | 0.47±0.11 | 6 | 2.3 ±0.6 | 3 | 4.3 ±1.5 |
| Liver (μg/g wet wt) | 9 | 0.05±0.01 | 6 | 0.93±0.26 | 3 | 1.24±0.20 |
| Rumen ingesta (μg/g dry wt) | 4 | 0.09±0.06 | 5 | 1.3 ±0.4 | 3 | 1.3 ±0.9 |

[a]From Hoekstra, Pope and Phillips (78).

### 3. *Cobalt and Vitamin $B_{12}$ Requirements*

Under grazing conditions, lambs are the most sensitive to cobalt deficiency, followed by mature sheep, calves, and mature cattle in that order (4). Field experience suggests that species differences among ruminants in cobalt requirements are small. Early evidence from Australia (49,105) and New Zealand (112) indicated that 0.07 or 0.08 ppm Co in the dry diet was just adequate for sheep and cattle. This level of dietary cobalt, therefore, became accepted as the minimum requirement for these species. Later studies placed the minimum level of "pasture-associated" cobalt required by growing lambs appreciably higher, namely 0.11 ppm on dry basis (11). This is in accord with the minimum level suggested for Scottish conditions (156a). In a later study of a marginally cobalt-deficient area in New Zealand, Andrews (5) assessed the position as follows: "Mean (pasture) values of 0.11 ppm Co or more would probably exclude the likelihood of cobalt deficiency. Mean values approaching 0.08 ppm would suggest but not prove actual or potential existence of the disease." More precise estimates of minimum cobalt requirements, applicable under all grazing conditions, are difficult because of the influence of many variables such as seasonal changes in herbage cobalt concentrations, selective grazing habits, and soil contamination. However, Lee and Marston (98), in a recently published critical study, provide evidence that for sheep grazing grossly Co-deficient pastures, the total intake of cobalt to ensure optimum growth and hemoglobin production is 0.08 mg/day when supplementary cobalt is given 3 times each week. In growing lambs the requirement was stated to be higher.

The results of pen-feeding experiments with sheep consuming purified diets, carried out by Somers and Gawthorne (153), point to a minimum cobalt requirement closer to the earlier level quoted above, namely 0.07-0.08 ppm, than to a level of 0.11 ppm. These figures were obtained with sheep that were 18 months of age at the beginning of the experiment, designed for other purposes, which lasted only 39 weeks. Clinical signs of Co deficiency would almost certainly have appeared later in the sheep consuming 0.06 ppm Co, since their serum vitamin $B_{12}$ concentrations had already fallen to a critically low level. It should be emphasized that natural rations high in cereal grains do not necessarily supply adequate cobalt. Growth responses to supplementary cobalt have, in fact, been demonstrated in steers fed a fattening ration based on barley grain (131) and on sorghum grain and silage (119).

The requirements of ruminants for parenterally administered or absorbed vitamin $B_{12}$ are of the same order as those found for the rat, the pig, and the chick but are higher than those of man [see Smith and Loosli (151)]. A recent careful assessment places the minimum total requirement of sheep for parenteral or absorbed cobalamin at $11 \pm 2$ $\mu$g/day (146). Injections of 150 $\mu$g once every 2 weeks were found to be adequate for lambs fed a cobalt-deficient diet (149), and lambs grazing a Co-deficient pasture were found to grow as well with injections of 100 $\mu$g of vitamin $B_{12}$ at weekly intervals as they did with ample oral cobalt. By contrast, the oral requirements of ruminants for vitamin $B_{12}$ are substantially higher than those of other species. Dairy calves have been reported to require between 20 and 40 $\mu$g vitamin $B_{12}$ per kilogram dry matter consumed (90). Andrews and Anderson (7) fed crystalline vitamin $B_{12}$ to lambs grazing a Co-deficient pasture at the rate of 1000 $\mu$g/week for 16 weeks. The growth response was much smaller than that obtained from either oral cobalt or injected vitamin $B_{12}$. Extrapolation from these experiments suggests an oral requirement for growing lambs of some 200 $\mu$g/day, which is about 10 times the reported oral requirement of other species per unit of food intake. The likely reasons for this high requirement are discussed below when considering the mode of action of cobalt in ruminants.

### 4. *Diagnosis of Cobalt-Vitamin $B_{12}$ Deficiency*

The milder forms of cobalt deficiency in ruminants are impossible to diagnose with certainty on the basis of clinical and pathological observations alone. The only evidence of the deficiency is a state of unthriftiness and there is usually no sign of anemia. A secure diagnosis of cobalt deficiency can only be achieved in these circumstances by measuring the response in temperament, appetite, and live weight that follows cobalt feeding or vitamin $B_{12}$ injections. However, if the ration or grazing consistently contains less than 0.08 ppm Co, cobalt deficiency can be predicted with confidence.

The concentration of cobalt in the livers of sheep and cattle is sufficiently responsive to changes in cobalt intake to have some value in the diagnosis of cobalt deficiency. For the reasons given earlier, liver vitamin $B_{12}$ concentration is a more sensitive and reliable criterion than liver cobalt concentration. The criteria for sheep, shown in the following tabulation, were suggested by Andrews *et al.* (9) and can tentatively be applied to cattle:

| Condition of animal | Vitamin $B_{12}$ concentration ($\mu$g/g fresh liver) |
|---|---|
| Severe cobalt deficiency | $<0.07$ |
| Moderate cobalt deficiency | 0.07–0.10 |
| Mild cobalt deficiency | 0.11–0.19 |
| Cobalt sufficiency | $>0.19$ |

The diagnostic value of liver vitamin $B_{12}$ levels, and still more of kidney vitamin $B_{12}$ levels, may be reduced if the cobalt deficiency is coexistent with other diseases or conditions resulting in loss of appetite (8). From an examination of numerous controlled trials with sheep, Andrews (5) emphasized the limitations of liver vitamin $B_{12}$ assays as an aid to diagnosis of mild or intermittent forms of cobalt deficiency, but stated that "values of 0.10 $\mu$g/g or less for livers from individual sheep can be accepted as clearly diagnostic of cobalt deficiency disease.".

Serum vitamin $B_{12}$ assays have obvious advantages over liver or kidney determinations. With the onset of cobalt deficiency, serum vitamin $B_{12}$ levels fall markedly and the levels can be related to the amount of cobalt ingested (10, 38, 78, 84, 153). Dawbarn *et al* (38) concluded from their pen-feeding experiments with growing sheep that "signs of cobalt deficiency may be expected to supervene when the mean vitamin $B_{12}$ activity in the plasma falls to about 0.2 m$\mu$g/ml." Andrews and Stephenson (10) reached similar conclusions from experiments with grazing ewes and lambs, although considerable individual variability was observed. Incipient stages of cobalt deficiency were associated with mean serum vitamin $B_{12}$ values of 0.26 m$\mu$g/ml for ewes and 0.30 m$\mu$g/ml for lambs. When the cobalt deficiency had become marked in ewes and acute in lambs, mean serum $B_{12}$ concentrations fell in both cases to levels less than 0.20 m$\mu$g/ml. Similar conclusions were reached by Somers and Gawthorne (153), although they showed that some sheep can maintain plasma vitamin $B_{12}$ concentrations of 0.2 m$\mu$g/ml, or less, for some weeks without the appearance of any clinical signs of deficiency.

### 5. *Prevention and Treatment of Cobalt Deficiency*

Cobalt deficiency in ruminants can be controlled either by treatment of the pastures with cobalt-containing fertilizers, by direct oral administration of cobalt to the animal, or by injections of vitamin $B_{12}$. The method of choice depends upon a number of factors, discussed below, but treatment with vitamin $B_{12}$ is of little practical significance for reasons of cost and convenience. The doses required were considered in the previous section. Vitamin $B_{12(b)}$ (hydroxycobalamin) is as effective as vitamin $B_{12}$ itself (cyanocobalamin) in correcting cobalt deficiency in lambs when injected at the rate of 100-150 $\mu$g/week (87), but folinic acid at levels of 71 $\mu$g, 5 mg, or 15 mg daily produces no such response (84). Large oral doses of dried or fresh whole liver were also shown to be highly curative (47), a finding which led Filmer and Underwood (49) as early as 1937 to suggest prophetically that "the potency of liver may be due to the presence of a stored factor and that cobalt may function through the production of this factor within the body." Subsequent attempts to cure cobalt deficiency with liver were unsuccessful (84, 108), presumably because of differences in either the vitamin $B_{12}$ or the cobalt contents of the liver preparations employed.

The most economical and widely practiced means of ensuring continuous and adequate supplies of cobalt to sheep and cattle grazing cobalt-deficient areas is by treatment or "topdressing" of the pastures with fertilizers to which a small proportion of cobalt salts or ores has been added. In this way the concentration of cobalt in the herbage can be raised for extended periods to levels adequate for the requirements of the animal. In some areas improvement in the growth of the legume component of the pasture can also be achieved by such treatment (123, 130) but this is not common. As little as 100-150 g of cobalt sulfate per acre, applied annually or biennially, suffices on most deficient pastures (14, 138). A single dressing of 500 to 600 g of cobalt sulfate per acre has maintained satisfactory pasture cobalt levels for 3 to 6 years (3). The most efficient quantity and frequency of topdressing with cobalt depends upon the degree of deficiency and the soil type and has to be determined for each area. Cobalt uptake by plants on highly calcareous soils, or on soils high in manganese which fixes the cobalt in unavailable forms (1), can be so low as to render treatment with "cobaltized" fertilizers inefficient or ineffective. Other procedures must therefore be adopted in such areas.

Direct administration of cobalt to stall-fed animals is usually achieved by incorporating cobalt oxide or salts into the mineral mixtures normally fed as supplements. The inclusion of cobalt in such mixtures is now common, even where there is no clear evidence that the unsupplemented rations are cobalt-deficient. Cobalt may also be supplied successfully to either stall-fed or grazing sheep and cattle in the form of cobalt-containing salt "licks," but variable consumption of the lick and, therefore, of cobalt is a disability inherent in this form of treatment. Oral dosing or "drenching" of animals with solutions of cobalt salts is practiced in many areas, and can be completely successful if the dosing is frequent enough.

Dosing sheep with 2 mg Co twice weekly or 7 mg Co weekly (94) is completely adequate, even under acutely cobalt-deficient conditions. Since individual drenching is a tedious and labor-demanding operation the effect of larger doses at longer intervals has been investigated. Lee (94) found that 35 mg cobalt administered once in 5 weeks "merely delays the onset of symptoms." Filmer (48) observed that monthly doses of 140 mg cobalt kept ewes and lambs alive on Co-deficient land, but the sheep did not thrive as well as those treated twice weekly or weekly with the usual small doses. Stewart *et al.* (157) found that monthly doses of 250 mg cobalt to lambs grazing Co-deficient hill pastures in Scotland "is a most useful method of preventing the condition of cobalt deficiency and producing healthy lambs." However, the dosed sheep gained appreciably less than did others grazing cobalt-topdressed pastures. This problem was later reexamined by Andrews and co-workers (12) in New Zealand. Monthly doses of 300 mg Co, either as soluble cobalt sulfate or insoluble cobalt oxide, were given to lambs grazing Co-deficient pastures for 5 months. Their growth rates and vitamin $B_{12}$ levels in blood and liver were compared with those of untreated lambs and of lambs dosed with 7 mg Co weekly. The two forms of cobalt were equally effective but neither of the monthly treatments prevented cobalt deficiency entirely. Both permitted growth at suboptimal rates and greatly reduced mortality but were less effective than the weekly dosing. Lee and Marston (98) showed that the equivalent of 1 mg Co/day is as effective for 2 to 3 years when given weekly as when given more frequently. Over longer periods, the body weight response is slightly but significantly in favor of the more frequent administration.

The necessity for regular and frequent dosing arises from the fact that cobalt, unlike iron and copper, is not readily stored in the liver or elsewhere in the body. The cobalt which is present in the tissues does not easily pass into the rumenoreticular region of the digestive tract where it is needed for vitamin $B_{12}$ synthesis. It is for this latter reason that injections of cobalt are largely ineffective. Where large amounts are injected some improvement in the condition of Co-deficient sheep has been observed (84, 132), presumably as a consequence of small amounts of cobalt reaching the rumen in the saliva or via the rumen wall. Studies with radioactive cobalt indicate that the element only reaches the rumen, reticulum, and omasum when large amounts are injected and even then only in very small quantities (32). Cobalt injections are capable of increasing the vitamin $B_{12}$ content of the cecum and large intestine of Co-deficient lambs, probably as a result of bacterial synthesis following the excretion of cobalt into the duodenum via the bile. The vitamin so produced is apparently not absorbed since blood and body tissue vitamin $B_{12}$ levels are not improved (84). Phillipson and Mitchell (127) placed cobalt salts directly into the abomasum or duodenum, by means of appropriate fistulas, and found them to be much less effective than cobalt given by mouth, despite evidence of some movement of cobalt into the rumen. All the above findings combine to show clearly that any treatment with cobalt, if it is to be fully successful, must be capable of maintaining adequate cobalt concentrations in the rumen fluid.

Many of the disadvantages of other forms of treatment can be avoided by the use of cobalt pellets or "bullets," first devised by Dewey and coworkers (40) in 1958. The small dense pellets composed of cobalt oxide and finely divided iron (Spec. Grav. 5.0), when delivered into the esophagus with a balling gun, lodge in the reticulorumen where they usually remain to yield a steady supply of cobalt to the rumen liquor. The usefulness of cobalt pellet therapy in the prevention of cobalt deficiency has been established in sheep and cattle (8, 9, 41, 134, 143). Two problems have arisen with this form of treatment—rejection of the pellets by regurgitation in a proportion of animals, and surface coating of the pellets with calcium phosphate so that the rate of cobalt release into the rumen is reduced. Millar and Andrews (114) prepared radioactive cobaltic oxide pellets so that they could be detected in live sheep and followed their retention in grazing ewes and lambs over a 16-month period. By the end of the period about one-third of the ewes and the lambs had rejected their pellets and a proportion of those retained were coated with calcium phosphate. Very similar results were obtained by Poole and Connolly (128) in Ireland. Furthermore, both groups of workers reported that the administration of two pellets or one pellet plus a 0.5 in. grub-screw grinder was ineffective in reducing the calcium phosphate coating. This is contrary to Australian experience in which abrasion between the two objects has been quite successful in keeping the pellet surface clean and maintaining a steady supply of cobalt to the rumen of sheep for more than 5 years (41).

A cobaltic oxide pellet has been designed and tested by Connolly and Poole (33) which is claimed to eliminate these drawbacks. The pellet consists of a $1.75 \times 0.5$-in. steel rod covered with layers of cotton gauze bandage impregnated with 5 g cobaltic oxide. The cobalt is released during the slow digestion of the cotton by the microorganisms of the rumen. In an experiment with lambs lasting 73 days, none of the new type pellets was lost, there was no evidence of coating, and live-weight gains and serum vitamin $B_{12}$ levels were satisfactory.

## 6. *Mode of Action of Cobalt*

When the concentration of cobalt in the bacteria-free rumen fluid falls below a critical level, placed at 20 $\mu\mu$g/ml (106), the rate of vitamin $B_{12}$ synthesis by the rumen organisms is reduced below the animal's needs; when the stores of vitamin $B_{12}$ laid down mainly in the liver become depleted below a critical level close to 0.10 $\mu$g/g wet weight (5) and when the plasma vitamin $B_{12}$ levels fall below a critical level of 0.2 m$\mu$g/ml, as a result of inadequate production and absorption and of inability to obtain supplies from body stores, characteristic signs of cobalt deficiency begin to appear in most lambs and calves. The length of time taken for clinical signs of deficiency to become apparent depends primarily upon the degree of dietary cobalt deficiency, the magnitude of the vitamin $B_{12}$ body stores previously built up, and the age of the animal. There is no evidence that vitamin $B_{12}$ synthesis is possible within the body tissues (84) or that the vitamin $B_{12}$ synthesized by the bacteria of the cecum and large intestine can be absorbed. The ruminant is, therefore, ultimately dependent upon the synthetic capacity of its rumen organisms.

In Co-deficient lambs the total concentrations of rumen bacteria and the principal types of these organisms are reduced below normal (52). No direct evidence has been obtained implicating particular organisms in vitamin $B_{12}$ synthesis in the rumen nor demonstrating a preferential diminution in the numbers of any such organisms as a consequence of lack of dietary cobalt. However, Dryden *et al.* (44) established that several pure strains of rumen bacteria each produced a mixture of vitamin $B_{12}$ and its analogs when cultured in a Co-adequate medium. Ample *in vivo* evidence has been obtained that the rumen organisms normally produce many Co-containing vitamin $B_{12}$-like compounds, with no physiological activity in the body tissues, in addition to true vitamin $B_{12}$ (50, 58, 76, 129).

Comparative assays with *Escherichia coli* or *Lactobacillus leichmannii* and with *Ochromonas mahlemensis,* which responds only to vitamin $B_{12}$ itself, have shown that a considerable proportion of the total activity of the rumen contents, as measured by the former assay organisms, is normally contributed by the inactive analogs. Under cobalt deficiency conditions, both the total and the true vitamin $B_{12}$ activity fall markedly, the former more than the latter (76). These findings have been confirmed and extended by Gawthorne (58) who developed a bioautographic ionophoretic technique capable of estimating millimicrogram amounts of the individual cobamides and cobinamides in rumen contents (57). Using this technique this worker found the total concentrations of these compounds to fall significantly, and the proportion present as true vitamin $B_{12}$ (5,6-dimethylbenzimidazolyl cobamide cyanide or DMBC) to rise significantly as dietary cobalt was reduced from 0.34 to 0.04 ppm. The proportion of the total due to the physiologically inactive 2-methyladenyl cobamide cyanide (2MAC) remained unaltered at about 44%. These two compounds together constituted over 80% of the total concentration, irrespective of cobalt intake and, therefore, of the magnitude of the total concentration of cobamides and cobinamides. It was shown further that the total vitamin $B_{12}$ activity of the blood plasma of sheep can be almost entirely (99%) accounted for as DMBC even when the rumen concentrations of its analogs are high. From these data it appears that the Co-deficient sheep converts at least 60% of its limited supply of dietary cobalt into compounds that it cannot absorb and cannot use. At higher cobalt intakes, a still larger proportion is so converted. In the experiments of Smith and Marston (146), an even more inefficient production of cobalamin from cobalt was observed. This was assessed at 13±5% under cobalt deficiency and at about 3% under cobalt sufficiency conditions.

The above findings provide a partial explanation of the higher cobalt requirement of the ruminant than of the nonruminant animal. A further factor contributing to this high cobalt requirement is the limited ability of the ruminant to absorb the vitamin $B_{12}$ that is produced. This is apparent from a study with sheep using vitamin $B_{12}$ containing $^{60}Co$ (125) and can be deduced from the studies of Kercher and Smith (84). These workers found that the effective oral dose of the vitamin in Co-deficient lambs was about 35 times the parenteral dose and calculated that about 3% of the orally administered vitamin must have been absorbed. This calculation involves the assumption that cobalamin injected in relatively high doses is retained as effectively as cobalamin absorbed from the alimentary tract. There is evidence that it is not. Thus Dawbarn and Hine (37) found that cobalamin injected into sheep at the level of 50 $\mu$g/day was 30-36% excreted in the urine. Subsequently, Smith and Marston (146) found injected labeled cobalamin to be 70-90% retained by cobalt-deficient sheep, with excretion in the urine and feces, and with evidence of substantial secretion into the duodenum. About half the cobalamin produced in the rumen was lost during passage through the alimentary tract and only about 5% was absorbed. The ruminant clearly makes extremely inefficient use of its dietary cobalt, both in respect to production of cobalamin in the rumen and to absorption of the vitamin from the alimentary tract.

The marked failure of appetite that is such a conspicuous feature of cobalt-vitamin $B_{12}$ deficiency in ruminants is not nearly so noticeable in vitamin $B_{12}$ deficiency in man or other species. This difference can probably be related to the means by which the two types of animals derive their energy. The main source of energy to ruminants is not glucose but acetic and propionic acids, together with smaller quantities of butyric and other fatty acids produced by fermentation in the rumen. Following the discovery that methylmalonyl-CoA isomerase, a vitamin $B_{12}$-requiring enzyme, catalyzes the conversion of methylmalonyl-CoA to succinyl-CoA (17) and that the activity of this isomerase is severely depressed in the livers of vitamin $B_{12}$-deficient rats (67, 147), Marston and associates (106) investigated the possibility that (a) a breakdown in propionate utilization at this point in the metabolic pathway might be the primary defect in Co-vitamin $B_{12}$ deficiency in ruminants, and (b) the depression of appetite might be due to an increased level in the blood of propionic acid or of some other metabolic product stemmed back by the reduced capacity of the $B_{12}$-deficient animal to metabolize propionate. The production and absorption of propionic and other fatty acids were found to proceed more or less normally in the Co-deficient sheep. As the deficient state progressed, the rate of disappearance from the blood of injected propionate was shown to fall. Examination of liver homogenates from sheep revealed that in the $B_{12}$-deficient animal (a) there is a failure to convert propionate efficiently to succinate, (b) there is an accumulation of the intermediate methylmalonyl-CoA, and (c) this accumulation can be prevented by the addition of methylmalonyl-CoA isomerase.

These important findings were taken further by Somers (152) who found that both propionate and, in contrast to the observations of Marston *et al.* (106), acetate, clearance rates from the blood of sheep were increasingly adversely affected as vitamin $B_{12}$ deficiency intensified (see Fig. 11). These effects were greater than the effects of depressed feed intake alone, and both variables had a more pronounced effect on the rate of clearance of propionate than of acetate. Gawthorne (56) further demonstrated that the excretion of methylmalonic acid (MMA) in the urine of severely vitamin $B_{12}$-deficient sheep was 5–12 times greater than that of pair-fed vitamin $B_{12}$-injected controls, and was restored to normal within 3 weeks by intramuscular injections of the vitamin. A significant increase in urinary MMA excretion occurred only when the sheep were severely affected by the deficiency. It is apparent from these clearance and excretion studies that a significant depression in appetite can occur in the vitamin $B_{12}$-deficient sheep before there is a significant reduction in clearance rate of fatty acids absorbed from the gut or an increase in urinary MMA excretion. These findings suggest that the inappetence is

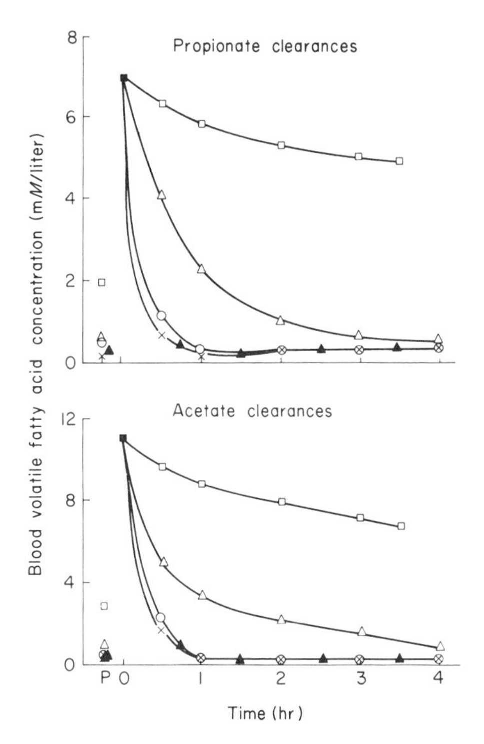


**Fig. 11.** Propionate and acetate clearance rates in sheep. P—preinjection values; X—no signs of deficiency; O—signs of incipient vitamin $B_{12}$ deficiency; □—acute state of vitamin $B_{12}$ deficiency; Δ—after 300 μg intramuscular vitamin $B_{12}$ therapy; ▲—all signs of deficiency remitted. [From Somers (152).]

not a direct consequence of the reduced capacity of the deficient animal to metabolize propionate. They do not exclude an early impairment in methylmalonyl-CoA isomerization, because the potential accumulation and augmented excretion of MMA could be offset by the reduced feed intake and, therefore, the amount of propionate presented for metabolism in the early stages of the deficiency. In the later stages it seems that methylmalonyl-CoA isomerization is so severely impaired by vitamin $B_{12}$ depletion that the sheep is unable to catabolize even the quantities of propionate produced by rumen fermentation of its severely reduced feed intake.

The extent to which metabolic pathways, other than that of propionate, may be impaired in the vitamin $B_{12}$-deficient ruminant has been less explored. Increased concentrations of pantothenic acid or CoA occur in the livers of vitamin $B_{12}$-deficient sheep (39, 148). The cause of the increase is unknown but is apparently not primarily due to accumulation of methylmalonyl-CoA. Thus Smith *et al.* (148) found the molar concentrations of CoA in the livers of ewes to be more than 3 times those of methylmalonic acid, in both vitamin $B_{12}$-deficient and in pair-fed, vitamin $B_{12}$-sufficient animals.

Deficiencies of either vitamin $B_{12}$ or folic acid can cause an increase in the excretion of formiminoglutamic acid (FIGLU) in humans (27, 102), in rats (142, 158), and in chicks (155). It has been suggested that vitamin $B_{12}$, through participation in the methyltetrahydrofolate-homocysteine transmethylase reaction, regulates the availability of tetrahydrofolic acid and thereby indirectly affects FIGLU catabolism (25, 74). Gawthorne (56) showed that the excretion of FIGLU in the urine of severely vitamin $B_{12}$-deficient sheep is more than 30 times that of pair-fed, vitamin $B_{12}$-injected controls and that this excretion can be restored to normal by injections of vitamin $B_{12}$. In contrast to the position with methylmalonic acid excretion, urinary FIGLU excretion was markedly elevated even in the early stages of vitamin $B_{12}$ deficiency. Whether the increased FIGLU excretion in vitamin $B_{12}$-deficient sheep can be ascribed to a decrease in the activity of methyltetrahydrofolate-homocysteine transmethylase has not been established. However it is clear that the metabolic reactions involving this compound are impaired relatively early in the development of the deficiency syndrome. Increased FIGLU excretion thus appears to be a sensitive indicator of vitamin $B_{12}$ deficiency in the sheep.

### 7. *Cobalt and "Phalaris Staggers"*

In certain areas, sheep and to a lesser extent cattle, grazing pastures containing the perennial grass, *Phalaris tuberosa,* can develop either an acute form of disease from which they quickly die or a chronic form with nervous disorders characterized clinically by a marked incoordination of gait(staggers), muscular tremors, rapid breathing, and pounding of the heart, particularly when the animals are driven or disturbed. Phalaris staggers, which refers particularly to the chronic condition, was first described by McDonald (111) in South Australia and has been observed elsewhere in Australia (118, 154), in New Zealand (100), and apparently also in South Africa (169) where it is called "Ronpha staggers." Evidence has been obtained that the tryptamine alkaloids, closely related to serotonin, shown to be present in *P. tuberosa* (35) are responsible for the peracute syndrome of poisoning and for acute Phalaris staggers (53).

The relationship of tryptamine alkaloids to the acute forms of the disease is straightforward [see Gallagher *et al.* (53)] but how they produce chronic disease, with persistent neurological disorders, is not so clear.

Nor is it clear how cobalt protects the animal against chronic staggers. Lee and Kuchel (96) first demonstrated the association of this condition with incipient cobalt deficiency and secured its complete prevention in sheep grazing affected pastures by regular oral dosing with cobalt salts. This finding has been confirmed (97,154,169) and extended to include the protective action of cobalt pellets. As cobalt does not appear to be effective against the acute forms of the disease, it seems that cobalt must in some way be concerned with preventing the development of degenerative, structural changes in the nervous system. In this respect the action of cobalt appears to be different from, and additional to, its action in preventing cobalt deficiency in ruminants. This is because (a) administration of vitamin $B_{12}$ is ineffective against Phalaris staggers and (b) a high toxic potential of *Phalaris* pastures increases the level of cobalt required for protection. Where chronic staggers does not develop in animals grazing *Phalaris* pastures, it is contended that the soils maintain sufficient concentrations of cobalt in the pastures to meet the normal requirements of ruminants for this element plus the extra requirements needed to prevent the degenerative changes in the nervous system.

## V. Cobalt in the Nutrition of Man and Other Nonruminants

### 1. *Man*

Conclusive evidence of the existence of a dietary deficiency of cobalt per se, affecting human health, has not been produced, even in areas where cobalt deficiency in ruminants is severe. However, an interesting case was reported of a child, living in a Co-deficient area in Scotland, with marked geophagia which responded rapidly to oral cobalt therapy (141). The accompanying anemia only responded to subsequent treatment with iron. The significance of cobalt in human health and nutrition is confined, so far as is now known, to its presence in vitamin $B_{12}$. The problem of vitamin $B_{12}$ in the health and nutrition of man lies outside the scope of this text. All ordinary diets supply much more cobalt than can be accounted for as vitamin $B_{12}$, and no relationship necessarily exists between their cobalt and their vitamin $B_{12}$ contents.

Earlier studies (21, 70) estimated that an average, good quality adult diet supplies only 5-8 $\mu$g of total cobalt daily. More recently Tipton *et al.* (164), by emission spectroscopy, obtained the much higher mean daily intakes of 160 and 170 $\mu$g Co for the diets of two individuals measured over 30-day periods. Subsequently, Schroeder *et al.* (140), using both atomic absorption and emission spectroscopy, obtained a still higher range of 140 to 580 $\mu$g Co/day for the diets studied, and gave an estimated figure for a typical North American adult diet of 300 μg Co/day.

There seems little doubt that discrepancies of this magnitude reflect analytical uncertainties, but the actual daily cobalt intakes by man must be influenced greatly by differences in the amounts and proportions of different foods composing the diets. The green leafy vegetables are the richest and the most variable sources of cobalt and dairy products and cereals, especially refined cereals, are among the poorest (80, 159). Typical values taken from these earlier investigations are spinach, 0.4-0.6; cabbage and lettuce, 0.2; corn (maize) seed, 0.01; and white flour, 0.003 ppm Co on dry basis. Comparable levels of cobalt were obtained for some foods by Schroeder and associates (140). Many other food items, notably milk and cereals, were found to be so much higher than earlier recorded values obtained by colorimetric methods that the whole question of the validity of different analytical procedures for cobalt needs critical examination.

Apart from its relation to vitamin $B_{12}$, cobalt has received sporadic attention in human medicine. It has been used as a nonspecific erythropoietic stimulant in the treatment of the anemias of nephritis and of infection (54, 172) and several reports of responses to cobalt, in addition to iron, in the treatment of iron deficiency anemia in children and in pregnant women have appeared (30, 55, 69, 82, 160). The amounts of cobalt required to elicit the hemopoietic response are large (20-30 mg daily) so that serious toxic manifestations, including thyroid hyperplasia, myxedema, and congestive heart failure in infants can occur (89, 137). Cobalt, therefore, occupies a very restricted place in the clinical management of human anemias.

Clinical and experimental evidence of effects of varying intakes of cobalt upon the thyroid has been obtained in other studies. These are considered in relation to iodine and endemic goiter in Chapter II. The possible value of cobalt therapy in the treatment of hypertension has also been investigated. Since vasodilatation, with flushing has been observed following injection of cobalt salts (99), Perry and Schroeder (126) treated 9 hypertensive patients with 50 mg oral cobalt chloride per day for 10 to 65 days. Marked lowering of blood pressure to normal in 3 of these patients was observed, with a 14-72% reduction in requirements for antihypertensive drugs in 5. No toxic effects were apparent in any of the treated patients.

### 2. *Nonruminant Animals*

Cobalt deficiency has never been demonstrated in monogastric animals. Rabbits have been successfully maintained on diets reported to supply only 0.1 μg Co/day (162), and rats were found to grow as well on semisynthetic diets supplying 0.6 μg Co/day (166) or 0.3 μg/day (79), as when those diets were supplemented with cobalt. Horses thrive on pastures so low in cobalt that sheep and cattle dependent upon such pastures soon waste and die. Some of the earlier studies with laboratory animals are now known to have supplied appreciable amounts of dietary vitamin $B_{12}$, but it is evident that monogastric species have lower cobalt requirements than ruminants.

Monogastric animals consuming all-plant rations, which supply little or no vitamin $B_{12}$, obviously need some dietary cobalt so that their intestinal flora can synthesize sufficient vitamin $B_{12}$ to meet the metabolic needs of the tissues. The diets consumed generally contain sufficient cobalt for this purpose. However, Dinusson *et al.* (43) reported a small but significant increase, from additions of cobalt, in the rate of gain and efficiency of feed use of pigs fed a corn-soybean meal diet. In some experiments a similar improvement was observed from supplements of vitamin $B_{12}$ or meat scraps, with a response from cobalt even when those rations contained 5% meat scraps. Although the level of dietary cobalt does not appear to be a significant factor, it must be emphasized that intestinal synthesis of vitamin $B_{12}$ is not always adequate for the needs of growing pigs and poultry consuming all-plant rations. Additional supplies of this vitamin can be obtained by coprophagy and by the consumption of litter and refuse in which bacterial fermentation by the bacteria voided in the feces, or contaminated from the soil, has occurred. Where these adventitious sources of vitamin $B_{12}$ are denied to the animals, marked growth responses can be obtained from supplements rich in the vitamin. The "animal protein factor" effect with pigs and poultry receiving all-plant diets is due partly and in some cases largely to the vitamin $B_{12}$ supplied. Cobalt, of itself, is not important in this respect.

## VI. Cobalt Polycythemia and Cobalt Toxicity

### 1. *Polycythemia*

Since the time of the original demonstration that cobalt in large amounts induces polycythemia in rats (170), this effect has been demonstrated in mice, rabbits, guinea pigs, dogs, pigs, ducks, and chickens and in man [see Grant and Root (63)]. Cobalt polycythemia does not occur in mature ruminants (18,61,81), but is one of the first effects of excessive consumption in calves treated prior to rumen development (83).

It is a true polycythemia, accompanied by hyperplasia of the bone marrow, reticulocytosis and increased blood volume, and the red cells contain normal hemoglobin of normal survival time (22). The condition can be enhanced or suppressed by various means that do not act in the same way in all species. Thus the feeding of liver and choline or the injection of liver extracts inhibits the polycythemic effect of cobalt in dogs (36). In rats, choline has no such effect and liver powder or extract may accentuate a cobalt polycythemia (2). In rats and calves, oral supplementation with methionine or cystine or injections of methionine, cystine, cysteine, or histidine reduce the toxicity and the polycythemic effect of cobalt (62,65,122). This reduction apparently arises as a consequence of the capacity of these amino acids to form stable association complexes with cobalt. When fed with cobalt to rats, EDTA also prevents the development of polycythemia, presumably by combining with the metal to form a nonionized, poorly absorbed compound (28).

The effectiveness of cobalt as an erythropoietic stimulant has been much more clearly established than has its mechanism of action. Cobalt polycythemia can be summated with the polycythemia of altitude (29) and with the marrow stimulation that follows hemorrhage (86). Its usefulness in clinical medicine has been mentioned earlier. Apparently cobalt does not induce polycythemia by way of vitamin $B_{12}$, since injections of large doses of this vitamin do not significantly increase hematopoiesis (101). The most acceptable explanations are those ascribing the polycythemia to increased production of erythropoietins and to an anoxic action of cobalt (63). The latter hypothesis receives strong support from a study of cobalt polycythemia in rabbits (175). Tissue hypoxia, with a marked depression of oxidative phosphorylation, was observed and shown to be an important part of the stimulating effect of cobalt upon erythropoiesis. It was suggested further that an additional mechanism exists. This involves a secondary action upon erythropoiesis of the erythrocyte breakdown products arising from an initial destruction of erythrocytes.

### 2. *Cobalt Toxicity*

Cobalt has a fairly low order of toxicity in all species studied. According to Becker and Smith (18), daily doses of about 3 mg/kg body weight, which is roughly equivalent to a dietary intake of 150 ppm Co, can be tolerated by sheep for periods of at least 8 weeks without harmful effects. Doses of 4 or 10 mg Co/kg body weight result in severely depressed appetite and weight loss, with anemia and some deaths at the higher level. Andrews (6), from his study of acute cobalt toxicity, estimated that a single dose of 300 mg Co/kg body weight as a soluble salt would usually be lethal to sheep and that much smaller single doses equivalent to 40 to 60 mg/kg would occasionally be fatal. It appears from the work of MacLaren *et al.* (103) that cattle are less tolerant of high-cobalt intakes than are sheep.

## REFERENCES


1. S. N. Adams, J. L. Honeysett, K. G. Tiller, and K. Norrish, *Aust. J. Soil Res.* **7**, 29 (1969).
2. H. D. Anderson, E. J. Underwood, and C. A. Elvehjem, *Amer. J. Physiol.* **130**, 373 (1940).
3. E. D. Andrews, *N. Z. J. Sci. Technol., Sect. A* **35**, 301 (1953).
4. E. D. Andrews, *N. Z. J. Agr.* **92**, 239 (1956).
5. E. D. Andrews, *N. Z. J. Agr. Res.* **8**, 788 (1965).
6. E. D. Andrews, *N. Z. Vet. J.* **13**, 101 (1965).
7. E. D. Andrews and J. P. Anderson, *N. Z. J. Sci. Technol., Sect. A.* **35**, 483 (1954).
8. E. D. Andrews and L. I. Hart, *N. Z. J. Agr. Res.* **5**, 403 (1962).
9. E. D. Andrews, L. I. Hart, and B. J. Stephenson, *N. Z. J. Agr. Res.* **2**, 274 (1959); **3**, 364 (1960).
10. E. D. Andrews and B. J. Stephenson, *N. Z. J. Agr. Res.* **9**, 491 (1966).
11. E. D. Andrews, B. J. Stephenson, J. P. Anderson, and W. C. Faithful, *N. Z. J. Agr. Res.* **1**, 125 (1958).
12. E. D. Andrews, B. J. Stephenson, C. E. Isaacs, and R. H. Register, *N. Z. Vet. J.* **14**, 191 (1966).
13. J. G. Archibald, *J. Dairy Sci.* **30**, 293 (1947).
14. H. O. Askew, *N. Z. J. Sci. Technol., Sect. A* **28**, 37 (1946).
15. H. O. Askew and J. Watson, *N. Z. J. Sci. Technol., Sect. A* **25**, 81 (1943).
16. B. C. Aston, *N. Z. J. Agr.* **28**, 38 and 301; **29**, 14 and 84 (1924).
17. W. S. Beck, M. Flavin, and S. Ochoa, *J. Biol. Chem.* **229**, 997 (1957); W. S. Beck and S. Ochoa, *ibid.* **232**, 931 (1958).
18. D. E. Becker and S. E. Smith, *J. Anim. Sci.* **10**, 266 (1951).
19. D. E. Becker, S. E. Smith, and J. K. Loosli, *Science* **110**, 71 (1949).
20. K. C. Beeson, V. A. Lazar, and S. G. Boyce, *Ecology* **36**, 155 (1955).
21. R. Belz, *Voeding* **6**, 236 (1960).
22. L. Berk, J. H. Burchenal, and W. B. Castle, *N. Engl. J. Med.* **240**, 754 (1949).

22a. G. Bertrand and M. Machebouef, *C. R. Acad. Sci.* **180**, 1380 and 1993 (1925); **182**, 1504 (1926); **183**, 5 (1926).

23. G. Bertrand and H. Nakamura, *C. R. Acad. Sci.* **185**, 321 (1927); **186**, 1480 (1928).
24. R. Braude, A. A. Free, J. E. Page, and E. L. Smith, *Brit. J. Nutr.* **3**, 289 (1949).
25. J. M. Buchanan, *Medicine (Baltimore)* **43**, 697 (1964).
26. E. M. Butt, R. E. Nusbaum, T. C. Gilmour, and S. L. DiDio, *in* "Metal-Binding in Medicine" (M. J. Seven and L. A. Johnson, eds.), p. 43. Lippincott, Philadelphia, Pennsylvania. 1960.
27. I. Chanarin, M. C. Bennett, and V. Berry, *J. Clin. Pathol.* **15**, 269 (1962).
28. G. P. Child, *Science* **114**, 466 (1951).
29. E. W. Cohn and F. E. D'Amour, *Amer. J. Physiol.* **166**, 394 (1951).
30. B. L. Coles, *Arch. Dis. Childhood* **30**, 121 (1955).
31. C. L. Comar, *Nucleonics* **3**, 30 (1948).

32. C. L. Comar, G. K. Davis, and R. F. Taylor, *Arch. Biochem.* **9**, 149 (1946); C. L. Comar and G. K. Davis, *ibid.* **12**, 257 (1947).
33. J. F. Connolly and D. B. R. Poole, *Ir. J. Agr. Res.* **6**, 229 (1967).
34. R. Correa, *Arq. Inst. Biol.* (*Sao Paulo*) **24**, 199 (1957).
35. C. C. Culvenor, R. DalBon, and L. W. Smith, *Aust. J. Chem.* **17**, 1301 (1964).
36. J. E. Davis, *Amer. J. Physiol.* **127**, 322 (1939).
37. M. C. Dawbarn and D. C. Hine, *Aust. J. Exp. Biol. Med. Sci.* **33**, 335 (1955).
38. M. C. Dawbarn, D. C. Hine, and J. Smith, *Aust. J. Exp. Biol. Med. Sci.* **35**, 273 (1957).
39. M. C. Dawbarn, H. Forsyth, and D. Kilpatrick, *Aust. J. Exp. Biol. Med. Sci.* **41**, 1 (1963).
40. D. W. Dewey, H. J. Lee, and H. R. Marston, *Nature* (*London*) **181**, 1367 (1958).
41. D. W. Dewey, H. J. Lee, and H. R. Marston, *Aust. J. Agr. Res.* **20**, 1109 (1969).
42. C. G. Dickinson, *Aust. Vet. J.* **3**, 82 (1927); R. B. Becker, W. M. Neal, and A. L. Shealy, *Fla., Agr. Exp. Sta., Bull.* **231** (1931).
43. W. E. Dinusson, E. W. Klosterman, E. L. Lasley, and M. L. Buchanan, *J. Anim. Sci.* **12**, 623 (1953).
44. L. P. Dryden, A. M. Hartman, M. P. Bryant, J. M. Robinson, and L. A. Moore, *Nature* (*London*) **195**, 201 (1962).
45. P. Dutoit and C. Zbinden, *C. R. Acad. Sci.* **188**, 1628 (1929); **190**, 172 (1929).
46. G. H. Ellis and J. F. Thompson, *Ind. Eng. Chem., Anal. Ed.* **17**, 254 (1945).
46a. R. W. Engel, N. O. Price, and R. F. Miller, *J. Nutr.* **92**, 197 (1967).
47. J. F. Filmer, *Aust. Vet. J.* **9**, 163 (1933).
48. J. F. Filmer, *N. Z. J. Agr.* **63**, 287 (1941).
49. J. F. Filmer and E. J. Underwood, *Aust. Vet. J.* **10**, 84 (1934); **13**, 57 (1937).
50. J. E. Ford, S. K. Kon, and J. W. G. Porter, *Chem. Ind.* (*London*) **22**, 495 (1952).
51. H. M. Fox and H. Ramage, *Proc. Roy. Soc., London, Ser. B* **108**, 157 (1931).
52. L. S. Gall, S. E. Smith, D. E. Becker, C. N. Stark, and J. K. Loosli, *Science* **109**, 468 (1949).
53. C. H. Gallagher, J. H. Koch, R. M. Moore, and J. D. Steel, *Nature* **204**, 542 (1964).
54. F. H. Gardner, *J. Lab. Clin. Med.* **41**, 56 (1953).
55. J. E. Gardner and F. Tevetoglu, *J. Pediat.* **51**, 667 (1957).
56. J. M. Gawthorne, *Aust. J. Biol. Sci.* **21**, 789 (1968).
57. J. M. Gawthorne, *Aust. J. Exp. Biol. Med. Sci.* **47**, 311 (1969).
58. J. M. Gawthorne, *Aust. J. Exp. Biol. Med. Sci.* **48**, 285 and 293 (1970).
59. J. M. Gawthorne, M. Somers, and H. J. Woodliff, *Aust. J. Exp. Biol. Med. Sci.* **44**, 585 (1966).
60. C. F. Gessert, D. T. Berman, J. Kastelic, O. G. Bentley, and P. H. Phillips, *J. Dairy Sci.* **35**, 696 (1952).
61. R. P. Geyer, I. W. Rupel, and E. B. Hart, *J. Dairy Sci.* **28**, 291 (1945).
62. W. C. Grant, *Fed. Proc., Fed. Amer. Soc. Exp. Biol.* **14**, 61 (1955).
63. W. C. Grant and W. S. Root, *Physiol. Rev.* **32**, 449 (1952).
64. J. R. Greig, H. Dryerre, W. M. Godden, A. Chrichton, and W. G. Ogg, *Vet. J.* **89**, 99 (1933).
65. W. H. Griffith, C. L. Pavcek, and D. J. Mulford, *J. Nutr.* **23**, 603 (1942).
66. R. E. R. Grimmett and F. B. Shorland, *Trans. Roy. Soc. N. Z.* **64**, 191 (1934).
67. S. Gurnani, S. P. Mistry, and B. C. Johnson, *Biochim. Biophys. Acta* **38**, 187 (1960).
68. E. G. Hallsworth, S. B. Wilson, and E. A. Greenwood, *Nature* (*London*) **187**, 79 (1960).
69. H. G. Hamilton, *S. Med. J.* **49**, 1056 (1956).

70. M. J. Harp and F. I. Scoular, *J. Nutr.* **47**, 67 (1952).
71. A. E. Harper, R. M. Richard, and R. A. Collins, *Arch. Biochem. Biophys.* **31**, 328 (1951).
72. L. I. Hart and E. D. Andrews, *Nature* (*London*) **184**, 1242 (1959).
73. A. M. Hartman and L. P. Dryden, *Arch. Biochem. Biophys.* **40**, 310 (1952).
74. V. Herbert and R. Zalusky, *J. Clin. Invest.* **41**, 1263 (1962).
75. A. Heyrovsky, *Cas. Lek. Cesk.* **91**, 680 (1952) (translation supplied by Commonwealth Bur. Animal Nutr., Aberdeen, Scotland).
76. D. C. Hine and M. C. Dawbarn, *Aust. J. Exp. Biol. Med. Sci.* **32**, 641 (1954).
77. D. C. Hodgkin, J. Pickworth, J. H. Robertson, K. N. Trueblood, R. J. Prosen, and J. G. White, *Nature* (*London*) **176**, 325 (1955); R. Bonnett, J. R. Cannon, A. W. Johnson, L. Sutherland, A. R. Todd, and E. L. Smith, *ibid.* p. 328.
78. W. G. Hoekstra, A. L. Pope, and P. H. Phillips, *J. Nutr.* **48**, 421 (1952).
79. A. E. Houk, A. W. Thomas, and H. C. Sherman, *J. Nutr.* **31**, 106 (1946).
80. G. Hurwitz and R. C. Beeson, *Food Res.* **9**, 348 (1944).
81. S. W. Josland, *N. Z. J. Sci. Technol.* **19**, 31 (1937).
82. K. Kato, *J. Pediat.* **11**, 385 (1937).
83. H. A. Keener, G. P. Percival, K. S. Morrow, and G. H. Ellis, *J. Dairy Sci.* **32**, 527 (1949).
84. C. J. Kercher and S. E. Smith, *J. Anim. Sci.* **14**, 458 and 878 (1955); **15**, 550 (1956).
85. M. Kirchgessner, *Z. Tierphysiol., Tierernaehr. Futtermittelk.* **14**, 270 (1959).
86. W. Kleinberg, A. S. Gordon, and H. A. Charipper, *Proc. Soc. Exp. Biol. Med.* **42**, 119 (1939).
87. B. A. Koch and S. E. Smith, *J. Anim. Sci.* **10**, 1017 (1951).
88. H. J. Koch, E. R. Smith, N. F. Shimp, and J. Connor, *Cancer* **9**, 499 (1956).
89. J. P. Kriss, W. H. Carnes, and R. T. Gross, *J. Amer. Med. Ass.* **157**, 117 (1955).
90. C. A. Lassiter, G. M. Ward, C. F. Huffman, C. W. Duncan, and H. D. Webster, *J. Dairy Sci.* **36**, 997 (1953).
91. G. W. Leddicotte, *Int. Comm. Radiol. Protect., Rep. Permissible Doses Intern. Radiat., 1958.*
92. C. C. Lee and L. F. Wolterink, *Amer. J. Physiol.* **183**, 173 (1955); *Poultry Sci.* **34**, 764 (1955).
93. C. C. Lee and L. F. Wolterink, *Science* **123**, 6 (1956).
94. H. J. Lee, *Aust. Vet. J.* **26**, 152 (1950).
95. H. J. Lee, *Brit. Commonwealth Offic. Sci. Conf., Australia, 1949* p. 262 (1951).
96. H. J. Lee and R. E. Kuchel, *Aust. J. Agr. Res.* **4**, 88 (1953).
97. H. J. Lee, R. E. Kuchel, and R. F. Trowbridge, *Aust. J. Agr. Res.* **7**, 333 (1956).
98. H. J. Lee and H. R. Marston, *Aust. J. Agr. Res.* **20**, 1109 (1969).
99. J. M. LeGoff, *J. Pharm. Exp. Ther.* **38**, 1 (1930).
100. H. D. LeSouef, *Aust. Vet. J.* **24**, 12 (1948).
101. S. Levy and J. M. Orten, *J. Nutr.* **45**, 487 (1951).
102. A. L. Luhby, J. M. Cooperman, and D. N. Teller, *Proc. Soc. Exp. Biol. Med.* **101**, 350 (1959).
103. A. P. C. MacLaren, W. G. Johnston, and R. C. Voss, *Vet. Rec.* **76**, 1148 (1964).
104. H. R. Marston, *J. Counc. Sci. Ind. Res.* (*Aust.*) **8**, 111 (1935); E. W. Lines, *ibid.* p. 117.
105. H. R. Marston, *Physiol. Rev.* **32**, 66 (1952).
106. H. R. Marston, S. H. Allen, and R. M. Smith, *Nature* (*London*) **190**, 1085 (1961).
107. H. R. Marston and H. J. Lee, *Nature* (*London*) **164**, 529 (1949).
108. H. R. Marston and H. J. Lee, *Nature* (*London*) **170**, 791 (1952).

109. H. R. Marston, H. J. Lee, and I. W. McDonald, *J. Agr. Sci.* **38**, 222 (1948).
110. H. R. Marston, R. G. Thomas, D. Murnane, E. W. Lines, I. W. McDonald, and L. B. Bull, *Commonwealth Counc. Sci. Ind. Res. (Aust.), Bull.* **113** (1938).
111. I. W. McDonald, *Aust. Vet. J.* **17**, 165 (1942); **22**, 91 (1946).
112. K. J. McNaught, *N. Z. J. Sci. Technol., Sect. A* **20**, 14 (1938).
113. K. J. McNaught, *N. Z. J. Sci. Technol., Sect. A.* **30**, 26 (1948).
114. K. R. Millar and E. D. Andrews, *N. Z. Vet. J.* **12**, 9 (1964).
115. M. Moinuddin, A. L. Pope, P. H. Phillips, and G. Bohstedt, *J. Anim. Sci.* **12**, 497 (1953).
116. R. A. Monroe, H. E. Sauberlich, C. L. Comar, and S. L. Hood, *Proc. Soc. Exp. Biol. Med.* **80**, 250 (1952).
117. M. L. Montgomery, G. E. Sheline, and I. L. Chaikoff, *J. Exp. Med.* **78**, 151 (1943).
118. R. M. Moore, G. W. Arnold, R. J. Hutchins, and H. W. Chapman, *Aust. J. Sci.* **24**, 88 (1961).
119. J. G. Morris and R. J. W. Gartner, *J. Agr. Sci.* **68**, 1 (1967).
120. M. W. O'Halloran and K. D. Skerman, *Brit. J. Nutr.* **15**, 99 (1961).
121. J. B. Orr and A. Holm, "Mineral Content of Pastures," 6th Rep. *Econ. Advis. Counc., Gt. Brit.*, 1931.
122. J. M. Orten and M. C. Bucciero, *J. Biol. Chem.* **176**, 961 (1948).
123. P. G. Ozanne, E. A. Greenwood, and T. C. Shaw, *Aust. J. Agr. Res.* **14**, 39 (1963).
124. R. Paulais, *Ann. Pharm. Fr.* **4**, 110 (1946).
125. P. B. Pearson, L. Struglia, and I. L. Lindahl, *J. Anim. Sci.* **12**, 213 (1953).
126. H. M. Perry, Jr. and H. A. Schroeder, *Amer. J. Med. Sci.* **228**, 396 (1954).
127. A. T. Phillipson and R. L. Mitchell, *Brit. J. Nutr.* **6**, 176 (1952).
128. D. B. R. Poole and J. F. Connolly, *Ir. J. Agr. Res.* **6**, 281 (1967).
129. J. W. G. Porter, *Proc. Nutr. Soc.* **12**, 106 (1953).
130. J. K. Powrie, *Aust. J. Sci.* **23**, 198 (1960).
131. N. S. Raum, G. L. Stables, L. S. Pope, O. F. Harper, G. R. Waller, R. Renbarger, and A. D. Tillman, *J. Anim. Sci.* **27**, 1695 (1968).
132. S. N. Ray, W. C. Weir, A. L. Pope, G. Bohstedt, and P. H. Phillips, *J. Anim. Sci.* **7**, 3 (1948).
133. H. M. Reisenhauer, *Nature (London)* **186**, 375 (1960).
134. D. Richardson, W. S. Tsien, B. A. Koch, J. K. Ward, and E. F. Smith, *J. Anim. Sci.* **19**, 910 (1960).
135. E. L. Rickes, N. G. Brink, F. R. Koniusky, T. R. Wood, and K. Folkers, *Science* **108**, 134 (1948).
136. T. Rigg and H. O. Askew, *Emp. J. Exp. Agr.* **2**, 1 (1934).
137. J. S. Robey, P. M. Veazey, and J. D. Crawford, *N. Engl. J. Med.* **255**, 955 (1956).
138. R. C. Rossiter, D. H. Curnow, and E. J. Underwood, *J. Aust. Inst. Agr. Sci.* **14**, 9 (1948).
139. P. Rothery, J. M. Bell, and J. W. T. Spinks, *J. Nutr.* **49**, 173 (1953).
139a. F. C. Russell and D. Duncan, *Commonwealth Bur. Anim. Nutr. Tech. Commun.* No. 15 (1956).
140. H. A. Schroeder, A. P. Nason, and I. H. Tipton, *J. Chronic Dis.* **20**, 869 (1967).
141. V. S. Shuttleworth, R. S. Cameron, G. Alderman, and H. T. Davies, *Practitioner* **186**, 760 (1961).
142. M. Silverman and A. J. Pitney, *J. Biol. Chem.* **233**, 1179 (1958).
143. K. D. Skerman, A. K. Sutherland, M. W. O'Halloran, J. M. Bourke, and B. L. Munday, *Amer. J. Vet. Res.* **20**, 977 (1959).

144. E. L. Smith, *Nature (London)* **162**, 144 (1948).
145. E. L. Smith, "Vitamin $B_{12}$," Monograph. Methuen, London, 1960.
146. R. M. Smith and H. R. Marston, *Brit. J. Nutr.* (1970) (in press).
147. R. M. Smith and K. J. Monty, *Biochem. Biophys. Res. Commun.* **1**, 105 (1959).
148. R. M. Smith, W. S. Osborne-White, and G. R. Russell, *Biochem. J.* **112**, 703 (1969).
149. S. E. Smith, D. E. Becker, J. K. Loosli, and K. C. Beeson, *J. Anim. Sci.* **9**, 221 (1950).
150. S. E. Smith, B. A. Koch, and K. L. Turk, *J. Nutr.* **44**, 455 (1951).
151. S. E. Smith and J. K. Loosli, *J. Dairy Sci.* **40**, 1215 (1957).
152. M. Somers, *Aust. J. Exp. Biol. Med. Sci.* **47**, 219 (1969).
153. M. Somers and J. M. Gawthorne, *Aust. J. Exp. Biol. Med. Sci.* **47**, 227 (1969).
154. W. H. Southcott, *Aust. Vet. J.* **32**, 225 (1956).
155. M. R. Spivey-Fox and W. J. Ludwig, *Proc. Soc. Exp. Biol. Med.* **108**, 703 (1961).
156. F. J. Stare and C. A. Elvehjem, *J. Biol. Chem.* **99**, 473 (1933).
156a. J. Stewart, *Brit. Commonwealth Offic. Sci. Conf., Australia, 1949* p. 281 (1951).
157. J. Stewart, I. W. Mitchell, and F. J. Young, *Vet. Rec.* **67**, 755 (1955).
158. E. L. R. Stokstad, R. E. Webb, and E. Shah, *J. Nutr.* **88**, 225 (1966).
159. N. D. Sylvester and L. H. Lampitt, *J. Soc. Chem. Ind., London* **59**, 57 (1940).
160. F. Tevetoglu, *J. Pediat.* **49**, 46 (1956).
161. R. E. Thiers, J. F. Williams, and J. H. Yoe, *Anal. Chem.* **27**, 1725 (1955).
162. J. F. Thompson and G. H. Ellis, *J. Nutr.* **34**, 121 (1947).
163. I. H. Tipton and M. J. Cook, *Health Phys.* **9**, 103 (1963).
164. I. H. Tipton, P. L. Stewart, and P. G. Martin, *Health Phys.* **12**, 1683 (1966).
165. E. J. Underwood, *Aust. Vet. J.* **10**, 87 (1934).
166. E. J. Underwood and C. A. Elvehjem, *J. Biol. Chem.* **124**, 419 (1938).
167. E. J. Underwood and J. F. Filmer, *Aust. Vet. J.* **11**, 84 (1935).
168. E. J. Underwood and R. J. Harvey, *Aust. Vet. J.* **14**, 183 (1938).
169. F. J. Van der Merwe, *Farming S. Afr.* **35**, 44 (1959).
170. K. Waltner and K. Waltner, *Klin. Wochenschr.* **8**, 313 (1929).
171. D. A. Webb, *Sci. Proc. Roy. Dublin Soc.* **21**, 505 (1937).
172. C. Wilbert, *Muench. Med. Wochenschr.* **92**, 1373 (1950).
173. H. Wolff, *Klin. Wochenschr.* **28**, 280 (1950).
174. N. Yamagata, S. Murata, and T. Torii, *J. Radiat. Res.* **3**, 4 (1962).
175. A. P. Yastrebov, *Fed. Proc., Fed. Amer. Soc. Exp. Biol.* **25**, Trans. Suppl., 630 (1966).

# 6

# NICKEL

## I. Sources of Nickel

Nickel occurs regularly in soils and plants in concentrations substantially higher than those normally present in animal tissues and fluids. Common pasture plants contain 0.5-3.5 ppm Ni on dry basis (15, 17, 24), so that a grazing sheep consuming 1 kg dry matter per day would ingest an average of about 2 mg Ni daily. Human adults consuming a good mixed diet ingest substantially less. Kent and McCance (10) give an estimate of 0.3 to 0.5 mg Ni/day, and Schroeder *et al.* (23) give a similar calculated range, namely 0.3-0.6 mg Ni/day. The latter workers indicate that actual nickel intakes vary greatly with the amounts and proportions of the different classes of foodstuffs consumed. They point out that vegetarian diets will supply much more nickel than nonvegetarian diets because of the higher nickel content of foods of plant origin than of those of animal origin. One of their diets, high in refined foods and in fat, and low in vegetables, was calculated to supply only 0.003-0.01 mg Ni/day, whereas another diet of similar caloric and protein value but containing oysters and rich in whole grain foods and vegetables, and low in fat, would supply 0.7-0.9 mg Ni daily.

Data on the nickel content of foods are meager. Over 40 years ago, Bertrand and Macheboeuf (2) reported levels of 1.5 to 3.0 ppm Ni on dry basis for the green leafy vegetables, and much lower values ranging from 0.15 to 0.35 ppm for fruits, tubers, and grains. Concentrations of this order have been largely confirmed in the more recent study of Schroeder and associates (23). The individual variability in this investigation was large and some items were found to be exceptionally rich in nickel, e.g., tea, 7.6; buckwheat seed, 6.4; herring, 6.8; and oysters, 6.0 ppm Ni on dry basis. Whether such values are generally representative of these foods will only be known when more systematic analysis are undertaken, but very low levels of nickel in muscle meats, eggs, and milk were consistently obtained.

## II. Nickel Metabolism

Nickel is poorly absorbed from ordinary diets and excreted mostly in the feces. This is apparent from the early studies of Drinker *et al.* (5) in man and of Tedeschi and Sunderman in dogs (31). The latter reported that normal dogs excreted 90% of ingested nickel in the feces and 10% in the urine, with no significant retention in the body. Poor absorption and predominant excretion in the feces are further implied by the results of several studies of urinary nickel excretion in man. Perry and Perry (21) found the urine of 24 healthy adults in the United States to contain 10-70 $\mu$g Ni/liter (mean 20±2.6), with an average total excretion of about 30 $\mu$g/day. Imbrus *et al.* (9), in a study of 154 normal adults, obtained a lower mean value of 10 $\mu$g Ni/liter, with the very wide range of 1 to 81 $\mu$g/liter. Sunderman (29), using an atomic absorption procedure developed specially for the quantitative measurement of nickel in biological materials, found the nickel concentrations of 24-hr collections of urine from 17 normal adults to range from 4 to 31 $\mu$g/liter, with a mean of 18 $\mu$g/liter. The mean total daily urinary excretion of nickel in this study was 19.8 $\mu$g (range 7.2-37.6). If it is assumed that the individuals in these studies ingested 0.3-0.6 mg Ni/day from the diet, it is clear that the absorption of dietary nickel must be very low, probably about 3 to 6%. Even at very high intakes of nickel from different sources, rats were shown to absorb nickel poorly and to excrete it mainly in the feces (22).

Studies with radioactive nickel ($^{63}$Ni) indicate that this element is widely distributed in very low concentrations in the tissues and rapidly eliminated. This has been shown in mice following intraperitoneal injections of $^{63}$Ni as nickel chloride (34) and in rats given intravenous injections of smaller doses of the same compound (25). In the latter study, 61% of a single injection of $^{63}$Ni was excreted via the urine within 72 hr and 5.9% appeared in the feces, presumably from the bile. Within 48 hr the $^{63}Ni$ activity had completely disappeared from the blood, with no accumulation in the red cells. The injected nickel was distributed throughout the tissues in a manner directly related to the blood volume of the particular organ or tissue and decreased rapidly in all tissues, so that after 72 hr only the kidney contained significant amounts of $^{63}Ni$.

Several spectrographic studies of human tissues (3, 16, 32) also indicate that the body does not readily retain nickel and does not accumulate this element with age in any organ so far examined, other than the lungs. Koch *et al.* (16) reported the following mean levels in human tissues: small intestine, 2; bladder, 0.5; lung, muscle, and heart, 0.4; and liver, 0.12 ppm Ni on wet tissue basis. These values are higher than those obtained much earlier by Bertrand (2) for animal tissues but are very similar to those of Tipton and Cook (32).

The level of nickel in human blood plasma has been investigated by several groups with unusually uniform results. The levels reported were 0.01-0.06, mean 0.04 (18); 0.01-0.09, mean 0.03 (16); 0.00-0.18, mean 0.02 (20); and 0.00-0.27, mean 0.06 $\mu$g Ni/g blood (8). The levels obtained for the nickel content of red cells in the study of Herring *et al.* (8) were almost identical with those obtained for plasma, namely, 0.00-0.31, mean 0.053 $\mu$g Ni/g. The spectrochemical method used in this investigation was claimed to be sensitive to 0.007 $\mu$g Ni/g. Much higher levels of nickel than those just quoted, i.e., within the range of 0.5 to 2 $\mu$g/g, have been reported in the blood serum of a series of myocardial infarction patients, although only 2 out of 20 healthy controls contained such high levels (4). It was concluded that high concentrations of serum nickel are associated with infarction but whether they are a consequence of the condition or are involved in its etiology is not known. The nickel content of the hair of human females is significantly higher than that of males. The mean concentration of 25 females was 3.96±1.055 ppm and that of 79 males, 0.97±0.147 ppm Ni in the washed fat-free hair (23a). Of the 8 elements studied, only nickel, copper, and cobalt showed this sex difference.

The average level of nickel in cow's milk and colostrum has been reported to be 0.03 and 0.10 ppm, respectively (14). Lower values were found by Archibald (1), who was unable to detect any nickel in some samples when suitable precautions were used to avoid contamination.

## III. Nickel As an Essential Element

Completely convincing evidence that nickel is essential for plants or animals has not yet appeared. Schroeder (23) argues that it behaves like an essential trace metal because (*1*) it is ubiquitous in the biosphere, (*2*) it shows biological activity *in vitro,* affecting certain enzyme systems, (*3*) it has a low molecular weight and two interchangeable valencies, and (*4*) it is nontoxic to mammals orally, except in astringent doses. Early indications that nickel could partially replace cobalt in the treatment of cobalt deficiency in sheep (6) have apparently not been confirmed.

Rats consuming highly purified diets containing 80 ppb Ni were found to grow just as well, over a period of 55 days in an all-plastic environment "isolator" system, as those given supplemental nickel intraperitoneally at the rate of 70 $\mu$g/day (25a). Similar results were obtained with Japanese quail reared to adulthood in plastic isolators and consuming diets containing either 0.08 ppm or 1.72 ppm Ni (38). No significant differences were observed in weight gains, feed efficiency, mortality, egg production, or hatchability. Nielsen (18a) has produced evidence pointing to a physiological role for nickel in the chick. Chicks maintained in an all-plastic controlled environment system and consuming diets containing either 79 ppb or less than 40 ppb Ni, developed the following signs of deficiency within 3 weeks: bright orange leg color instead of pale yellow-brown; slightly enlarged hocks; legs slightly thickened, especially near the joints; edematous type of dermatitis on the shank skin; and a less friable liver. In addition, the bone, liver, kidney, and aorta, but not the blood, muscles, or feathers, retained more $^{63}$Ni than these tissues did in Ni-supplemented chicks. None of these abnormalities were apparent in the chicks receiving the same diet supplemented with nickel at the rate of 3 to 5 ppm.

Nickel activates several enzyme systems, including arginase (7), carboxylase (26), acetyl coenzyme A synthetase (37), and trypsin (28). It inhibits acid phosphatase under certain conditions (19) and catalyzes the nonenzymic decarboxylation of oxalacetic acid (26). These activities are not specific to nickel, and no functional significance has yet been ascribed to them in the living organism. Two further findings of interest and possible significance in relation to a physiological function for nickel may be cited. It is consistently present in ribonucleic acid (RNA) from diverse sources in concentrations many times higher than in the native material from which the RNA is isolated (29, 33) and may play a part in maintaining the configuration of the protein molecules of crystalline complexes of ribonuclease (13). Indirect but suggestive evidence that nickel and other metals may play a role in pigmentation in several species of animals, fish, birds, and insects has been obtained by Kikkawa and co-workers (11, 12) who put forward the hypothesis, based on the strong affinity of melanin and its precursors for metals, that color depends upon specific metals which are transmitted as genes.

## IV. Nickel Toxicity

Many of the early studies on the toxicity of nickel to man are unreliable, due mainly to inadequate analytical methods, but they leave no doubt that this element, like zinc, manganese, and chromium, is relatively nontoxic and that nickel contamination does not present a serious health hazard. Acid foods take up nickel from nickel vessels during cooking but it is poorly absorbed and causes no detectable damage (5,22). A high incidence of respiratory tract neoplasia and nickel dermatitis among exposed workers in nickel refineries has been reported (27), and nickel has been implicated as a pulmonary carcinogen in tobacco smoke (30). The reported elevation of serum nickel in patients with myocardial infarction has been mentioned earlier.

The low toxicity of nickel when orally ingested has been established in rats, mice, monkeys, and chickens. In a study with rats, levels of 250, 500, and 1000 ppm in three different forms in the diet did not significantly affect growth rate or reproduction, and no signs of toxicity were observed even after 3 to 4 months of continuous feeding (22). Adult monkeys maintained their weights and were in perfect health after 6 months on similar nickel intakes (22). The feeding of 1600 ppm of nickel, as the acetate, to young growing mice resulted in a growth reduction and lowered food consumption in both males and females, whereas 1100 ppm was sufficient to induce these changes in the females but not in the males (36). Apparent digestion coefficients for energy, fat, and protein were unaffected, but there were significant reductions in the activities of cytochrome oxidase and isocitric dehydrogenase in the liver, of malic dehydrogenase in the kidneys, and of cytochrome oxidase and malic dehydrogenase in the heart of the nickel-fed groups. Dietary nickel levels of 1100 and 1600 ppm did not influence the body weights of adult mice or their litter size, but the numbers of pups weaned were reduced at the 1600 ppm level.

Experiments with growing chicks given adequate diets, to which nickel was added as the sulfate or the acetate at levels up to 1300 ppm, indicate a lower tolerance for nickel in this species. Growth of chicks to 4 weeks of age was significantly depressed at 700 ppm Ni and above. This was associated with a reduction in food intake, an impairment of energy metabolism, and a marked reduction in nitrogen retention (35). In a subsequent experiment, designed to differentiate between the effects on food consumption and on nickel toxicity per se, no significant differences in growth rate were obtained with 1100 ppm Ni, neither as the sulfate nor as the acetate, compared with pair-fed controls. However nitrogen retention was depressed at this level of nickel intake. It is apparent from these studies that soluble nickel salts at dietary levels of 700 ppm Ni or higher affect both feed intake and nitrogen retention in chicks. The mode of action of nickel upon these processes remains to be determined.

## REFERENCES


1. J. G. Archibald, *J. Dairy Sci.* **32**, 877 (1949).
2. G. Bertrand and M. Macheboeuf, *C. R. Acad. Sci.* **180**, 1380 and 1993 (1925); **182**, 1504 (1926); **183**, 5 (1926).
3. E. M. Butt, R. E. Nusbaum, T. C. Gilmour, and S. L. DiDio, *in* "Metal-Binding in Medicine" (M. J. Seven and L. A. Johnson, eds.), p. 43. Lippincott, Philadelphia, Pennsylvania, 1960.
4. C. A. D'Alonzo and S. Pell, *Arch. Environ. Health* **6**, 381 (1963).
5. K. R. Drinker, L. T. Fairhall, G. B. Ray, and C. K. Drinker, *J. Ind. Hyg.* **6**, 307 (1924).
6. J. F. Filmer and E. J. Underwood, *Aust. Vet. J.* **13**, 57 (1937).
7. L. Hellerman and M. E. Perkins, *J. Biol. Chem.* **112**, 175 (1935).
8. W. B. Herring, B. S. Leavell, L. M. Paixao, and J. H. Yoe, *Amer. J. Clin. Nutr.* **8**, 846 (1960).
9. H. R. Imbrus, J. Cholak, L. H. Miller, and T. Sterling, *Arch. Environ. Health* **6**, 286 (1963).
10. N. L. Kent and R. A. McCance, *Biochem. J.* **35**, 837 and 877 (1941).
11. H. Kikkawa, *Jap. Med. Congr., Proc., 14th, 1955* Part II, p. 25 (1956); *J. Jap. Biochem. Soc.* **27**, 427 (1955).
12. H. Kikkawa, Z. Ogita, and S. Fujito, *Science* **121**, 43 (1955).
13. M. U. King, *Nature* (*London*) **201**, 918 (1964).
14. M. Kirchgessner, *Schriftenreihe Mangelkrankh.* **6**, 61 and 105 (1955).
15. M. Kirchgessner, G. Merz, and W. Oelschläger, *Arch. Tierernaehr.* **10**, 414 (1960).
16. H. J. Koch, E. R. Smith, N. F. Shimp, and J. Connor, *Cancer* **9**, 499 (1956).
17. R. L. Mitchell, *Soil Sci.* **60**, 63 (1945).
18. R. Monacelli, H. Tanaka, and J. H. Yoe, *Clin. Chim. Acta* **1**, 557 (1956).
18a. F. H. Nielsen, private communication (1970).
19. P. Ohlmeyer, *Hoppe-Seyler's Z. Physiol. Chem.* **282**, 1 (1945).
20. L. M. Paixao and J. H. Yoe, *Clin Chim. Acta* **4**, 507 (1959).
21. H. M. Perry, Jr. and E. F. Perry, *J. Clin. Invest.* **38**, 1452 (1959).
22. S. S. Phatak and V. N. Patwardhan, *Indian J. Sci. Ind. Res. A* **9**, 70 (1950); **11**, 172 (1952).
23. H. A. Schroeder, J. J. Balassa, and I. H. Tipton, *J. Chronic Dis.* **15**, 51 (1961).
23a. H. A. Schroeder and A. P. Nason, *J. Invest. Dermatol.* **53**, 71 (1969).
24. W. E. Seay and L. E. Demumbrum, *Agron, J.* **50**, 237 (1958).
25. J. C. Smith and B. Hackley, *J. Nutr.* **95**, 541 (1968).
25a. J. C. Smith, Jr., *Proc. 2nd Mo. Conf. Trace Substances Environ. Health*, p. 223 1968.
26. J. F. Speck, *J. Biol. Chem.* **178**, 315 (1949).


26. J. F. Speck, *J. Biol. Chem.* **178**, 315 (1949).


27. G. A. Stephens, *Med. Press* **187**, 216 (1933); **194**, 283 (1934).
28. K. Sugai, *J. Biochem.* (*Tokyo*) **36**, 91 (1944).
29. F. W. Sunderman, Jr., *Amer. J. Clin. Pathol.* **44**, 182 (1965).

30. F. W. Sunderman and F. W. Sunderman, Jr., *Amer. J. Clin. Pathol.* **35**, 203 (1961).
31. R. E. Tedeschi and F. W. Sunderman, *A. M. A. Arch. Ind. Health* **16**, 486 (1957).
32. I. H. Tipton and M. J. Cook, *Health Phys.* **9**, 103 (1963).
33. W. E. C. Wacker and B. L. Vallee, *J. Biol. Chem.* **243**, 3257 (1959).
34. A. W. Wase, D. M. Goss, and J. M. Boyd, *Arch. Biochem. Biophys.* **51**, 1 (1954).
35. C. W. Weber and B. L. Reid, *J. Nutr.* **95**, 612 (1968).
36. C. W. Weber and B. L. Reid, *J. Anim. Sci.* **28**, 620 (1969).
37. L. T. Webster, Jr., *J. Biol. Chem.* **240**, 4164 (1965).
38. R. H. Wellenreiter, D. E. Ullrey, and E. R. Miller, *Proc. 1st Int. Symp. Trace Element Metab. Anim., 1969* p. 52 (C. F. Mills, ed.) Livingstone, Edinburgh, 1970.

# 7

# MANGANESE

## I. Introduction

More than half a century ago, manganese was shown to be a constant component of plant and animal tissues (17). Early attempts to demonstrate an essential role for this element in laboratory animals were unconvincing. The purified diets employed were so deficient in other essential nutrients that the animals made little growth or survived for only short periods, even when manganese was added (18, 114). The first clear evidence that manganese is an essential element in animal nutrition was obtained in 1931. In that year, Hart and associates (92, 166) found the metal to be necessary for growth in mice and for normal ovarian activity in mice and rats, and Orent and McCollum (128) showed that it is required to prevent testicular degeneration in male rats and for ovarian function in female rats. Five years later, a great stimulus was given to nutritional studies with manganese by the discovery that the two diseases of poultry—perosis or "slipped tendon" (171) and nutritional chondrodystrophy (107)—were caused by inadequate intakes of manganese and could be prevented by manganese supplementation of the diet.

The importance of manganese in practical poultry feeding was so great that extensive investigations were undertaken of its content in different feeds, of the factors affecting its dietary requirements, and of its mode of action within the body of birds and mammals. As a result of these studies, the particular relation of manganese to growth, reproduction, skeletal development, and to carbohydrate and lipid metabolism became established.

The identification of a specific biochemical role for manganese proved extremely elusive for many years. Many enzymes were shown to be activated by manganese *in vitro*, but this property was usually shared with other bivalent ions, notably magnesium. A specific role for manganese in the synthesis of the mucopolysaccharides of cartilage was then demonstrated by Leach and Muenster (102) in 1962. This important discovery was subsequently confirmed and extended by others (101, 104, 161), and the manganese was shown to catalyze glucosamine-serine linkages (64, 135). An additional biochemical role for manganese emerged when pyruvate carboxylase was found to be a manganese metalloprotein (148), and indications were obtained that the element functions in the transcarboxylation part of the pyruvate carboxylase reaction (118).

The importance of manganese in carbohydrate metabolism was further enhanced when Everson and Shrader (53) demonstrated that this element is involved in glucose utilization.

Manganese was first shown to be required by plants and microorganisms by McHargue (113) in 1923. Manganese deficiency diseases affecting a range of crop and pasture plants were subsequently recognized in many parts of the world. Unequivocal evidence of uncomplicated manganese deficiency in grazing livestock has not been produced, despite reports of responses to manganese supplements in several areas.

Manganese deficiency has not been observed in man but the problem of chronic manganese poisoning affecting men engaged in the mining of manganese ores has been greatly illuminated by Cotzias and associates (35, 37, 115). These workers conducted highly successful studies of the metabolic movements of manganese which established the existence of specific and efficient homeostatic control mechanisms in animals and man.

Interest in the biological significance of manganese was further widened by the discovery of Weinberg (167) that certain bacteria require manganese in excess of that needed for growth, in order to synthesize antibiotics, bacteriophage, protective antigens and several enzymes, and for the formation of endospores and longevity of cultures. No other transitional metal of atomic numbers 24-30 was able to satisfy these requirements of the organism. The concept emerging from these studies, namely that manganese requirements for optimal function are in excess of those needed for growth, applies equally to mammals, as will be seen.

## II. Manganese in Animal Tissues and Fluids

### 1. *Total Content and Distribution*

The body of a normal 70-kg man is estimated to contain a total of 12 to 20 mg Mn (35). This quantity is only one-fifth of the estimated total of copper and one-hundredth of that of zinc. Manganese is distributed widely throughout the body tissues and fluids in concentrations that are highly characteristic for particular organs. They vary comparatively little within or among species or with age (147). Manganese tends to be higher in tissues rich in mitochondria and is associated with the presence of melanins. It is more concentrated within the mitochondria than in the cytoplasm or other organelles of the cell (111, 157). The pigmented portions of the eye are not exceptionally rich in manganese, as they are in zinc and copper, although the retina appears to be richer in this metal than most body tissues (155). The pigmented melanin-containing parts of the conjunctiva are higher in manganese than the nonpigmented parts, and dark hair and feathers are higher than lighter-colored hair and feathers (39, 163).

The average concentrations of manganese in the tissues of adult men, as reported by Tipton and Cook (159), and of rabbits and a range of animal species as reported by Fore and Morton (56), are presented in Table 23. Additional relevant data have been published by Schroeder *et al.* (147). The bones, liver, kidney, pancreas, and pituitary gland normally carry higher manganese concentrations (1-3 ppm on fresh basis) than do other organs. The skeletal muscles are among the lowest in manganese (0.1-0.2 ppm) of the tissues of the body. The levels in the bones can be raised or lowered by substantially varying the manganese intake of the animal. This has been demonstrated in the rat (103), rabbit (48), pig (88), and chick (87, 109). Thus levels of 0.8 and 0.6 ppm Mn were found in the bones of Mn-deficient rats and chicks, respectively, compared with 2.2 to 2.0 ppm in the bones of normal animals (58, 103). More recently, Mathers and Hill (109) reported a skeletal Mn concentration of 2.54 ppm on dry, fat-free basis for pullets fed a low-Mn diet from 18 weeks of age to after 6 to 7 months of egg production, compared with 7.83 ppm for similar birds fed a high-Mn diet for this period. The skeletal manganese constituted about 25% of total body manganese, irrespective of differences in dietary Mn. The proportions of the body total present in the skin and feathers and in the muscles were of a similar order, with only about 10% present in the liver, kidney, ovary, and oviduct. Of particular interest is their finding that skeletal manganese does not constitute an important mobilizeable store of this element. Birds

TABLE 23
*Concentrations of Manganese in Animal Tissues*[a]

| Tissue | Man[b] (ppm Mn) | Rabbit[c] (ppm Mn) | Av. figures[c] for a range of animal species (ppm Mn) |
|---|---|---|---|
| Bones (long) | – | – | 3.3 |
| Adrenals | 0.20 | 0.67 | 0.40 |
| Aorta | 0.19 | – | – |
| Brain | 0.34 | 0.36 | 0.40 |
| Heart | 0.23 | 0.28 | 0.34 |
| Kidney | 0.93 | 1.2 | 1.2 |
| Liver | 1.68 | 2.1 | 2.5 |
| Lung | 0.34 | – | – |
| Muscle | 0.09 | 0.13 | 0.18 |
| Ovaries | 0.19 | 0.60 | 0.55 |
| Pancreas | 1.21 | 1.6 | 1.9 |
| Pituitary | – | 2.4 | 2.5 |
| Prostate | 0.24 | – | – |
| Spleen | 0.22 | 0.22 | 0.40 |
| Testes | 0.19 | 0.36 | 0.50 |
| Hair | – | 0.99 | 0.80 |

[a]Values given are on fresh basis.
[b]From Tipton and Cook (159).
[c]From Fore and Morton (56).

given a high-Mn diet to the point of lay, and a low-Mn diet during egg-laying, decreased in body manganese by about 100 μg. This decrease came largely from the liver (one-third) and the skin and feathers (one-third) and most of the remainder from the muscle and remaining tissue. Only 8 μg was contributed by the skeleton.

The storage capacity of the liver for manganese is limited, compared with the great capacity of this organ to accumulate iron and copper. Increasing the manganese intake of pigs from 12 to 172 ppm raised the liver manganese levels from 8.8 to only 10.3 ppm on dry basis (88). A 200-fold increase in the level of manganese in the diet of mice resulted in a doubling of the manganese concentration of the liver and the diaphragm (1). Raising the manganese content of a low-Mn diet (6-7 μg/g) by 100 μg/g, only increased the Mn concentration in the liver from 4.8 to 9.3 ppm on dry, fat-free basis, even when fed to pullets for several months (109).

In further contrast to the position with iron and copper, reserve stores of manganese do not occur in the livers of newborn rats, rabbits, guinea pigs, pigs, cattle, or humans (29, 105, 150, 153, 163). This is surprising because milk is relatively just as deficient in manganese as it is in iron and copper. In fact, the manganese concentration of the liver varies little with age or with species. Human livers from healthy individuals of all ages contain about 6 to 8 ppm Mn on dry basis (29, 150, 159); the livers of sheep (162), cattle, (61, 163), and hens (109, 145), 8-10 ppm; and those of normal rats, rabbits, and guinea pigs, 6-8 ppm (105, 139).

Unlike many other trace metals, manganese does not accumulate significantly in the lungs with age (147, 159). In a recent study of human lungs, a mean manganese concentration of 0.22±0.126 ppm was reported (120). Examination of tissues from patients who had died from cancer of the lung disclosed significantly lower manganese levels in the tumor than in the uninvolved lung or in the rest of the involved lung. This latter finding conforms with those of other workers (158).

### 2. *Manganese in Hair, Wool, and Feathers*

The level of manganese in the skin and its appendages varies with the individual, with the color, and with the manganese status of the diet. Van Koetsveld (163) reported that the hair of healthy adult cows contained 8-15 ppm Mn (mean 12), whereas the hair of cows showing signs of manganese deficiency revealed levels below 8 ppm. At very high intakes of manganese, concentrations as high as 80 ppm were observed. Meyer and Engelbartz (116) found the manganese content of the hair of 351 cattle to range from 3.9 to 49.9 ppm, with a mean of 15.8. O'Mary *et al.* (127) observed the hair of Hereford cows and calves to vary from 6 to 104 ppm Mn. Red hair was consistently and significantly higher in manganese than white hair from the same animals. The manganese content of the black hair of Friesian cattle has also been shown to be significantly higher than the white hair (24). Anke and Groppel (5) found the hair of mature goats to reflect the manganese dietary supply better than any other parts of the body studied. The level in the hair of a group receiving a low-Mn diet averaged 3.5 ppm Mn, in contrast to 11.1 ppm in a comparable group receiving adequate manganese. The manganese levels in wool and in feathers have been shown to vary similarly with dietary manganese intakes. Thus Lassiter and Morton (99) reported a mean concentration of 6.1 ppm Mn in the wool of lambs fed a very low-Mn diet for 22 weeks, compared with 18.7 in the wool of control lambs. Mathers and Hill (109) observed a mean level of 1.2 ppm Mn in the skin and feathers of pullets fed a low-Mn diet for several months, whereas comparable birds similarly fed a high-Mn diet revealed a mean concentration of 11.4 ppm Mn in their skin and feathers.

### 3. *Manganese in Blood*

Widely varying values have been reported for the manganese concentration in blood. Bowen (25), using neutron activation analysis, found normal human blood to average 24±8 μg Mn/liter, divided fairly equally between cells and plasma. Cotzias and co-workers (38, 129), also using activation analysis, reported considerably lower levels in two separate studies. In the first study, the levels were 8.44±2.73 μg/liter for whole blood and only 0.59±0.18 μg/liter for plasma. In the second, the following mean concentrations were obtained: whole blood, 9.84±0.4; serum, 1.42±0.2; and red cells, 23.57±1.2 μg/liter. Similar low values for human blood have been reported by two other groups (19, 54). Following acute coronary occlusion, serum manganese levels are almost invariably elevated. Hedge *et al.* (70a) found this elevation to be a better diagnostic criterion of myocardial infarction than were serum glutamic oxalacetic transaminase (SGOT) levels. Cotzias *et al.* (41) observed significantly elevated concentrations of manganese in the red cells of rheumatoid arthritis patients, with no such change in serum manganese (see Table 24).

Manganese levels of approximately 20 μg/liter have been reported for the whole blood of calves (70). Slightly lower values have been obtained by other workers (14,20) for bovine blood and for avian blood (21), but the whole question of the concentration, distribution between cells and plasma, and forms of manganese in the blood of different species requires reexamination employing the sensitive and specific methods of analysis now available. In human serum, manganese is selectively and almost totally bound by a $\beta_1$-globulin (40, 55).

### 4. *Manganese in Milk*

Normal cow's milk averages 30 μg Mn/liter (range 10-40), with concentrations of 130 to 160 μg/liter in colostrum (7, 95, 143). The level in milk can be increased two- to fourfold by feeding manganese sulfate to the cows in amounts equivalent to 10 or 13 g Mn/day (7,163). The provi-

TABLE 24
*Blood Manganese Levels in Rheumatoid Arthritis*[a]

| Blood | Controls[b] | | | Rheumatoid[b] | | |
|---|---|---|---|---|---|---|
| | $n_1$ | $n_2$ | Mean | $n_1$ | $n_2$ | Mean |
| Whole blood | 14 | 63 | 9.84 ± 0.4 | 6 | 16 | 14.99 ± 2.4 |
| Red cells | 14 | 46 | 23.57 ± 1.2 | 6 | 14 | 33.60 ± 5.0 |
| Serum | 14 | 55 | 1.42 ± 0.2 | 6 | 16 | 1.69 ± 0.4 |

[a]From Cotzias *et al.* (41).
[b]Concentrations of manganese in micrograms/liter of whole blood, red cells and serum. $n_1$, No. of patients; $n_2$, No. of analyses.

sion of high-manganese feeds (28), or feeding supplements providing only 3 g Mn/day (93), results in smaller increases. Limited data obtained with sows indicate that subnormal dietary intakes of manganese result in subnormal manganese levels in the milk, but this question does not appear to have been critically studied in any species.

The manganese content of milk is so responsive to differences in dietary manganese content that real species differences are difficult to establish, although there seems little doubt that human milk is significantly lower in manganese than the milk of cows, sheep, or goats (28, 63). In a study of the trace element concentrations of the milk of several species, Grebennikov *et al.* (63) found mare's milk to be slightly richer in manganese than human milk. Cow's milk was reported to be twice as high, and ewe's milk more than 4 times higher in manganese than human milk. All the values obtained by these latter workers are substantially higher than most of those reported by others.

### 5. *Manganese in the Avian Egg*

On normal rations, hens produce eggs containing 10–15 μg of manganese. The amount can vary widely with the level of manganese in the diet. Thus, in the experiments of Hill and Mather (73), eggs from pullets fed a low-Mn diet contained about 4–5 μg Mn, whereas those from pullets fed a high-Mn diet contained some 10–15 μg. In earlier studies it was found that raising the manganese level in the hen's diet from 13 to 1000 ppm increased the amount in the yolk from 4 to 33 μg (59), and supplementing a Mn-deficient diet with 40 ppm Mn increased the concentration in the whole egg from 0.5 to 0.9 ppm on dry basis (107). Virtually none of this manganese is present in the shell, but, unlike zinc and iron, appreciable amounts occur in the egg white. The concentration in the yolk is 4–5 times that of the white.

### 6. *Forms of Manganese in the Tissues*

The only known manganese-containing metalloprotein, with a fixed amount of the metal per molecule of protein, is pyruvate carboxylase isolated from chicken mitochondria (148). However, liver arginase has been claimed to contain manganese as an essential component (10, 47). A firmly bound manganese compound, which is nondialyzable, nonexchangeable and not available for chelation with EDTA, occurs in human and rabbit erythrocytes, and the suggestion has been made that this is a manganese-porphyrin compound (22). In human blood serum, manganese occurs almost totally bound to a $\beta_1$-globulin (15, 40, 55). Complexes of manganese and other bivalent cations with RNA and deoxyribonucleic acid (DNA) have also been identified (169).

In the mammalian liver most of the manganese is present in the arginase extract, indicating that in this organ the metal exists largely in combination with protein (47). By dialysis of liver and kidney homogenates, Fore and Morton (56) showed that about 20% of the manganese in these tissues is loosely held, whereas in retinal tissue most of the manganese is nondialyzable. In bone the manganese is largely associated with the inorganic portion, with a little bound to the organic matrix. It is apparent from the work of Borg and Cotzias (23) and of Kato (90) with radioactive manganese that a high proportion of the manganese of the tissues exists in highly labile intracellular combinations. In human spinal fluid, which is low in protein, the manganese occurs very largely as micromolecular chelates (40).

## III. Manganese Metabolism

The absorption, distribution, and excretion of manganese have been studied in recent years by several groups of workers, among which Cotzias has been particularly prominent. The early studies of Greenberg (65) with radiomanganese indicated that only 3-4% of an orally administered dose is absorbed in rats. The absorbed manganese quickly appeared in the bile and was excreted in the feces. Experiments since that time with several species, including man, indicate that manganese is almost totally excreted via the intestinal wall by several routes. These routes are interdependent and combine to provide the body with an efficient homeostatic mechanism regulating the manganese levels in the tissues (16, 90, 129). The relative stability of manganese concentrations in the tissues, to which earlier reference was made, is due to such controlled excretion rather than to regulated absorption (27).

Under ordinary conditions the bile flow is the main route of excretion of manganese and constitutes the principal regulatory mechanism. The total concentration of manganese in the bile fluid is highly correlated with its bilirubin content (112), and can be increased tenfold or more by the administration of large amounts of manganese to the animal (134a). Excretion also occurs via the pancreatic juice (30). When the hepatic (biliary) route is blocked or when overloading with manganese occurs, excretion by this route increases (129). Excretion of manganese also takes place into the duodenum and jejunum and to a smaller extent into the terminal ileum (16). These may be regarded as auxiliary routes which can participate significantly in the regulatory and excretory processes when the enterohepatic circulation of manganese is saturated by overloading (16).

Very little manganese is excreted in the urine, even when stable or radioactive manganese is injected, or added to the diet (94, 112). The administration of chelating agents produces a marked rise in urinary manganese (96, 112). Since manganese is carried in the plasma bound to protein, the urinary manganese excretion of nephritics should be higher than normal, as is urinary iron, copper, and zinc excretion in such patients.

Injected radiomanganese disappears rapidly from the bloodstream (23, 90). Borg and Cotzias (23) have resolved this clearance into three phases. The first and fastest of these is identical with the clearance rate of other small ions, suggesting the normal transcapillary movement; the second can be identified with the entrance of manganese into the mitochondria of the tissues; and the third and slowest component could indicate the rate of nuclear accumulation of the element. These interpretations are supported by radiomanganese studies demonstrating early and preferential accumulation in the mitochondria-rich organs of the body (35, 90), localization of manganese in the mitochondria of the cell, and high mitochondrial and low nuclear manganese turnover rates (35). The kinetic patterns for blood clearance and for liver uptake of manganese are almost identical, indicating that the two manganese pools—blood manganese and liver mitochondrial manganese—rapidly enter equilibrium. A high proportion of the body manganese must, therefore, be in a dynamic, highly mobile state.

Evidence for the existence of a specific intracellular manganese pathway through the body has been produced by Cotzias (27, 36). Loading the body with stable manganese, but not with other elements, rapidly elutes manganese from the body and redistributes it within the tissues. The turnover of parenterally administered $^{54}Mn$ has been directly related to the level of stable manganese in the diet of mice over a wide range (27). A linear relationship between the rate of excretion of the tracer and the level of manganese in the diet was observed, and the concentration of $^{54}Mn$ in the tissues was directly related to the level of stable manganese in the diet. This provides further support for the contention that variable excretion rather than variable absorption regulates the concentration of this metal in tissues. An inverse relationship between dietary manganese and the percentage of $^{54}Mn$ taken up by liver, kidney, spleen, and muscle has been further demonstrated in chicks (149). Mahoney and Small (108) showed that the disappearance of radiomanganese from the human body can be described by a curve having two exponential components. An average of 70% of the injected manganese was eliminated by a "slow" pathway with an average half-life of 39 days. The half-time of the "fast" component was 4 days. Preloading two subjects with manganese greatly decreased the fraction eliminated by the slow pathway.

The administration of exogenous glucocorticoid hormones markedly affects the tissue distribution of $^{54}Mn$ in the mouse. There is a shift in the partition within the body from the liver to the carcass (78). Stimulation of the animal's own adrenal cortices with ACTH was then shown to result in similar changes (79). However, adrenalectomy did not alter the manganese concentrations in the liver and diaphragm, except in animals receiving high-manganese intakes. In this case the levels in these two organs increased following adrenalectomy. The constancy of manganese concentrations in the tissues cannot, therefore, be ascribed solely to the function of the adrenocortical hormones (79).

Little is known of the mechanism of absorption of manganese from the gastrointestinal tract, or of the means by which excess dietary calcium and phosphorus reduces manganese availability. In birds the effect of high dietary levels of calcium phosphate in aggravating manganese deficiency is believed to be due to a reduction in soluble manganese through adsorption by solid mineral (144,172). Such adsorption of manganese by phosphates and carbonates has been demonstrated *in vitro* at acid reactions similar to those in the absorptive regions of the intestine. Nevertheless, chemical forms as diverse as the oxide, carbonate, sulfate, and chloride, with widely varying solubilities, are equally valuable as sources of manganese in poultry rations and presumably, therefore, are equally well absorbed (144). Exceptions are provided by one carbonate ore (rhodochrosite) and a silicate ore (rhodomite) which are relatively unavailable (59,144).

The effect of variations in dietary calcium and phosphorus on the metabolism of $^{54}Mn$ in rats has been studied further by Lassiter and associates (100). These workers found that the fecal excretion of parenterally administered $^{54}Mn$ was much higher, and the liver retention much lower, on a 1.0% calcium diet than on a 0.6% calcium diet. It appears, therefore, that calcium can influence manganese metabolism by affecting retention of absorbed manganese as well as by affecting manganese absorption. Variations in dietary phosphorus had no comparable effects on the excretion of intraperitoneally administered $^{54}Mn$, but the absorption of orally administered $^{54}Mn$ was impaired.

## IV. Manganese Deficiency and Functions

Manganese deficiency has been demonstrated experimentally in mice, rats, rabbits, guinea pigs, pigs, poultry, cattle and sheep and occurs naturally on certain diets composed of normal feeds fed to pigs and poultry. Unequivocal evidence of manganese deficiency, absolute or conditioned, has not been obtained in man. The cardinal manifestations of manganese deficiency, namely, impaired growth, skeletal abnormalities, disturbed or depressed reproductive function, and ataxia of the newborn, are similar in all species, but their actual expression varies with the degree and duration of the deficiency imposed, with the age or stage of growth of the animal, and with the rate at which the deficiency develops.

Hemoglobin levels do not appear to be significantly affected by lack of manganese (154, 165). The growth inhibition results partly from reduced food consumption and partly from an impaired efficiency of utilization of the ingested food (26), but severe inappetence is not a conspicuous feature of manganese deficiency, as it is in zinc and cobalt deficiencies.

### 1. *Manganese and Bone Growth*

Skeletal abnormalities, ranging from gross and crippling deformities to mild rarefaction of bone, occur in all species, depending upon the degree and duration of manganese deficiency imposed and upon the age of the animal. The following deformities, according to species, have been observed: in mice, rats, and rabbits, retarded bone growth with shortening and bowing of the forelegs (11,151,154); in pigs, lameness, enlarged hock joints and crooked and shortened legs (119,124); in cattle, leg deformities with "overknuckling" (62); in sheep, joint pains with poor locomotion and balance (99); in goats, tarsal joint excrescences and ataxia (5); in chicks, poults, and ducklings, the disease perosis or "slipped tendon" (171); and in chick embryos, nutritional chondrodystrophy (107). Perosis is characterized by enlargement and malformation of the tibiometatarsal joint, twisting and bending of the tibia and the tarsometatarsus, thickening and shortening of the long bones, and slipping of the gastrocnemius tendon from its condyles. With increasing severity of the condition the chicks are reluctant to move, squat frequently, walk upon their hocks, and soon die. The severity of the deformity of the hock joint is aggravated by weight applied to the leg (42).

Nutritional chondrodystrophy is characterized by shortened and thickened legs and wings, "parrot beak" resulting from a disproportionate shortening of the lower mandible, globular contour of the head due to anterior bulging of the skull, and high mortality. A similar congenital defect in bone development, emphasizing the need for manganese in embryonic life, occurs in the offspring of Mn-deficient rats. A severe shortening of the radius, ulna, tibia, and fibula is evident in these young at birth (82). In addition, a marked epiphyseal dysplasia at the proximal end of the tibia has been observed in such animals and a shortening and doming of the skull with an anomalous ossification of the inner ear (85, 86). The skulls of deficient young are significantly shorter, wider, and higher than those of controls, especially at birth. An increased ratio of skull height to skull length is also found in adult Mn-deficient rats.

The length, density, breaking strength, and ash content are significantly reduced in manganese deficiency but not the volume of the bone or the gross composition of the bone ash (2,31,154). Skeletal maturation is retarded due to an inhibition of endochondral osteogenesis at the epiphyseal cartilages (57,124,174). Vitamin D utilization is not affected (31), and the bone changes differ distinctly from those of calcium, phosphorus, vitamin D, or copper deficiencies (31,101,154). Deficient bones appear normal to X-ray examination and silver nitrate staining (31,58), despite some reduction in bone ash content. Furthermore, changes in dietary manganese content do not significantly alter the amount or location of deposition of $^{45}Ca$ or $^{32}P$ in chick tibia (130).

These findings combine to show that an impairment of the calcification process per se is not a primary causal factor in the bone abnormalities of manganese deficiency. The emphasis placed upon this process by the earlier workers is understandable in the light of the belief that bone phosphatase plays a major role in calcification (137), and the discovery that blood and bone phosphatase activity is reduced in Mn-deficient chicks (170) and rabbits (153). Subnormal alkaline phosphatase activity has also been demonstrated in the serum of Mn-deficient lambs (99) and in the serum, heart, kidney, and liver of Mn deficient ducks (164). However, alkaline phosphatase activity in blood, bone, and other tissues is not always subnormal in manganese deficiency (52,81,125,165). Nor is the manganese concentration of bone invariably reduced below normal, although this is usually so. In Mn-deficient pigs, normal whole bone manganese levels were reported in the presence of highly significant reductions in soft tissue manganese concentrations (134).

Attention was then focused upon a possible involvement of manganese in the synthesis of the organic matrix of cartilage. Leach and Muenster (102) discovered that radiosulfate uptake is lowered in the cartilage of the Mn-deficient chick and that the total concentration of hexosamines and hexuronic acid is reduced in this tissue. A significant but less pronounced reduction was observed in the hexosamine content of other tissues. The relative amounts of glucosamine and galactosamine in epiphyseal cartilage were changed, with most of the decrease in total hexosamine content occurring in the galactosamine component (Table 25). Everson and associates (51,152,161) demonstrated a significant reduction in the concentration of acid mucopolysaccharides (AMPS) in rib and epiphyseal cartilage and in the otoliths of Mn-deficient, newborn

TABLE 25
*Effect of Manganese on the Hexosamine Content of Chick Epiphyseal Cartilage*[a]

| Parameters affected | Level of added manganese | | | |
|---|---|---|---|---|
| | 0 mg/kg | 10 mg/kg | 20 mg/kg | 100 mg/kg |
| 4-Week weight (g) | 369 | 471 | 547 | 562 |
| Enlarged hock (%) | 100 | 93 | 39 | 14 |
| Hexosamine content (%)[b] | 1.98 | 3.64 | 4.16 | 4.51 |
| Glucosamine | 0.72 | 1.02 | 1.02 | 0.88 |
| Galactosamine | 1.26 | 2.62 | 3.14 | 3.69 |

[a]From Leach (101).
[b]Dry, lipid-free cartilage.

guinea pigs. Hyaluronic acid and heparin were also significantly lowered in these tissues in the newborn Mn-deficient animals.

The above data, together with the histological changes in the cartilage observed by Leach (101), indicate that a severe reduction in cartilage chondroitin sulfate content is associated with manganese deficiency. These changes are apparently specific for this deficiency, since they are not observed in other perosislike bone abnormalities. The importance of the chondroitin sulfate-protein complex to the maintenance of the rigidity of connective tissue indicates further that the bone abnormalities of manganese deficiency almost certainly arise as a consequence of these changes. The critical sites of manganese function in chondroitin sulfate synthesis have recently been identified by Leach *et al.* (102a). These are the two enzyme systems: (*a*) polymerase enzyme, which is responsible for the polymerization of uridine diphosphate (UDP)-*N*-acetyl galactosamine to UDP-glucuronic acid to form the polysaccharide and (*b*) galactotransferase, an enzyme which incorporates galactose from UDP-galactose into the galactose-galactose-xylose trisaccharide which serves as the linkage between the polysaccharide and the protein associated with it. These findings provide a biochemical explanation for the observed effects of manganese deficiency upon the composition and histopathology of connective tissue.

Manganese is similarly involved in eggshell formation. Many years ago several groups of workers (32,67,106) showed that eggshell strength and ash content are reduced in manganese deficiency. More recently, Hill and associates (73,104) have shown that shell thickness is depressed by a low-Mn diet when given before laying but not when given from the point of lay. It was concluded that a lack of manganese during the period of sexual maturity limits an essential physiological component of the process of shell formation but once development of this component is complete shell formation can proceed normally. It was found further that low-Mn diets are not associated with impaired formation of shell matrix as a whole but they markedly depress the acid mucopolysaccharide content of the matrix as measured by uronic acid (104).

2. *Manganese and Reproductive Function*

Defective ovulation, testicular degeneration, and infant mortality were reported in the earliest studies demonstrating the essentiality of manganese in the diet of rats (92,128,166). The mortality was at first thought to be due to deficient lactation, but the young from normal rats were found to grow normally when placed with Mn-deficient mothers (46). Subsequently, Shils and McCollum (151) showed that Mn-deficient rats experience no loss of ability to suckle normal young and no lack of maternal interest. The latter finding contrasts greatly with the recent discovery that maternal behavior is seriously affected in Zn-deficient rats (6). In the female, three stages of manganese deficiency can be recognized. In the least severe stage the animals give birth to viable young, some or all of which exhibit ataxia; in the second and more severe stage the young are born dead or die shortly after birth; in the third, acute stage of deficiency, estrous cycles are absent or irregular, the animals will not mate, and sterility results. A delay in the opening of the vaginal orifice may also occur (26). A similar impairment of reproductive function occurs in poultry, since lowered egg production and decreased hatchability have been reported, even in the absence of signs of perosis and chondrodystrophy (8,145).

The omission of manganese from the maternal diet of guinea pigs results in a decrease in litter size and an increase in the percentage of young born dead or delivered prematurely (52). Experimental and field evidence is accumulating that manganese is necessary for normal fertility in cattle. Cows (14,138) and goats (5) exhibit delayed estrus and conception on low-Mn rations, and manganese supplements have been reported by several workers to improve the fertility of dairy cows in parts of Europe (62,72,122,168,173). These manifestations of manganese deficiency may be conditioned by the presence of unidentified environmental factors that limit manganese absorption or utilization by the animals, because the manganese intakes from the grazing are normal or high by standards elsewhere, where no such Mn-responsive functional infertility exists.

The results of two studies with pigs provide further evidence of the relation of manganese to estrus incidence and ability to conceive. In one of these the Mn-deficient animals required more services per conception and farrowed fewer piglets per litter than normal sows (66). In the other study, the Mn-deficient sows exhibited either no estrus or depressed or irregular estrus, with some evidence of fetal death and resorption. They also farrowed fewer piglets per litter than similar pigs receiving added manganese (134).

No indication of the precise locus or mode of action of manganese in male or female reproductive function has emerged from any of these investigations. Manganese is not present in exceptionally high concentrations in the ovaries, nor in the male sex glands, as is zinc. In the Mn-deficient male rat and rabbit, the sterility and absence of libido is associated with seminal tubular degeneration, complete lack of spermatozoa, and accumulation of degenerating cells in the epididymis (26,154). This suggests that manganese is specifically involved in spermatogenesis, but these observations provide no clue to the nature of its action.

### 3. *Manganese and Neonatal Ataxia*

Ataxia in the offspring of Mn-deficient animals was first observed in the chick by Caskey and Norris (33), who showed that ataxic chicks could not be cured by manganese administration after hatching. Subsequently, Shils and McCollum (151) reported ataxia, incoordination, and poor equilibrium in many of the young from Mn-deficient rats. These observations were followed by an illuminating series of studies by Hurley and Everson and their associates with rats and guinea pigs establishing that manganese deficiency during pregnancy produces an irreversible congenital defect in the young, characterized by ataxia and marked loss of equilibrium and often also by head retraction and tremors, increased susceptibility to stimuli, and delayed development of the body-righting reflexes (52,80,81). The defect responsible for these disturbances arises relatively late in gestation—in the rat between the fourteenth and eighteenth days. There is a critical need for manganese on the fifteenth and sixteenth days of gestation because manganese supplementation, begun at that time, results in an improvement of survival time and of the incidence of ataxia in the young, whereas supplementation begun on the eighteenth day is completely ineffective in preventing ataxia (81) (see Table 26).

Histological examination of the brain and spinal cord has not disclosed any lesions which could account for the ataxia (34,74,81,151), such as occur in the ataxia of copper-deficient lambs. Nor have assays of the liver, brain, and other tissues for a number of enzymes (164), including acetylcholinesterase, in the brain of rats and guinea pigs (52,81), revealed any relevant biochemical defects. Cerebrospinal fluid pressures

TABLE 26
*Effect of Manganese Supplementation at Various Times during Gestation*[a]

| Initiation of supplementation (day of gestation)[b] | No. of litters | Young born | | Survival to 28 days | |
|---|---|---|---|---|---|
| | | Total | Per litter | Live young (%) | Ataxic (%) |
| 7-12 | 14 | 105 | 7.5 | 53 | 0 |
| 14 | 6 | 42 | 7.0 | 87 | 0 |
| 15 | 8 | 60 | 7.5 | 36 | 48[c] |
| 16 | 8 | 54 | 6.8 | 44 | 46[c] |
| 18 | 8 | 65 | 8.1 | 26 | 100 |

[a]From Hurley *et al.* (81).
[b]Day of finding sperm considered first day of gestation.
[c]Mild.

do not differ significantly in normal and deficient young (85), but studies of the threshold and pattern of electroshock seizures in ataxic Mn-deficient rats indicate increased brain excitability and convulsability in these animals (83).

The importance of a structural defect in the inner ear as a causative factor in the ataxia and postural defects of Mn-deficient rats and guinea pigs then emerged (80,86,152). These symptoms were shown to arise from impaired vestibular function, itself a reflection of a specific effect of lack of manganese upon cartilage mucopolysaccharide synthesis and, hence, bone development of the skull, particularly of the otoliths. A small proportion of Mn-deficient guinea pigs also exhibit abnormal curvatures of the semicircular canals and misshapen ampullae (152). The defects of motion and posture characteristic of the Mn-deficient young thus result primarily from faulty otolith development during fetal life.

A specific congenital ataxia in mice, resulting from defective development of the otoliths and caused by the presence of mutant genes, has been shown by Hurley and associates (49) to be completely and permanently rectified by appropriately timed manganese supplementation of the diet of the pregnant, mutant mice. This remarkable discovery suggests an interaction between the mutant genes and manganese metabolism. Various teratogenic agents are known to mimic, or phenocopy, the developmental effects of known genetic mutants, but this finding is the first demonstration of a phenocopy-inducing agent, manganese, having a reciprocal effect upon the mutant itself.

### 4. *Manganese and Carbohydrate Metabolism*

The possibility that manganese plays a part in gluconeogenesis through its presence in the metalloenzyme, pyruvate carboxylase, has already been mentioned. Direct evidence that manganese is involved in glucose utilization has recently been produced by Everson and Shrader (53). Newborn guinea pigs severely affected with manganese deficiency exhibited aplasia, or marked hypoplasia, of all the cellular components of the pancreas. Where hypoplasia occurred, islet population was reduced but islet size was increased. Islets contained fewer and less intensely granulated beta cells than the islets of control newborn guinea pigs. Young, adult Mn-deficient guinea pigs also had decreased numbers of pancreatic islets, with less intensely granulated beta cells and more alpha cells. When glucose was administered orally and intravenously the Mn-deficient guinea pigs revealed a decreased capacity to utilize glucose and exhibited a diabeticlike curve in response to glucose loading (Fig. 12). Manganese supplementation completely reversed the reduced glucose utilization.

Other evidence possibly implicating manganese in glucose utilization has been reported. The administration of this element to diabetic subjects has a hypoglycemic effect (12,140), and both pancreatectomy and diabetes have been correlated with decreased manganese levels in blood and tissues (9,97). Whether manganese is involved in insulin synthesis, or insulin need, is not yet known, but it is perhaps pertinent that decreased concentrations of stainable mucopolysaccharides occur in the skin of young rats born to diabetic mothers (175). It has also been suggested that insulin may regulate the utilization of glucose in the synthesis of mucopolysaccharides (146). It is, therefore, conceivable that the im-

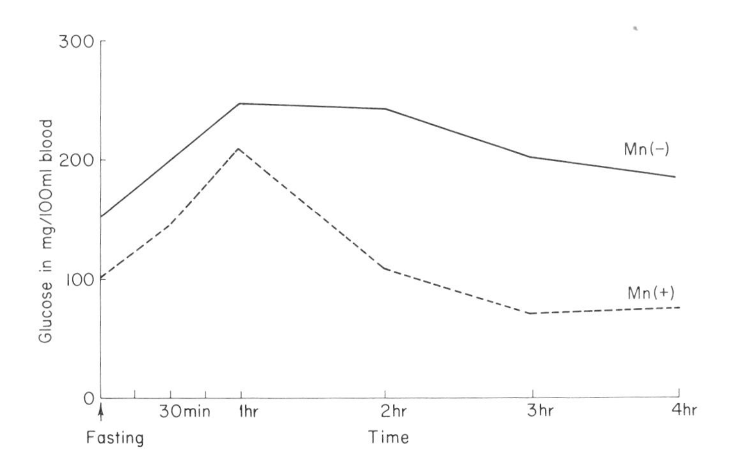


**Fig. 12.** Glucose tolerance curves of manganese-deficient and control guinea pigs (glucose administered orally). [From Everson and Shrader (53).]

pairment of glucose utilization found in the Mn-deficient guinea pig is related to the connective tissue defect that occurs in such animals.

### 5. *Manganese and Lipid Metabolism*

Evidence of a metabolic interaction between manganese and choline in skeletal development was obtained some years ago (50,75,89). Subsequently, manganese and choline supplements were both shown to reduce liver and bone fat in Mn-deficient rats (3), and manganese supplementation of a Mn-deficient diet reduced fat deposition and back fat thickness in pigs (134). The nature of the choline-manganese interaction and of the lipotropic action of manganese is little understood. Manganese stimulates the hepatic synthesis of cholesterol and fatty acids in rats (44), and manganous ion is a necessary cofactor for the conversion of mevalonic acid to squalene by mevalonic kenase (1). A phosphorylated derivative of mevalonic acid necessary for this reaction requires manganese for its synthesis (156). The above findings clearly implicate manganese in lipid metabolism but their significance in the living animal remains to be determined.

### 6. *Manganese and Arginase Activity*

A marked reduction in liver arginase activity has been demonstrated in Mn-deficient rats (26,151) and rabbits (154). The arginase activity of liver preparations from these animals was greatly increased by adding manganese *in vitro*, but the *in vivo* importance of this activation is doubtful. The proportion of the total urinary nitrogen excreted as urea is only slightly reduced and the proportions as ammonia, amino acid, and uric acid are unchanged in Mn-deficient rats (26). Moreover, increasing the burden on the urea-forming system does not accentuate the manifestations of manganese deficiency. No significance can be attached to a possible manganese-arginine relationship in poultry, since their tissues contain little or no arginase and they do not synthesize urea.

## V. Manganese Requirements

The minimum dietary requirements of manganese depend upon the species, the criteria of adequacy employed, the chemical form in which the element is ingested, and the nature of the rest of the diet. Birds have higher manganese requirements than mammals. Species and breed differences also exist among mammals and birds. The minimum dietary intakes of manganese compatible with maximum body growth, normal bone development, and normal reproductive performance are not necessarily identical.

1. *Laboratory Animals*

Mice, rats, and rabbits are unable to grow normally on milk diets containing 0.1-0.2 ppm Mn or on synthetic diets containing 0.2-0.3 ppm of the dry diet (151,154,165). A level of 1 ppm is inadequate to promote normal fetal development in pregnant rats (86). Holtkamp and Hill (76), in a study of the effect of various levels of manganese in the diet, concluded that 2 mg/day at 30 days of age, which is equivalent to about 40 ppm of the dry diet, is optimal for growth of young rats. Rabbits fed dried milk containing 0.14 ppm Mn, plus iron and copper, develop normal bones when 0.3 mg Mn/day as manganese chloride is added. For normal growth 1-4 mg/day is reported to be necessary (153). High intakes of phosphorus relative to calcium increase the manganese requirements of rats. This is evident from experiments in which lowering the Ca:P ratio from 1:0.85 to 1:2.55 aggravated the signs of manganese deficiency on a series of diets supplying only 0.5 $\mu$g Mn/rat/day (165).

2. *Pigs*

The minimum manganese requirements of pigs for normal growth, skeletal development, and fertility cannot be given with any precision. The requirements of this species for satisfactory reproductive performance are substantially higher than those for somatic growth. A high-corn, high-mineral ration containing 11-14 ppm Mn was found satisfactory for the growth of pigs. This diet resulted in leg stiffness and lameness which could be prevented, but not cured, by additional manganese (91). In a later experiment, a diet containing 12 ppm Mn was found adequate for growth and skeletal development, but slight indications of improved reproductive performance from additional manganese were obtained (66). The results of two further experiments (88,134) indicate that satisfactory growth can be obtained in young pigs with diets containing only 0.5 or 1.0-1.5 ppm Mn. However, marked tissue manganese depletion was observed. The growth rate and feed efficiency of pigs fed such low-manganese diets from normal weaning weights to varying market weights were as good as those of similar pigs fed the same basal rations plus 40 ppm Mn. When female pigs were weaned at 3 weeks of age and fed throughout the growing, gestation, and lactation periods on a diet containing 0.5 ppm Mn, skeletal abnormalities, impaired reproduction, increased fatness, and ataxia in the newborn piglets occurred. All of these manifestations of manganese deficiency were prevented by supplemental manganese at a level of 40 ppm.

The experiments just described suggest that the bare requirements of manganese for growth of pigs are extremely low, in fact well below the levels ordinarily found in practical swine rations, including those high in corn. Growth on such low-Mn rations is maintained at the expense of tissue manganese and if prolonged into the reproductive life of the animal, or if begun early enough in life, especially *in utero*, typical signs of manganese deficiency will arise. Whether 40 ppm of dietary manganese is necessary, or whether some lower level is adequate for all functional purposes, can only be answered when more definitive experiments are undertaken.

### 3. *Sheep, Goats, and Cattle*

The minimum dietary manganese requirements of sheep are unknown and those of cattle and goats are not known with any precision. The requirements of cattle for body growth appear to be substantially lower than the requirements for normal bone growth and for fertility. The level of calcium and phosphorus in the diet affects requirements, and on the basis of field evidence with cattle, other unidentified factors exist which are capable of producing "conditioned" manganese deficiency. Hawkins *et al.* (70) maintain that the minimum requirements of young calves for growth are probably not more than 1 ppm Mn and that they are increased by high dietary intakes of Ca and P. Bentley and Phillips (14) state that approximately 10 ppm is adequate for growth in heifers but is marginal for optimum reproductive performance in dairy cows. Heifers grew just as well and cows produced as much milk on rations containing 7-10 ppm Mn as similar animals consuming the same rations supplied with 30, 40, or 60 ppm Mn. The former animals were slower to exhibit estrus and to conceive, and more of their calves were born with weak legs and pasterns at first calving. Anke and Groppel (5) reported that calves fed maize-milk diets containing 8 ppm Mn develop a general ataxia, as well as joint abnormalities. Rojas *et al.* (138) concluded that the manganese requirements of Hereford cows for maximum fertility are in excess of 16 ppm of the dry diet. Anke and Groppel (5) reported no growth differences in female goats fed a ration containing 20 ppm Mn in the first year and 6 ppm in the second year, compared with controls consuming the same rations supplemented with manganese to 100 ppm. However, the former animals exhibited depressed estrus, 23% aborted in the third to fifth month of pregnancy, the birth weights of the young were significantly reduced, and there was a higher male:female sex ratio, compared with the controls.

Earlier claims (62,163) that manganese deficiency in cattle, characterized by poor growth, leg deformities, and subnormal fertility, occurs under field conditions in parts of Holland on pastures containing 44 to 190 ppm Mn on dry basis are not supported by later work. Hartmans (69) reports no such disabilities in cattle grazing pastures containing 25-30 ppm Mn, and no improvement in growth, productivity, or fertility from manganese supplements equivalent to 150 ppm of the dry ration. Furthermore, in subsequent experiments with identical twins consuming the same low-Mn pastures, no evidence of manganese deficiency was evident when the calcium levels were raised from about 0.75 to 1.2% and the phosphorus levels from about 0.35 to 0.8%. It was concluded that 25 ppm Mn is adequate to meet the full requirements of cattle under these conditions, even where calcium and phosphorus intakes are higher than normal.

### 4. *Poultry*

The minimum requirements of dietary manganese are higher for poultry than they are for mammals. These approximate 40 ppm under normal dietary conditions and are similar for the growth of chicks and for normal egg production and hatchability in hens. A total intake of 50 ppm Mn is recommended to provide a margin of safety and to cope with variations in calcium and phosphorus intakes. Atkinson and co-workers (8) have produced evidence that the National Research Council's recommended manganese allowance (123) is inadequate for optimal reproductive performance in turkey hens. These workers showed that for this purpose an ordinary diet containing 3.22% Ca and 0.78% inorganic P needs to be supplemented with between 54 and 108 ppm Mn.

Excess dietary calcium and phosphorus increases manganese requirements by reducing manganese availability (145). For example, 64% of the chicks fed a ration containing 3.2% Ca, 1.6% P, and 37 ppm Mn developed perosis, whereas no perosis was apparent when a diet containing the same level of manganese but only 1.2% Ca and 0.9% P was fed. Freedom from perosis was achieved by omission of the bone meal without additional manganese or by retaining the bone meal and increasing the manganese to levels of 62 and 124 ppm (145).

The heavier breeds of poultry have slightly higher requirements than the lighter breeds but the requirement does not increase greatly with a higher rate of egg production. If it is assumed that 5% of dietary manganese is absorbed, a hen with a 50% egg production, i.e., averaging one every second day, would require at most 0.2 mg/day, additional to that required by a similar nonlaying hen. The corresponding addition for a hen with a 90% egg production is 0.4 mg Mn/day. These amounts of manganese represent only an additional 2-4 ppm Mn at normal feed intakes by laying hens.

The higher requirements of birds than of mammals for manganese arise partly from poor absorption of ingested manganese from the gut. It is doubtful if this is sufficient explanation because absorption is also low in mammals. Injections of manganese into chicks in amounts equivalent to 6 to 10 ppm of the diet can completely prevent perosis, compared with 4 to 5 times these quantities ingested with the feed (59,172).

Two factors are responsible for the special importance of manganese in poultry nutrition. The first is their relatively high requirement. The second is the low manganese content of corn (maize) compared with other cereal grains and seed products. Of the common grains, wheat and oats are the richest in manganese, maize is the poorest, and barley is intermediate. Sorghum grain is very similar to barley in manganese content (60). Poultry rations based on corn, and to a lesser extent on barley or sorghum, are deficient in manganese unless supplemented with manganese salts or ores, or with manganese-rich feeds such as wheaten bran or middlings in which the manganese of the wheat grain is concentrated. Rations based on wheat or oats, by contrast, are likely to be adequate in this respect, unless the diet contains calcium and phosphorus in excess of normal requirements. Soybean meal, an important protein supplement for poultry, contains about 30 ppm Mn (145). It would, therefore, improve the manganese content of a corn or a barley based diet but would not eliminate the necessity for additional manganese. Protein supplements of animal origin such as tankage, fish meal, meat meal or skim milk are normally so low in manganese that they would decrease the manganese content of high-corn and other diets.

## VI. Manganese in Human Health and Nutrition

The common foods in human dietaries are exceedingly variable in manganese content. Peterson and Skinner (133) listed twelve major food groups in descending order of their manganese concentration on the fresh basis. These groups included: nuts, cereals, dried fruits, roots, tubers and stalks, fresh fruits, nonleafy vegetables, animal tissues, poultry and poultry products, fish and seafoods (see Table 27). The average concentrations ranged from 20 to 23 ppm for the first three groups to as low as from 0.2 to 0.5 ppm for the last three groups. The leafy vegetables would, of course, rank much higher if expressed on a dry weight basis. Similar but highly variable average manganese values have been reported for fresh vegetables obtained from different parts of the United

TABLE 27
*Manganese Content of Groups of Principal Foodstuffs*[a,b]

| Class of food | No. of samples | Minimum (ppm Mn) | Maximum (ppm Mn) | Average (ppm Mn) |
|---|---|---|---|---|
| Nuts | 10 | 6.3 | 41.7 | 22.7 |
| Cereals and their products | 23 | 0.5 | 91.1 | 20.2 |
| Dried legume seeds | 4 | 10.7 | 27.7 | 20.0 |
| Green leafy vegetables | 18 | 0.8 | 12.6 | 4.5 |
| Dried fruits | 7 | 1.5 | 6.7 | 3.3 |
| Roots, tubers, and stalks | 12 | 0.4 | 9.2 | 2.1 |
| Fresh fruits (including blueberries) | 26 | 0.2 | 44.4 | 3.7 |
| Fresh fruits (excluding blueberries) | 25 | 0.2 | 10.7 | 2.0 |
| Nonleafy vegetables | 5 | 0.8 | 2.4 | 1.5 |
| Animal tissues | 13 | 0.08 | 3.8 | 1.0 |
| Poultry and poultry products | 6 | 0.30 | 1.1 | 0.5 |
| Dairy products | 7 | 0.03 | 1.6 | 0.5 |
| Fish and seafoods (including oysters) | 7 | 0.12 | 2.2 | 0.5 |
| Fish and seafoods (excluding oysters) | 6 | 0.12 | 0.4 | 0.25 |

[a]From Peterson and Skinner (133).
[b]Manganese content of fresh edible portion.

States (77). More recently Schroeder *et al.* (147), using the microanalytical method of Sandell (142), have confirmed and extended these findings. They reported the highest manganese concentrations in nuts and whole cereals, variable amounts in vegetables, and low concentrations in meats, fish, and dairy products. Tea and cloves were found to be exceptionally rich in manganese. In fact, one cup of tea was estimated to contain 1.3 mg Mn, compared with only 0.15 mg in a cup of coffee. Much lower values of 0.29 and 0.05 mg Mn for a cup of tea and of coffee, respectively, were reported by North *et al.* (126).

The wide range for manganese in cereal grains and their products is due partly to plant species differences and partly to the effects of the milling processes which separate the manganese-rich from the manganese-poor parts of the grain. When whole wheat containing 31 ppm Mn was milled it yielded 160 ppm in the germ, 119 ppm in the bran, and only 5 ppm in low-grade flour (145). The degree of concentration of manganese in the outer layers of the grain is revealed by a later comprehensive study of the minerals of wheat, flour, and bread (45). Patent flours, averaging 72% extraction, were found to contain lower proportions of all the minerals investigated than the original commercial hard wheat blends from which the flours were made. With manganese, this reduction was quite striking, the whole wheats averaging 46 ppm and the patent flours, 6.5 ppm Mn.

It is apparent from the figures quoted above that the daily intakes of manganese from human dietaries will vary greatly with the nature of the dietary components and particularly with the amounts of unrefined cereals, green leafy vegetables, and tea consumed. For example, the average daily intakes of 2 adults consuming a diet in which 40-50% of the calories came from white flour were 2.2-2.7 mg Mn, whereas the corresponding manganese intakes for 2 individuals consuming a diet in which the same proportion of the calories came from 92% extraction flour were 8.5-8.8 mg (94). The latter figures are close to the 7 mg Mn daily calculated for adults on a typical English winter diet, although in this case no less than 3.3 mg of the total was estimated to come from tea (121). Adults consuming a typical Dutch diet were estimated to consume an average of 2.3 mg Mn/day (13), which is similar to the 2.4 mg Mn daily calculated by Schroeder *et al.* (147) for adults on an institutional diet in the United States. The mean manganese intake of 9 college women in that country was reported as 3.7 mg daily (126), which is very much lower than the 6.4-7.5 mg measured for 2 adults by Tipton *et al.* (160) in their extensive study of trace elements in diets and excreta in man or than the mean daily intake of 7 mg Mn by men consuming a vegetarian diet reported by Lang *et al.* (98).

From the daily manganese intakes cited above, it can be calculated that the overall manganese concentration in the dry matter of ordinary human diets, excluding tea, generally lies between 5 and 10 ppm. This is presumably adequate since manganese deficiency has never been observed or demonstrated in man. However, diets high in milk, sugar, and refined cereals and low in fruits and vegetables could contain less than 5 ppm Mn. The possibility that such diets supply insufficient manganese cannot be excluded. This possibility certainly warrants investigation with growing children and pregnant women, in view of the special involvement of manganese in skeletal development in embryonic and early life and in the reproductive processes. In this connection, Schroeder (147) has suggested that the following human disorders require examination from the standpoint of possible manganese deficiency: conditions in pregnancy involving nervous instability and convulsions; disorders of bony and cartilaginous growth in infants and children; and certain types of sterility in males and females. In the light of evidence presented earlier, diabetes can now be added to the above list. In addition, abnormal manganese metabolism, exemplified by slow turnover rates of the metal, has been demonstrated in rheumatoid arthritis (41).

Limited evidence that manganese depletion, or an induced manganese deficiency, may occur in hydralazine disease and in lupus erythematosis disseminatus in man has been produced by Comens (34a). Hydralazine, a chelating agent, is known to bind manganese *in vitro* (132), and an interaction between these substances occurs in rats (34a,84). However, disseminated lupus has not been reported among many patients taking large doses of phenothiazines, which also bind manganese (117).

## VII. Manganese Toxicity

Manganese is among the least toxic of the trace elements to mammals and birds. The growth of rats is unaffected by dietary manganese intakes as high as 1000-2000 ppm, although larger amounts interfere with phosphorus retention (59). Hens tolerate 1000 ppm without ill-effects (59), but 4800 ppm Mn is highly toxic to young chicks (71). Growing pigs are less tolerant since 500 ppm Mn retards growth and depresses appetite (66). Depressed feed intake and lowered body weight gains were observed in calves fed a low-manganese basal ration supplemented with manganese sulfate at levels of 2460 and 4920 ppm Mn. When supplemented at a dietary level of 820 ppm, no effect on growth or appetite was apparent (43). The adverse effects of excess manganese on growth are mainly a reflection of depressed appetite. When the feed intakes of the calves were equalized for each treatment, no significant effects upon daily gain or feed efficiency were found from manganese supplements of 1000, 2000, or even 3000 ppm. However, a small but significant depression in hemoglobin levels was observed with increasing manganese (43).

A relationship between manganese, iron metabolism, and hemoglobin formation is apparent from studies with lambs (68), cattle (136), rabbits, and pigs (110). Hemoglobin regeneration was greatly retarded and serum iron depressed in anemic lambs fed diets containing 1000 or 2000 ppm Mn. In normal lambs, higher manganese levels, up to 5000 ppm, produced similar effects and also brought about decreased concentrations of iron in the liver, kidney, and spleen. A similar depression in hemoglobin formation was observed in mature rabbits and baby pigs fed diets containing 1250 and 2000 ppm Mn. The depressing effect was overcome by a dietary supplement of 400 ppm of iron (110). At these levels, manganese appears to interfere with iron absorption rather than with hematopoiesis. Lower-manganese levels are reported to have some adverse effects. In fact, Matrone and co-workers (110) estimate that the minimum level of dietary manganese capable of affecting hemoglobin formation is only 45 ppm for anemic lambs and lies between 50 and 125 ppm for mature rabbits and baby pigs. These findings suggest that the Mn:Fe dietary ratio is of wider significance than commonly conceived, especially as most livestock diets contain manganese in concentrations as high, or higher, than those just quoted.

Over 30 years ago the toxic effects of excess manganese in rats were reported to be counteracted by additional thiamine and vice versa (131). The depletion periods of rats on a thiamine-deficient, Mn-adequate diet were also found to be shortened by supplementary manganese (141). Later studies by Anderson and Parker (4) disclosed no evidence of a manganese-thiamine reaction in this species.

Chronic manganese poisoning occurs among miners working with manganese ores. The manganese enters the lungs as oxide dust from the air and also enters the body via the gastrointestinal tract [see Cotzias (35)]. The disease is characterized by a severe psychiatric disorder (locura manganica) resembling schizophrenia, followed by a permanently crippling neurological (extrapyramidal) disorder clinically similar to Parkinson's disease. Comparative studies of a normal population of "healthy" manganese miners and of patients suffering from chronic manganese poisoning, revealed significant differences in the rate of loss of injected $^{54}Mn$ from the whole body and from an area representing the liver (37,115). The healthy miners showed faster losses of the isotope from these areas than did the other two groups and carried higher tissue concentrations of manganese than those with chronic manganese poisoning. The presence of elevated tissue concentrations is, therefore, not necessary for the continuance of the neurological manifestations of the disease, and metal chelation therapy is unlikely to be effective in their remission. The possible effectiveness of chelation therapy during the early psychiatric phases of chronic manganese poisoning has yet to be investigated.

## REFERENCES


1. B. H. Amdur, H. Rilling, and K. Bloch, *J. Amer. Chem. Soc.* **79**, 2646 (1957).
2. M. O. Amdur, L. C. Norris, and G. F. Heuser, *Proc. Soc. Exp. Biol. Med.* **59**, 254 (1945).
3. M. O. Amdur, L. C. Norris, and G. F. Heuser, *J. Biol. Chem.* **164**, 783 (1946).
4. B. M. Anderson and H. E. Parker, *J. Nutr.* **57**, 55 (1955).
5. M. Anke and B. Groppel, *Proc. 1st Int. Symp. Trace Element Metab. Anim., 1969* p. 133 (C. F. Mills, ed.) Livingstone, Edinburgh, 1970.
6. J. Apgar, *Amer. J. Physiol.* **215**, 160 (1968).
7. J. G. Archibald and J. G. Lindquist, *J. Dairy Sci.* **26**, 325 (1943).
8. R. L. Atkinson, J. W. Bradley, J. R. Couch, and J. H. Quisenberry, *Poultry Sci.* **46**, 472 (1967).
9. G. O. Babenko and Z. V. Karplyuk, *Ukr. Biokhim. Zh.* **35**, 732 (1963); cited in *Chem. Abstr.* **60**, 5973f (1964).
10. S. J. Bach and D. B. Whitehouse, *Biochem. J.* **57**, PXXI (1954).
11. L. L. Barnes, G. Sperling, and L. A. Maynard, *Proc. Soc. Exp. Biol. Med.* **46**, 562 (1941).

12. P. M. Belyaev, *J. Physiol. USSR* **25**, 741 (1938).
13. R. Belz, *Voeding* **21**, 236 (1960).
14. O. G. Bentley and P. H. Phillips, *J. Dairy Sci.* **34**, 396 (1951).
15. A. J. Bertinchamps and G. C. Cotzias, *Fed. Proc., Fed. Amer. Soc. Exp. Biol.* **18**, 469 (1959).
16. A. J. Bertinchamps, S. T. Miller, and G. C. Cotzias, *Amer. J. Physiol.* **211**, 217 (1966).
17. G. Bertrand and F. Medigreceanu, *Ann. Inst. Pasteur, Pasteur* **27**, 282 (1913); G. Bertrand and M. Rosenblatt, *ibid* **35**, 815 (1921); **36**, 23 (1922).
18. G. Bertrand and H. Nakamura, *C. R. Acad. Sci.* **186**, 480 (1928).
19. W. F. Bethard, P. A. Olehy, and R. A. Schmidt, "The Use of Neutron Activation Analysis for Quantitation of Selected Cations in Human Blood." General Dynamics Corp., U.S.A., 1962.
20. F. Blakemore, J. A. Nicholson, and J. Stewart, *Vet. Rec.* **49**, 415 (1937).
21. W. Bolton, *Brit. J. Nutr.* **9**, 170 (1955).
22. D. C. Borg and G. C. Cotzias, *Nature* (*London*) **182**, 1677 (1958).
23. D. C. Borg and G. C. Cotzias, *J. Clin. Invest.* **37**, 1269 (1958).
24. S. Bosch, J. van der Grift, and J. Hartmans, *Versl. Landbouwk. Onderzoek.* 666 (1966).
25. H. J. M. Bowen, *J. Nucl. Energy* **3**, 18 (1956).
26. P. D. Boyer, J. H. Shaw, and P. H. Phillips, *J. Biol. Chem.* **143**, 417 (1942).
27. A. A. Britton and G. C. Cotzias, *Amer. J. Physiol.* **211**, 203 (1966).
28. A. Brock and L. K. Wolff, *Acta Brevia Neer. Physiol., Pharmacol., Microbiol.* **5**, 80 (1935).
29. G. Bruckmann and S. G. Zondek, *Biochem. J.* **33**, 1845 (1933).
30. W. T. Burnett, R. R. Bigelon, A. W. Kimbol, and C. W. Sheppard, *Amer. J. Physiol.* **168**, 520 (1952).
31. C. D. Caskey, W. D. Gallup, and L. C. Norris, *J. Nutr.* **17**, 407 (1939).
32. C. D. Caskey and L. C. Norris, *Poultry Sci.* **17**, 433 (1938).
33. C. D. Caskey and L. C. Norris, *Proc. Soc. Exp. Biol. Med.* **44**, 332 (1940).
34. C. D. Caskey, L. C. Norris, and G. F. Heuser, *Poultry Sci.* **23**, 516 (1944).
34a. P. Comens, *Amer. J. Med.* **20**, 944 (1956); *in* "Metal-Binding in Medicine" (M. J. Seven and L. A. Johnson, eds.), p. 312. Lippincott, Philadelphia, Pennsylvania, 1960.
35. G. C. Cotzias, *Physiol. Rev.* **38**, 503 (1958); *in* "Mineral Metabolism: An Advanced Treatise" (C. L. Comar and F. Bronner, eds.), Vol. 2, Part 2B, p. 403. Academic Press, New York, 1962.
36. G. C. Cotzias and J. J. Greenough, *J. Clin. Invest.* **37**, 1298 (1958).
37. G. C. Cotzias, K. Horiuchi, S. Fuenzalida, and I. Mena, *Neurology* **18**, 376 (1968).
38. G. C. Cotzias, S. T. Miller, and J. Edwards, *J. Lab. Clin. Med.* **67**, 836 (1966).
39. G. C. Cotzias, P. S. Papavasiliou, and S. T. Miller, *Nature* (*London*) **201**, 1228 (1964).
40. G. C. Cotzias and P. S. Papavasiliou, *Nature* (*London*) **195**, 823 (1962).
41. G. C. Cotzias, P. S. Papavasiliou, E. R. Hughes, L. Tang, and D. C. Borg, *J. Clin. Invest.* **47**, 992 (1968).
42. R. D. Creek, H. E. Parker, S. M. Hauge, F. N. Andrews, and C. W. Carrick, *Poultry Sci.* **39**, 96 (1960).
43. G. N. Cunningham, M. B. Wise, and E. R. Barrick, *J. Anim. Sci.* **25**, 532 (1966).
44. G. L. Curran, *J. Biol. Chem.* **210**, 765 (1954).
45. C. P. Czerniejewski, C. W. Shank, W. G. Bechtel, and W. B. Bradley, *Cereal Chem.* **41**, 65 (1964).

46. A. L. Daniels and G. J. Everson, *J. Nutr.* **9**, 191 (1935).
47. S. Edelbacher and H. Bauer, *Naturwissenchaften* **26**, 26 (1938).
48. G. H. Ellis, S. E. Smith, and E. M. Gates, *J. Nutr.* **34**, 21 (1947).
49. L. Erway, L. S. Hurley, and A. Fraser, *Science* **152**, 1766 (1966).
50. R. J. Evans, M. Rhian, and C. I. Draper, *Poultry Sci.* **22**, 88 (1943).
51. G. J. Everson, W. DeRafols, and L. S. Hurley, *Fed. Proc., Fed. Amer. Soc. Exp. Biol.* **23**, 448 (1964).
52. G. J. Everson, L. S. Hurley, and J. F. Geiger, *J. Nutr.* **68**, 49 (1959).
53. G. J. Everson and R. E. Shrader, *J. Nutr.* **94**, 89 (1968); R. E. Shrader and G. J. Everson, *ibid.* p. 269.
54. A. A. Fernandez, C. Sobel, and S. L. Jacobs, *Anal. Chem.* **35**, 1721 (1962).
55. A. C. Foradori, A. J. Bertinchamps, J. M. Gulebon, and G. C. Cotzias, *J. Gen. Physiol.* **50**, 2255 (1967).
56. H. Fore and R. A. Morton, *Biochem. J.* **51**, 594, 598, 600, and 603 (1952).
57. G. Frost, C. W. Asling, and M. M. Nelson, *Anat. Rec.* **134**, 37 (1959).
58. W. D. Gallup and L. C. Norris, *Science* **87**, 18 (1938).
59. W. D. Gallup and L. C. Norris, *Poultry Sci.* **18**, 76 and 83 (1939).
60. R. J. W. Gartner and J. O. Twist, *Aust. J. Exp. Agr. Anim. Husb.* **8**, 210 (1968).
61. G. F. Gessert, D. T. Berman, J. Kastelic, O. G. Bentley, and P. H. Phillips, *J. Dairy Sci.* **35**, 696 (1952).
62. J. Grashuis, J. J. Lehr, L. L. E. Beuvery, and A. Beuvery-Asman, *Meded. "De Schothorst"* **S40** (1953).
63. E. P. Grebennikov, V. R. Soroka, and E. V. Sabadash, *Vop. Pitan.* **22**, 87. Fed. Proc. **23**, Trans. Suppl. 461 (1964).
64. E. E. Grebner, C. W. Hall, and E. F. Neufeld, *Arch. Biochem. Biophys.* **116**, 391 (1966); *Biochem. Biophys. Res. Commun.* **22**, 672 (1966).
65. D. M. Greenberg and W. A. Campbell, *Proc. Nat. Acad. Sci. U.S.* **26**, 448 (1940); D. M. Greenberg, H. D. Copp, and E. M. Cuthbertson, *J. Biol. Chem.* **147**, 749 (1943).
66. R. H. Grummer, O. G. Bentley, P. H. Phillips, and G. Bohstedt, *J. Anim. Sci.* **9**, 170 (1950).
67. M. S. Gutowska and R. T. Parkhurst, *Poultry Sci.* **21**, 277 (1942).
68. R. H. Hartman, G. Matrone, and G. H. Wise, *J. Nutr.* **57**, 429 (1955).
69. J. Hartmans, personal communication (1969).
70. G. E. Hawkins, G. H. Wise, G. Matrone, and P. K. Waugh, *J. Dairy Sci.* **38**, 536 (1955).
70a. B. Hedge, G. C. Griffith, and E. M. Butt, *Proc. Soc. Exp. Biol. Med.* **107**, 734 (1961).
71. V. H. Heller and R. Penquite, *Poultry Sci.* **16**, 243 (1937).
72. S. L. Hignett, *Proc. 3rd Int. Congr. Anim. Reprod., Artif. Insem., 1956* p. 116 (1956).
73. R. Hill and J. W. Mather, *Brit. J. Nutr.* **22**, 625 (1968).
74. R. M. Hill, D. E. Holtkamp, A. R. Buchanan, and E. K. Rutledge, *J. Nutr.* **41**, 359 (1950).
75. A. G. Hogan, L. R. Richardson, H. Patrick, and H. L. Kempter, *J. Nutr.* **21**, 327 (1941).
76. D. E. Holtkamp and R. M. Hill, *J. Nutr.* **41**, 307 (1950).
77. H. Hopkins and J. Eisen, *J. Agr. Food Chem.* **7**, 633 (1959).
78. E. R. Hughes and G. C. Cotzias, *Amer. J. Physiol.* **201**, 1061 (1961).
79. E. R. Hughes, S. T. Miller, and G. C. Cotzias, *Amer. J. Physiol.* **211**, 207 (1966).
80. L. S. Hurley and G. J. Everson, *Proc. Soc. Exp. Biol. Med.* **102**, 360 (1959).

81. L. S. Hurley, G. J. Everson, and J. F. Geiger, *J. Nutr.* **66**, 309 (1958); **67**, 445 (1959).
82. L. S. Hurley, G. J. Everson, E. Wooten, and C. W. Asling, *J. Nutr.* **74**, 274 (1961).
83. L. S. Hurley, D. E. Woolley, and P. S. Timiras, *Proc. Soc. Exp. Biol. Med.* **106**, 343 (1961).
84. L. S. Hurley, D. E. Woolley, F. Rosenthal, and P. S. Timiras, *Amer. J. Physiol.* **204**, 493 (1963).
85. L. S. Hurley, E. Wooten, and G. J. Everson, *J. Nutr.* **74**, 282 (1961).
86. L. S. Hurley, E. Wooten, G. J. Everson, and C. W. Asling, *J. Nutr.* **71**, 15 (1960).
87. W. M. Insko, Jr., M. Lyons, and J. H. Martin, *Poultry Sci.* **17**, 12 and 264 (1958).
88. S. R. Johnson, *J. Anim. Sci.* **2**, 14 (1943).
89. T. H. Jukes, *J. Biol. Chem.* **134**, 789 (1940); *J. Nutr.* **20**, 445 (1940).
90. M. Kato, *Quart. J. Exp. Physiol.* **48**, 355 (1963).
91. T. B. Keith, R. C. Miller, W. T. S. Thorp, and M. A. McCarthy, *J. Anim. Sci.* **1**, 120 (1942).
92. A. R. Kemmerer, C. A. Elvehjem, and E. B. Hart, *J. Biol. Chem.* **92**, 623 (1931).
93. A. R. Kemmerer and W. R. Todd, *J. Biol. Chem.* **94**, 317 (1931).
94. N. L. Kent and R. A. McCance, *Biochem. J.* **35**, 877 (1941).
95. M. Kirchgessner, *Mangelkrankheiten* **6**, 61 and 105 (1955).
96. M. F. Kosai and A. G. Boyle, *Ind. Med. Surg.* **25**, 1 (1956).
97. L. G. Kosenko, *Klin. Med. (Moscow)* **42**, 113 (1964); *Fed. Proc., Fed. Amer. Soc. Exp. Biol.* **24**, Trans. Suppl. T237 (1965).
98. V. M. Lang, B. B. North, and L. M. Morse, *J. Nutr.* **85**, 132 (1965).
99. J. W. Lassiter and J. D. Morton, *J. Anim. Sci.* **27**, 776 (1968).
100. J. W. Lassiter, J. D. Morton, and W. J. Miller, *Proc. 1st Int. Symp. Trace Element Metab. Anim., 1969* p. 430 (C. F. Mills, ed.) Livingstone, Edinburgh, 1970.
101. R. M. Leach, Jr., *Fed. Proc., Fed. Amer. Soc. Exp. Biol.* **26**, 118 (1967).
102. R. M. Leach, Jr. and A. M. Muenster, *J. Nutr.* **78**, 51 (1962).
102a. R. M. Leach, Jr., A. M. Muenster, and E. M. Wien, *Arch. Biochem. Biophys.* **133**, 22 (1969).
103. C. W. Lindow, W. H. Peterson, and H. Steenbock, *J. Biol. Chem.* **84**, 419 (1929).
104. M. Longstaff and R. Hill, *Proc. 1st Int. Symp. Trace Element Metab. Anim. 1969* p. 137 (Mills, C. F., ed.) Livingstone, Edinburgh, 1970.
105. E. J. Lorenzen and S. E. Smith, *J. Nutr.* **33**, 143 (1947).
106. M. Lyons, *Arkansas, Agr. Exp. Sta., Bull.* **374** (1939).
107. M. Lyons and W. M. Insko, Jr., *Ky., Agr. Exp. Sta., Bull.* **371** (1937).
108. J. P. Mahoney and W. J. Small, *J. Clin. Invest.* **47**, 643 (1968).
109. J. W. Mathers and R. Hill, *Brit. J. Nutr.* **22**, 635 (1968).
110. G. Matrone, R. H. Hartman, and A. J. Clawson, *J. Nutr.* **67**, 309 (1959).
111. L. S. Maynard and G. C. Cotzias, *J. Biol. Chem.* **214**, 489 (1955).
112. L. S. Maynard and S. Fink, *J. Clin. Invest.* **35**, 83 (1956).
113. J. S. McHargue, *J. Agr. Res.* **24**, 781 (1923).
114. J. S. McHargue, *Amer. J. Physiol.* **77**, 245 (1926).
115. I. Mena, O. Marin, S. Fuenzalida, and G. C. Cotzias, *Neurology* **17**, 128 (1967).
116. H. Meyer and T. Engelbartz, *Deut. Tieraerztl. Wochenschr.* **67**, 124 (1960).
117. L. Meyler, "Side Effects of Drugs," 4th ed. Excerpta Med. Found., Amsterdam, 1964.
118. A. S. Mildvan, M. C. Scrutton, and M. F. Utter, *J. Biol. Chem.* **241**, 3488 (1966).
119. R. C. Miller, T. B. Keith, M. A. McCarthy, and W. T. S. Thorp, *Proc. Soc. Exp. Biol. Med.* **45**, 50 (1940).

120. M. M. Molokhia and H. Smith, *Arch. Environ. Health* **15**, 745 (1967).
121. G. W. Monier-Williams, "Trace Elements in Foods." Chapman & Hall, London, 1949.
122. I. B. Munro, *Vet. Rec.* **69**, 125 (1957).
123. National Research Council, "Nutrient Requirements for Domestic Animals, No. 1. Nutrient Requirements for Poultry. Natl. Acad. Sci.—Natl. Res. Council, Washington, D.C., 1962.
124. G. M. Neher, L. P. Doyle, D. M. Thrasher, and M. P. Plumlee, *Amer. J. Vet. Res.* **17**, 121 (1956).
125. H. M. Nielsen and D. E. Madsen, *Poultry Sci.* **21**, 500 (1942).
126. B. B. North, J. M. Leichsenring, and L. M. Norris, *J. Nutr.* **72**, 217 (1960).
127. C. C. O'Mary, W. T. Butts, R. A. Reynolds, and M. C. Bell, *J. Anim. Sci.* **28**, 268 (1969).
128. E. R. Orent and E. V. McCollum, *J. Biol. Chem.* **92**, 651 (1931).
129. P. S. Papavasiliou, S. T. Miller, and G. C. Cotzias, *Amer. J. Physiol.* **211**, 211 (1966).
130. H. E. Parker, F. N. Andrews, C. W. Carrick, R. D. Creek, and S. M. Hauge, *Poultry Sci.* **34**, 1154 (1955).
131. D. Perla, *Science* **89**, 132 (1939); D. Perla and M. Sandberg, *Proc. Soc. Exp. Biol. Med.* **41**, 522 (1939).
132. H. M. Perry, Jr. and H. A. Schroeder, *Amer. J. Med. Sci.* **228**, 405 (1954).
133. W. H. Peterson and J. T. Skinner, *J. Nutr.* **4**, 419 (1931).
134. M. P. Plumlee, D. M. Thrasher, W. N. Beeson, F. N. Andrews, and H. E. Parker, *J. Anim. Sci.* **13**, 996 (1954); **15**, 352 (1956).
134a. C. K. Reiman and A. S. Minot, *J. Biol. Chem.* **45**, 133 (1920).
135. H. C. Robinson, A. Telser, and A. Dorfman, *Proc. Nat. Acad. Sci. U.S.* **56**, 1859 (1966).
136. N. W. Robinson, S. L. Hansard, D. M. Johns, and G. L. Robertson, *J. Anim. Sci.* **19**, Proc. 1290 (1960).
137. R. Robison, "The Significance of Phosphoric Esters in Metabolism," p. 40. N. Y. U. Press, New York, 1932.
138. M. A. Rojas, I. A. Dyer, and W. A. Cassatt, *J. Anim. Sci.* **24**, 664 (1965).
139. H. Roth, *Biochem. Z.* **333**, 361 (1960).
140. A. H. Rubenstein, N. W. Levin, and G. A. Elliott, *Nature (London)* **194**, 188 (1962).
141. M. Sandberg, D. Perla, and O. M. Holly, *Proc. Soc. Exp. Biol. Med.* **42**, 368 (1939).
142. E. B. Sandell, "Colorimetric Determination of Traces of Metals," p. 619. Wiley (Interscience), New York, 1959.
143. M. Sato and K. Murata, *J. Dairy Sci.* **15**, 461 (1932).
144. P. J. Schaible and S. L. Bandemer, *Poultry Sci.* **21**, 8 (1942).
145. P. J. Schaible, S. L. Bandemer, and J. A. Davidson, *Mich., Agr. Exp. Sta., Tech. Bull.* **159** (1938).
146. S. Schiller and A. Dorfman, *J. Biol. Chem.* **227**, 625 (1957).
147. H. A. Schroeder, J. J. Balassa, and I. H. Tipton, *J. Chronic Dis.* **19**, 545 (1966).
148. M. C. Scrutton, M. F. Utter, and A. S. Mildvan, *J. Biol. Chem.* **241**, 3480 (1966).
149. E. A. Settle, F. R. Mraz, C. R. Douglas, and J. K. Bletner, *J. Nutr.* **97**, 141 (1969).
150. J. H. Sheldon, *Brit. Med. J.* **2**, 869 (1932).
151. M. E. Shils and E. V. McCollum, *J. Nutr.* **26**, 1 (1943).
152. R. E. Shrader and G. J. Everson, *J. Nutr.* **91**, 453 (1967).
153. S. E. Smith and G. H. Ellis, *J. Nutr.* **34**, 33 (1947).
154. S. E. Smith, M. Medlicott, and G. H. Ellis, *Arch. Biochem.* **4**, 281 (1944).
155. F. W. Tauber and A. C. Krause, *Amer. J. Ophthalmol.* **26**, 260 (1943).

156. T. T. Tchen, *J. Amer. Chem. Soc.* **79**, 6344 (1957).
157. R. E. Thiers and B. L. Vallee, *J. Biol. Chem.* **226**, 911 (1957).
158. N. W. Tietz, E. F. Hirsch, and B. Neyman, *J. Amer. Med. Ass.* **165**, 2187 (1957).
159. I. H. Tipton and M. J. Cook, *Health Phys.* **9**, 103 (1963).
160. I. H. Tipton, P. L. Stewart, and P. G. Martin, *Health Phys.* **12**, 1683 (1966).
161. H. Tsai and G. J. Everson, *J. Nutr.* **91**, 447 (1967).
162. E. J. Underwood and D. H. Curnow, unpublished data (1944).
163. E. E. Van Koetsveld, *Tijdschr. Diergeneesk.* **83E**, 229 (1958).
164. R. Van Reen and P. B. Pearson, *J. Nutr.* **55**, 225 (1955).
165. L. W. Wachtel, C. A. Elvehjem, and E. B. Hart, *Amer. J. Physiol.* **140**, 72 (1943).
166. J. Waddell, H. Steenbock, and E. B. Hart, *J. Nutr.* **4**, 53 (1931).
167. E. D. Weinberg, *Appl. Microbiol.* **12**, 436 (1964).
168. A. Werner and M. Anke, *Arch. Tierernaehr.* **10**, 142 (1960).
169. J. S. Wiberg and W. F. Newman, *Arch. Biochem. Biophys.* **72**, 66 (1957).
170. A. C. Wiese, B. C. Johnson, C. A. Elvehjem, E. B. Hart, and J. G. Halpin, *J. Biol. Chem.* **127**, 411 (1939).
171. H. S. Wilgus, Jr., L. C. Norris, and G. F. Heuser, *Science* **84**, 252 (1936); *J. Nutr.* **14**, 155 (1937).
172. H. S. Wilgus, Jr. and A. R. Patton, *J. Nutr.* **18**, 35 (1939).
173. J. G. Wilson, *Vet. Rec.* **64**, 621 (1952); **79**, 562 (1966).
174. S. B. Wolbach and D. M. Hegsted, *A. M. A. Arch. Pathol.* **56**, 437 (1953).
175. V. P. Zhuk, *Bull. Exp. Biol. Med. (USSR)* **54**, 1272 (1964).

# 8

# ZINC

## I. Introduction

Just over a century ago, Raulin (189) showed that zinc is essential in the nutrition of *Aspergillus niger*. Not until 1926 was the essentiality of zinc for the higher forms of plant life clearly established (223). Before this zinc had been shown to be a constituent of hemosycotypin, the respiratory pigment of the snail *Sycotypus* (119).

Following the original demonstration of the occurrence of zinc in living tissues (97), many investigations revealed the regular presence of this element in plants and animals in concentrations often comparable with those of iron and usually much greater than those of most other trace elements. Particularly high concentrations were reported in serpent venom (38), in many marine organisms, notably oysters (17), and in the iridescent layer of the choroid of the eye (274).

The first indications of a function for zinc in the higher animals came from the work of Birckner (14) carried out over 50 years ago. Early attempts to demonstrate such a function using semipurified diets met with limited success (12,83,118) until 1934 when Todd and associates (249) produced the first indisputable evidence that zinc is a dietary essential for the rat. Twenty years passed before Tucker and Salmon (251) made the important discovery that zinc cures and prevents parakeratosis in pigs, although Raper and Curtin (188) had previously reported that a combined supplement of cobalt and zinc prevents "dermatitis" in pigs receiving corn-cotton seed meal rations. Subsequently O'Dell and co-workers (148) showed that zinc is required for growth, feathering, and skeletal development in poultry. Interest in zinc in human nutrition followed the exciting discoveries of Pories and Strain (178) on the relation of zinc to wound healing, and of Prasad and co-workers (180,181) on the relation of this element to the occurrence of dwarfism and hypogonadism in boys in parts of the Middle East. About this time it also became evident that zinc deficiency can occur in cattle under natural conditions in some areas (42,62,98).

Concurrently with these nutritional investigations, intensive studies were undertaken of zinc metabolism and functions in living cells. As early as 1939, Keilin and Mann (87) showed that zinc is a constituent of the enzyme carbonic anhydrase. Many other zinc-containing metalloenzymes were subsequently discovered, including pancreatic carboxypeptidase, alkaline phosphatase, alcohol, malic, lactic, and glutamic dehydrogenases, and tryptophan desmolase (153,255). In addition, zinc was found to act as a cofactor in a variety of enzyme systems, including arginase, enolase, several peptidases, oxalacetic decarboxylase, and carnosinase (153). It became apparent that zinc is involved in a wide range of cellular activities and is vitally concerned with the fundamental process of RNA and protein synthesis and metabolism in plants (208), microorganisms (142), and the higher animals (55,203,221).

## II. Zinc in Animal Tissues and Fluids

### 1. *Total Content and Distribution in the Body*

The whole body of a 70-kg man is estimated to contain 1.4-2.3 g of zinc (278). This is about half the amount of iron, some 10-15 times that of copper, and more than 100 times that of manganese. The mammalian newborn does not consistently carry higher concentrations of total body zinc than mature animals of the same species, and there is little fetal storage of zinc (228). In fact, the zinc concentration of the human fetus changes little throughout fetal life. During the suckling period, whole body zinc concentration rises substantially from newborn levels in the rat and the pig but not in the guinea pig. In the kitten the levels of zinc in the liver and spleen fall during suckling (228). In mice a pronounced reduction in whole body zinc was achieved by denying them access to zinc-rich colostrum (145).

Typical normal values for zinc concentration in the organs of several species are given in Table 28. These levels can be raised in some tissues, notably the bones and the hair, by high dietary intakes of zinc (109) and are reduced to some extent in zinc deficiency, notably in the pancreas, the male sex glands, and the hair, bones, and blood plasma. Because of the relatively high-zinc concentrations (150-250 ppm) present in the bones and teeth (21,35,41) and their large weight, an appreciable proportion of total body zinc resides in these structures. In animals covered with hair, fur, or wool, a considerable proportion is also present in these tissues. In the rat and the hedgehog, 38% of the whole body zinc is situated in the skin and hair or the skin and bristles (228). In man, 20% of total body zinc is estimated to be present in the skin (175).

The zinc content of the pancreas is of interest because of evidence that this element plays a part in insulin production and function *in vivo* (59,186,215). There is no significant difference in the zinc content of the total, fat-free pancreatic tissue of diabetics and nondiabetics (46), despite earlier claims to this effect (212). Recent work by Shevchuk (215) indicates that the zinc present in the pancreas of rats and guinea pigs is concentrated in the islets of Langerhans. It was suggested that the concentration of zinc in the pancreas varies with the functional state of the islet cells and that zinc plays a role in the function of the beta cells that elaborate insulin.

The wide distribution of zinc throughout the cells and tissues of the body is evident from many studies employing radioautographic and his-

TABLE 28
*Typical Zinc Concentrations of Normal Tissues*[a]

| Tissue | Man (μg/g) | Monkey (μg/g) | Rat (μg/g) | Pig (μg/g) |
|---|---|---|---|---|
| Adrenal | 12 | 16 | – | 33 |
| Brain | 14 | – | 18 | – |
| Heart | 33 | 22 | 21 | – |
| Kidney | 55 | 29 | 23 | 40 |
| Liver | 55 | 51 | 30 | 40 |
| Lung | 15 | 19 | 22 | – |
| Muscle | 54 | 24 | 13 | – |
| Pancreas | 29 | 48 | 33 | 45 |
| Prostate | 102 | – | 223 | – |
| Spleen | 21 | 21 | 24 | 28 |
| Testis | 17 | 17 | 22 | – |

[a]Fresh tissue. Values obtained from following sources: man, Tipton and Cook (247); monkey, Macapinlac *et al.* (110); rat, Gilbert and Talyor (57) and Mawson and Fischer (115); pig, Hoekstra (75a).

tochemical techniques. Millar *et al.* (123) showed that $^{65}Zn$ is not localized in any particular region of the liver of the rat. It is more concentrated in the cortex than in the medulla of the kidney and the adrenals, and in the epidermis rather than in the dermis. In all these sites the zinc occurs largely in combination with protein.

The concentration of zinc in different muscles varies with their color and their functional activity. Swift and Berman (240) found a threefold variation in the zinc concentration of eight bovine muscles, and Cassens *et al.* (29), a fourfold variation in the same eight porcine muscles. The mean zinc concentration of the longissimus dorsi muscle, which is light-colored and shows little activity, was 69 ppm on dry, fat-free basis, whereas that of the serratus ventralis, which is a dark, highly active muscle, was 247 ppm (29). In the lobster (*Homaris vulgaris*) the comparison was made between fast-contracting and slow-contracting muscles (22). The overall mean zinc concentration of the latter was some 6 times higher than that of the fast-contracting muscles. Further studies by Cassens and co-workers (30) indicate that such differences are minimal in newborn pigs and develop with use of the muscles, so that marked differences between red and white muscle are apparent by 8 weeks of age. The higher zinc content of the red muscle resides entirely in the heavy subcellular fraction, composed mainly of myofibrils and nuclei.

In liver and mammary cells, zinc is present in the nuclear, mitochondrial, and supernatant fractions, with the highest concentrations in the supernatant and the microsomes (8,245). Experiments with radiozinc injections in mice indicate that about one-sixth of the $^{65}Zn$ in these tissues is firmly bound to protein and cannot be removed by long dialysis or with EDTA (8). This fraction probably consists of the zinc metalloenzymes and the zinc bound to nucleic acids. The remainder, which is exchangeable with zinc ions or removable by EDTA, is bound to the imidazole or sulfhydryl groups of the proteins (272). Becker and Hoekstra (9) also observed a pool of very firmly bound zinc in rat liver cells, together with three further pools of cellular zinc bound with differing intensities. The binding pattern of liver zinc, like the total zinc concentration, was unaltered in zinc deficiency.

### 2. *Zinc in Nails, Hair, and Wool*

The skin and its appendages have been known to be relatively rich in zinc since the early studies of Lutz (109). Smith (217) found the zinc content of the nails of 18 normal human subjects to range from 93 to 292 ppm and to average 151 ppm. The hair of 46 subjects ranged similarly from 92 to 255, with a mean of 173 ppm. This is very close to the mean

167±5 ppm for healthy males and 172±9 ppm for females reported by Schroeder and Nason (210). Strain *et al.* (234) obtained a lower overall mean of 119.6±4.6 ppm Zn for the hair of north American males. The corresponding mean level for the hair of normal Egyptians was 103.3±4.4 ppm Zn and of untreated "zinc-deficient" dwarfs, 54.1±5.5 ppm. Oral zinc sulfate therapy increased the level of zinc in the hair to 121.1±4.8 ppm and alleviated the clinical signs of zinc deficiency. Reinhold *et al.* (193) reported that the hair of Iranian villagers, in a region where zinc deficiency had been suspected, averaged 127 ppm, compared with 220 ppm for control subjects whose diet was diversified and abundant.

The zinc content of hair reflects dietary zinc intakes in rats (192), pigs (101), and cattle and goats (128,138), although the changes in zinc concentrations do not necessarily reflect the severity of the metabolic effects of zinc deficiency as manifested by impaired growth rates in rats (192). Individual variability is high, and there is some variation with age and part of the body (138). O'Mary and co-workers (152) found no significant effects of color or season on the zinc content of Hereford cattle hair. Most of the values fell within the range 115-135 ppm.

### 3. *Zinc in Eye Tissues*

The highest concentrations of zinc known to exist normally in living tissues occur in the choroid of the eye. Individual and species differences are large and the levels in this tissue are higher in freshwater fish than in saltwater fish (99) and, among mammals, in carnivorous than in herbivorous species (18,273,274). Eye tissues other than the choroid carry normal concentrations of zinc, although high values can occur in the iris (18,273). Thus 436 and 277 ppm Zn have been reported for the iris and choroid, respectively, of sheep's eyes, and 246 and 139 ppm for these tissues in the eyes of cattle (18). The choroids of the dog, fox, and marten contain very much higher concentrations, namely 14,600, 69,000, and 91,000 ppm on dry basis, respectively (273). Further dissection of the choroid disclosed that the zinc is concentrated in the iridescent layer (tapetum lucidum) with values reaching the extraordinary high levels of up to 8.5% (85,000 ppm) in the dog and of up to 13.8% (138,000 ppm) in the fox (274). The function of zinc in eye tissues is unknown.

### 4. *Zinc in Male Sex Organs and Secretions*

Attention was first drawn to the high-zinc content of the male sex organs by Bertrand and Vladesco (13). These workers reported high concentrations of this element in horse testes, the prostate gland of the bull and man, the epididymis of the rat, and the seminal vesicles of the pig. The mammalian testis is not consistently exceptionally high in zinc, compared with several other tissues, including hair and red muscle (45,114,221,254). Thus, Underwood and Somers (254) reported a mean concentration of 105±4.4 ppm Zn (dry weight) for the testes of normal rams, compared with 74±5.0 ppm for the testes of severely Zn-deficient rams in which spermatogenesis had ceased. Prasad and co-workers (183) obtained values of 176±12 ppm Zn and 132±16 ppm for the testes of control rats and of Zn-deficient rats, respectively. Parizek *et al.* (160) determined the zinc content of the testes of rats varying in age from 7 to 58 days. During the first 30–35 days the concentration remained fairly constant at about 120 ppm dry weight, with a considerable increase to close to 200 ppm during the second month of life. This coincides with the time when spermatids are transformed into spermatozoa.

Higher levels than those just cited occur in the normal prostate gland of the rabbit and man, the dorsolateral prostate of the rat, and in human seminal fluid and spermatozoa. The following average values, in parts per million Zn on dry basis have been reported: human prostate, 859; dorsolateral rat prostate, 891; human semen, first fraction of ejaculate, 2930, second fraction, 1400, third fraction, 910; human spermatozoa, 1990 (115). Sperm-rich boar semen is intermediate in zinc concentration between bull semen (approximately 10 μg/ml) (275) and human semen (50–200 μg/ml) (13). The concentration in sperm and seminal plasma is approximately equal when expressed on a dry weight basis (115). Boar spermatozoa accumulate zinc when stored *in vitro* in contact with boar seminal plasma (275). The zinc content of spermatozoa can, therefore, be affected by time of sampling after ejaculation, conditions of storage, and membrane integrity.

Little is known of the chemical forms in which these high concentrations of zinc exist. A very small proportion, of the order of 5%, is present as carbonic anhydrase in the rat dorsolateral prostate and in human semen, and a variable proportion is dialyzable (113,115). In ram (113) and rat (123) sperms, the zinc is concentrated in the tail, and in the rat prostate in the luminal edge of the acinar cells (123,194). The level of zinc in the dorsolateral prostate is reduced by castration and the rate of accumulation in young rats is greatly increased by injections of testosterone or gonadotropin (60,121,122). Chorionic gonadotropin increases the weight and $^{65}$Zn uptake in all the other accessory reproductive structures in male rats, as well as the dorsolateral prostate, whereas follicle-stimulating hormone (FSH) reduces this uptake per organ and per gram of tissue (198).

The administration of stilbestrol lowers the zinc content of human prostatic cancerous tissue, which even before treatment is lower in zinc concentration than that of normal or of hyperplastic but nonmalignant, prostatic tissue (56,73a,248). The range of zinc concentration is so wide that zinc determinations have little diagnostic value. In one study the zinc concentration of 28 normal prostate glands taken from accident victims ranged from 186 to 1340 ppm (mean 732 ppm Zn on dry basis); 33 hyperplastic but not malignant glands taken by surgery ranged from 190 to 2900 ppm (mean 972 ppm Zn) (248). Very similar values were obtained for normal human prostates in another study, but the zinc concentration in benign prostatic cancer was generally lower, with a range of 30 to 884 ppm and a mean of 486 ppm Zn of dry tissue (56). Whitmore (277) states that the concentration of zinc in tumor tissue localized in the prostate gland is approximately one-third that of the normal prostate.

## 5. *Zinc in Blood*

Zinc is present in the plasma and in the erythrocytes, leukocytes, and platelets. About one-third of the plasma zinc is loosely associated with the serum albumins and the remainder is more firmly bound to the globulins (283). A specific zinc-protein transport compound has not been identified in plasma. Almost the whole of the zinc in erythrocytes occurs as carbonic anhydrase (78), together with a very small fraction associated with other zinc enzymes known to be present (257). Carbonic anhydrase cannot be detected in plasma or leukocytes so that the zinc in erythrocytes must account for the whole of the activity of this enzyme in blood (100). Human leukocytes contain a zinc metalloprotein with no known enzymic activity (255,258). The polymorpholeukocytes of rabbit blood are almost free of zinc, whereas the zinc content of mononuclear lymphocytes parallels their phosphatase activity (112). Calculations from data obtained by Vallee and Gibson (256) indicate that 75-88% of the total zinc of normal human blood is contained in the red cells, 12-22% in the plasma, and 3% in the leukocytes. The individual leukocyte contains 25 times as much zinc as the individual erythrocyte. Human blood platelets contain zinc in amounts ranging from 0.2 to 0.45 $\mu g/10^9$ platelets, with a mean of 7.1 $\mu g$ of platelet zinc in 100 ml of whole blood (50). The zinc content of serum is consistently higher than that of plasma, by an average of 16%. Of this increase, 44% was demonstrated as being derived from platelets disintegrating during clotting, 39% from a slightly greater dilution in plasma, and 4.0% from hemolysis (50).

In all species studied the zinc content of the erythrocytes is lower than in man. This is evident from the following figures for zinc in erythrocytes reported by Smirnov (216): man, 13; rat, 10; dog and rabbit, 9; and goose, 6.5 μg/g. In the adult rabbit, average zinc levels are 2.5 μg/ml of whole blood, 2.7 μg/ml of plasma, and 9 μg/ml of erythrocytes (10,112). In young growing pigs, 0.6 μg Zn/ml of plasma and 7 μg/ml of erythrocytes can be considered typical normal levels (77,125). Vallee and Gibson (256), using a dithizone method, reported the following values for normal human blood: whole blood, 8.8±0.2; plasma, 1.21±0.19; and erythrocytes, 14.4±2.7 μg/ml. Prasad *et al.* (182) later confirmed these values by an atomic absorption method.

Numerous studies have established the normal range of zinc in human blood plasma and red cells and the changes that occur in various disease states (64,70a,91,95,182,256,260,266). Individual variability is high, differences due to age, race, or sex are small or nonexistent, and there appear to be no significant seasonal or diurnal variations. Significant regional differences in the United States in the plasma zinc of normal males were reported by Kubota and co-workers (95). These are of particular interest in view of the growing evidence that the supply of zinc to the human population may be marginal. In women taking oral contraceptives, and in the third trimester of pregnancy, plasma zinc levels are significantly lower than normal. Using a simplified method of plasma zinc determinations by atomic absorption spectrometry (63), Halsted *et al.* (65) reported mean values of 0.96±0.13 μg Zn/ml for control women, 0.60±0.11 μg/ml for 25 women in the third trimester of pregnancy, and 0.65±0.08 μg/ml for 10 women who had been taking oral contraceptives for 1 month to 4 years. In newborn infants the zinc content of the erythrocytes is only one-quarter of the adult value, rising progressively over the first 12 years of life (10). This would be expected from the low levels of carbonic anhydrase present in the red cells of newborn and premature infants (231).

The zinc content of blood is responsive to changes in dietary zinc intake. Large oral doses of zinc greatly increase whole blood and plasma zinc concentrations in rats, rabbits, cats, and pigs (10,77,109) and in lambs (156) and cattle (154). Cattle given incremental zinc supplements, to provide total dietary zinc intakes ranging from 18 to 189 ppm, revealed increases in serum zinc from means of 1.5 and 1.8 μg/ml to means of 2.5 and 2.7 μg/ml (167). Subnormal plasma zinc levels have not been observed in all studies of zinc deficiency in animals (78,148). With the use of improved analytical methods, and at more severe deficiency states, a decline in plasma zinc is invariably evident. Mills *et al.* (139)

reported a marked fall from normal levels of 0.8–1.2 μg Zn/ml to below 0.4 μg/ml in the serum of severely Zn-deficient lambs and calves. Significant falls in whole blood and plasma zinc occur in less acutely Zn-deficient lambs, calves, and goats (127,128). A mean serum zinc concentration of 0.22 μg/ml was reported in Zn-deficient baby pigs, compared with 0.98 μg/ml in pair-fed controls (125). Pigs receiving adequate zinc but a low-protein diet have also been shown to carry serum zinc concentrations as low as 0.36 μg/ml associated with very low plasma protein levels (171). Mean serum zinc levels in rats fed a severely Zn-deficient diet (0.8 ppm Zn) were shown by Luecke *et al.* (105) to rise steadily from less than 0.4 μg/ml to a plateau close to 1.4 μg/ml with incremental increases in supplementary zinc up to 12.5 ppm. It was concluded that a serum zinc level of 1.3 μg/ml indicates dietary zinc adequacy with respect to growth in this species.

Profound changes in the zinc content of the plasma and cellular elements occur in various diseases. Subnormal plasma zinc levels have been reported in patients with atherosclerosis (268), with malignant tumors (1,266), with myocardial infarctions in one study (269) but not in another (37), with chronic and acute infections (266), with untreated pernicious anemia (70a,242,256), and with postalcoholic cirrhosis of the liver and other liver diseases (64,86,182,260). Such changes indicate abnormalities in zinc metabolism but their mechanisms are not understood. In patients with anemia, other than pernicious anemia, the zinc and carbonic anhydrase levels in the blood are lowered in parallel fashion so that the decreases are proportional to the decreases in hemoglobin and red cell counts, and the zinc and enzyme values per unit of red blood cells (RBC) remain in the normal range. In pernicious anemia, Vallee and Gibson (256) found the zinc and carbonic anhydrase values per unit of RBC to be significantly above normal, even when the increased red cell size was eliminated as a contributing factor. A significant decrease in erythrocyte carbonic anhydrase activity occurs in sheep given sulfanilamide orally at the rate of 4 g/kg body weight (243) and in hemolytic anemia (2). Significant breed differences in erythrocyte carbonic anhydrase activity also occur in this species (2).

Marked changes in leukocyte zinc occur in patients with chronic leukemia. The concentration of zinc in the peripheral leukocytes is greatly reduced below normal and cannot be raised by injections of zinc gluconate. A rise to normal levels occurs in clinical remission and under therapy with X-rays or urethan, accompanying the falling leukocyte count (56). The zinc content of the leukocytes also decreases in patients with a variety of neoplastic diseases and this difference has been suggested as a diagnostic test for cancer (241). In refractory anemia, accompanied by leukopenia with white cell counts below 2000/mm$^3$, a three- to tenfold *elevation* of the zinc content of the leukocytes has been observed. The physiological significance of these changes in zinc concentration is unknown.

### 6. *Zinc in Milk*

The concentration of zinc in milk is relatively high. A high proportion of reported values for cow's, ewe's, and human milk lie between 3 and 5 μg Zn/ml (6,10,31,90,161,205). The level of zinc in normal sow's milk appears to be somewhat higher; Earle and Stevenson (43) reported a mean of 6.9 μg Zn/g for milk obtained on the thirty-fifth day of lactation. The actual levels depend upon the zinc status of the diet, since the zinc concentration in milk is raised by high intakes of zinc (6,10,43,132). Very little of the zinc in cow's milk is associated with the fat. About 12% is present in dissolved ultrafilterable form and the remainder is bound to casein (161).

The concentration of zinc in colostrum is 3-5 times that of true milk in cows (90), sows (43), and women (10).

### 7. *Zinc in Avian Egg*

The normal hen's egg contains 0.7-1.0 mg Zn, almost all of which is present in the yolk (14). The zinc is associated with the protein component (vitellin) of the lipovitellin (252). The amount of zinc present is reduced in the eggs of Zn-deficient hens, resulting in reduced hatchability and embryonic abnormalities.

## III. Zinc Metabolism

### 1. *Absorption*

Zinc absorption occurs mainly in the small intestine, predominantly in the duodenum. Little is yet known of the mechanisms involved. It is poorly absorbed from ordinary diets by rats (48) and by steers (49). Absorption is affected by the level of intake of the element, by the amounts and proportions of several other elements and dietary components, and by the chemical form in which the zinc is ingested. There is some slight evidence that zinc absorption decreases with age in rats (61) and cattle (126), although there is some doubt whether this is a true maturity effect (129).

Much of the evidence on the factors that can influence zinc absorption is inferred from the way these factors mitigate or aggravate signs of zinc deficiency in the animal. Such changes need not necessarily arise from effects on zinc absorption, since zinc utilization and excretion could also be affected. High calcium intakes potentiate the zinc deficiency syndrome in pigs (101,103,230,252), in dogs (195), and in birds (89,196). The major but not the only site of the calcium-zinc interaction is at the intestinal level, resulting in reduced zinc absorption (71,72,73). Such a calcium-zinc antagonism could not be demonstrated in man (226). The average intestinal absorption of radiozinc was 35.7% in 5 normal subjects during a period of low Ca intake, with no significant change in $^{65}Zn$ absorption when their Ca intake was increased six- to tenfold. High inorganic phosphorus intakes can aggravate zinc deficiency in rats. Whether this results from an interference with zinc absorption is not clear (26). Judging by capacity to promote growth in chicks fed a Zn-deficient diet, zinc is equally well absorbed as the oxide, carbonate, sulfate, or metal, whereas the zinc in sphalerite (mostly zinc sulfide) and in franklinite (oxides of Zn, Fe, and Mn) is largely unabsorbed (44).

Direct evidence of an effect of high levels of copper upon zinc absorption has been obtained by Van Campen (262). This worker placed $^{65}Zn$ into isolated duodenal segments of the rat and administered copper either intraduodenally or intraperitoneally. Zinc absorption was reduced with the former treatment but not with the latter, indicating that copper interference with zinc uptake is mediated at the intestinal level. Taken in conjunction with earlier work (263), in which high intakes of zinc were similarly shown to depress copper absorption, a mutual antagonism is apparent between zinc and copper in the absorptive process, taking place in or on the intestinal epithelium. The *in vitro* studies of Sahagian and co-workers (201) suggest that the transmural passage of a metal by the rat intestinal wall depends upon two separate steps. The first is the uptake and binding of the metal by extracellular and intracellular surfaces, and the second is the transport of the metal across cellular membranes and accumulation in the solution bathing the opposite surface of the wall.

Cadmium aggravates zinc deficiency in poultry (236), pigs (69), and calves (179), and zinc can prevent or delay certain toxic effects of injection of a cadmium salt (159,225). It appears that cadmium competes with zinc at important cellular binding sites. Further aspects of zinc-cadmium antagonism are considered in Chapter 10.

Different protein and amino acid sources vary in their effects upon the zinc needs of rats, pigs, chicks, and poults. O'Dell and Savage (149) attributed these differences to the presence or absence of phytates which bind the zinc in a form from which it is not readily released and absorbed. This has been amply confirmed, and the addition of phytic acid to casein diets was shown to reduce zinc retention to that in soybean protein diets (52,146,172). The availability of zinc for absorption from combinations with phytic acid is increased by autoclaving or by treatment with EDTA (92,149,218). However, phytic acid is only one of the factors concerned. Lease and Williams (96) found the availability of zinc from various oilseed meals to be highly variable and not related to their phytate contents. Scott and Zeigler (214) have presented evidence that certain natural feeds, such as casein and liver extract, contain chelates which improve zinc absorption and utilization. Several natural and synthetic chelating agents with suitable stability constants improve the availability of zinc to chicks (143) and poults (267) consuming soybean protein diets. In rye grass, and presumably other forages, a proportion of the zinc is present in the form of anionic complexes (19). The influence of such complexing upon the availability of zinc to the animal is unknown.

Published results of a possible effect of vitamin D on zinc absorption and utilization are confusing. From the work of Becker and Hoekstra (9) with rats, it appears that the increased absorption of zinc attributed to vitamin D is not a direct effect of the vitamin but results from a homeostatic response to the increased need for zinc which accompanies enhanced skeletal growth and calcification. Homeostatic control of zinc absorption also operates in respect to the dietary level of zinc. Absorption of $^{65}Zn$ is enhanced in zinc deficiency and declines at higher zinc intakes (120,129,130).

## 2. *Excretion*

Zinc leaves the body largely by way of the feces (47,49,117,130). Fecal zinc consists mostly of unabsorbed dietary zinc, with a small amount of endogenous origin. Endogenous zinc is secreted mainly into the small intestine, chiefly via the pancreatic juice, except in the pig (162). Only small amounts are secreted into the bile and into the cecum and colon (120). Injected zinc is also mostly excreted in the feces, with little appearing in the urine (117,130,162). Thus about 70% of radiozinc administered orally to steers was recovered in the feces and 0.3% in the urine. When the zinc was given intravenously, 20% appeared in the feces and 0.25% in the urine (49). This pattern of excretion was followed on diets normal in zinc and on those high in zinc. Miller and associates (128-130) showed that endogenous excretion of fecal $^{65}Zn$ and stable zinc is significantly reduced in calves and goats on low-Zn diets, so contributing substantially to homeostatic control of this element. Urinary excretion of zinc was increased somewhat in the Zn-deficient animals, but the total excretion of $^{65}Zn$ in the urine over 13 days was still less than 0.3% of the dose.

The quantity of zinc excreted in the urine of healthy human adults is very small (0.1-0.7 mg/day), compared with the 10-15 mg/day normally ingested. The amounts so excreted do not vary appreciably with the level of zinc in the diet and are not significantly increased following zinc injections. The normal kidney has clearly a very limited ability to excrete zinc. However, Cavell and Widdowson (31) found the urine of 10 1-week old, breast-fed infants, 9 of whom were in negative zinc balance, to average 1 $\mu$g Zn/ml, which is about 5 times the concentration of normal, adult urine.

Urinary excretion of zinc is well above normal in nephrosis, postalcoholic hepatic cirrhosis, and hepatic porphyria (168,182,235). Six patients with albuminuria averaged 2.1 mg Zn/day in their urine (range 1.0-3.8), whereas healthy individuals averaged 0.3 mg/day (range 0.1-0.5) (117). Heavy losses of zinc in the urine of albuminurics can be explained by the inability of the nephrotic kidney to provide an effective barrier against the protein-bound zinc of the plasma. The pronounced zincuria of postalcoholic cirrhosis of the liver is less easily understood. Vallee and co-workers (261) found sufferers from this disease to excrete from 1 mg Zn/day in the urine, compared with under half this amount by noncirrhotic controls.

Increased urinary excretion of zinc occurs in total starvation in man (224), in hypertensive patients (209), and during the administration of EDTA (166). In hypercholesteremic patients, parenteral administration of $Na_2Ca$ EDTA induces a tenfold increase in urinary zinc levels and a significant fall in serum cholesterol. Oral administration of this compound caused a fourfold increase in urinary zinc but no change in serum cholesterol levels (165). The relatively long biological half-life of $^{65}Zn$ in man (322 days), its wide distribution due to stratospheric fallout, and prolonged retention in the tissues, have stimulated studies of means of accelerating its removal from the body. Considerable enhancement of urinary $^{65}Zn$ excretion in man has been demonstrated by Spencer and Rosoff (225) by administration of both EDTA and diethylenetriaminepentaacetic acid (DTPA) particularly the latter. These compounds and two other chelating agents with high stability constants for zinc, were found highly effective at relatively long-term intervals after the administration of the $^{65}Zn$ dose.

Significant quantities of zinc can be lost in the sweat, especially in hot climates where the volume of sweat may rise to 5 or more liters/day (184a). Prasad *et al.* (184a) found the sweat of normal individuals to average $1.15 \pm 0.30$ $\mu$g Zn/ml, most of which was present in the aqueous phase, i.e., not associated with the cellular elements, as is the case with iron. In Zn-deficient patients the mean zinc level in the whole sweat was reduced to $0.6 \pm 0.27$ μg/ml. On this basis, a normal individual secreting 5 liters of sweat per day could lose 5 mg Zn/day and a Zn-deficient individual, about 2 mg/day. In temperate climates the losses of zinc by this route would, of course, be substantially smaller.

### 3. *Intermediary Metabolism*

Absorbed or injected zinc is incorporated at differing rates into different tissues which reveal varying rates of zinc turnover. Uptake of zinc by the bones and the central nervous system (CNS) is relatively slow, and the zinc remains firmly bound for long periods. Zinc entering the hair is not available to the tissues and is only lost as the hair is shed (11,57). The most rapid accumulation and turnover rate of retained zinc occurs in the pancreas, liver, kidney, and spleen (49,66). In the rat, the retained radiozinc accumulates most rapidly in the pancreas and the dorsolateral prostate, but the turnover rate in the latter is slower than it is in other tissues (270). Uptake of $^{65}$Zn by tumor tissue located in organs that normally have a low concentration of zinc is similar or greater than that of the tissue in which the tumor is located (199).

The zinc in these tissues and the more slowly exchanging zinc in the muscles and the red cells constitute a soft tissue zinc pool composed of compartments of varying exchange rates. Further compartments, also with varying exchange rates, exist within the cells of particular tissues (8,32). Cotzias *et al.* (32,33) showed that the various tissues of mice, as well as various intracellular organelles of mouse liver cells, show distinct individuality relative to the time course of $^{65}$Zn distribution after intraperitoneal injection. The feeding of a zinc salt accelerated the loss of $^{65}$Zn without affecting the characteristic partition of the isotope among either the organs or the intracellular organelles. The specificity of the zinc pathway through the body was disturbed when the stable zinc was injected and normal absorption was bypassed. The sensitive homeostatic control of zinc metabolism acting at the sites of absorption and of excretion appears, therefore, to operate only when the absorptive and excretory mechanisms are functioning together (33).

Injected $^{65}$Zn combines initially with the plasma proteins to form a plasma zinc pool and more slowly with the intracellular proteins of the cellular components. Clearance of plasma zinc, but not of erythrocyte zinc, is rapid to the soft tissues of the body and to the feces via the small intestine (162). The $^{65}$Zn is also rapidly transported from the plasma through the placenta, so that 2 hr after injection into pregnant mice, a high content is present in fetal bones and liver (11).

## IV. Zinc Deficiency and Functions

Zinc deficiency has been demonstrated in mice, rats, pigs, guinea pigs, birds, sheep, cattle, goats, monkeys, and man. The deficiency syndrome is characterized clinically by growth retardation or failure, by inappetence, by lesions of the skin and its appendages, and by impaired reproductive development and function. Gross bone disorders further characterize zinc deficiency, particularly in birds. The sequence in which these manifestations appear, and their relative prominence, differ with the age and sex of the animal, and with the degree of deficiency imposed.

Naturally occurring zinc deficiency has been reported in pigs (251), in cattle (42,62,98), and probably in young lambs (169). However, zinc deficiency is not a major "area" problem in grazing stock in the way that copper, cobalt, and selenium deficiencies are. Marginal zinc deficiency in man may be more widespread, as will be shown below.

### 1. *Growth and Food Utilization*

Retardation of growth in young animals was observed in the original demonstration of zinc deficiency in rats (249) and has been a conspicuous feature of this deficiency in all subsequent studies. In experiments with lambs and calves fed a grossly Zn-deficient diet (average 1.2 ppm Zn), weight gain ceased abruptly and growth arrest occurred within 2 weeks (139). At higher but still severely deficient intakes of zinc (2.4 ppm) the typical skin lesions appeared several weeks before growth and appetite became significantly depressed (254). The growth inhibition results partly from reduced food intake and partly from an impaired utilization of the food that is consumed and absorbed (109a,125,135,183, 184,222). The apparent digestibility of the food is unaffected in zinc deficiency, at least in ruminants (137,222). Somers and Underwood (222) found the fecal excretion of nitrogen and sulfur to be similar in zinc-deficient ram lambs and in pair-fed control lambs consuming and digesting the same amounts of the same basal diet. The urinary excretion of nitrogen and of sulfur was greatly elevated in the zinc-deficient animals (see Table 29). Of great interest, also, is the evidence obtained by Oberleas and Prasad (147) that a relationship exists between zinc intakes and the utilization of seed proteins and that these proteins can be equal in quality to animal proteins for growth promotion when supplemented with zinc.

### 2. *Zinc and Keratogenesis*

Alopecia accompanied by gross epithelial lesions, especially of the skin, were observed in the early investigations of zinc deficiency in rats

TABLE 29
*Effects of Zinc Deficiency on Feed, Digestibility, and Nitrogen and Sulfur Balances in Ram Lambs*[a]

| Parameters affected | Zn-deficient group[b] | Pair-fed controls[b] |
|---|---|---|
| Dry matter digestibility (%) | 64.5±1.2 | 66.8±0.6 |
| Nitrogen in feces (g/day) | 4.6±0.2 | 4.5±0.2 |
| Sulfur in feces (g/day) | 0.4±0.04 | 0.4±0.04 |
| Nitrogen in urine (g/day) | 5.8±0.30 | 3.7±0.21[c] |
| Sulfur in urine (g/day) | 0.63±0.06 | 0.38±0.02[c] |
| Nitrogen balance (g/day) | +2.0±0.08 | +4.2±0.16[c] |
| Sulfur balance (g/day) | +0.25±0.02 | +0.49±0.04[c] |

[a]From Somers and Underwood (222).
[b]Each group consisted of four animals.
[c]Means were significantly different at the 0.01 level.

and mice. Histological studies revealed a condition of parakeratosis, i.e., a thickening or hyperkeratinization of the epithelial cells of the skin and esophagus (51). The changes of parakeratosis involve excessive keratinization, retention of the nuclei, and exfoliation with some exudate containing red cells. The many histochemical similarities between the lesions of parakeratosis in pigs, and those of psoriasis in man, have recently been revealed (3). In more severe zinc deficiency in rats, scaling and cracking of the paws, with deep fissures, develop, in addition to coarseness and loss of hair and to dermatitis (53). In pigs, parakeratosis can appear on any part of the body subjected to mild trauma (134). It mostly occurs around the eyes and mouth and on the scrotum and lower parts of the legs. A similar distribution of the parakeratotic lesions occurs in zinc-deficient ruminants, with the legs becoming tender, easily injured, and often raw and bleeding (15,254). In the zinc-deficient squirrel monkey (*Saimiri sciureus*), in addition to alopecia, a unique histopathological lesion, parakeratosis of the tongue, has been reported (7,110). The healing response of the parakeratotic lesions to supplemental zinc, whether administered orally or injected, is rapid and dramatic.

The influence of zinc deficiency upon keratogenesis is particularly evident in the sheep, through changes in the wool and horns. In horned lambs the normal ring structure disappears from new horn growths. These are ultimately shed leaving soft spongy outgrowths that continually hemorrhage (139,254). In breeds which normally show no horn growth, prolonged zinc depletion leads to the development of keratinous outgrowth or "buds" in the positions normally occupied by horns in other breeds (139). Changes in the structure of the hooves can also occur and the effects of zinc deficiency upon wool growth in lambs are particularly striking (139,254). The wool fibers lose their crimp and become thin and loose and are readily shed. Sometimes the whole fleece is shed and no further wool growth occurs until additional zinc is supplied when regrowth is immediate.

In the Zn-deficient chick (148), turkey poult (93), and pheasant (213), feathering is poor and abnormal and dermatitis is usual. In the Japanese quail (227), there is no dermatitis but feathering is severely affected (see Fig. 13). The poor feather development has been explained on the basis of a degeneration of the feather follicles resulting from hyperkeratosis (148). Further evidence of the involvement of zinc in keratogenesis comes from studies of chick embryos from the eggs of severely Zn-deficient hens (16, 89). Gross disturbances of the integument, as well as of the skeleton, were observed.

### 3. *Zinc and Bone Growth*

Bowing of the hind legs and stiffness of the joints in Zn-deficient calves (127), rapidly reversible by zinc feeding, and skeletal malformations in the fetuses from Zn-deficient rats (84) have been observed. In a study of zinc deficiency in baby pigs, Miller and co-workers (125) found a reduction in the size and strength of the femur. Comparisons with pair-fed controls indicated that these changes were due to the reduced food intake. Bone growth was affected in direct proportion to the effect of zinc deficiency upon body growth, and maximum strength of the femur was directly related to bone size.

**Fig. 13.** Wings from zinc-supplemented (upper pair) and zinc-deficient (two lower pairs) quail, 4 weeks of age. (Photo by courtesy of Dr. M. R. Spivey Fox.)

Skeletal abnormalities are a regular and conspicuous feature of zinc deficiency in growing birds. This was first observed in chicks by O'Dell (148) and has been confirmed in further studies with chicks, poults, pheasants, and quail (93, 213, 227, 284). The long bones are shortened and thickened in proportion to the degree of zinc deficiency (285). Changes and disproportions occur in other bones, giving rise to a perosis histologically similar to that of manganese deficiency (148, 284). In chick embryos from the eggs of hens fed severely Zn-deficient diets, gross skeletal and other deformities arise, including agenesis of all or part of the limbs, dorsal curvature of the spine, and shortened and fused thoracic and lumbar vertebrae (16, 89).

The exact mechanism of action of zinc in bone formation is still not known. Decreased osteoblastic activity in the bony collar of the long bones has been noted and a reduction or failure in chondrogenesis associated with an increase in the amount of cartilage matrix (148, 284). The ash content of the bones is sometimes reduced in zinc deficiency (24, 227, 239, 284), whereas, in other instances, there is no such change (187, 229). Bone alkaline phosphatase activity is invariably reduced, even in the absence of any significant reduction in bone ash (229). The significance of the reduction in phosphatase activity is, therefore, not clear. In all cases the zinc concentration of the bones in zinc deficiency is below normal, whether expressed on a basis of fresh weight, dry fat-free weight, or bone ash. Zinc clearly plays a part in the events leading to calcification but at what point remains to be discovered.

An intriguing feature of the effects of zinc deficiency upon bone is that the leg defects do not develop in Zn-deficient chicks on all types of diet, when other signs of the deficiency are apparent (143). Furthermore, these "perosislike" or "arthritislike" abnormalities can be alleviated by histamine, histidine, and various antiarthritic agents, without affecting the other manifestations of zinc deficiency (144). Nielsen *et al.* (144) found that histidine at 1.0 and 2.0%, histamine at 0.2%, aspirin at 0.5-1.0%, phenylbutazone at 0.2%, cortisone acetate at 0.1%, and indomethacin at 0.025% of the diet protected against the leg disorder of chicks fed a Zn-deficient soybean protein diet. Only supplementary zinc prevented this abnormality and also all other signs of the deficiency. The nature of the interaction between zinc and these agents is unknown. They do not act by increasing the availability of dietary zinc.

### 4. *Zinc and Wound Healing*

The first indication that zinc might play a significant role in wound healing came from the studies of Pories and Strain and their associates.

They observed an accelerated rate of wound healing in rats fed an accidentally contaminated diet and later identified the beneficial contaminant as zinc (174,178). Subsequent investigations of patients suffering from major burns revealed subnormal levels of zinc in the hair. This suggested that such patients develop serious zinc deficits and that zinc deficiency might be one of the causes of the difficult healing problems often seen in this condition (178). A quantitative method of measuring wound healing in man was then developed (177), and the rate of wound closure was shown to be directly proportional to hair zinc levels (178). The addition of zinc sulfate to the diet of young men, at the rate of 50 mg Zn, 3 times daily, following surgery for pilonidal sinuses, increased significantly the rate of wound healing, compared with that of unmedicated controls (175, 176) (see Table 30). The normal diets being consumed by these men could thus be regarded as zinc deficient, if rapid rate of wound healing is considered a criterion of adequacy.

Wound healing is impaired in Zn-deficient calves (134) and rats (204). In the calves the rate of regeneration of new skin, following removal of biopsy specimens, was similarly impaired by feed restriction alone and was not improved by adding zinc to a practical diet containing 30 ppm Zn (131). The latter point was not tested in the experiment with rats, but the tensile strength of a healing surgical incision in the integument was significantly lower, 12 days after operation, in the Zn-deficient than in pair-fed control animals. The precise mode of action of zinc in tissue repair is unknown. Zinc is preferentially concentrated in healing tissues in rats, both in skin and muscle wounds (206), and in bone fractures (28). This suggests a heightened metabolic demand for zinc for tissue synthesis during the healing process.

## 5. *Zinc and Reproduction*

Zinc deficiency affects spermatogenesis and the development of the primary and secondary sex organs in the male, and every phase of the

TABLE 30

*Influence of Oral Zinc Sulfate on Wound Healing in Young Men*[a]

| Patients (No.) | Average age | Wound vol. (ml) | Days for healing | Healing rate (ml/day) |
|---|---|---|---|---|
| Controls (10) | 25.0 | 32.3 | 76.9±11.3 | 0.45±0.09 |
| Zinc-supplemented (10)[b] | 24.6 | 54.5 | 45.8±2.6 | 1.25±0.30 |
| | | | $P<0.02$ | $P<0.01$ |

[a]From Pories *et al.* (174).

[b]Oral zinc sulfate (50 mg Zn) 3 times per day.

reproductive process, from estrus to parturition, in the female.

Atrophic seminiferous tubules were observed in the earliest study of the histopathology of zinc deficiency in male rats (51). Mawson and coworkers (115, 121) later reported retarded development of the testes, epididymes, prostate and pituitary glands, with atrophy of the testicular germinal epithelium. Rapid growth and development of the dorsolateral prostate was induced in the Zn-deficient rat by gonadotropin or testosterone treatment, without increasing the zinc concentration in this gland to its normally high level (122). Evidence from experiments involving restriction of food intake to the level of Zn-deficient rats, and treatment of such rats with gonadotropin or male sex hormone, suggests that the impaired development of the accessory sex glands is secondary to the severe inanition of zinc deficiency, and that this inanition results in a reduced gonadotropin output and a consequent fall in androgen production (122). Direct evidence to support this hypothesis involving actual measurements of the output of these hormones in zinc deficiency has not appeared. The testicular atrophy and failure of spermatogenesis appears to be due directly to lack of zinc. The availability of sufficient zinc for incorporation of high concentrations into sperm during the final stages of maturation seems to be essential for the maintenance of spermatogenesis, and for the survival of the germinal epithelium (122, 254, 276).

Impaired development and functioning of the male gonads is apparent in other species. Hypogonadism, with suppression in the development of the secondary sexual characteristics, is a conspicuous feature of the conditioned zinc deficiency observed in young men in parts of Iran and Egypt (181). Hypogonadism also occurs in Zn-deficient bull calves (128, 170), kids (136), and ram lambs (254). In an experiment with ram lambs, testicular growth was significantly impaired (254). Spermatogenesis ceased, within a period of 20 to 24 weeks, in the lambs receiving a diet containing 2.4 ppm Zn. Testicular weights and sperm production of comparable lambs consuming the same basal diet, plus zinc sulfate to bring the total dietary zinc intake to 32.4 ppm but restricted to the total feed intake of the Zn-deficient animals, were significantly increased, without affecting body-weight gains. Further evidence of the particular importance of zinc to male sex gland development and function was obtained by a comparison of two groups of ram lambs receiving diets containing 17.4 and 32.4 ppm Zn, each with unrestricted feeding. Growth and food consumption were similar in the two groups but testicular growth and sperm production were significantly greater in the lambs receiving the larger zinc supplement. These differences were further reflected in the histological ratings given to the testes of the animals from the different treatments. The fructose concentration of the seminal plasma, which provides an index of testosterone production by the gonads (113), was similar for all dietary treatments. This suggests that testosterone output is not impaired in zinc deficiency in ram lambs, as was suggested for rats (122). Remission of all signs of zinc deficiency and recovery of testicular size, structure, and sperm production were achieved during a zinc repletion period lasting 20 weeks, indicating that the tissues had not been permanently damaged by the severe zinc deficiency imposed. Complete recovery was similarly obtained by Pitts *et al.* (170) with Zn-deficient calves.

The nature of the effects of zinc deficiency upon reproduction in female rats depends upon the severity, the duration, and the timing of the deficiency. When Hurley and Swenerton (84) fed a virtually Zn-free, soybean protein diet to female rats from weaning to maturity, they did not grow and showed severe disruption of the estrous cycles. In most cases no mating took place and the animals were completely infertile. With similar rats maintained on a marginally Zn-deficient diet (9 ppm) throughout the period from weaning to maturity and then mated, estrous cycles were not disrupted and mating took place normally. When these rats were given the severely Zn-deficient diet during pregnancy, less than half had living young at term and 98% showed gross congenital malformations. In the deficient females, 99% of the implantation sites either showed a resorbed conceptus or gave rise to a malformed fetus. These abnormalities apparently arose from a direct effect of lack of zinc in the fetal tissues, since the fetuses from the deficient females contained less zinc than their controls. Similar effects of severe zinc deficiency upon mating and fetal development in rats were obtained by Apgar (5) with an egg-albumin diet. After 4 weeks on the low-Zn diet ( $< 1$ ppm) the number of female rats that would mate was greatly reduced, compared with those receiving the same diet plus 2-3 ppm Zn in the drinking water. When this source of zinc was removed on the day sperm were found, only half the females were pregnant on day 8 and all of them had resorbed their fetuses by day 21. No congenital abnormalities and no reduction in the zinc content of the fetuses occurred in an earlier study in which the zinc deficiency was less severe (85).

It has also been shown by Apgar (4) that severe zinc deficiency, imposed upon normal adult female rats from the first day of gestation, can markedly affect the female herself, as well as the fetuses. The zinc content of the fetuses was reduced at day 22, and 75% of the young died within 2 hr of birth. Marked differences in maternal behavior before and after parturition and in the time of emergence of a fetus were observed between the Zn-deficient females and those from either the pair-fed or the unrestricted-fed control groups. The Zn-deficient females delivered their litters with extreme difficulty, suffered excessive bleeding, and failed to consume afterbirths or to prepare a nest site (see Fig. 14). A feature of this notable demonstration of the effect of zinc deficiency upon maternal behavior at parturition is the speed with which it developed. Body stores of zinc were apparently too small and too inaccessible to meet the needs of either the mother or her fetuses during pregnancy. This effect of zinc upon parturition has since been confirmed (27). No such behavioral disturbances arise when deficiencies of copper and manganese are similarly imposed upon pregnant rats (4).

Comparable studies of the effects of severe zinc deficiency upon the female reproductive processes have not been carried out with farm animals other than poultry. The decreased hatchability of eggs and the grossly impaired development and high mortality of chick embryos from the eggs of Zn-deficient hens (16, 89) were noted earlier. Suboptimal dietary levels of zinc fed to sows reduced the size of litters and the zinc

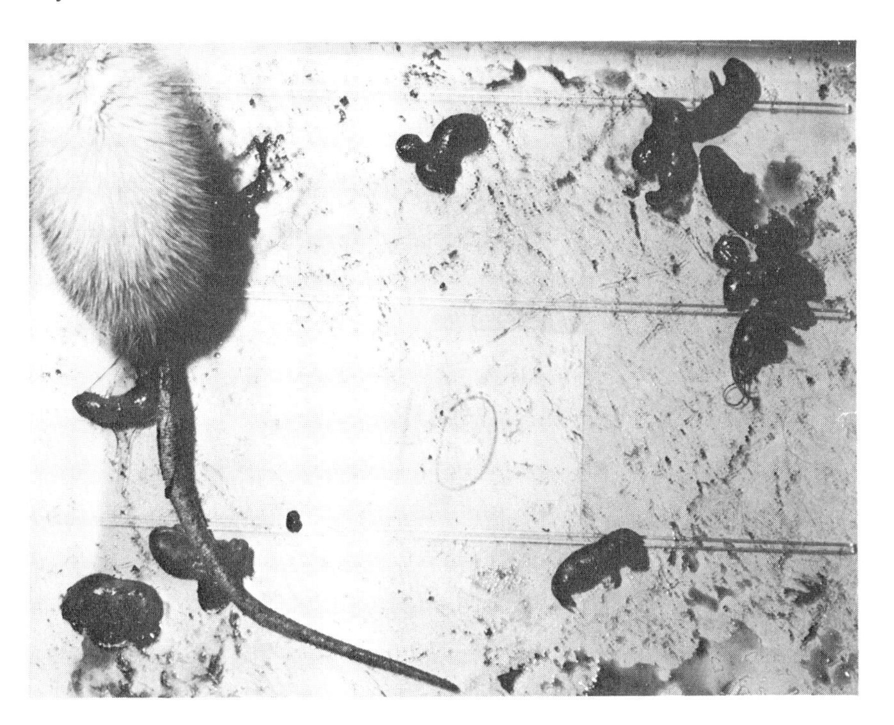

**Fig. 14.** Cage of zinc-deficient female rats 2 hr after parturition. Note excessive bleeding, failure to consume afterbirths, and lack of a nest site. (Photo by courtesy of Dr. J. Apgar.)

content of some of the tissues of the young but no abnormalities in fetal development or maternal behavior were reported (76).

### 6. *Zinc and Learning Behavior*

Further evidence of the profound influence of zinc on behavior potential has been obtained by Caldwell and Oberleas (27). These workers compared the behavioral pattern of the surviving offspring of mildly Zn-deficient mothers with similar young rats from Zn-supplemented mothers. The latter were found to be significantly superior both in the Lashley III water maze test and in a platform avoidance conditioning test. In addition, the Zn-supplemented rats revealed significantly greater conditioned responses, versus escape responses, compared with the Zn-deficient animals. This superior learning ability was accompanied by a larger "activity score" which was interpreted as indicating a reduced emotionality. It was emphasized that these effects of zinc deficiency upon behavior were obtained on soybean protein diets apparently adequate in protein.

### 7. *Zinc and Protein Metabolism*

Nucleic acid and protein synthesis was first shown to be impaired in Zn-deficient microorganisms (142, 271, 281) and plants (88, 208). Impaired *in vivo* synthesis of DNA in the liver of Zn-deficient rats, compared with pair-fed controls, has been demonstrated in several studies (23, 55, 203). The incorporation of $^{32}P$ into rat liver nucleotides (280) and of tritiated thymidine into the DNA of liver, testis, kidney, and spleen (203, 279) are depressed, and the oxidation of $^{14}C$-labeled lysine and leucine is increased (244), in zinc deficiency. Decreased RNA concentration in the liver and decreased DNA concentration in the thymus also occur in Zn-deficient baby pigs (94). In addition, zinc is essential for the incorporation of glycine into glutathione (81), of methionine into rat tissue protein (80), and is involved in the metabolism of cysteine (82). These defects in protein and nucleic acid metabolism respond rapidly to supplemental zinc. Thus Sandstead and Rinaldi (203) found a single injection of 100 $\mu g$ Zn, administered 24 hr before injection of labeled thymidine, to be sufficient to stimulate the synthesis of DNA in Zn-deficient rats. Quantitative evidence of defective protein utilization has been obtained in Zn-deficient lambs (222). Highly significant increases in urinary N and S excretion were observed, compared with pair-fed controls (see Table 29). Zinc has also been shown to enhance the utilization of soybean protein by rats (147).

Evidence of a different nature implicating zinc in nucleic acid and protein metabolism comes from two further studies with rats. Macapinlac and co-workers (110) found the total protein and RNA contents of the testes of Zn-deficient rats to be reduced, although the incorporation of $^{14}C$-labeled leucine into testicular protein and adenine into testicular RNA remained unaltered. Somers and Underwood (221) found that the testes of their more severely Zn-deficient rats contained lower concentrations of zinc, RNA, DNA, and protein and higher nonprotein N concentrations and ribonuclease activity, than either the pair-fed or the unrestricted-fed controls (see Table 31). It seems that one of the functions of zinc may be to control ribonuclease activity at the cellular level. Support for such a hypothesis comes from the fact that yeast ribonuclease activity can be completely inhibited by zinc at a concentration of $10^{-4}$ *M* (151). Furthermore, the ribonuclease activity of Zn-deficient citrus leaves is higher than that of normal leaves (88).

Many years ago, Hove and co-workers (78) found the proteolytic activity of the pancreas to be reduced in the Zn-deficient rat. Vallee and Neurath (259) later discovered that pancreatic carboxypeptidase is a zinc metalloenzyme. These findings stimulated Mills and co-workers (140) to investigate the effects of zinc deficiency on pancreatic carboxypeptidase activity. They found that carboxypeptidase activty was appreciably reduced in the Zn-deficient rats and returned rapidly to normal with zinc therapy. A similar reduction in pancreatic carboxypeptidase activity in Zn-deficient rats was observed by Hsu *et al.* (79). No evidence was obtained by Mills that this reduction had limited protein digestion or absorption. This particular defect is, therefore, unlikely to be directly responsible for the growth arrest and impaired food utilization of zinc deficiency.

### 8. *Zinc and Carbohydrate Metabolism*

Carbohydrate metabolism is impaired in Zn-deficient fungi (54) and plants (191). The addition of zinc to insulin solutions delays the physiological action of the hormone and prolongs the hypoglycemia. Convincing evidence that zinc plays a part in the normal production or action of insulin *in vivo* has only recently appeared. Quarterman and co-workers (186) observed a serious impairment of blood glucose homeostasis in Zn-deficient rats. This was evident from the glucose tolerance curves following intraperitoneal injections of glucose. The deficient rats also revealed a greatly increased resistance to insulin coma. These data suggest that zinc is necessary for normal insulin secretion and for some further undefined function involving sensitivity to insulin.

TABLE 31

*Composition of Testes from Zinc-Deficient Rats and of Two Zinc-Supplemented Control Groups*[a]

| Group | Testes Zn ($\mu$g/g) | Testes RNA (mg/g) | Testes DNA (mg/g) | Testes Total N (mg/g) | Testes protein N (mg/g) | Testes nonprotein N[b] (mg/g) | Testes ribonuclease activity[c] |
|---|---|---|---|---|---|---|---|
| 1. Zn-deficient | 21.0 | 3.2 | 3.0 | 14.0 | 8.3 | 5.7 | 0.21 |
| | ±0.83 | ±0.16 | ±0.15 | ±0.40 | ±0.32 | ±0.13 | ±0.009 |
| 2. Pair-fed controls | 32.5 | 5.0 | 4.5 | 14.5 | 10.8 | 3.8 | 0.12 |
| | ±1.08 | ±0.14 | ±0.24 | ±0.36 | ±0.28 | ±0.10 | ±0.005 |
| 3. Ad lib-fed controls | 36.5 | 5.3 | 4.7 | 15.0 | 11.6 | 3.4 | 0.13 |
| | ±1.68 | ±0.19 | ±0.13 | ±0.41 | ±0.38 | ±0.21 | ±0.003 |
| LSD 0.05 | 3.47 | 0.47 | 0.51 | 1.11 | 0.93 | 0.43 | 0.02 |
| 00.01 | 4.67 | 0.64 | 0.68 | 1.49 | 1.25 | 0.59 | 0.03 |

[a] From Somers and Underwood (221).
[b] Nonprotein nitrogen calculated by difference (total N−protein N).
[c] Absorbancy at 260 m$\mu$ due to nucleotide production from RNA by 100 mg of testes.

From evidence of *in vitro* studies, it appears that zinc participates in a further aspect of carbohydrate metabolism. Quarterman (185) showed that glucose uptake by adipose tissue is increased by zinc and by insulin, and even more by a combination of the two. It was suggested that decreased glucose uptake by adipose tissue may explain the high-fasting free fatty acid levels in the blood and the virtual absence of adipose tissue in animals with advanced zinc deficiency.

9. *Zinc and Atherosclerosis*

Indications that zinc therapy can be beneficial in some cases of atherosclerosis have recently been obtained. In one study by Pories and associates (175), 13 patients with advanced vascular disease were given zinc sulfate orally for up to 29 months. Twelve of these 13 showed marked clinical improvement, 9 returned to their previously normal activity, and 7 had a return of previously absent pulses. In another study, Henzel and associates (70) administered zinc sulfate to 14 patients with extensive, inoperable symptomatic atherosclerosis and to 2 who had inabling vasospastic disease for 3 to 11 months. Six of these were excluded from the evaluation of zinc therapy because of changes in their habits or voluntary dietary restriction. Four of the remaining 10 experienced no improvement, 2 of whom had rapid progression of their disease. Six of the 10 experienced clinical improvement which could be ascribed to the zinc sulfate medication.

The mode of action of zinc in atherosclerosis is far from clear. Henzel *et al.* (70) suggest that it may act as a pharmacological agent which in some way allows increased blood volume to reach areas of the body where diseased blood vessels have restricted arterial perfusion. Serial arteriograms, after a year of zinc therapy, show no apparent diminution of sclerotic deposits in the great vessels, suggesting a possible "small vessel" effect (175). Hair and plasma zinc levels are usually significantly subnormal in patients with atherosclerosis and myocardial infarctions (175, 268, 269), and the aortic wall has an extremely active turnover of $^{65}Zn$ (232). This becomes even more active when the arterial wall is injured, tripling during the first week of healing, in a manner similar to that seen in healing skin, muscle, and bone. Since atherosclerosis is believed to begin with some form of arterial trauma, the possibility arises that atherosclerosis may be, in part, an expression of inadequate arterial repair (175).

10. *Changes in the Blood in Zinc Deficiency*

The hematocrit values are slightly elevated in the Zn-deficient rat (109a), baby pig (125), and Japanese quail (227). Leukopenia has been reported in the Zn-deficient rat (109a). Total white cell counts do not change significantly, compared with pair-fed controls, but marked changes occur in the differential white cell count. In the Zn-deficient baby pig (125) and weanling rat (109a), there is a relative and absolute reduction in lymphocytes. In male weanling rats fed a low-Zn diet for 14 days the percentage of polymorphonuclear (P) leukocytes markedly increased and of lymphocytes (L) markedly decreased (40). Similar changes in the P:L ratio were observed in normal adult female rats fed a low-Zn diet during pregnancy but not in adult nonpregnant rats fed a similar diet for 3 weeks (40). The physiological significance of these intriguing changes in the numbers and proportions of different white cells and their precise relationship to zinc deficiency are obscure.

A decline in plasma protein levels has been observed in zinc deficiency in some investigations (78, 157, 187) and not in others (109a, 125, 227). The electrophoretic patterns of these proteins are greatly altered in the Zn-deficient pig (125) and, after fasting, in quail (227). The relationship of these changes in the plasma proteins to the wider problem of protein synthesis in zinc deficiency has yet to be evaluated.

Alkaline phosphatase activity is reduced in the blood serum of Zn-deficient pigs (103, 125), calves (135), and rats (104, 109a). Comparisons of the levels of this enzyme in the serum of pair-fed controls indicate that in the baby pig this reduction is due to zinc deficiency and not to reduced feed intake (104, 125, 184). Carbonic anhydrase and lactic dehydrogenase activities remain within normal limits in the blood of parakeratotic pigs and in Zn-deficient rats and chicks. Subnormal levels of the former enzyme occur in Zn-deficient calves (127) and in the blood of the Middle East dwarfs suffering from conditioned zinc deficiency (202).

### 11. *Enzyme Activities in the Tissues in Zinc Deficiency*

Changes occur in the activities of various zinc-containing or zinc-dependent enzymes in the tissues of the Zn-deficient animal. In a comprehensive study carried out by Prasad and co-workers (183), histochemical determinations revealed reduced activities of certain enzymes, accompanied by reduced concentrations of zinc, in the testes, bones, esophagus, and kidneys of Zn-deficient rats. Compared with restricted-fed controls, decreased enzyme activity was noted in the deficient rats as follows: in the testes, lactic dehydrogenase (LDH), malic dehydrogenase (MDH), alcohol dehydrogenase (ADH), and NADH diaphorase; in the bones, LDH, MDH, ADH, and alkaline phosphatase; in the esophagus, MDH, ADH, and NADH diaphorase; and in the kidneys, MDH and alkaline phosphatase. No changes in succinic dehydrogenase were apparent. In a zinc-repleted group of rats the levels of zinc in the testes and bones increased significantly and the activities of the various enzymes increased in all the tissues examined. Essentially similar results were obtained with baby pigs (184). Liver alcohol dehydrogenase activity is unaffected in the Zn-deficient rat (79, 109a) and the baby pig (125). Liver glutamic dehydrogenase activity is also unaffected in the Zn-deficient baby pig (125). Intestinal alkaline phosphatase is reduced in Zn-deficient rats compared with pair-fed controls (104).

The relationship of these particular changes in tissue enzyme activities to the clinical manifestations of zinc deficiency in the animal is not clear. Since they are all proteins, and it is believed that RNA synthesis and, hence, DNA and protein synthesis, are primary defects in zinc deficiency, it seems reasonable to expect that the synthesis of these enzymes by the cell would also be impaired. The greater susceptibility of certain zinc-containing enzymes to loss of activity in zinc deficiency presumably relates to the firmness of binding of zinc to the enzyme protein, compared with others in which the zinc is more firmly bound. Thus pancreatic carboxypeptidase, in which the zinc is not as firmly bound as it is in liver alcohol dehydrogenase, is readily reduced in activity, whereas liver alcohol dehydrogenase activity is largely unaffected. Prasad (183) has speculated on this question in the following terms:

> In a growing cell there is a series of apoenzymes, each present at some low concentration and, with normal levels of zinc, the apoenzymes combine with metal, forming functional zinc enzymes. At limiting levels of zinc, the apoenzymes behave as a series of ligands, each competing for the available zinc ions according to their stability constants. According to this view, the apoenzymes or ligands with high stability constants will be satisfied first, whereas those with low stabilities will not form functional zinc complexes. The concentration of zinc ions will decrease progressively as zinc is diluted by continuing growth, but among apoenzymes or other macromolecular ligands, there will be first one and then perhaps others whose decreasing activities will limit growth. Even when growth stops altogether, we can expect that some of the ligands with the tightest affinities for zinc might still be completely satisfied.

### 12. *Zinc and Hormones*

Evidence exists that zinc is involved in the production and function of several hormones and that zinc metabolism is affected by the sex hormones. The data of Quarterman *et al.* (186), discussed earlier, suggest strongly that zinc is required for normal insulin secretion and those of Millar *et al.* (122) suggest similarly that gonadotropin output is reduced in the Zn-deficient rat. In neither case were actual hormone outputs measured and compared with those of pair-fed controls. The endocrine abnormalities of the Zn-deficient male dwarfs have been described as those of hypopituitarism (202). However, significant weight gains were not observed in Zn-deficient rats treated with bovine growth hormone

(109a). Furthermore, Prasad *et al.* (183) showed with hypophysectomized and nonhypophysectomized rats that the effects of zinc and growth hormone on growth are independent of each other.

The maternal behavioral abnormalities of Zn-deficient rats are difficult to explain, except in terms of some disturbance in output or activity of sex hormones. In estrous, pregnant, and pseudopregnant rabbits the activity of endometrial carbonic anhydrase is directly controlled by progesterone (107). A marked increase in the concentration of zinc occurs in the endometrium and in the uterine tissue as a whole in the progestational phase of pregnancy in the rabbit (108). This means that zinc must have been mobilized from body depots and transferred to the uterus, presumably to meet the requirements of the endometrium-mediated phase of embryonic development. Since control over uterine events in early pregnancy in the rabbit is exercised primarily by progesterone, it seems that this hormone controls the changes in uterine zinc and carbonic anhydrase. In this connection, Apgar (5) found the administration of progesterone and estrone to be partially effective in maintaining pregnancy to term in Zn-deficient rats. The effects of chorionic gonadotropin (198) and of testosterone upon zinc uptake by the dorsolateral prostate and other accessory sex glands of the male rat (121, 122) have previously been mentioned.

## V. Zinc Requirements

In all species there is a series of minimum zinc requirements depending upon the nature of the diet being consumed and the age and functional activity of the animal. The effect of diet is a reflection of the amounts and proportions, relative to those of zinc, of many organic and inorganic factors that influence zinc absorption, utilization, and excretion. The effects of age and of functional activity are shown most clearly by the substantially higher zinc requirements of the young growing and the reproducing male and female animal than of nonreproducing adult animals of the same species. Minimum zinc requirements may also be influenced by ambient temperatures where these cause profuse sweating with consequent severe losses of zinc in the sweat and by parasitic infestation with its attendant blood and, hence, zinc losses. The criteria of adequacy employed can also be important.

### 1. *Rats*

The zinc requirements of young rats for growth and health were studied by Forbes and Yohe (53). These workers reported markedly subnormal growth and other signs of zinc deficiency on diets supplying 7 ppm Zn. When these diets contained casein or egg white as the protein source an additional 5 ppm as $ZnCO_3$ was found adequate for satisfactory growth and health. On a soybean protein diet, 11 ppm Zn from this source was necessary. This indicates a minimum zinc requirement for growing rats of 12 to 18 ppm, depending upon the protein source and, by implication, the phytate content of the diet. On the basis of concurrent zinc balance studies, the minimum requirement in terms of "absorbable" zinc was estimated to be 8-9 ppm. Forbes and Yohe found that increasing the calcium level of the diet from 0.8 to 1.6% significantly depressed weight gains. The requirements must, therefore, be raised beyond the 12-18 ppm just stated as minimal, if the calcium intake is substantially above normal. The more recent experiments of Luecke *et al.* (105) indicate that on an egg white basal diet, the minimum zinc requirement of the rat for growth is close to 11 ppm, although higher serum zinc levels occurred at an intake of 13 ppm Zn.

The minimum zinc requirements for normal reproductive performance cannot be given with precision for either males or females. In the male rat, Swenerton and Hurley (239) have produced evidence that 60 ppm Zn is inadequate to prevent long-term testicular changes. They state that rat diets containing isolated soybean protein should contain at least 100 ppm of zinc if extraneous sources of the element are minimal.

## 2. *Pigs*

A soybean protein purified diet containing 12 ppm Zn was found by Miller *et al.* (125) to be inadequate to promote growth in baby pigs, compared with that of control pigs receiving the same diet supplemented with 100 ppm Zn. An interesting feature of these studies was that comparable pigs, consuming a similar purified diet in which casein was the protein source and which contained only 10 ppm Zn, maintained growth and appetite similar to those consuming the soybean protein diet supplemented to provide 100 ppm Zn, over the 4-week period of the experiment. They also showed no signs of parakeratosis, as did the pigs on the 12 ppm soybean protein diet. It is unlikely that the 10 ppm casein diet would have sustained adequate growth for much longer because the pigs on this diet revealed significantly lowered serum zinc and alkaline phosphatase levels by the end of the 4-week period. Subsequently, these workers (124) obtained evidence that the dietary zinc requirement is higher for the male infant pig than for the female. With purified diets containing isolated soybean protein, 45 ppm Zn was adequate for the female baby pig but not for the male.

In an earlier trial, Smith *et al.* (220) fed weanling pigs a basal soybean protein diet containing 0.66% Ca, 0.47% P, and 16 ppm Zn, supplemented with zinc oxide to provide total dietary zinc levels of 21, 26, 31,

36, 41, and 46 ppm. Complete freedom from parakeratosis was not achieved until the 41-ppm level was reached. A further increase to 46 ppm improved growth rate. It seems that the requirement of zinc for growth is slightly higher than the requirement for the prevention of any clinical evidence of parakeratosis and that an optimum zinc requirement of 46 ppm can be assigned to pigs of this age consuming diets of the type used. Essentially similar results were obtained by Miller *et al.* (124), with no differences between male and female weanling pigs. Diets with casein or egg albumin as the main protein source and, therefore, lower in phytic acid would be expected to impose somewhat lower zinc requirements. On the other hand, the diet used by Smith and co-workers, cited above, conform more to most practical-type rations, so that 45-50 ppm Zn appears adequate to meet the normal nutritional needs of growing pigs.

Raising the dietary calcium level to about twice normal, or higher, increases the zinc requirements of growing pigs, as shown by the increased incidence and severity of parakeratosis on Zn-deficient diets and the signs of zinc deficiency induced on otherwise marginal or adequate diets (74, 102, 103, 230, 251). The level of this dietary factor must, therefore, always be taken into account in assessing zinc requirements. High dietary intakes of copper also raise zinc requirements. Copper supplements of 125 to 250 ppm can aggravate signs of zinc deficiency in pigs (75, 150) and significantly increase zinc requirements on normal rations above those found adequate without the copper supplement (237).

The zinc requirements of breeding sows appear to be slightly lower than those of growing pigs. Thus Pond and Jones (173) reported no benefit upon reproductive performance from adding 50 ppm zinc to a corn-soybean meal diet containing 35 ppm Zn and 1.4% Ca. Hennig (68) similarly observed no effect upon the reproductive performance of sows from zinc supplementation of barley-fish meal diets containing 36-44 ppm Zn, even when given excess calcium up to 1.5-2.2% of the diet. However, Hoekstra *et al.* (76) carried out three trials with gilts fed a high-calcium (1.6%) corn-soybean meal diet containing 30-34 ppm Zn. In two of these trials the addition of 100 ppm zinc significantly increased the number of live pigs per litter, without affecting birth or weaning weights. On this evidence, corn-soybean meal rations high in calcium must be considered marginal in zinc for reproduction in swine.

### 3. *Poultry*

The minimum zinc requirements of chicks for growth and health, when they are maintained in lacquered batteries and fed a diet with soybean protein as the principal protein source and containing 1.6% Ca, 0.7% P, and 15 ppm Zn, are estimated by O'Dell (148) to be 35 ppm. The requirement was slightly decreased by lowering the calcium to 1.1% and was unaffected by 2.1% Ca. This has been substantially confirmed in subsequent investigations (141, 196, 284), but, with casein or egg white as the protein source, lower total zinc requirements have been found (141, 163). The minimum dietary level of zinc for growth and health in chicks thus appears to be 35-40 ppm on soybean protein diets and 25-30 ppm on casein or egg white diets. Furthermore, the chick is less vulnerable than the pig to excess calcium (148, 163).

Under most circumstances a practical corn-soybean meal ration supplies sufficient zinc to meet the requirements of this element for chick growth and for egg production and hatchability. However, considerable variation from sample to sample can occur in the zinc content of cereal grains and of other protein concentrates used in poultry feeding. The zinc content of fish meal, whale meal, and meat meal is usually much higher than that of soybean meal (106, 253), so that the inclusion of these materials should insure that the zinc requirements of chicks are fully supplied.

4. *Ruminants*

The minimum zinc requirements of sheep and cattle vary with the criteria of adequacy employed and apparently also with the type of diet consumed. Ott and co-workers (158) found that 18 ppm Zn did not support maximal growth in lambs and suggested a requirement between 18 and 33 ppm. Mills *et al.* (139), using a diet in which 60% of the nitrogen was supplied by urea, found that 7 ppm Zn was adequate for growth of lambs, whereas 15 ppm was necessary to maintain normal plasma zinc levels. Underwood and Somers (254), also employing a diet in which 60-70% of the nitrogen came from urea and most of the remainder from egg albumin, observed that ram lambs grew just as well when the diet supplied 17 ppm Zn as when it supplied 32 ppm. However, testicular development and spermatogenesis were markedly greater at the higher zinc intake. It was concluded that 17 ppm Zn is adequate for body growth and appetite on this type of diet but is inadequate for normal testicular growth and function.

The experiments of Mills *et al.* (139) and of Miller *et al.* (133) indicate that a dietary concentration of 8 to 9 ppm is adequate for the growth of calves, although the former workers estimate that 10-14 ppm is necessary to maintain normal plasma zinc levels. These estimates of zinc requirements based on semisynthetic diets are substantially lower than would be suggested from observations of zinc deficiency in cattle under field conditions and of growth responses to zinc in the feed lot. Thus Perry *et al.* (167) obtained significant increases in daily gain from supplementary zinc in two out of four experiments with cattle fed fattening rations based on corn and soybean meal and either corn cobs or corn silage, containing 18 and 29 ppm Zn. Raum and co-workers (190) observed a slight improvement in the growth of steers from zinc (and cobalt) supplementation of barley rations containing 29-33 ppm Zn. Furthermore, signs of zinc deficiency, responsive to zinc, occur in cattle where the pasture or fodder contains 18-42 ppm (98), 19-83 ppm (42), and estimates of 28 to 50 ppm Zn (62). Since herbage zinc concentrations of this magnitude are common in areas where clinical zinc deficiency in cattle is unknown, factors must be present which reduce zinc absorption or impair its utilization in the animal. Limited evidence with ruminants suggests that neither phytic acid nor calcium can be incriminated in this respect (139, 157). The presence of anionic complexes of zinc in the herbage, of unknown availability, such as have been shown by Bremner and Knight (19) to exist in rye grass, may prove to be significant.

## VI. Zinc in Human Nutrition

Common foods vary greatly in zinc content. White sugar, pome, and citrus fruits are among the lowest in zinc (less than 1 ppm of fresh edible portion). Wheat germ and bran (40-120 ppm) and oysters, which may contain over 1000 ppm Zn, are among the richest sources of this element (211). Between these extremes, in ascending order, lie roots and tubers, white flour and bread, milk, leafy vegetables, meat, fish, eggs, whole cereals, nuts, and leguminous seeds. Species differences among the cereal grains are small (39, 253), although zinc is similarly concentrated in the germ and branny layers. White flours milled from North American hard wheat blends averaged 7.8 ppm Zn, compared with 35 ppm in the original wheats (36). Similar flour from Australian soft wheats averaged only 5 ppm Zn, compared with 16 ppm in the wheats from which the flour was made (253).

Adult human diets were shown to supply from 5 to 22 mg Zn/day in one study (117), but most mixed diets supply from 12 to 15 mg/day (202, 211). Little is known of the relative availability to man of the zinc in different foods. High dietary intakes of whole wheat bread and legumes, both high in phytates, have been incriminated as factors contributing to the zinc deficiency syndrome in Middle Eastern dwarfs. Whether phytates appreciably affect the availability of zinc from good mixed diets is more doubtful. The studies of Oberleas and Prasad (147) on zinc and protein nutrition in rats indicate that the utilization of plant seed proteins for growth is greatly affected by the level of zinc intake. They suggest that zinc supplementation of cereal diets may improve the growth and well-being of large segments of the human population in many areas. The balance studies of Engel *et al.* (47) with young girls indicate that zinc is equally well retained from diets in which the protein is supplied mainly by cereals and legumes as by meat and milk. Little is known of the significance of chelates in human dietaries. Disodium EDTA and calcium sodium EDTA are used in some canned foods to reduce deterioration from heavy metals and may affect the availability of dietary zinc. Certain drugs, such as D-penicillamine, which is used in the treatment of Wilson's disease, may also affect zinc availability, since they increase the excretion of zinc as well as of copper (25, 116).

Tribble and Scoular (250) found the average zinc intake of college women aged 17–27 years to be 12 mg/day, of which 6.6 mg/day was retained. Engel and co-workers (47) observed daily zinc intakes ranging from 4.6 to 9.3 mg/day, and suggested that 6 mg Zn/day is adequate for the normal needs of preadolescent girls. Normal infants receiving breast milk were shown to be in negative zinc balance at 1 week of age, in some cases severely so, on daily zinc intakes ranging from 0.2 to 1.2 mg/kg body weight (31). The significance of such losses of body zinc will depend on how long they continue into later infancy. In this connection, decreases have been reported in the zinc content of the hair of 10 of 12 infants studied, particularly between 4 and 12 months of age (233).

Balance studies give useful information on zinc intakes and retentions over short periods of time, but they do not provide a satisfactory answer to the question of the optimal zinc needs of man through the life-span and to meet periods of stress. Indeed, if a rapid rate of wound healing is taken as a criterion of adequacy, many dietaries must be considered inadequate or at best marginal in zinc, even though other signs of zinc deficiency, as seen in animals, are not apparent. Zinc intakes, like iron intakes, may be falling in many communities because of declining calorie consumption and because of a decrease in opportunities for contamination due to greater cleanliness and more hygienic food handling and processing practices. A lesser use of galvanized iron pipes and containers may also be of significance. Already there is evidence of regional differences in plasma zinc levels in human adults in the United States (95), which presumably reflect differences in dietary zinc intakes. Whether such differences arise from variations in dietary habits or from local variations in the zinc content of the foods and beverages consumed is not known. However, there is ample evidence that the zinc content of plants can be influenced by the soil type and the fertilizer treatment. In a limited Australian study the zinc content of wheaten grain was markedly influenced by fertilizer treatment (253).

The only evidence of zinc deficiency in man of sufficient severity to produce frank deficiency symptoms comes from studies of male dwarfs living in restricted areas in the Middle East. The condition in these patients is characterized clinically by markedly retarded growth and sexual development, pathologically by hepatosplenomegaly, delayed epiphyseal closure of the long bones, and anemia and biochemically by subnormal concentrations of zinc in the blood, hair, urine, and sweat and of iron and alkaline phosphatase in the blood serum, together with high plasma zinc turnover rates [see Prasad (180)]. The diets consist mainly of whole wheat or corn bread and beans which supply approximately the same amount of zinc (15 mg/day) as a typical North American diet (202). The availability of the zinc is presumed to be low because of the high phytate content of these diets, and, in Iran, probably also because of the prevalence of clay eating. Blood loss due to intestinal parasitism is a further factor with the Egyptian dwarfs, and in both areas sweat losses are high, with losses of zinc of about 2 mg/day or more contributing to the zinc drain (184a).

The anemia responds slowly to iron medication. Transfer to good quality diets produces a slow response in growth and sexual development. When an additional 20 mg Zn/day is supplied as zinc sulfate the response is dramatic in body weight gain and height and particularly in the development of the external genitalia and the secondary sexual characteristics (202). The dwarfism and hypogonadism syndrome thus appears to be an expression of zinc deficiency conditioned by a combination of factors which conspire to affect adversely zinc absorption and retention and which operate throughout the growing period when zinc requirements are high. In a followup study of Iranian patients with this condition, those who were able to consume a good diet had lost the abnormalities or they were ameliorated, whereas with those who had not changed their diets the syndrome continued unabated (197).

The relation of zinc to wound healing in man and to various human disorders such as postalcoholic hepatic cirrhosis and atherosclerosis was considered earlier. Postalcoholic cirrhosis can be regarded as a conditioned zinc deficiency in view of the low plasma and liver zinc levels, the zincuria, and the responses, in some cases, to zinc therapy (261).

## VII. Zinc Toxicity

Zinc is relatively nontoxic to birds and mammals and a wide margin of safety exists between normal intakes and those likely to produce deleterious effects. Rats, pigs, poultry, sheep, and cattle exhibit considerable tolerance to high intakes of zinc, the extent of the tolerance depending upon the composition of the basal diet, particularly its content of minerals known to affect zinc absorption and utilization, such as copper, iron, and cadmium.

Dietary zinc intakes of 0.25%, or 2500 ppm, have no discernible effects on rats whether ingested as the metal, chloride, or carbonate. At twice this level of zinc, growth is severely depressed, with heavy mortality in young animals when ingested as the chloride, and is slightly depressed with little mortality when ingested as the oxide (67). Intakes of 5000 or 10,000 ppm Zn, as the carbonate, produce severe anemia in young rats in addition to subnormal growth, anorexia, and, at the higher rate, heavy mortality (238). Adult female rats fed a diet containing 0.2% (2000 ppm) Zn, as zinc oxide, beginning either 21 days before breeding or from the first day of gestation maintained normal pregnancies with no anatomical malformations in the fetuses. Rats similarly fed 0.4% Zn revealed variable degrees of death and resorption of the fetuses, with 100% resorption in the 15- and 16-day-old fetuses of mothers fed this diet from 21 days before breeding and during pregnancy (207).

Weanling pigs fed for several weeks on diets containing 1000 ppm Zn, either as the sulfate (101) or as the carbonate (20) suffered no ill-effects. At higher levels of zinc, depressed growth and appetite, arthritis, and internal hemorrhages were observed and at 4000 and 8000 ppm mortality was high (20). Broilers and layer hens exhibit a similar tolerance to zinc at levels of 1200 to 1400 ppm of the diet and a similar growth and appetite depression when the level is raised to 3000 ppm (164).

Studies carried out by Ott and co-workers (155) with lambs and feeder cattle indicate that these species are somewhat less tolerant of high-zinc intakes than are rats, pigs, and poultry. Consumption by lambs, over a 10-week period, of diets containing more than 1500 ppm Zn (as the oxide) caused depressed feed consumption, whereas 1000 ppm Zn caused reduced gains, decreased feed efficiency, and increased mineral consumption. No other external signs of toxicity were observed. Dietary zinc levels of 500 ppm or less had no detrimental effects on steers and heifers, but 900 ppm caused reduced gains and lowered feed efficiency, whereas 1700 ppm induced, in addition, a depraved appetite characterized by excessive salt and other mineral consumption and wood chewing.

The growth depression that characterizes zinc toxicity is due largely, but not entirely, to reduced food consumption, probably as a result of unpalatability of the high-zinc diets (58, 155). In the zinc-intoxicated rat, an anemia of the hypochromic, microcytic type develops (34, 58), accompanied by high levels of zinc in the liver and other tissues and subnormal levels of copper, iron, cytochrome oxidase, catalase and δ -aminolevulinate dehydratase (34, 58, 111, 264, 265, 282). The anemia and the accompanying biochemical changes can be largely overcome by supplements of copper (58, 219, 264) and completely overcome by supplements of copper plus iron (34, 111). It appears, therefore, that the anemia of zinc toxicity results first from an induced copper deficiency and, second, from an induced iron deficiency brought about by an interference, by the high-zinc intakes, with the absorption and utilization of these metals.

The studies of tissue changes in zinc toxicity in sheep and cattle carried out by Ott and associates (154, 156) disclose certain differences from those just described for rats. Subnormal liver copper levels were observed, the anemia was much less marked, and the iron concentration in the liver was actually increased in both species. Defective development and mineralization of bone were not apparent, as reported by Sadavisan (200) for zinc-toxic rats; nor was there any evidence of reduced calcium and phosphorus absorption as found for lambs (246). At the higher levels of zinc intake, changes in rumen metabolism, evidenced by a reduction in the volatile fatty acids (VFA) concentration and the acetic acid-propionic acid ratio, occurred in the lambs, probably through a toxic effect of the zinc on the rumen microorganisms (156). It was suggested that an effect on these microorganisms may account for the lower-zinc toxicity threshold in ruminants than in nonruminants.

## REFERENCES

1. N. W. H. Addink, *Nature (London)* **186**, 253 (1960); N. W. H. Addink and L. J. P. Frank, *Cancer* **12**, 544 (1959).
2. N. S. Agar, J. Roberts, and J. V. Evans, *J. Comp. Biochem. Physiol.* **35**, 639 (1970).
3. J. W. Anderson, G. A. Cooper, and W. G. Hoekstra, *J. Invest. Dermatol.* **48**, 521 (1967).
4. J. Apgar, *Amer. J. Physiol.* **215**, 160 and 1478 (1968).
5. J. Apgar, *J. Nutr.* **100**, 470 (1970).
6. J. G. Archibald, *J. Dairy Sci.* **27**, 257 (1944).
7. G. H. Barney, M. P. Macapinlac, W. N. Pearson, and W. J. Darby, *J. Nutr.* **93**, 511 (1967).
8. M. E. Bartholemew, R. Tupper, and A. Wormall, *Biochem. J.* **73**, 256 (1959).
9. W. M. Becker and W. G. Hoekstra, *J. Nutr.* **90**, 301 (1966); **94**, 455 (1968).
10. R. Berfenstam, *Acta Paediat. (Stockholm)* **41**, 389, Suppl. 87, 105 (1952).
11. R. Bergman and R. Söremark, *J. Nutr.* **94**, 6 (1968).
12. G. Bertrand and R. Benson, *C. R. Acad. Sci.* **175**, 289 (1922).
13. G. Bertrand and R. Vladesco, *C. R. Acad. Sci.* **173**, 176 (1921).
14. V. Birckner, *J. Biol. Chem.* **38**, 191 (1919).

15. D. M. Blackmon, W. J. Miller, and J. D. Morton, *Vet. Med.* **62**, 265 (1967).
16. D. L. Blamberg, U. B. Blackwood, W. C. Supplee, and G. F. Combs, *Proc. Soc. Exp. Biol. Med.* **104**, 217 (1960).
17. H. Bodansky, *J. Biol. Chem.* **44**, 399 (1920); H. Severy, *ibid.* **55**, 79 (1923).
18. J. M. Bowness and R. A. Morton, *Biochem. J.* **51**, 530 (1950); **53**, 620 (1953).
19. I. Bremner and A. H. Knight, *Brit. J. Nutr.* **24**, 279 (1970).
20. M. F. Brink, D. E. Becker, S. W. Terrill, and A. H. Jensen, *J. Anim. Sci.* **18**, 836 (1959).
21. F. Brudevold, L. T. Steadman, M. A. Spinelli, B. H. Amdur, and P. Gron, *Arch. Oral Biol.* **8**, 135 (1963).
22. G. W. Bryan, *Nature (London)* **213**, 1043 (1967).
23. P. J. Buchanan and J. M. Hsu, *Fed. Proc., Fed. Amer. Soc. Exp. Biol.* **27**, 483 (1968).
24. E. M. Butters and M. L. Scott, *Poultry Sci.* **35**, 1135 (1954).
25. D. A. Buyski, W. Sterling, and E. Peets, *Ann. N.Y. Acad. Sci.* **135**, 711 (1966).
26. C. A. Cabell and I. P. Earle, *J. Anim. Sci.* **24**, 800 (1965).
27. D. F. Caldwell and D. Oberleas, *Pan-Amer. Health Org. Sci. Publ. No. 185,* 1969.
28. N. R. Calhoun and J. C. Smith, *Lancet* **2**, 682 (1968).
29. R. G. Cassens, E. J. Briskey, and W. G. Hoekstra, *J. Sci. Food Agr.* **6**, 427 (1963).
30. R. G. Cassens, W. G. Hoekstra, E. C. Falton, and E. J. Briskey, *Amer. J. Physiol.* **212**, 688 (1967).
31. P. A. Cavell and E. M. Widdowson, *Arch. Dis. Childhood* **39**, 496 (1964).
32. G. C. Cotzias, D. C. Borg, and B. Selleck, *Amer. J. Physiol.* **202**, 359 (1962).
33. G. C. Cotzias and P. S. Papavasiliou, *Amer. J. Physiol.* **206**, 787 (1964).
34. D. H. Cox and D. L. Harris, *J. Nutr.* **70**, 514 (1960).
35. D. B. Cruickshank, *Brit. Dent. J.* **61**, 530 (1936); **63**, 395 (1937); **68**, 257 (1940).
36. C. P. Czerniejewski, C. W. Shank, W. G. Bechtel, and W. B. Bradley, *Cereal Chem.* **41**, 65 (1964).
37. C. A. D'Alonzo and S. Pell, *Arch. Environ. Health* **6**, 381 (1963).
38. C. Delezenne, *Ann. Inst. Pasteur, Paris* **33**, 68 (1919).
39. W. A. Dewar, *J. Sci. Food Agr.* **18**, 68 (1967).
40. I. E. Dreosti, S-H. Tsao, and L. S. Hurley, *Proc. Soc. Exp. Biol. Med.* **128**, 169 (1968).
41. R. E. Drinker and E. S. Collier, *J. Ind. Hyg.* **8**, 257 (1926).
42. P. Dynna and G. N. Havre, *Acta Vet. Scand.* **4**, 197 (1963).
43. I. P. Earle and J. W. Stevenson, *J. Anim. Sci.* **24**, 325 (1965).
44. H. M. Edwards, Jr., *J. Nutr.* **69**, 306 (1959).
45. W. G. Eggleton, *Biochem. J.* **34**, 991 (1940).
46. J. Eisenbrand and M. Sienz, *Hoppe-Seyler's Z. Physiol. Chem.* **268**, 1 (1938).
47. R. W. Engel, R. F. Miller, and N. O. Price, *in* "Zinc Metabolism" (A. S. Prasad, ed.), p. 326. Thomas, Springfield, Illinois, 1966.
48. J. P. Feaster, S. L. Hansard, J. T. McCall, and G. K. Davis, *Amer. J. Physiol.* **181**, 287 (1955).
49. J. P. Feaster, S. L. Hansard, J. T. McCall, F. H. Skipper, and G. K. Davis, *J. Anim. Sci.* **13**, 781 (1954).
50. B. Foley, S. A. Johnson, B. Hackley, J. C. Smith, and J. A. Halsted, *Proc. Soc. Exp. Biol. Med.* **128**, 265 (1968).
51. R. H. Follis, Jr., H. G. Day, and E. V. McCollum, *J. Nutr.* **22**, 223 (1941).
52. R. M. Forbes, *Fed. Proc., Fed. Amer. Soc. Exp. Biol.* **19**, 643 (1960).
53. R. M. Forbes and M. Yohe, *J. Nutr.* **70**, 53 (1960).
54. J. W. Foster and F. W. Denison, *Nature (London)* **166**, 833 (1950).
55. M Fujioka and I. Lieberman, *J. Biol. Chem.* **239**, 1164 (1964).

56. J. G. Gibson, B. L. Vallee, R. G. Fluharty, and J. E. Nelson, *Proc. 6th Int. Congr. Cancer Res.*, p. 1102 (1954).
57. I. G. F. Gilbert and D. M. Taylor, *Biochim. Biophys. Acta* **21**, 546 (1956).
58. D. R. Grant-Frost and E. J. Underwood, *Aust. J. Exp. Biol. Med. Sci.* **36**, 339 (1958).
59. G. M. Grodsky and P. H. Forsham, *Annu. Rev. Physiol.* **28**, 364 (1966).
60. S. A. Gunn and T. C. Gould, *Endocrinology* **58**, 443 (1956); *Amer. J. Physiol.* **193**, 505 (1958).
61. S. A. Gunn, T. C. Gould, and W. A. D. Anderson, *Radiat. Res.* **20**, 504 (1963).
62. S. Haaranen, *Nord. Veterinaer Med.* **14**, 265 (1962); **15**, 536 (1963).
63. B. Hackley, J. C. Smith, and J. A. Halsted, *Clin. Chem.* **14**, 1 (1968).
64. J. A. Halsted, B. Hackley, C. Rudzki, and J. C. Smith, *Gastroenterology* **54**, 1098 (1968).
65. J. A. Halsted, B. Hackley, and J. C. Smith, *Lancet* **2**, 278 (1968).
66. J. C. Heath and J. Liquier-Milward, *Biochim. Biophys. Acta* **5**, 404 (1950).
67. V. G. Heller and A. D. Burke, *J. Biol. Chem.* **74**, 85 (1927).
68. A. Hennig, *Arch. Tierernaehr.* **15**, 331, 345, 353, 363, and 377 (1965).
69. A. Hennig, M. Anke, and W. Kracht, *Jahrb. Tierernaehr. Futterung.* **5**, 272, 286, and 297 (1964-1965).
70. J. H. Henzel, B. Holtman, F. W. Keitzer, M. S. DeWeese, and E. Lichti, *Proc. 2nd Conf. Trace Substances Environ. Health, 1968*, p. 83 (1968).
70a. W. B. Herring, B. S. Leavell, L. M. Paixao, and J. H. Yoe, *Amer. J. Clin. Nutr.* **8**, 846 (1960).
71. D. A. Heth, W. M. Becker, and W. G. Hoekstra, *J. Nutr.* **88**, 331 (1966).
72. D. A. Heth and W. G. Hoekstra, *J. Nutr.* **85**, 367 (1965).
73. D. A. Heth, M. L. Sunde, and W. G. Hoekstra, *Poultry Sci.* **45**, 75 (1966).
73a. R. Hoare, G. E. Delory, and D. W. Penner, *Cancer* **9**, 721 (1956).
74. J. A. Hoefer, E. R. Miller, D. E. Ullrey, H. D. Ritchie, and R. W. Luecke, *J. Anim. Sci.* **19**, 249 (1960).
75. W. G. Hoekstra, *Fed. Proc. Fed. Amer. Soc. Exp. Biol.* **23**, 1068 (1964).
75a. W. G. Hoekstra, Private Communication (1961).
76. W. G. Hoekstra, E. C. Faltin, C. W. Lin, H. F. Roberts, and R. H. Grummer, *J. Anim. Sci.* **26**, 1348 (1967).
77. W. G. Hoekstra, P. K. Lewis, Jr., P. H. Phillips, and R. H. Grummer, *J. Anim. Sci.* **15**, 752 (1956).
78. E. Hove, C. A. Elvehjem, and E. B. Hart, *Amer. J. Physiol.* **119**, 768 (1937); **124**, 750 (1938); *J. Biol. Chem.* **136**, 425 (1940).
79. J. M. Hsu, J. K. Anilane, and D. E. Scanlan, *Science* **153**, 882 (1966).
80. J. M. Hsu and W. L. Anthony, *Fed. Proc., Fed. Amer. Soc. Exp. Biol.* **28**, 762 (1969); J. M. Hsu, W. L. Anthony, and P. J. Buchanan, *J. Nutr.* **99**, 425 (1969).
81. J. M. Hsu, W. L. Anthony, and P. J. Buchanan, *Proc. Soc. Exp. Biol. Med.* **127**, 1048 (1968).
82. J. M. Hsu, W. L. Anthony, and P. J. Buchanan, *Proc. 1st Int. Symp. Trace Element Metab. Anim. 1969*, p. 151 (C. F. Mills, ed.) Livingstone, Edinburgh, 1970.
83. R. B. Hubbell and L. B. Mendel, *J. Biol. Chem.* **75**, 567 (1927).
84. L. S. Hurley and H. Swenerton, *Proc. Soc. Exp. Biol. Med.* **123**, 692 (1966).
85. L. S. Hurley, H. Swenerton, and J. T. Eichner, *Fed. Proc., Fed. Amer. Soc. Exp. Biol.* **23**, 292 (1964).
86. A. M. Kahn, H. W. Helwig, A. G. Redeker, and T. B. Reynolds, *Amer. J. Clin. Pathol.* **44**, 426 (1965).

87. D. Keilin and T. Mann, *Nature* (*London*) **144**, 442 (1939).
88. B. Kessler and S. P. Monselise, *Physiol. Plant.* **12**, 1 (1959).
89. E. W. Kienholz, D. E. Turk, M. L. Sunde, and W. G. Hoekstra, *J. Nutr.* **75**, 211 (1961).
90. M. Kirchgessner, *Z. Tierphysiol., Tierernaehr. Futtermittelk.* **14**, 270 (1959).
91. H. J. Koch, E. R. Smith, N. F. Shimp, and J. Connor, *Cancer* **9**, 499 (1956).
92. F. H. Kratzer, J. B. Allred, P. N. Davis, B. J. Marshall, and P. Vohra, *J. Nutr.* **68**, 313 (1959).
93. F. H. Kratzer, P. Vohra, J. B. Allred, and P. N. Davis, *Proc. Soc. Exp. Biol. Med.* **98**, 205 (1958).
94. P. K. Ku, D. E. Ullrey, and E. R. Miller, *Proc. 1st Int. Symp. Trace Element Metab. Anim., 1969* p. 158 (C. F. Mills, ed.) Livingstone, Edinburgh, 1970.
95. J. Kubota, V. A. Lazar, and F. L. Losee, *Arch. Environ. Health* **16**, 788 (1966).
96. J. G. Lease and W. P. Williams, Jr., *Poultry Sci.* **46**, 233, and 242 (1967).
97. G. Lechartier, F. Bellamy, F. Raoult, and H. Breton, *C. R. Acad. Sci.* **84**, 867; **85**, 40 (1887).
98. S. P. Legg and L. Sears, *Nature* (*London*) **186**, 1061 (1960).
99. M. Leiner and G. Leiner, *Biol. Zentralbl.* **64**, 293 (1944).
100. H. D. Lewis and M. D. Altschule, *Blood* **4**, 442 (1949).
101. P. K. Lewis, Jr., W. G. Hoekstra, and R. H. Grummer, *J. Anim. Sci.* **16**, 578 (1957).
102. P. K. Lewis, Jr., W. G. Hoekstra, R. H. Grummer, and P. H. Phillips, *J. Anim. Sci.* **15**, 741 (1956).
103. R. W. Luecke, J. A. Hoefer, W. S. Brammell, and F. Thorp, *J. Anim. Sci.* **16**, 3 (1957).
104. R. W. Luecke, M. E. Holman, and B. V. Baltzer, *J. Nutr.* **94**, 344 (1968).
105. R. W. Luecke, B. E. Rukson, and B. V. Baltzer, *Proc. 1st Int. Symp. Trace Element Metab. Anim., 1969* p. 471 (C. F. Mills, ed.) Livingstone, Edinburgh, 1970.
106. G. Lunde, *J. Sci. Food Agr.* **19**, 432 (1968).
107. C. Lutwak-Mann and C. E. Adams, *J. Endocrinol.* **15**, 43 (1957).
108. C. Lutwak-Mann and J. E. A. McIntosh, *Nature* (*London*) **221**, 1111 (1969).
109. R. E. Lutz, *J. Ind. Hyg.* **8**, 177 (1926).
109a. M. P. Macapinlac, W. N. Pearson, and W. J. Darby, *in* "Zinc Metabolism" (A. S. Prasad, ed.), p. 142. Thomas, Springfield, Illinois, 1966.
110. M. P. Macapinlac, W. N. Pearson, G. H. Barney, and W. J. Darby, *J. Nutr.* **93**, 511 (1967); **95**, 569 (1968).
111. A. C. Magee and G. Matrone, *J. Nutr.* **72**, 233 (1960).
112. M. Mager and F. Lionetti, *Fed. Proc., Fed. Amer. Soc. Exp. Biol.* **13**, 258 (1954).
113. T. Mann, "The Biochemistry of the Semen and the Male Reproductive Tract," 2nd ed. Butler & Tanner, London, 1964.
114. C. A. Mawson and M. I. Fischer, *Nature* (*London*) **167**, 859 (1951).
115. C. A. Mawson and M. I. Fischer, *Biochem. J.* **55**, 696 (1953); *Can. J. Med. Sci.* **30**, 336 (1952); *Arch. Biochem. Biophys.* **36**, 485 (1952).
116. J. T. McCall, N. P. Goldstein, R. V. Randall, and J. B. Gross, *Amer. J. Med. Sci.* **254**, 13 (1967).
117. R. A. McCance and E. M. Widdowson, *Biochem. J.* **36**, 692 (1942).
118. J. S. McHargue, *Amer. J. Physiol.* **77**, 245 (1926).
119. L. B. Mendel and H. C. Bradley, *Amer. J. Physiol.* **14**, 313 (1905); **17**, 167 (1906).
120. A. H. Methfessel and H. Spencer, *Fed. Proc., Fed. Amer. Soc. Exp. Biol.* **25**, 483 (1966) (abstr.).
121. M. J. Millar, P. V. Elcoate, and C. A. Mawson, *Can. J. Biochem. Physiol.* **35**, 865 (1957).

122. M. J. Millar, M. I. Fischer, P. V. Elcoate, and C. A. Mawson, *Can. J. Biochem. Physiol.* **36**, 557 (1958); **38**, 1457 (1960).
123. M. J. Millar, N. R. Vincent, and C. A. Mawson, *J. Histochem. Cytochem.* **9**, 111 (1961).
124. E. R. Miller, D. O. Liptrap, and D. E. Ullrey, *Proc. 1st Int. Symp. Trace Element Metab. Anim., 1969* p. 377 (C. F. Mills, ed.) Livingstone, Edinburgh, 1970.
125. E. R. Miller, R. W. Luecke, D. E. Ullrey, B. V. Baltzer, B. L. Bradley, and J. A. Hoefer, *J. Nutr.* **95**, 278 (1968).
126. J. K. Miller and R. G. Cragle, *J. Dairy Sci.* **48**, 370 (1965).
127. J. K. Miller and W. J. Miller, *J. Dairy Sci.* **43**, 1854 (1960); *J. Nutr.* **76**, 467 (1962).
128. W. J. Miller, D. M. Blackmon, R. P. Gentry, G. W. Powell, and H. E. Perkins, *J. Dairy Sci.* **49**, 1446 (1966).
129. W. J. Miller, D. M. Blackmon, R. P. Gentry, W. J. Pitts, and G. W. Powell, *J. Nutr.* **92**, 71 (1967).
130. W. J. Miller, D. M. Blackmon, G. W. Powell, R. P. Gentry, and J. M. Hiers, *J. Nutr.* **90**, 335 (1966).
131. W. J. Miller, D. M. Blackmon, J. M. Hiers, P. R. Fowler, C. M. Clifton, and R. P. Gentry, *J. Dairy Sci.* **50**, 715 (1967).
132. W. J. Miller, C. M. Clifton, and P. R. Fowler, *J. Anim. Sci.* **23**, 885 (1964).
133. W. J. Miller, C. M. Clifton, and N. W. Cameron, *J. Dairy Sci.* **46**, 715 (1963).
134. W. J. Miller, J. D. Morton, W. J. Pitts, and C. M. Clifton, *Proc. Soc. Exp. Biol. Med.* **118**, 427 (1965).
135. W. J. Miller, W. J. Pitts, C. M. Clifton, and J. D. Morton, *J. Dairy Sci.* **48**, 1329 (1965).
136. W. J. Miller, W. J. Pitts, C. M. Clifton, and S. C. Schmittle, *J. Dairy Sci.* **47**, 556 (1964).
137. W. J. Miller, G. W. Powell, and J. M. Hiers, *J. Dairy Sci.* **49**, 1012 (1966).
138. W. J. Miller, G. W. Powell, W. J. Pitts, and H. F. Perkins, *J. Dairy Sci.* **48**, 1091 (1965).
139. C. F. Mills, A. C. Dalgarno, R. B. Williams, and J. Quarterman, *Brit. J. Nutr.* **21**, 751 (1967).
140. C. F. Mills, J. Quarterman, R. B. Williams, A. C. Dalgarno, and B. Panic, *Biochem. J.* **102**, 712 (1967).
141. M. W. Moeller and H. M. Scott, *Poultry Sci.* **37**, 1227 (1958).
142. A. Nason, N. D. Kaplan, and H. A. Oldewurtel, *J. Biol. Chem.* **201**, 435 (1953).
143. F. H. Nielsen, M. L. Sunde, and W. G. Hoekstra, *J. Nutr.* **89**, 24 and 35 (1966).
144. F. H. Nielsen, M. L. Sunde, and W. G. Hoekstra, *Proc. Soc. Exp. Biol. Med.* **116**, 256 (1964); *J. Nutr.* **86**, 89 (1965).
145. H. Nishimura, *J. Nutr.* **49**, 79 (1953).
146. D. Oberleas, M. E. Muhrer, and B. L. O'Dell, *J. Anim. Sci.* **19**, 1280 (1960).
147. D. Oberleas and A. S. Prasad, *Amer. J. Clin. Nutr.* **22**, 1304 (1969).
148. B. L. O'Dell and J. E. Savage, *Poultry Sci.* **36**, 489 (1957); B. L. O'Dell, P. M. Newberne, and J. E. Savage, *J. Nutr.* **65**, 503 (1958).
149. B. L. O'Dell and J. E. Savage, *Proc. Soc. Exp. Biol. Med.* **103**, 304 (1960).
150. P. J. O'Hara, A. P. Newman, and R. Jackson, *Aust. Vet. J.* **36**, 225 (1960).
151. Y. Ohtaka, K. Uchida and T. Sakai, *J. Biochem.* (*Tokyo*) **54**, 322 (1963).
152. C. C. O'Mary, W. T. Butts, R. A. Reynolds, and M. C. Bell, *J. Anim. Sci.* **28**, 268 (1969).
153. J. M. Orten, *in* "Zinc Metabolism" (A. S. Prasad, ed.), p. 38. Thomas, Springfield, Illinois, 1966.

154. E. A. Ott, W. H. Smith, R. B. Harrington, H. E. Parker, and W. M. Beeson, *J. Anim. Sci.* **25**, 432 (1966).
155. E. A. Ott, W. H. Smith, R. B. Harrington, and W. M. Beeson, *J. Anim. Sci.* **25**, 414 and 419 (1966).
156. E. A. Ott, W. H. Smith, R. B. Harrington, M. Stob, H. E. Parker, and W. M. Beeson, *J. Anim. Sci.* **25**, 432 (1966).
157. E. A. Ott, W. H. Smith, M. Stob, and W. M. Beeson, *J. Nutr.* **82**, 41 (1964).
158. E. A. Ott, W. H. Smith, M. Stob, H. E. Parker, R. B. Harrington, and W. M. Beeson, *J. Nutr.* **87**, 459 (1965).
159. J. Parizek, *J. Reprod. Fert.* **1**, 294 (1960).
160. J. Parizek, J. C. Boursnell, M. F. Hay, A. Babicky, and D. M. Taylor, *J. Reprod. Fert.* **12**, 501 (1966).
161. S. Parkash and R. Jenness, *J. Dairy Sci.* **50**, 127 (1967).
162. J. C. Pekas, *Amer. J. Physiol.* **211**, 407 (1966); *J. Anim. Sci.* **27**, 1559 (1958).
163. J. M. Pensack, J. N. Henson, and P. D. Bogdonoff, *Poultry Sci.* **37**, 1232 (1958).
164. J. M. Pensack and R. C. Klussendorff, *Poultry Nutr. Conf., 1956,* Atlantic City, New Jersey.
165. H. M. Perry, Jr. and S. H. Camel, *in* "Metal-Binding in Medicine" (M. J. Seven and L. A. Johnson, eds.), p. 209. Lippincott, Philadelphia, Pennsylvania, 1960.
166. H. M. Perry, Jr. and H. A. Schroeder, *Amer. J. Med.* **22**, 168 (1957).
167. T. W. Perry, W. M. Beeson, W. H. Smith, and M. T. Mohler, *J. Anim. Sci.* **27**, 1674 (1968).
168. H. A. Peters, *in* "Metal-Binding in Medicine" (M. J. Seven and L. A. Johnson, eds.), p. 190, Lippincott, Philadelphia, Pennsylvania, 1960.
169. R. E. Pierson, *J. Amer. Vet. Med. Ass.* **149**, 1279 (1966).
170. W. J. Pitts, W. J. Miller, O. T. Fosgate, J. D. Morton, and C. M. Clifton, *J. Dairy Sci.* **49**, 455 (1966).
171. B. S. Platt and W. Frankul, *Proc. Nutr. Soc.* **21**, vii (1962).
172. M. P. Plumlee, D. R. Whitaker, W. H. Smith, J. H. Conrad, H. E. Parker, and W. M. Beeson, *J. Anim. Sci.* **19**, 1285 (1960).
173. W. G. Pond and J. R. Jones, *J. Anim. Sci.* **23**, 1057 (1964).
174. W. J. Pories, J. H. Henzel, and J. A. Hennessen, *Proc. 1st Annu. Conf. Trace Substances Environ. Health, 1967* p. 114 (1968).
175. W. J. Pories, J. H. Henzel, C. G. Rob, and W. H. Strain, *Lancet* **2**, 121 (1967).
176. W. J. Pories, J. H. Henzel, C. G. Rob, and W. H. Strain, *Ann. Surg. 165,* **432** (1967).
177. W. J. Pories, E. W. Schear, D. R. Jordan, J. Chase, G. Parkinson, R. Whittaker, W. H. Strain, and C. G. Rob, *Surgery* **59**, 821 (1966).
178. W. J. Pories and W. H. Strain, *in* "Zinc Metabolism" (A. S. Prasad, ed.), p. 378, Thomas, Springfield, Illinois, 1966.
179. G. W. Powell, W. J. Miller, J. D. Morton, and C. M. Clifton, *J. Nutr.* **84**, 205 (1964).
180. A. S. Prasad, *in* "Zinc Metabolism" (A. S. Prasad, ed.), p. 250. Thomas, Springfield, Illinois, 1966.
181. A. S. Prasad, J. A. Halsted, and M. Nadimi, *Amer. J. Med.* **31**, 532 (1961); A. S. Prasad, A. Miale, Z. Farid, H. H. Sandstead, A. Schulert, and W. J. Darby, *Arch. Intern. Med.* **111**, 407 (1963).
182. A. S. Prasad, D. Oberleas, and J. A. Halsted, *J. Lab. Clin. Med.* **66**, 508 (1965).
183. A. S. Prasad, D. Oberleas, P. Wolf, and J. P. Horwitz, *J. Clin. Invest.* **46**, 549 (1967); *J. Lab. Clin. Med.* **73**, 486 (1969).
184. A. S. Prasad, D. Oberleas, P. Wolf, J. P. Horwitz, E. R. Miller, and R. W. Luecke, *Amer. J. Clin. Nutr.* **22**, 628 (1969).

184a. A. S. Prasad, A. Schulert, H. H. Sandstead, A. Miale, and Z. Farid, *J. Lab. Clin. Med.* **62**, 84 (1963).
185. J. Quarterman, *Biochem. J.* **102**, 41P (1967).
186. J. Quarterman, C. F. Mills, and W. R. Humphries, *Biochem. Biophys. Res. Commun.* **25**, 354 (1966).
187. M. M. Rahman, R. E. Davies, C. W. Deyoe, B. L. Reid, and J. R. Couch, *Poultry Sci.* **40**, 195 (1961).
188. J. T. Raper and L. V. Curtin, *Proc. 3rd Conf. Processing as Related to Nutritive Value of Cottonseed Meal,* 1953.
189. J. Raulin, *Ann. Sci. Nat. Bot. Biol. Vegetale* **11**, 93 (1869).
190. N. S. Raum, G. L. Stables, L. S. Pope, O. F. Harper, G. R. Waller, R. Renbarger, and A. D. Tillman, *J. Anim. Sci.* **27**, 1695 (1968).
191. H. S. Reed, *Amer. J. Bot.* **33**, 778 (1946).
192. J. G. Reinhold, G. A. Kfoury, and M. Arslanian, *J. Nutr.* **96**, 519 (1968).
193. J. G. Reinhold, G. A. Kfoury, M. A. Ghalambor, and J. C. Bennett, *Amer. J. Clin. Nutr.* **18**, 294 (1966).
194. R. H. Rixon and J. F. Whitfield, *J. Histochem. Cytochem.* **7**, 262 (1959).
195. B. T. Robertson and M. J. Burns, *Amer. J. Vet. Res.* **24**, 997 (1963).
196. R. H. Robertson and P. J. Schaible, *Poultry Sci.* **37**, 1321 (1958); **39**, 837 (1960).
197. H. A. Ronaghy, P. G. Moe, and J. A. Halsted, *Amer. J. Clin. Nutr.* **21**, 709 (1968).
198. B. Rosoff and C. Martin, *Fed. Proc., Fed. Amer. Soc. Exp. Biol.* **25**, 316 (1966) (abstr.).
199. B. Rosoff and H. Spencer, *Nature* (*London*) **207**, 652 (1965).
200. V. Sadavisan, *Biochem. J.* **48**, 527 (1951); **49**, 186 (1951); **52**, 452 (1952).
201. B. M. Sahagian, I. Harding-Barlow, and H. M. Perry, Jr., *J. Nutr.* **90**, 259 (1966); **93**, 291 (1967).
202. H. H. Sandstead, A. S. Prasad, A. S. Schulert, Z. Farid, A. Miale, S. Bassily, and W. J. Darby, *Amer. J. Clin. Nutr.* **20**, 422 (1967).
203. H. H. Sandstead and R. A. Rinaldi, *J. Cell. Physiol.* **73**, 81 (1969).
204. H. H. Sandstead and G. H. Shepard, *Proc. Soc. Exp. Biol. Med.* **128**, 687 (1968).
205. M. Sato and K. Murata, *J. Dairy Sci.* **15**, 451 (1932).
206. E. D. Savlov, W. H. Strain, and F. Huegin, *J. Surg. Res.* **2**, 209 (1962).
207. S. A. Schlicker and D. H. Cox, *J. Nutr.* **95**, 287 (1968).
208. E. Schneider and C. A. Price, *Biochim. Biophys. Acta* **55**, 406 (1962).
209. H. A. Schroeder, "Mechanisms of Hypertension." Thomas, Springfield, Illinois, 1957.
210. H. A. Schroeder and A. P. Nason, *J. Invest. Dermatol.* **53**, 71 (1969).
211. H. A. Schroeder, A. P. Nason, I. H. Tipton, and J. J. Balassa, *J. Chronic Dis.* **20**, 179 (1967).
212. D. M. Scott and A. M. Fisher, *J. Clin. Invest.* **17**, 725 (1938).
213. M. L. Scott, E. R. Holm, and R. E. Reynolds, *Poultry Sci.* **38**, 1344 (1959).
214. M. L. Scott and T. R. Zeigler, *J. Agr. Food Chem.* **11**, 123 (1963).
215. I. A. Shevchuk, *Fed. Proc., Fed. Amer. Soc. Exp. Biol.* **24**, Trans. Suppl. 48 (1965).
216. A. A. Smirnov, *Biokhimiya* **13**, 79; *Chem. Abstr.* **42**, 8302 (1948).
217. H. Smith, *Forensic Sci. Soc., J.* **7**, 97 (1967).
218. I. D. Smith, R. H. Grummer, W. G. Hoekstra, and P. H. Phillips, *J. Anim. Sci.* **19**, 586 (1960).
219. S. E. Smith and E. J. Larson, *J. Biol. Chem.* **163**, 29 (1946).
220. W. H. Smith, M. P. Plumlee, and W. M. Beeson, *Science* **128**, 1280 (1960).
221. M. Somers and E. J. Underwood, *Aust. J. Biol. Sci.* **22**, 1229 (1969).
222. M. Somers and E. J. Underwood, *Aust. J. Agr. Res.* **20**, 899 (1969).

223. A. L. Somner and C. B. Lipman, *Plant Physiol.* **1**, 231 (1926); A. L. Somner, *ibid.* **3**, 231 (1928).
224. H. Spencer and J. Samachson, *Proc. 1st Int. Symp. Trace Element Metab. Anim., 1969* p. 312 (C. F. Mills, ed.) Livingstone, Edinburgh, 1970.
225. H. Spencer and B. Rosoff, *Health Phys.* **12**, 475 (1966).
226. H. Spencer, V. Vankinscott, I. Lewin, and J. Samachson, *J. Nutr.* **86**, 169 (1965).
227. M. R. Spivey-Fox and B. N. Harrison, *Proc. Soc. Exp. Biol. Med.* **116**, 256 (1964); *J. Nutr.* **86**, 89 (1965).
228. C. M. Spray and E. M. Widdowson, *Brit. J. Nutr.* **4**, 361 (1951); E. M. Widdowson and C. M. Spray, *Arch. Dis. Childhood* **26**, 205 (1951).
229. B. Starcher and F. H. Kratzer, *J. Nutr.* **79**, 18 (1963).
230. J. W. Stevenson and I. P. Earle, *J. Anim. Sci.* **15**, 1036 (1956).
231. S. S. Stevenson, *J. Clin. Invest.* **22**, 403 (1943).
232. W. H. Strain, F. Huegin, C. A. Lankau, W. P. Beiliner, R. K. McEvoy, and W. J. Pories, *Int. J. Radiat. Isotopes* **15**, 231 (1964).
233. W. H. Strain, A. Lascari, and W. J. Pories, *Proc. 7th Int. Congr. Nutr.*, Vol. 5, p. 759 (1966).
234. W. H. Strain, L. T. Steadman, C. A. Lankau, W. P. Beiliner, and W. J. Pories, *J. Lab. Clin. Med.* **68**, 244 (1966).
235. J. F. Sullivan and H. E. Lankford, *Amer. J. Clin. Nutr.* **10**, 153 (1962).
236. W. C. Supplee, *Science* **139**, 119 (1963).
237. N. F. Suttle and C. F. Mills, *Brit. J. Nutr.* **20**, 135 (1966).
238. W. R. Sutton and V. E. Nelson, *Proc. Soc. Exp. Biol. Med.* **36**, 211 (1937).
239. H. Swenerton and L. S. Hurley, *J. Nutr.* **95**, 8 (1968).
240. C. E. Swift and M. D. Berman, *Food Technol.* **13**, 365 (1969).
241. S. Szmigielski and J. Litwin, *Cancer* **17**, 1381 (1964).
242. T. R. Talbot and J. F. Ross, *Lab. Invest.* **9**, 174 (1964).
243. E. A. Teekell, M. C. Bell, and D. L. Anderson, *Amer. J. Vet. Res.* **23**, 226 (1962).
244. R. C. Theuer and W. G. Hoekstra, *J. Nutr.* **89**, 448 (1966).
245. R. E. Thiers and B. L. Vallee, *J. Biol. Chem.* **226**, 911 (1957).
246. A. Thompson, S. L. Hansard, and M. C. Bell, *J. Anim. Sci.* **18**, 187 (1959).
247. I. H. Tipton and M. J. Cook, *Health Phys.* **9**, 103 (1963).
248. J. R. Todd, private communication (1969).
249. W. R. Todd, C. A. Elvehjem, and E. B. Hart, *Amer. J. Physiol.* **107**, 146 (1934).
250. H. M. Tribble and F. I. Scoular, *J. Nutr.* **52**, 209 (1954).
251. H. F. Tucker and W. D. Salmon, *Proc. Soc. Exp. Biol. Med.* **88**, 613 (1955).
252. R. Tupper, R. W. E. Watts, and A. Wormall, *Biochem. J.* **57**, 254 (1954).
253. E. J. Underwood, *Proc. 12th World's Poultry Congr., 1962*, p. 216 (1962).
254. E. J. Underwood and M. Somers, *Aust. J. Agr. Res.* **20**, 889 (1969).
255. B. L. Vallee, *Physiol. Rev.* **39**, 443 (1959).
256. B. L. Vallee and J. G. Gibson, *J. Biol. Chem.* **176**, 445 (1948); *Blood* **4**, 455 (1949).
257. B. L. Vallee, F. L. Hoch, S. J. Adelstein, and W. E. C. Wacker, *J. Amer. Chem. Soc.* **78**, 5879 (1956).
258. B. L. Vallee, F. L. Hoch, and W. L. Hughes, *Arch. Biochem. Biophys.* **48**, 347 (1954).
259. B. L. Vallee and H. Neurath, *J. Biol. Chem.* **217**, 253 (1955).
260. B. L. Vallee, W. E. C. Wacker, A. F. Bartholomay, and F. L. Hoch, *Ann. Intern. Med.* **50**, 1077 (1959).
261. B. L. Vallee, W. E. C. Wacker, A. F. Bartholomay, and E. D. Robin, *N. Engl. J. Med.* **255**, 403 (1956); **257**, 1055 (1957).
262. D. R. Van Campen, *J. Nutr.* **97**, 104 (1969).

263. D. R. Van Campen, and P. U. Scaife, *J. Nutr.* **91**, 473 (1967).
264. R. Van Reen, *Arch. Biochem. Biophys.* **46**, 337 (1953).
265. R. Van Reen and P. B. Pearson, *Fed. Proc., Fed. Amer. Soc. Exp. Biol.* **12**, 283 (1953).
266. I. Vikbladh, *Scand. J. Clin. Lab. Invest.* **2**, 143 (1950); Suppl. 2 (1951).
267. P. Vohra and F. H. Kratzer, *J. Nutr.* **82**, 249 (1964).
268. N. F. Volkov, *Fed. Proc., Fed. Amer. Soc. Exp. Biol.* **22,** Trans. Suppl. 897 (1963).
269. W. E. C. Wacker, D. D. Ulmer, and B. L. Vallee, *N. Engl. J. Med.* **255**, 449 (1956).
270. J. C. N. Wakeley, B. Moffat, A. Crook, and J. R. Mallard, *Appl. Radiat.* 7, 225 (1960).
271. W. S. Wegner and A. H. Romano, *Science* **142**, 1669 (1963).
272. G. Weitzel, *Angew. Chem.* **68**, 566 (1966).
273. G. Weitzel and A. M. Fretzdorff, *Hoppe-Seyler's Z. Physiol. Chem.* **292**, 221 (1953).
274. G. Weitzel, F. J. Strecker, U. Roester, E. Buddecke, and A. M. Fretzdorff, *Hoppe-Seyler's Z. Physiol. Chem.* **296**, 19 (1954).
275. N. Westmoreland, N. L. First, and W. G. Hoekstra, *J. Reprod. Fert.* **13**, 223 (1968).
276. B. Wetterdal, *Acta Radiol.* Suppl. 156 (1958).
277. W. F. Whitmore, Jr., *Nat. Cancer Inst., Monogr.* **12** (1962).
278. E. M. Widdowson, R. A. McCance, and C. M. Spray, *Clin. Sci.* **10**, 113 (1951).
279. R. B. Williams and J. K. Chesters, *Proc. 1st Int. Symp. Trace Element Metab. Anim., 1969* p. 64 (C. F. Mills, ed.) Livingstone, Edinburgh, 1970.
280. R. B. Williams, C. F. Mills, J. Quarterman, and A. C. Dalgarno, *Biochem. J.* **95**, 29P (1965).
281. E. Winder and J. M. Denneny, *Nature (London)* **184**, 742 (1959).
282. I. J. Witham, *Biochim. Biophys. Acta* **11**, 509 (1963).
283. H. Wolff, *Deut. Arch. Klin. Med.* **197**, 263 (1950); *Klin. Wochenschr.* **34**, 409 (1956).
284. R. J. Young, H. M. Edwards, and M. B. Gillis, *Poultry Sci.* **37**, 1100 (1958).
285. T. R. Zeigler, M. L. Scott, R. McEvoy, R. H. Greenlaw, F. Huegin, and W. H. Strain, *Proc. Soc. Exp. Biol. Med.* **109**, 239 (1962).

# 9

# CHROMIUM

## I. Introduction

It is almost 60 years since chromium was first shown to be taken up by plants from soils (33). Various reports of a stimulating effect of chromates upon crop growth have since appeared (14). Chromium stimulation of growth and biotin synthesis in *Aerobacter aerogenes* has also been observed (49), but there is no conclusive evidence that this element is essential for either microorganisms or the higher plants. Indications have been obtained that excess chromium in certain soils is responsible for various disease conditions in plants (29, 71).

For many years biological interest in chromium in the higher animals was confined to its toxic properties in man. In 1954, Curran (13) showed that the synthesis of chloresterol and fatty acids from acetate by rat liver was enhanced in the presence of chromium ions, and, in 1959, Wacker and Vallee (74) reported high concentrations of chromium in a ribonucleoprotein fraction from beef liver. High concentrations of this element were also demonstrated in an ascidian (34). The physiological significance of these aggregations of chromium is not yet clear. Chromium has been further reported to be a constituent of proteolytic enzymes and to be essential for their function (6).

Interest in the nutritional physiology of chromium was raised to a new plane of significance in 1959 when Schwarz and Mertz (65) showed that trivalent chromium increases the glucose tolerance of rats subsisting on certain diets. Later work by these workers, and by Schroeder, established chromium as a cofactor with insulin, necessary for normal glucose utilization and for growth and longevity in rats and mice. Indications of its possible importance in human health and nutrition and of deficient or marginal chromium intakes from diets high in refined carbohydrates have recently appeared. A comprehensive and critical review of the occurrence and function of chromium in biological systems has recently been prepared by Mertz (40).

## II. Chromium in Animal Tissues and Fluids

Chromium is widely distributed in human tissues in extremely low and variable concentrations (32, 67, 69). How much of this variation is a reflection of analytical difficulties is unknown. Most adult human tissues contain chromium levels of the order of 0.02 to 0.04 ppm on dry basis, without special concentration in any known tissue or organ. The total content in the body of adult man is estimated to be less than 6 mg (51). Unlike most trace elements, tissue levels of chromium decline with age. The chromium concentration in the human fetus increases throughout pregnancy and is followed by a fall at birth (46, 50). Human stillborn and infant tissues carry chromium concentrations several times higher than those of adults (68). In the heart, lung, aorta, and spleen the levels decline rapidly in the first decade of life, while in the liver and kidney the neonatal levels are maintained until the second decade when a significant decline occurs (61). The lung is the only organ in which chromium concentrations are known to rise in later life (61, 68). Substantial variations in human liver and kidney chromium levels have been observed in different geographical regions (61). These presumably reflect regional differences in chromium intakes.

Acceptable data on the chromium content of the tissues of domestic animals are exceedingly meager. Limited evidence for a range of tissues from normal rats (37, 61) and from various domestic and wild animals (61) suggests that the levels are higher than those of the same tissues of most adult humans.

Highly divergent results have been reported for the chromium content of human blood and its distribution between cells and plasma. Herring *et al.* (24), employing a spectrochemical method described as an "accurate procedure," reported that the chromium level in 109 samples of normal human plasma ranged from 0.009 to 0.055 $\mu$g/g, with a mean of 0.027

μg/g. The corresponding range for red cells was very similar, namely 0.005-0.054 μg/g with a mean of 0.021 μg/g. The values for chromium were found to be consistently normal in a range of hematological diseases, both in the plasma and the red cells. Feldman and co-workers (18), using atomic absorption, found healthy human adult plasma to contain 0.011-0.065 μgCr/ml. This range is comparable with the 0.01-0.10 μg/ml observed in a small group of infants suffering from kwashiorkor, which were stated to be "within normal limits" (9). Higher mean values have been reported by Bala and Lifshits (3) employing a "fractional spectrographic" method. These workers also found the greater part of the chromium to be in the red cells, with a significant reduction in the levels in whole blood and erythrocytes but not in plasma, in all the types of anemia studied. Gray and Sterling (22), using a colorimetric method, found normal human serum to average 0.14 μg Cr/ml, a level very close to that of Hopkins and Schwarz (26). The latter workers observed further that the chromium of the plasma is very largely bound to the siderophilin (transferrin) component of the β-globulin fraction of the plasma proteins, and suggested that siderophilin is involved in the normal transport of chromium. The affinity of chromium(III) for siderophilin was shown to approach that of iron, with the two trivalent elements appearing to compete with each other for the metal-binding site.

Chromium is bound by other compounds within the tissues. Wacker and Vallee (74) isolated a nucleoprotein from beef liver which contained 1080 ppm Cr, representing a 20,000-fold aggregation of the metal over the 0.05 ppm found in whole liver. A similar but smaller aggregation of chromium, equally difficult to remove by dialysis or by chelating agents, was found in RNA preparations from diverse sources. The function of this chromium and of other metals present in RNA is unknown—they may play a role in the maintenance of the configuration of the molecule, perhaps linking purine or pyrimidine through covalent bonds. Chromium is known to be effective in serving as a cross-linking agent for collagen (23) and for conarachin (48).

Kirchgessner (30) reported a mean level of 0.057 μg Cr/g for cow's colostrum and of 0.013 μg/g for normal cow's milk. The latter value conforms well with the figure of 0.01 μg/g given by Schroeder *et al.* (61) but is appreciably lower than the range of 0.04 to 0.08 μg/g reported for human breast milk from 4 individuals (9).

## III. Chromium Metabolism

The absorption, distribution, and excretion of chromium by animals and man at levels well beyond those normally ingested have been well investigated (2). Less is known of the metabolism of chromium in physiological amounts. At these amounts, orally administered trivalent chromium is poorly absorbed, to the extent of only about 1% or less, regardless of dose and of dietary chromium status (15, 43, 72). The mechanism of transfer across the intestinal wall is unknown. Even soluble chromates are poorly absorbed and appear mainly in the feces in insoluble complex form (11). Chromic oxide is so insoluble that it has found wide application as a "marker" for determining the digestibility of components of the diet and of feed intakes by grazing stock (28).

Studies with $^{51}Cr$ indicate that hexavalent chromium is better absorbed than trivalent chromium. Mackenzie *et al.* (36) observed a three- to fivefold greater blood radioactivity following intestinal administration of hexavalent $^{51}Cr$ than after trivalent $^{51}Cr$. Donaldson and Barreras (15) similarly obtained a consistently greater intestinal uptake of chromium from $Na_2\,^{51}CrO_4$ than from $^{51}CrCl_3$ in man and in rats. Of interest was their further observation that patients with pernicious anemia and achlorhydria absorbed significantly more chromium from the former compound than did controls. This can probably be explained by the fact that acid gastric juice reduces hexavalent chromium ions to poorly absorbable $Cr^{3+}$ ions.

Absorbed anionic hexavalent chromium readily passes through the membrane of the red blood cells and becomes bound to the globin fraction of the hemoglobin. Cationic trivalent chromium is unable to pass this membrane and combines with the $\beta$-globulin fraction of the plasma proteins and, in physiological quantities, is transported to the tissues bound to siderophilin (22, 26). These observations led to the development of effective techniques for labeling erythrocytes, platelets, and plasma proteins to determine their life-span or survival time (1, 12, 22).

Absorbed or injected chromium is excreted mainly in the urine, with small amounts being lost in the bile and small intestine and possibly through the skin (25, 38). Tissue uptake is quite rapid, and the plasma is cleared of a dose of $^{51}Cr$ within a few days (25, 72). Whole body radioactivity disappears much more slowly and can be expressed by at least three components, with half-lives of 0.5, 6, and 83 days, respectively (43). Hopkins (25) injected $^{51}Cr$ trichloride into rats at levels of 0.01 and 0.10 $\mu$g Cr/100 g body weight and found little difference in blood clearance, tissue distribution, or excretion due to dose level, previous diet, or sex. The bones, spleen, testis, and epididymis retained more $^{51}Cr$ after 4 days than the heart, lung, pancreas, or brain. The fact that various tissues retain chromium much longer than the plasma suggests that there is no equilibrium between tissue stores and circulating chromium and, therefore, that plasma chromium levels may not be good indicators of body chromium status. Tissue uptake of chromate is markedly affected by age in mice (73). When older mice were used, the concentration of intraperitoneally injected $^{51}Cr$ in liver, stomach, epididymal fat pad, thymus, kidney, and especially the testes, declined to almost half the values observed in young animals. As Mertz (40) has pointed out, "these observations may offer one explanation for the declining tissue chromium levels with age, detected in a survey of the U.S. population" (61).

The chromium entering the tissues is distributed among the subcellular fractions in rather unusual proportions. Edwards and co-workers (16) found that 49% was concentrated in the nuclear fraction, 23% occurred in the supernatant, and the remainder was divided equally between the mitochondria and the microsomes.

Chromium is mobilized from body stores in response to glucose administration. The plasma chromium levels of 5 healthy young subjects were shown to rise sharply parallel to glucose levels and to fall as the glucose levels declined (19). Rises in circulating chromium have been induced in rats by injections of insulin, as well as of glucose (43). In the fasting state, plasma chromium levels appear to correlate poorly with glucose tolerance but, as has been suggested (19), the absence of a rise in plasma chromium following a glucose load could indicate a deficiency of available tissue chromium stores. In diabetic patients, urinary chromium is also increased after a glucose load, as shown in Table 32.

The placental transfer of chromium presents problems of particular metabolic interest. Several studies with rats have failed to detect any significant transfer of $^{51}Cr$ from the mother into the fetus, regardless of chemical binding and valence state (42, 52, 72), and despite abundant evidence of the presence of measurable amounts of stable chromium in the newborn (52, 61) and in human embryos (46, 50) and newborn infants (61). Furthermore, newborn rats from mothers fed an ordinary commercial diet contained significantly more chromium than those from

TABLE 32

*Mean Urinary Chromium Concentrations after Glucose Tolerance Test in Diabetic Patients*[a]

| Chromium State | No. of tests | Fasting (μg/liter) | 1 Hr (μg/liter)[b] | 2 Hr (μg/liter) | 2-Hr. Increase (%) |
|---|---|---|---|---|---|
| No chromium | 14 | 22 | 20 | 33 | 50 |
| Chromium-fed | 5 | 13 | 12 | 38 | 190 |
| Mean | | 19 | 16 | 36 | 90 |

[a]From Schroeder (58).

[b]Values for 1/2 hr and 1 hr combined.

mothers fed a torula yeast diet, or this diet supplemented with chromium acetate in the drinking water, in amounts at least equal to that supplied by the commercial diet (42). These results, and the further finding that trivalent $^{51}Cr$ extracted from brewer's yeast readily crossed the placenta, indicate that chromium in the form of a natural complex, but not as a simple salt, passes the placental barrier. Since the regulation of glucose tolerance responds fairly well to simple chromium salts, it appears that the rat has a more limited capacity to convert such salts into the biologically active form necessary for placental transport than it has to convert them into the glucose tolerance factor (GTF).

## IV. Chromium Deficiency and Functions

Chromium deficiency is characterized by impaired growth and longevity and by disturbances in glucose, lipid, and protein metabolism. Rats maintained on diets low in protein and in chromium (less than 0.1 ppm Cr) also develop a corneal lesion, grossly visible in one or both eyes, manifested as a pronounced opacity of the cornea and congestion of the iridal vessels (53). Chromium supplementation prevents the appearance of the lesion but does not cure the fully developed defect. The biochemical mechanism by which the low-chromium state leads to this pathological change is unknown.

### 1. *Glucose Metabolism*

In 1957, Schwarz and Mertz (66) observed impaired glucose tolerance in rats fed various diets. This was unrelated to the selenium-responsive necrotic liver degeneration that occurred on these diets. It was postulated that the condition was due to a deficiency of a new dietary agent, designated GTF. The active component was subsequently shown to be trivalent chromium (65). Various trivalent chromium compounds administered at dose levels of 20 to 50 $\mu$g Cr/100 g body weight were fully effective in restoring tolerance to injected glucose, whereas hexavalent chromium and certain forms of trivalent chromium were either inactive or less active. Fractionation studies with the glucose tolerance factor of brewer's yeast have yielded fractions with considerably greater biological activity *in vitro* than chromium as $CrCl_3$ (8). This suggested the existence of an unidentified chromium-containing complex, GTF, analogous to the potent Factor 3 form of selenium.

Other systems involving carbohydrate metabolism, including glucose uptake by the isolated rat lens (17), glucose utilization for lipogenesis, and $CO_2$ production (45) (see Table 33), and glycogen formation from

TABLE 33
*Production of $^{14}CO_2$ from Glucose-1-$^{14}C$ by Rat Adipose Tissue in Vitro*[a,b]

| Supplement | No. of rats | 0 | Insulin ($\mu$U/flask): 200 | 500 | 1000 |
|---|---|---|---|---|---|
| None | 5 | 10.7±0.8 | 17.9±2.5 | 15.4±1.1 | 18.7±2.0 |
| 5 ppm Cr | 5 | 19.9±1.3 | 19.9±7.7 | 33.9±8.0 | 38.9±7.0 |

[a] From Mertz *et al.* (44).
[b] Values are expressed as mmoles $CO_2$/100 mg tissue.

glucose (55), were then shown to respond to chromium plus insulin, with little or no response in the absence of the hormone. These results and those of a polarographic study of chromium-insulin-mitochondrial interaction (10), led Mertz to the hypothesis that Cr(III) acts as a cofactor with insulin at the cellular level, through the formation of a ternary complex with membrane sites, insulin, and chromium (39).

In these experiments the fasting levels of serum glucose remained normal, suggesting that the chromium deficiency (0.1 ppm Cr) was insufficient to impair normal glucose utilization. Schroeder and co-workers (44, 57) induced a more severe chromium deficiency by strict control of environmental contamination. Under these conditions the glucose tolerance was worsened, and hyperglycemia and glycosuria were observed in rats and mice. The syndrome, resembling moderate diabetes mellitus, was rapidly reversed when 2 or 5 ppm Cr(III) was supplied in the drinking water. Schroeder (57) has also shown that the fasting serum glucose is relatively low in rats fed brown sugar (0.12–0.24 ppm Cr) and in females fed white sugar (0.02–0.03 ppm Cr) plus chromium.

Evidence is accumulating that chromium is involved in glucose tolerance in man. In a controlled clinical trial, 3 of 6 patients with mild diabetes, revealed a significant improvement of glucose tolerance during oral supplementation with chromium as $CrCl_3 \cdot 6H_2O$. Two more patients showed some improvement and in one the glucose tolerance was not influenced (20). In a study of the impaired glucose tolerance of old people, 10 subjects with diabetic glucose tolerance curves were given chromium supplements for 2 to 3 months. The tests of 4 of these improved to normal values, whereas those of the remaining 6 were unchanged (35). The group that responded to chromium had a much milder impairment than the nonresponders. This suggests that the low-chromium state of the responding patients had not yet been complicated by other factors. The supplementary chromium administered in these tests amounted to 150 $\mu$g/day, which increased the total intake from about 50 to 200 $\mu$g/day.

A severe impairment of glucose tolerance is characteristic of kwashiorkor and similar forms of malnourishment in infants (5). In some areas (27), but not in others (9), this condition can be markedly improved by chromium treatment. Thus Hopkins *et al.* (27) treated 12 such malnourished infants, 6 from Jordan and 6 from Nigeria, with 250 μg of chromium as $CrCl_3$. Within 18 hr after the Cr(III) administration the glucose removal rates of the Jordanian infants improved spectacularly from an average of 0.6 to 2.9%/min and those of the Nigerian infants from 1.2 to 2.9%/min. Further studies in Jordan elicited a clear-cut correlation between the geographical location of the family and a reduced glucose tolerance in the infant. This appeared to be related to the chromium content of the drinking water since the food sources of the families were similar. The chromium content of the water was significantly lower where the infants had impaired glucose tolerance than where no such impairment was evident.

Normal glucose tolerance tests in young healthy subjects are not influenced by chromium, just as glucose tolerance is not further improved in chromium-sufficient rats. The fact that not every malnourished infant or every diabetic or old person responds to chromium supplementation suggests either that the chromium status of the nonresponding individuals is adequate for this function or that one of the many other factors influencing carbohydrate metabolism is operating in these cases. Chromium can only be expected to correct that part of the condition which is caused by chromium deficiency. As Mertz (39) has cogently stated, "chromium is not considered a hypoglycemic agent, a substitute for insulin, nor a cure for diabetes."

### 2. *Lipid Metabolism*

Several lines of evidence collectively suggest that chromium may play a role in serum cholesterol homeostasis. The addition of chromium to a low-chromium diet suppressed serum cholesterol levels in rats and, in males, inhibited the tendency of these levels to increase with age. Further, a significant depression in serum cholesterol was achieved in male rats by feeding 1 μg Cr/ml in the drinking water, with a similar effect in females fed 5 μg Cr/ml in this way (60, 63). This effect was not specific to chromium since it was also observed in the rats receiving lead and cadmium in the drinking water. Confirmatory evidence implicating chromium was obtained in later experiments by Schroeder (59). Serum cholesterol levels were relatively elevated and increased with age in the rats receiving white sugar, very low in chromium, whereas those receiving brown sugar, or white sugar plus chromium, were low. The effects were similar in both sexes and younger rats fed raw sugar had lower levels than those fed white (59). A significant decline in serum cholesterol has also been reported in some institutionalized patients fed 2 mg Cr as the acetate daily for a period of 5 months. Other patients given similar treatment showed no such response (58).

Examination of the aortas of rats at the end of their natural lives revealed a significantly lower (2%) incidence of spontaneous plaques in the chromium-fed animals than in the chromium-deficient animals (19%). There were also lowered amounts of stainable lipids and of fluorescent material in the aorta (60). A satisfactory explanation of this effect of chromium has not yet appeared, although it should be noted that diabetes in man is associated with an increased incidence of vascular lesions (21). Further evidence of chromium involvement in lipid metabolism comes from the experiments of Mertz *et al.* (44), who found that chromium plus insulin significantly increased glucose uptake and incorporation of glucose carbon into epididymal fat in chromium-deficient rats (Table 33). Previously Curran (13) had observed that trivalent chromium enhances the synthesis of cholesterol and fatty acids from acetate in the livers of rats fed a commercial chow. The commercial chow used by Curran was probably deficient in chromium at that time.

### 3. *Protein Synthesis*

Rats fed diets deficient in chromium and protein have an impaired capacity to incorporate several amino acids into the protein of their hearts (54, 55). Slightly improved incorporation was achieved with insulin alone, which was significantly enhanced by Cr(III) supplementation. The amino acids affected by chromium were $\alpha$-amino isobutyric acid, glycine, serine, and methionine. No such effect of chromium was observed with lysine, phenylalanine, and a mixture of 10 other amino acids. Insulin *in vivo* also stimulated the cell transport of an amino acid analog to a greater degree in rats fed a low-protein, chromium-supplemented diet than it did in Cr-deficient controls (55). The hypothesis that chromium acts as cofactor for insulin can, therefore, also be applied to two insulin-responsive steps in amino acid metabolism which are independent of the action of insulin on glucose utilization.

### 4. *Growth and Longevity*

On a diet of rye, skim milk, and corn oil (containing 0.1 ppm Cr), with added vitamins and Zn, Cu, Mn, Co, and Mo in the water, male mice and rats receiving 2 or 5 ppm Cr(III) in the drinking water grew significantly better than their controls (62, 64). This effect was associated with decreased mortality. The last surviving male rats given chromium in water from the time of weaning lived to greater ages than those not so fed (62). The median age of male mice at death was 99 days longer when they were fed chromium than when they were not, and the mean age was 91 days longer (62). No such differences in longevity due to chromium were observed in female mice or rats. The chromium treatment had no effect upon the incidence of tumors but appeared to protect female rats against lung infection. Chromium was little accumulated in the organs, and evidence was obtained that the fetus is supplied with a considerable amount of chromium. This suggests that repeated pregnancies may deplete the tissue stores of the mother on a low-chromium diet (62). Mertz and Roginski (41) have since shown that raising rats in plastic cages on a low-protein, low-chromium diet results in a moderate depression of growth which can be alleviated by chromium supplementation. Subjecting the animals to controlled exercise or blood loss aggravated the low-chromium state.

## V. Sources of Chromium

Very little is known of the factors affecting the movement of chromium in the food chain from the soil, through plants and animals, to man. Saint-Rat (56) reported chromium concentrations in vegetable tissues ranging from 0.01 to 1 ppm, with the levels in most plants lying between 0.1 and 0.5 ppm. These values accord well with the levels obtained spectrographically for pasture grasses, namely 0.1-0.5 ppm on dry basis, obtained by Kirchgessner *et al.* (31) in Germany and by Mitchell in Scotland (47). They are also of the same order as those observed by Schroeder *et al.* (61) and by Carter *et al.* (9) in a rather heterogeneous range of tissues and human foods. The latter workers found 1, 1, 1-trichloro-2, 2-bis(p-chlorophenyl) ethane (DDT) and certain therapeutic preparations to be surprisingly high in chromium.

Chromium resembles most trace elements in being concentrated in the branny layers and germ of cereal grains. For instance, one sample of whole wheat was reported to contain 1.7 ppm Cr, compared with 0.23 ppm in patent flour and 0.14 ppm (oven-dry basis) in a sample of white bread (58). Chromium is also lost in the process of sugar refining, so that white sugars contain very little, compared with the amounts in brown or raw sugar (58, 59).

Dietary chromium intakes by man are greatly influenced by the amounts and proportions of refined carbohydrates that they supply. An institutional diet was shown to provide about 80 μg Cr per person per day in one study (61). In the two studies with diabetics and old people (20, 35), mentioned previously, in which some responses to chromium supplementation were obtained, the daily intake was estimated to be as low as 50 μg Cr/day. These intakes are very much lower than those reported for two adults by Tipton *et al.* (70), namely 330 and 400 μg Cr/day, in a comprehensive spectrographic study of trace elements in diets and excreta.

The drinking water can be a significant source of chromium in some areas. The municipal water supplies of twenty-four United States cities were found over 30 years ago to contain from 0.001 to 0.010 ppm Cr (4). Carter *et al.* (9) reported a level of 0.001 ppm in the tap water used in their studies of kwashiorkor in Egypt, and Hopkins *et al.* (27) reported a mean level of 0.0005 ppm Cr in the drinking water in one area in Jordan and 3 times this level (0.0016 ppm) in another area in that country. Marked differences in chromium content of potable water from different parts of the United States were also reported by Schroeder *et al.* (61).

All the above figures for the chromium content of foods and waters refer to total chromium. Little is known of the chemical forms in which chromium occurs in different foods. This is of particular importance because animals cannot readily convert the element into its specific biologically active complexes, as, for instance, they convert simple forms of zinc, copper, and iron into their active complexes. The animal, or at least the rat, is dependent to a large degree on dietary sources of GTF or similar organic forms of chromium. Foods rich in GTF such as brewer's yeast are, therefore, quantitatively superior, per unit of chromium, to others with less of their chromium in this form, and to simple salts of chromium. In respect to optimal response in systems concerned with carbohydrate metabolism, such differences in the forms of chromium ingested by the rat are real but relative. In respect to placental transport, the rat exhibits an absolute dietary requirement for GTF or for some organic complex capable of direct passage across the placental barrier. The definition of the nature of GTF, and of the chemical forms in which chromium occurs in foods and is transported across the placenta, presents a challenging problem for future research.

## VI. Chromium Toxicity

Chromium, particularly trivalent chromium, has a low order of toxicity. A wide margin of safety exists between the amounts ordinarily ingested and those likely to induce deleterious effects. Acute toxicity studies in rats with intravenously injected chromium(III) established the lethal dose for 50% of the animals at about 1 mg of the element per 100 g body weight (43). This is many times greater than the 0.05-0.10 $\mu g$ Cr/100 g required to correct the impairment of glucose metabolism *in vivo*. Hexavalent chromium is much more toxic than trivalent. Chronic exposure to chromate dust has been correlated with increased incidence of lung cancer (7), and oral administration of excessive levels (50 ppm) has been associated with growth depression and liver and kidney damage in experimental animals (37).

## REFERENCES

1. K. A. Aas and F. H. Gardner, *J. Clin. Invest.* **37**, 1257 (1958).
2. A. M. Baetjer, *in* "Chromium" (M. J. Udy, ed.), Amer. Chem. Soc. Mon. No. 132. Reinhold, New York, 1956.
3. Y. M. Bala and V. M. Lifshits, *Fed. Proc., Fed. Amer. Soc. Exp. Biol.* **25**, Trans. Suppl. 370 (1966).
4. M. M. Braideck and F. H. Emery, *J. Amer. Water Works Assoc.* **27**, 557 (1935); quoted by Schroeder (61).
5. H. A. Braig and J. C. Edozien, *Lancet* **2**, 662 (1965).
6. S. E. Bresler and N. N. Rozentsveig, *Biokhimiya* **16**, 84 (1951); cited by Mertz, (39).
7. H. P. Brinton, E. S. Fraiser, and A. L. Koven, *Pub. Health Rep.* **67**, 835 (1952).
8. J. N. Burkeholder and W. Mertz, *Fed. Proc., Fed. Amer. Soc. Exp. Biol.* **25**, 759 (1966).
9. J. P. Carter, A. Kattab, K. A. Abd-al-Hadi, J. T. Davis, A. E. Gholmy, and V. N. Patwardhan, *Amer. J. Clin. Nutr.* **21**, 195 (1968).
10. G. D. Christian, E. C. Knoblock, W. L. Purdy, and W. Mertz, *Biochim. Biophys. Acta* **66**, 420 (1963).
11. L. W. Conn, H. L. Webster, and A. H. Johnston, *Amer. J. Hyg.* **15**, 760 (1932).
12. M. Cooper and C. A. Owen, *J. Lab. Clin. Med.* **47**, 65 (1956).
13. G. L. Curran, *J. Biol. Chem.* **210**, 765 (1954).
14. G. K. Davis, *in* "Chromium" (M. J. Udy, ed.), Amer. Chem. Soc. Mon. No. 132. Reinhold, New York, 1956.
15. R. M. Donaldson and R. F. Barreras, *J. Lab. Clin. Med.* **68**, 484 (1966).
16. C. Edwards, K. B. Olson, G. Heggen, and J. Glenn, *Proc. Soc. Exp. Biol. Med.* **107**, 94 (1961).
17. T. G. Farkas and S. L. Roberson, *Exp. Eye Res.* **4**, 124 (1965).
18. F. J. Feldman, E. C. Knoblock, and W. C. Purdy, *Anal. Chim. Acta* **38**, 489 (1967).
19. W. H. Glinsmann, F. J. Feldman, and W. Mertz, *Science* **152**, 1243 (1966).
20. W. H. Glinsmann and W. Mertz, *Metab., Clin. Exp.* **15**, 510 (1966).
21. S. Goldenberg and H. T. Blumenthal, *in* "Diabetes Mellitus, Diagnosis and Treatment," p. 177. Am. Diabetes Assoc., New York, 1964.
22. S. J. Gray and K. Sterling, *J. Clin. Invest.* **29**, 1604 (1950).
23. K. H. Gustavson, *Nature* (*London*) **182**, 1125 (1958).
24. W. B. Herring, B. S. Leavell, L. M. Paixao, and J. H. Yoe, *Amer. J. Clin. Nutr.* **8**, 846 (1960).

25. L. L. Hopkins, Jr., *Amer. J. Physiol.* **209**, 731 (1965).
26. L. L. Hopkins, Jr. and K. Schwarz, *Biochim. Biophys. Acta* **90**, 484 (1964).
27. L. L. Hopkins, Jr., O. Ransome-Kuti, and A. S. Majaj, *Amer. J. Clin. Nutr.* **21**, 203 (1968).
28. E. A. Kane, W. C. Jacobson, and L. A. Moore, *J. Nutr.* **44**, 583 (1950); **47**, 263 (1952).
29. P. A. Keiller, *Tea Quart.* **12**, 96 (1939).
30. M. Kirchgessner, *Z. Tierphysiol., Tierernahr. Futtermittelk.* **14**, 270 and 278 (1959).
31. M. Kirchgessner, G. Merz, and W. Oelschläger, *Arch. Tierernahr.* **10**, 414 (1960).
32. H. J. Koch, E. R. Smith, N. F. Shimp, and J. Connor, *Cancer* **9**, 499 (1956).
33. P. Koenig, *Chem. Zentr.* **35**, 442 (1911); *Exp. Sta. Rec.* **28**, 730 (1913).
34. E. P. Levine, *Science* **133**, 1352 (1961).
35. R. A. Levine, D. H. Streeten, and R. J. Doisy, *Metab., Clin. Exp.* **17**, 114 (1968).
36. R. D. Mackenzie, R. Anwar, R. U. Byerrum, and C. A. Hoppert, *Arch. Biochem. Biophys.* **79**, 200 (1959).
37. R. D. Mackenzie, R. U. Byerrum, C. F. Deeker, C. A. Hoppert, and R. F. Langham, *A. M. A. Arch. Ind. Health* **18**, 232 (1958).
38. T. F. Mancuro and W. C. Hueper, *Ind. Med. Surg.* **20**, 358 (1951).
39. W. Mertz, *Fed. Proc., Fed. Amer. Soc. Exp. Biol.* **26**, 186 (1967).
40. W. Mertz, *Physiol. Rev.* **49**, 163 (1969).
41. W. Mertz and E. E. Roginski, *J. Nutr.* **97**, 531 (1969).
42. W. Mertz, E. E. Roginski, F. J. Feldman, and D. E. Thurman, *J. Nutr.* **99**, 363 (1969).
43. W. Mertz, E. E. Roginski, and R. Reba, *Amer. J. Physiol.* **209**, 489 (1965).
44. W. Mertz, E. E. Roginski, and H. A. Schroeder, *J. Nutr.* **86**, 107 (1965).
45. W. Mertz, E. E. Roginski, and K. Schwarz, *J. Biol. Chem.* **236**, 318 (1961).
46. A. Mikosha, *Nauk. Zap., Stanislav. Derzh. Med. Inst.* No. 3, 85 (1959); *Chem. Abstr.* **59**, 7969 (1963).
47. R. L. Mitchell, *Research (London)* **10**, 357 (1957).
48. W. E. F. Naismith, *Arch. Biochem. Biophys.* **73**, 255 (1958).
49. D. Perlman, *J. Bacteriol.* **49**, 167 (1945).
50. L. A. Pribluda, *Dokl. Akad. Nauk Beloruss. SSR* **7**, 135 (1963); *Chem. Abstr.* **59**, 3142 (1963).
51. Report of IRCP Committee II on Permissible Dose for Internal Radiation, *Health Phys.* **3**, 380 (1960).
52. E. E. Roginski, F. J. Feldman, and W. Mertz, *Fed. Proc., Fed. Amer. Soc. Exp. Biol.* **27**, 482 (1968).
53. E. E. Roginski and W. Mertz, *J. Nutr.* **93**, 249 (1967).
54. E. E. Roginski and W. Mertz, *Fed. Proc., Fed. Amer. Soc. Exp. Biol.* **26**, 301 (1967).
55. E. E. Roginski and W. Mertz, *J. Nutr.* **97**, 525 (1969).
56. L. D. Saint-Rat, *C. R. Acad. Sci.* **227**, 150 (1948).
57. H. A. Schroeder, *J. Nutr.* **88**, 439 (1966).
58. H. A. Schroeder, *Amer. J. Clin. Nutr.* **21**, 230 (1968).
59. H. A. Schroeder, *J. Nutr.* **97**, 237 (1969).
60. H. A. Schroeder and J. J. Balassa, *Amer. J. Physiol.* **209**, 433 (1965).
61. H. A. Schroeder, J. J. Balassa, and I. H. Tipton, *J. Chronic Dis.* **15**, 941 (1962).
62. H. A. Schroeder, J. J. Balassa, and W. H. Vinton, Jr., *J. Nutr.* **83**, 239 (1964); **86**, 51 (1965).
63. H. A. Schroeder, W. H. Vinton, Jr., and J. J. Balassa, *Proc. Soc. Exp. Biol. Med.* **109**, 859 (1962).

64. H. A. Schroeder, W. H. Vinton, Jr., and J. J. Balassa, *J. Nutr.* **80**, 39 and 48 (1963).
65. K. Schwarz and W. Mertz, *Arch. Biochem. Biophys.* **85**, 292 (1959).
66. K. Schwarz and W. Mertz, *Arch. Biochem. Biophys.* **72**, 515 (1957).
67. S. R. Stitch, *Biochem. J.* **67**, 97 (1957).
68. I. H. Tipton, *in* "Metal-Binding in Medicine" (M. J. Seven and L. A. Johnson, eds.), p. 27. Lippincott, Philadelphia, Pennsylvania, 1960.
69. I. H. Tipton and M. J. Cook, *Health Phys.* **9**, 103 (1963).
70. I. H. Tipton, P. L. Stewart, and P. G. Martin, *Health Phys.* **12**, 1683 (1966).
71. A. J. Van der Merwe and F. C. Anderson, *Farming S. Afr.* **12**, 439 (1937).
72. W. J. Visek, I. B. Whitney, U. S. G. Kuhn, and C. L. Comar, *Proc. Soc. Exp. Biol. Med.* **84**, 610 (1963).
73. P. V. Vittorio and E. W. Wright, *Can. J. Biochem. Physiol.* **41**, 1349 (1963).
74. W. E. C. Wacker and B. L. Vallee, *Fed. Proc., Fed. Amer. Soc. Exp. Biol.* **18**, 345 (1959); *J. Biol. Chem.* **234**, 3257 (1959).

# 10

# CADMIUM

## I. Introduction

Biological interest in cadmium is largely confined to its toxic properties, its possible relation to human hypertension, and its interactions with zinc and other essential metals. The similarity between the atomic structure and chemical behavior of cadmium and zinc, together with the high concentrations of cadmium that occur in the kidney, relative to those of other organs, led Vallee and associates (23,31) to suggest that it might occur as an integral part of a natural substance and, like zinc, perform some function in mammalian organisms. A protein compound containing as much as 5.9% Cd and 2.2% Zn was isolated by them from equine renal cortex and named metallothionein. This compound was also found in human kidney and liver, with varying proportions of cadmium and zinc and in sufficient amounts to account for all of the cadmium in these organs (23,59). An aggregation of cadmium of this size, higher by an order of magnitude than the metal content of any other known metalloprotein, and its specific association with a particular macromolecules, point to a functional role for this element. No such role has yet been demonstrated in living cells. Cotzias *et al.* (8) suggest that metallothionein may merely constitute a sequestering system for cadmium. No cadmium-containing metalloenzymes have yet been found, but *in vitro* studies have revealed an inhibition by cadmium of several mammalian enzymes (62). Increased esterase activity, accompanied by a parallel fall in peptidase activity, result from incubation of carboxypeptidase B with $Cd^{++}$ ions (14).

## II. Cadmium in Animal Tissues and Fluids

The total cadmium content of the body of the "standard American man" has been estimated to average 30 mg, of which some 10 mg occurs in the kidneys and 4 mg in the liver (60). Levels of 30 to 40 ppm Cd in the kidney and 2 to 3 ppm in the liver of human adults are common (78). The cadmium concentrations of all other organs and tissues studied are normally very much lower. Cadmium is virtually absent from the kidney of the newborn. It accumulates in these organs up to about the fifth decade and then declines slightly (55,65). Substantial variation in kidney cadmium concentration occurs with geographical location. Mongoloid subjects from several areas contained significantly higher levels of renal cadmium than their North American counterparts, with the Japanese averaging more than twice as much (55). Regional differences of this nature presumably reflect differences in cadmium intakes, either as a result of differences in food habits or in the cadmium contents of foods, water supplies, and the atmosphere, or from both together.

Normal human blood is very low and highly variable in cadmium content. Imbus *et al.* (21) reported a range of 0.3 to 5.4 μg Cd/100 ml, with a median concentration of 0.7 μg. Kubota *et al.* (28) determined the cadmium content of the blood of 243 adults from 19 cities in the United States. Less than one-half of the samples had detectable amounts of cadmium by the atomic absorption method used, and no consistent geographic pattern was apparent. More than half the samples had 0.5 μg Cd/100 ml or less, and 83% had less than 1 μg/100 ml. The median concentration was stated to be about 0.5 μg/100 ml.

Normal human urine is also very low and very variable in cadmium content. Perry and Perry (53) reported a range of $< 7$ to 22 μg Cd/liter, with an average $< 12.7$ μg/liter. Smith and Kench (74) found the urine of factory workers with no known exposure to cadmium to range similarly from 2 to 22 μg/liter, with a mean close to 10 μg/liter. Workers exposed to cadmium oxide dust disclosed much higher urinary cadmium levels, ranging from 15 to 420 μg/liter, with most values lying between 40 and 100 μg/liter. The urine of workers exposed to cadmium fumes varied from 40 to 410 μg Cd/liter. The initial site of deposition of cadmium was the lungs, from which site the metal was widely distributed throughout the body, accumulating in the liver and kidneys in concentrations as high as 300-400 ppm or more than 10-100 times normal (75). The high tissue concentrations of cadmium remained for many years after exposure to cadmium had ceased, indicating that excretion is slow. The urinary cadmium concentration of hypertensive individuals is also significantly higher than normal. In a comparison of 15 normal and 15 hypertensive patients, Perry and Schroeder (54) found the urine of the former group to average less than 1 μg Cd/liter, compared with nearly 50 μg/liter in the hypertensive subjects. The latter fell to approximately 5 μg/liter during therapy to control the blood pressure. Schroeder and Nason (68) reported the mean cadmium concentration in human hair to be 2.76±0.483 μg/g in males and 1.77±0.239 μg/g in females.

Little information is available on the normal cadmium content of animal tissues. Schroeder and Balassa (65) report some values for a small selection of wild animals which indicate generally lower levels in liver and kidneys than in adult man. Bunn and Matrone (3) report levels of 0.4 ppm Cd in the liver of rats on their basal diet, and 77.6 ppm when these animals were fed toxic (100-ppm) levels of cadmium. Corresponding figures for the liver of mice were 1.2-1.9 ppm, and 133-205 ppm for the cadmium-fed groups. Decker *et al* (11) observed comparable concentrations. In a study of cadmium toxicity in the bovine, Powell and co-workers (57) reported low (0.5-4.0 ppm on dry basis) tissue cadmium concentrations in a single control calf. When fed dietary cadmium levels of 640 and 2560 ppm, much higher concentrations were found in the liver, kidneys, spleen, incisors, and hair but not in the femur or the skin. The cadmium concentrations of the testicles of calves treated with 40 and 160 ppm Cd were not significantly increased beyond those of the controls (3 ppm). Calves receiving 640 and 2560 ppm of dietary Cd showed testicular cadmium concentrations of 14 and 16 ppm, respectively.

The cadmium content of cow's milk is low and apparently varies appreciably with the location of the cows. Thus Kubota *et al.* (28) found the milk from one area in Montana to range from 12 to 20 μg/liter with a mean of 15, whereas milk from other areas ranged from $< 1$ to 10 μg/liter and averaged 6 μg/liter. In a study of the effect of a high level of dietary cadmium on cows, Miller *et al.* (37) reported the cadmium content of the milk to be less than 1 ppm, which was the lower reliability limit of the method used. Less than 0.02% of the extra cadmium given appeared in the milk. Cadmium readily combined with the casein and whey fractions when added to milk *in vitro*, with the amount combined increasing linearly when from 1 to 25 ppm were added.

## III. Cadmium Metabolism and Interactions with Zinc and Other Metals

Little is known of the intestinal absorption of cadmium from ordinary foods or of the various routes of excretion of this element at physiological intakes. The metabolism of cadmium is known to be greatly affected by the relative intakes of zinc, copper, and other metals, and cadmium intakes, in turn, influence the metabolism of zinc and of iron and copper. Schroeder *et al.* (70) estimate that at a daily intake of 200 μg Cd by man and a urinary excretion of 40 μg, 20% appears in the urine, 0.9-1.8% is retained, and the remainder is lost in the feces. Judging by the figures given earlier for urinary cadmium this level of urinary excretion seems excessively high. How much of the fecal cadmium represents unabsorbed metal and how much comes from reexcretion directly into the bowel or via the bile or pancreatic juice, remain to be determined.

It appears that the body does not possess homeostatic control mechanisms for cadmium, such as exist for zinc and manganese. Cotzias *et al.* (8), in studies with injected $^{109}Cd$ in the mouse, found almost no total body turnover of this element, regardless of challenges with dietary loads of cadmium or zinc. Absorption occurred irrespective of the body's cadmium burden. The injected $^{109}Cd$ concentrated primarily in the soft tissues, rather than in the bones, and was distributed intracellularly in a manner similar to zinc, except for a lower uptake by the nuclear fraction. Limited tissue exchanges of cadmium and zinc were demonstrated in the rabbit and the changes in zinc metabolism in response to zinc loading and to cadmium loading were sufficiently similar to provide further evidence that cadmium is a zinc antimetabolite (8). In later experiments, intraperitoneally injected cadmium and cadmium feeding, in contrast to zinc feeding, influenced the intracellular distribution of $^{65}Zn$ in mouse liver cells (9). A partial replacement of zinc by cadmium in various tissues, as demonstrated in these experiments, is in accord with observations made in various other investigations (10,15,29,41,76). Such interchange implies a competition between cadmium and zinc for protein-binding sites, presumably including those of the zinc metalloenzymes. The fact that human renal metallothionein can vary in cadmium and zinc contents, but the sums of the two elements remain constant (59), suggests that cadmium and zinc compete for the protein-binding sites of this compound.

In studies with intact strips of rat intestine (61), cadmium uptake into the intestinal walls was demonstrated in the duodenum, jejunum, and ileum, with the least into the jejunum. These uptakes were initially rapid but remained constant and probably surface bound. Zinc uptake by the intestinal mucosa, measured in this way, was enhanced by cadmium or mercury, and mercury uptake was enhanced by cadmium or zinc. Cadmium uptake was enhanced by zinc at $10^{-5}$ to $10^{-3}$ $M$ but was depressed at higher zinc concentrations. A well-defined competition for transmural transport, i.e., across the intestinal wall, was observed between zinc and cadmium, and between mercury and cadmium. A severe depression by cadmium of $^{64}Cu$ uptake from ligated segments of rat intestine has also been demonstrated (79).

A mutual antagonism between cadmium and zinc, and important interactions between cadmium and copper and iron, are apparent from several whole animal studies. In the diet of the flour beetle, *Tribolium confusum,* additional copper slightly reduced the toxic effects of high levels of both zinc and cadmium, and high levels of zinc reduced the toxicity of cadmium. The zinc requirements of this beetle were thus considerably raised and the copper requirements slightly raised by cadmium feeding (34). When cadmium was added to the diets of chicks, in the presence of additional copper, zinc, or iron, alone or in combination, significant interactions between the metals were revealed affecting live-weight gains, mortality, red cell counts, hemoglobin levels, and cytochrome oxidase activity in the heart (20). Comparable interactions involving cadmium, zinc, copper, and iron were subsequently demonstrated in mice and rats (3). The reduced weight gains induced by cadmium at a dietary level of 100 ppm were largely overcome by additional zinc and copper. The mechanisms of these interactions are not fully understood but competition for protein-binding sites at the mucosal cell level, and elsewhere in the tissues, provides the most plausible explanation at present available. Cadmium feeding does not consistently reduce the zinc concentrations in the tissues nor does zinc feeding consistently reduce the cadmium concentrations, even though each element can ameliorate the toxicity effects of the other. This has been shown with mice and rats (3) and with pigs (18,19). However, Lease (29) demonstrated reduced absorption of $^{65}Zn$ in chicks by cadmium feeding, with reduced levels in the blood and tibia, but not in the liver, over 24 hr. This worker suggests that the effect of dietary cadmium on dietary zinc will depend on the absolute amounts of each, the proportions of one to the other, and the length of the feeding period.

Rabbits fed cadmium develop hyperplastic bone marrow and hypochromic, microcytic anemia, similar to that produced by iron deficiency (2). Japanese quail, similarly treated with cadmium, develop a severe anemia, associated with markedly increased concentrations of plasma transferrin and elevated transferrin-to-albumin ratios (22). The mechanism of this effect of cadmium is not yet known, although interference with iron metabolism is clearly indicated.

## IV. Cadmium and Hypertension

A relationship between cadmium and human hypertension has been postulated by Schroeder (64). It was noted that several effective antihypertensive drugs had the common characteristic of binding transition and related trace metals. The calcium disodium salt of the powerful chelating agent, EDTA, was then shown to lower the diastolic pressure of hypertensive but not of normotensive rats (71). Administration of this compound to human hypertensive patients did not have a marked or consistent effect on blood pressure, but it lowered circulating cholesterol and greatly increased the urinary excretion of several metals, particularly zinc, manganese, cadmium, and lead (52). Examination of the urine of a group of hypertensive patients revealed a higher than normal excretion of a range of metals, especially cadmium which was almost 50 μg/liter compared with an average of less than 1 μg/liter in the normal subjects (51) (Fig. 15). Hypertensive subjects were subsequently shown to have significantly more renal cadmium than similar normotensive individuals (64) (Table 34). A high cadmium-zinc ratio was also noted in the tissues of the hypertensive patients.

Epidemiological evidence obtained by Carroll (5) appeared to further implicate cadmium as a causal factor in hypertension. This worker found the average concentration of cadmium in the air of 28 U.S. cities to be markedly positively correlated with the death rates from hypertension and arteriosclerotic heart disease. This association could not be explained by the tendency of large industrialized cities to have both high

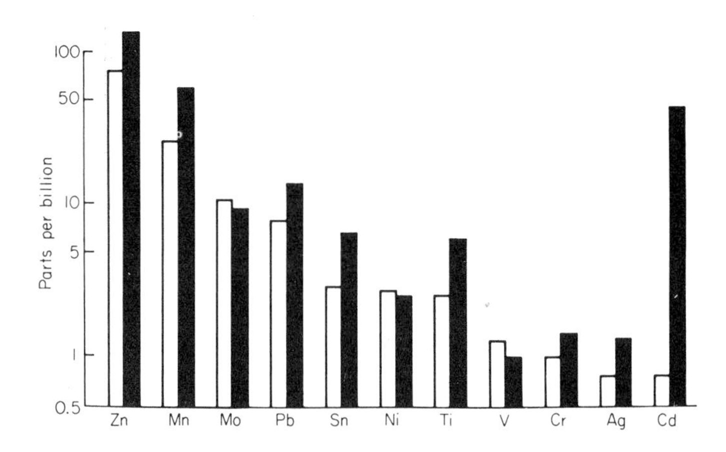


**Fig. 15.** Urinary excretion of metals in normotensive (white bars) and hypertensive (black bars) patients. [From Perry (51).]

TABLE 34
*Renal Cadmium in Hypertensive and Normotensive Patients*[a]

| Cause of death | No. of cases | Cadmium[b] (μg/g ash) |
|---|---|---|
| Accidents | 117 | 2940 ± 120 |
| Hypertension | 17 | 4220 ± 390 |

[a]From Schroeder (64).
[b]$P = <0.0005$.

heart disease rates and heavily polluted air, since no such correlation was observed with indices of air pollution in general. In a recent study of tissue cadmium concentration in man carried out by Morgan (38), no significant differences in kidney or liver cadmium levels were found between postmortem samples from a control group and from two groups of patients characterized by hypertensive cardiovascular disease and by ischemic cardiac or cerebrovascular disease. The zinc concentrations and the molar ratio of cadmium to zinc in these tissues also showed no significant group differences, although a further group characterized by neoplastic disease exhibited a wide and unpredictable variation in trace metal content. These findings do not conform with those obtained by Schroeder nor do they support the hypothesis that tissue cadmium concentration is related to hypertensive or degenerative vascular disease in man.

A progressively increasing incidence of systolic hypertension developed in rats exposed to 5 ppm cadmium in the drinking water from the time of weaning (63,72). The disturbance appears after a year or more of exposure and does not regress with age. Injection of cadmium intraperitoneally also induced hypertension in rats (67). The hypertension resulting from cadmium has been reversed by injection of disodium, zinc, trans-cyclohexane-1,2-diamine-tetraacetic acid ($Na_2Zn$-CDTA). This chelate of zinc has a somewhat higher stability constant for cadmium than for zinc. It depleted renal and hepatic cadmium and repleted renal and hepatic zinc, thus lowering the Cd-Zn ratio in these tissues (66). When the zinc chelate was given orally in the drinking water, or when deionized water was given for 200 days to cadmium hypertensive rats, both cadmium and zinc were reduced in the tissues and the hypertension regressed (69). It seems, therefore, that rats must not only be constantly exposed to cadmium, but must also have accumulated considerable quantities in their kidneys, if hypertension is to appear and to be sustained. The pathological changes associated with cadmium hypertension in rats, including renal arterial and arterioralar lesions and glomerular changes, have been shown to be indistinguishable from those accompanying benign hypertension from other causes (24). The mechanism of the chronic cadmium hypertension observed in these experiments with rats, and the significance of the relationship, if any, between cadmium and "essential" human hypertension, remain to be determined.

## V. Cadmium and Reproduction

Some reference to injurious effects of cadmium on the testes was made in early studies of cadmium toxicity (1,73). These appear to have been overlooked until 1956 when Parizek (41,50) dramatically revived interest in this subject. This worker showed that a single small injection of 0.02-0.04 m*M* $CdCl_2$/kg body weight induced selective hemorrhagic necrosis of the rat testes. This effect was specific for cadmium cations and was not shared by other metals studied, including mercury. Parizek's initial observation that both seminiferous tubules and interstitial tissue were acutely damaged by cadmium, resulting in permanent sterility (42), was soon confirmed by others (15,25,35), although interstitial tissue regenerates in time (15,41,42). The high sensitivity of the testes to cadmium develops gradually during the postnatal period with the successive differentiation of the seminiferous tubules (42). An analogous necrosis also occurs in the epididymis (16,43). Similar effects of cadmium upon the testes have been reported in the mouse (6,35), rabbit (4,42), hamster (42), and guinea pig (42). The species with cadmium-sensitive testes, other than the opossum, are scrotal, whereas those with cadmium-resistant testes (frogs and birds) have the male gonads in an abdominal location (6). The significance of this fact is obscure. The testes of all species so far examined are sensitive to cadmium when injected intratesticularly (7).

The mode of action of cadmium on the testis and the reason for its remarkable organ specificity are not fully understood. Evidence is accumulating which indicates that the necrosis is primarily caused by interference of cadmium with the testicular blood supply (16,25). Changes in the testicular vascular bed have been demonstrated by Chiquoine (6), whereas Niemi and Kormano (40), using a direct angiographic technique, observed closure of the intratesticular capillary network in rats, within a few hours of cadmium injection. A reestablishment of testicular vasculature was seen, beginning 4 weeks after the cadmium exposure. These findings establish the primary site of action of cadmium, but they throw little light on the mechanism involved or on the basis for the particular sensitivity to cadmium of the blood vessels of the testis and epididymis.

Several substances can protect the animal against the injurious effects of cadmium on the testes when injected concurrently with the cadmium. These are zinc (15,32,41), selenium (26,32), estrogens (17), and certain thiol compounds, notably cysteine, 2,3-dimercapto-1-propanol (BAL), and, to a lesser extent, glutathione (17). The latter finding suggests that cadmium may damage testicular vasculature by interference with essential sulfhydryl groups. The protective effect of zinc has excited particular interest because of the normal high-zinc concentrations in sperms and semen and the specific role of this element in testicular development and spermatogenesis and because of other zinc-cadmium interactions. It has been suggested that cadmium could displace zinc from key compounds, as has been demonstrated for certain zinc metalloenzymes (12), but the mechanism by which zinc protects the testes in cadmium intoxication remains unexplained (30). No decrease in testicular zinc concentration occurs at the time when rat testis first acquires its sensitivity to the necrotic action of cadmium at about 9 days of life (48).

The drastic toxic effects of injected cadmium on the male are paralleled, in certain circumstances, by comparable effects upon the female. During the original studies of cadmium-induced testicular necrosis in rats, no morphological abnormalities in the ovaries and no sterility were observed when similar doses of cadmium were administered to adult females (41,50). However, massive hemorrhagic necrosis occurred in the nonovulating ovaries of young rats maintained in a permanent state of estrus (43,44). This effect was specific for cadmium and occurred similarly in hypophysectomized animals, indicating that it is not mediated via the pituitary (44). Injection of cadmium salts into pregnant rats resulted in complete destruction of the *pars fetalis* of the placenta. Within a few hours following injection, this organ was transformed into an extensive blood clot with little remaining tissue. In all cases pregnancy was interrupted, with either resorption or delivery of the dead conceptus (43,45).

When cadmium is injected during the critical last 4 days of pregnancy, unusual signs of toxicity, with high mortality of the injected rats are observed (46). Visceral venous congestion, pulmonary congestion, hemorrhagic edema, hemorrhage in the adrenal cortex, and sometimes massive pleural effusion are apparent postmortem. These changes are specific for the last days of pregnancy. All rats injected in the seventeenth day of pregnancy, or earlier, survived even a higher dose of cadmium. The peculiar sensitivity of the maternal organism to cadmium is preserved if the fetuses are removed, leaving the placentae *in situ*. It is completely abolished by hysterectomy or by removal of the fetuses and the placentae from the uteri. It appears, therefore, that the presence of the placenta is responsible for this particular effect of cadmium. Parizek has put forward two alternative explanations of this phenomenon: (a) either the presence of the placenta could sensitize in some way the pregnant organism to cadmium or (b) the damage in other organs could be a secondary result of a primary destruction in the placenta.

The changes evoked by cadmium during the last part of pregnancy are similar to those which occur during pregnancy after feeding a diet high in highly unsaturated fatty acids or their peroxides and low in vitamin E (27,33,77). This observation stimulated Parizek to determine if this effect of cadmium could be prevented by antioxidants. Some protection was obtained with methylene blue, but selenium was found to prevent completely the cadmium-induced toxemia of pregnancy when injected as sodium selenite or selenate in a slightly higher than equimolar proportion to cadmium (49). Parenteral administration of salts of zinc also protect adult female rats against the toxic effects of cadmium (46a). With this treatment no symptoms of cadmium intoxication were observed and complete survival of mothers and fetuses and birth of living offspring were achieved. Such protective effect is specific for selenium, so that selenium emerges as a protective agent against cadmium toxicity effects on the testis, on the ovaries of prepubertal rats, or on rats in permanent estrus, as well as on pregnant rats. It seems logical to assume that selenium is acting through a similar mechanism in each case. Although the nature of this mechanism is not yet clear, Parizek and co-workers (47) have shown that selenium compounds similarly protect the organism against the toxic effects of bivalent mercury. These effects seem to be connected with a change in the chemical reactivity and distribution of cadmium and mercury in the animal. The level of mercury or cadmium in the blood of mothers injected with salts of these metals was increased by selenium treatment and the content of fetuses, milk, and sucklings was substantially decreased. Zinc protects against the toxicity of cadmium by a different mechanism. Rats protected by zinc salts show no reduction in blood cadmium comparable to that brought about by selenium, and the concentration of cadmium in the liver is actually increased (46a). This increase in the liver raises the possibility that zinc reduces cadmium toxicity by increasing the excretion of cadmium from the organism by kidneys, intestine, or biliary tract.

In addition to the highly selective toxic effects of injected cadmium on the reproductive organs and the effects of cadmium on hypertension in rats, dietary cadmium has been shown to be highly toxic to several monogastric species (1,13,19,56,73) and to be rather less toxic to ruminants (36,37). It should be noted that selenium compounds ameliorate the general toxic effects of cadmium as well as those related to reproduction (47). Part of the toxic effects of cadmium relates to the metabolic antagonism to zinc (19,22,36,57)—a fact not realized by some of the earlier workers. The livers and kidneys are the organs most severely affected in cadmium poisoning in rats (58) and in calves (57). Reduced red cell and hemoglobin concentrations are also common (3,19,22). The anemia induced in Japanese quail by feeding cadmium at 75 ppm of the diet is accompanied by significant changes in the plasma proteins, including a marked increase in plasma transferrin, as discussed earlier (22). In calves fed 40-160 ppm Cd from 9 to 20 weeks of age, few clinical signs of toxicity were apparent but at 640 ppm there was some mortality and at 2560 ppm there was 100% mortality, with deaths occurring in 2 to 8 weeks (57). The affected calves also revealed some of the typical signs of zinc deficiency, such as depressed appetite, growth, and feed efficiency, dry and scaly skin, loss of hair, and mouth and scrotal lesions.

## VI. Sources of Cadmium

Data on the sources of cadmium to man and domestic animals are exceedingly meager. The factors that affect the magnitude of normal intakes and the movements of cadmium from soils and plants to animals and man have also been little studied. This is a serious gap in knowledge, in view of the geographical differences in human renal cadmium levels and their increase with age, the toxicity of this element, and the possible association of cadmium with human hypertension. The interactions of cadmium with zinc, copper, iron, and selenium also indicate the need for more information on sources of cadmium to man and animals.

The data of Carroll (5) indicate that the atmosphere can be a significant source of cadmium in some cities. In the twenty-eight cities studied in the United States, the cadmium level in the air ranged from undetectable to as high as 0.062 $\mu$g Cd/m$^3$. The cadmium content of drinking water is also highly variable (65,70). Soft water remaining overnight in galvanized or black polyethylene pipes can take up 0.15-1.1 $\mu$g Cd/liter, although total intakes from these sources have been calculated to be no more than 1 or 2 $\mu$g/day (65).

Daily intakes of cadmium by human adults are estimated by Schroeder *et al.* (70) to be 200-500 $\mu$g, with considerable variation according to sources and types of food. Oysters are exceptionally rich in cadmium, as they are in zinc. Levels of 3 to 4 ppm Cd wet weight have been reported (70). No other foods or beverages examined approached this concentration (except canned anchovies). Vegetables, nuts, and fruits are poor sources of cadmium, with concentrations mostly ranging between 0.04 and 0.08 ppm. Cadmium is highly concentrated in the germ of wheat, but, unlike most other minerals, it appears to be more evenly distributed throughout the rest of the grain, so that there is a relatively small depletion of cadmium during the flour-milling process. In fact, Schroeder *et al.* (70) report a concentration of 0.26 ppm Cd for whole wheat, of 0.38 ppm for patent flour, and 1.11 ppm for the germ. It is not clear if the germ and flour were obtained from the whole grain actually analyzed. A similar enrichment of cadmium in milled rice has been demonstrated in Japan (39). From 79 to 89% of the cadmium in whole rice remained after polishing, even though 36-40% of the grain was removed and the ash content reduced by 18 to 26%. The cadmium content of the ash of the polished rice was 3.5-4 times that of the unpolished. These data suggest that the refining of cereals is not associated with significant losses of cadmium as occurs with most trace elements.

## REFERENCES

1. C. L. Alsberg and E. W. Schwartze, *J. Pharmacol. Exp. Ther.* **14**, 504 (1919-20).
2. B. Axelsson and M. Piscator, *Arch. Environ. Health* **12**, 374 (1966).
3. C. R. Bunn and G. Matrone, *J. Nutr.* **90**, 395 (1966).
4. E. Cameron and C. L. Foster, *J. Anat.* **97**, 269 (1963).
5. R. E. Carroll, *J. Amer. Med. Ass.* **198**, 177 (1966).
6. A. D. Chiquoine, *Anat. Rec.* **149**, 23 (1964).
7. A. D. Chiquoine and V. Suntzeff, *J. Reprod. Fert.* **10**, 455 (1965).
8. G. C. Cotzias, D. C. Borg, and B. Selleck, *Amer. J. Physiol.* **201**, 63 and 927 (1961).
9. G. C. Cotzias and P. S. Papavasilou, *Amer. J. Physiol.* **206**, 787 (1964).
10. D. H. Cox and D. L. Harris, *J. Nutr.* **70**, 514 (1962).
11. L. E. Decker, R. U. Byerrum, C. F. Decker, C. A. Hoppert, and R. J. Langham, *A. M. A. Arch. Ind. Health* **18**, 288 (1958).
12. R. Druyan and B. L. Vallee, *Fed. Proc., Fed. Amer. Soc. Exp. Biol.* **21**, 247 (1962).
13. O. G. Fitzhugh and T. H. Meiller, *J. Pharmacol. Exp. Ther.* **72**, 15 (1941).
14. J. E. Folk and J. A. Gardner, *Biochim. Biophys. Acta* **48**, 139 (1961).
15. S. A. Gunn, T. C. Gould, and W. A. D. Anderson, *Arch. Pathol.* **71**, 274 (1961).
16. S. A. Gunn, T. C. Gould, and W. A. D. Anderson, *Amer. J. Pathol.* **42**, 685 (1963).
17. S. A. Gunn, T. C. Gould, and W. A. D. Anderson, *Proc. Soc. Exp. Biol. Med.* **119**, 901 (1965); **122**, 1036 (1966).
18. A. Hennig and M. Anke, *Arch. Tierernahr.* **14**, 55 (1964).
19. A. Hennig, M. Anke, and W. Kracht, *Jahrb. Tierernahr. Futterung* **5**, 272, 286, and 297 (1964-1965).
20. C. H. Hill, G. Matrone, W. L. Payne, and C. W. Barber, *J. Nutr.* **80**, 227 (1963).
21. H. R. Imbrus, J. Cholak, L. H. Miller, and T. Sterling, *Arch. Environ. Health* **6**, 286 (1963).
22. R. M. Jacobs, M. R. Spivey-Fox, and M. H. Aldridge, *J. Nutr.* **99**, 119 (1969).
23. J. H. R. Kagi and B. L. Vallee, *J. Biol. Chem.* **235**, 3460 (1960); **236**, 2435 (1961).

24. M. Kanisawa and H. A. Schroeder, *Exp. Mol. Pathol.* **10**, 81 (1969).
25. A. B. Kar and R. P. Das, *Acta Biol. Med. Ger.* **5**, 153 (1960).
26. A. B. Kar and R. P. Das, *Proc. Nat. Inst. Sci. India, Part B* **29**, Suppl., 297 (1963).
27. M. A. Kenney and C. E. Roderuck, *Proc. Soc. Exp. Biol. Med.* **114**, 257 (1963).
28. J. Kubota, A. Lazar, and F. L. Losee, *Arch. Environ. Health* **16**, 788 (1968).
29. J. G. Lease, *J. Nutr.* **96**, 294 (1969).
30. Z. Lojda and J. Parizek, *Physiol. Bohemoslov.* **12**, 512 (1963).
31. M. Margoshes and B. L. Vallee, *J. Amer. Chem. Soc.* **79**, 4813 (1957).
32. K. E. Mason, J. O. Young, and J. E. Brown, *Anat. Rec.* **148**, 309 (1964).
33. D. G. McKay and T. Wong, *J. Exp. Med.* **115**, 1117 (1962).
34. J. C. Medici and M. W. Taylor, *J. Nutr.* **93**, 907 (1967).
35. E. S. Meek, *Brit. J. Exp. Pathol.* **40**, 503 (1959).
36. G. W. Miller, W. J. Miller, and D. M. Blackmon, *J. Nutr.* **93**, 203 (1967).
37. W. J. Miller, B. Lampp, G. W. Powell, C. A. Salotti, and D. M. Blackmon, *J. Dairy Sci.* **50**, 1404 (1967).
38. J. M. Morgan, *Arch. Intern. Med.* **123**, 405 (1969).
39. M. Moritsugu and J. Kobayashi, *Ber. Ohara Inst. Landwirt. Biol., Okayama Univ.* **12**, 145 (1964); cited by Schroeder *et al.* (70).
40. M. Niemi and M. Kormano, *Acta Pathol. Microbiol. Scand.* **63**, 513 (1965).
41. J. Parizek, *J. Endocrinol.* **15**, 56 (1957).
42. J. Parizek, *J. Reprod. Fert.* **1**, 294 (1960).
43. J. Parizek, *Cesk. Fysiol.* **11**, 466 (1962); **12**, 344 (1963).
44. J. Parizek, *Proc. 2nd Int. Congr. Endocrinol., 1964.*
45. J. Parizek, *J. Reprod. Fert.* **7**, 263 (1964).
46. J. Parizek, *J. Reprod. Fert.* **9**, 111 (1965).
46a. J. Parizek, I. Benes, J. Kalouskova, A. Babicky, and J. Lener, *Physiol. Bohemoslov.* **18**, 89 (1969).
47. J. Parizek, I. Benes, I. Ostadalova, A. Babicky, J. Benes, and J. Pitha, *in* "Mineral Metabolism in Paediatrics" (D. Barltrop and W. L. Burland, eds.), Blackwell, Oxford, 1969.
48. J. Parizek, J. C. Boursnell, M. F. Hay, A. Babicky, and D. M. Taylor, *J. Reprod. Fert.* **12**, 501 (1966).
49. J. Parizek, I. Ostadalova, I. Benes, and A. Babicky, *J. Reprod. Fert.* **16**, 507 (1968).
50. J. Parizek and Z. Zahor, *Nature (London)* **177**, 1036 (1956).
51. H. M. Perry, Jr., *Fed. Proc., Fed. Amer. Soc. Exp. Biol.* **20**, 254 (1961).
52. H. M. Perry, Jr. and G. H. Camel, *in* "Metal-Binding in Medicine" (M. J. Seven and L. A. Johnson, eds.), p. 209. Lippincott, Philadelphia, Pennsylvania, 1960.
53. H. M. Perry, Jr. and E. F. Perry, *J. Clin. Invest.* **38**, 1452 (1959).
54. H. M. Perry, Jr. and H. A. Schroeder, *Circulation* **12**, 758 (1955); *J. Lab. Clin. Med.* **46**, 936 (1955).
55. H. M. Perry, Jr., I. H. Tipton, H. A. Schroeder, R. L. Steiner, and M. J. Cook, *J. Chronic Dis.* **14**, 259 (1961).
56. W. G. Pond, P. Chapman, and E. Walker, *J. Anim. Sci.* **25**, 122 (1966).
57. G. W. Powell, W. J. Miller, J. D. Morton, and C. M. Clifton, *J. Nutr.* **84**, 205 (1964).
58. L. Prodan, *J. Ind. Hyg.* **14**, 174 (1932).
59. P. Pulido, J. H. R. Kagi, and B. L. Vallee, *Biochemistry* **5**, 1768 (1966).
60. Report of ICRP Committee II on Permissible Dose for Internal Radiation, *Health Phys.* **3**, 380 (1960).
61. B. M. Sahagian, I. Harding-Barlow, and H. M. Perry, Jr., *J. Nutr.* **90**, 259 (1966); **93**, 291 (1967).

62. H. A. Schroeder, "Mechanisms of Hypertension." Thomas, Springfield, Illinois, 1957.
63. H. A. Schroeder, *Amer. J. Physiol.* **207**, 62 (1964).
64. H. A. Schroeder, *J. Chronic Dis.* **18**, 647 (1965).
65. H. A. Schroeder and J. J. Balassa, *J. Chronic Dis.* **14**, 236 (1961).
66. H. A. Schroeder and J. Buckman, *Arch. Environ. Health* **14**, 693 (1967).
67. H. A. Schroeder, S. S. Kroll, J. W. Little, P. O. Livingstone, and M. A. G. Myers, *Arch. Environ. Health* **13**, 788 (1966).
68. H. A. Schroeder and A. P. Nason, *J. Invest. Dermatol.* **53**, 71 (1969).
69. H. A. Schroeder, A. P. Nason, and M. Mitchener, *Amer. J. Physiol.* **214**, 796 (1968).
70. H. A. Schroeder, A. P. Nason, I. H. Tipton, and J. J. Balassa, *J. Chronic Dis.* **20**, 179 (1967).
71. H. A. Schroeder and H. M. Perry, Jr., *J. Lab. Clin. Med.* **46**, 416 (1955).
72. H. A. Schroeder and W. H. Vinton, Jr., *Amer. J. Physiol.* **202**, 518 (1962).
73. E. W. Schwartze and C. L. Alsberg, *J. Pharmacol. Exp. Ther.* **21**, 1 (1923).
74. J. C. Smith and J. E. Kench, *Brit. J. Ind. Med.* **14**, 270 (1957).
75. J. P. Smith, J. C. Smith, and A. J. McCall, *J. Pathol. Bacteriol.* **80**, 287 (1960).
76. W. C. Supplee, *Science* **139**, 119 (1963).
77. F. W. Stamler, *Amer. J. Pathol.* **35**, 1207 (1959).
78. I. H. Tipton and M. J. Cook, *Health Phys.* **9**, 103 (1963).
79. D. R. Van Campen, *J. Nutr.* **88**, 125 (1966).

# 11

# IODINE

## I. Introduction

The ancient Greeks are reputed to have used burnt sponges successfully but quite empirically in the treatment of human goiter. Knowledge of this fact and the finding of iodine in abundance in sponges as early as 1819 (71) led the French physician, Coindet, to use salts of iodine therapeutically in the treatment of goiter in the following year (51). However, the first suggestion that goiter might be due to a deficiency of iodine does not appear to have been made until 1830 (181). Systematic investigation of this concept was later undertaken by the French botanist, Chatin, who between 1850 and 1876 determined the natural occurrence of iodine in air, water, soils, and foods from various localities and compared his results with the reported incidence of goiter (43). As a result of these studies, Chatin concluded that the occurrence of goiter was associated with a deficiency of environmental iodine and recommended that the water supply in goitrous districts should be enriched with this element. Certain anomalies in these observations discredited his conclusions, and the iodine deficiency theory lay almost forgotten for several decades.

During the second half of the 19th century, certain characteristics of endemic goiter were found to be linked with deficient thyroid function or experimental athyreosis and to be ameliorated by injections of thyroid gland extracts [see Harington (101)]. Furthermore, iodine was shown to be normally concentrated in the thyroid gland and to be reduced in concentration in this gland in endemic goiter (23). These findings were confirmed by Oswald (169) in 1899, and thyroglobulin was then identified. By 1919, Kendall (121) had isolated a crystalline compound from the thyroid containing 65% iodine which he claimed was the active principle and which he named "thyroxine." Subsequently Harington and Barger (102) showed that thyroxine is a tetraiodo derivative of a compound of phenol and tyrosine, or tetraiodothyronine, and accomplished its synthesis.

During the first quarter of this century, investigations in several countries established clearly that endemic goiter in man and in farm animals is associated primarily, but not exclusively, with a deficiency of iodine in the food and water supplies of the affected regions and can be largely controlled by raising individual iodine intakes. In some areas contributory factors may be involved (31,146), and iodine deficiency has not been incriminated in all endemic goiter areas (141). The first large-scale trial in man was carried out in the schools of Ohio between 1916 and 1920, although as early as 1915, Marine, a pioneer worker in this field, stated that "endemic goiter is the easiest known disease to prevent." Similar prophylactic measures were soon initiated in goitrous areas in other countries with such success that iodized salt soon became a widely recognized form of control.

Endemic goiter was subsequently shown to occur with varying intensity in many countries and in every continent (see Fig. 16[1]), with the actual number of goitrous individuals in the world estimated in 1960 at close to 200 million (120). Women and children are more affected than adult males (146). In certain countries, notably U.S.A., Switzerland, and New Zealand, the intensity of the disease declined markedly as a result of the use of iodized salt, although there is some evidence of a rise in goiter incidence in parts of U.S.A. in recent years, presumably associated with a decline in sales of iodized, as compared with noniodized salt. There is also evidence that the incidence of goiter has increased over the last 20 years in Ceylon, believed to be due to a lack of cooperation with the official program of iodine supplementation started in 1951 (166). In many Latin American countries, despite the simplicity of control measures, endemic goiter remains a serious public health problem, associated often with cretinism, feeblemindedness, and deaf-mutism (201).

[1]Map kindly supplied by Dr. F. C. Kelly, Iodine Educational Bureau, London.

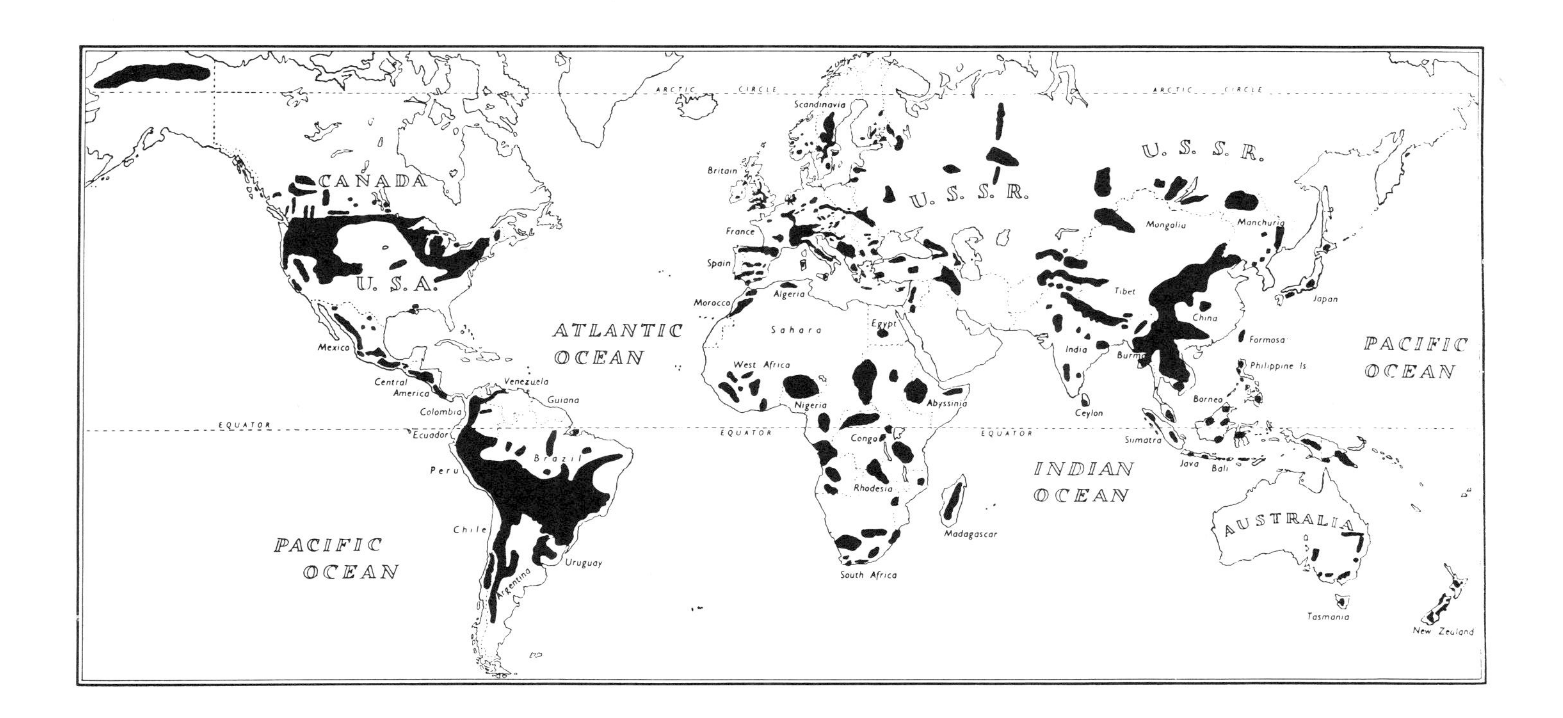


**Fig. 16.** World map showing occurrence of endemic goiter. (Black indicates areas where endemic goiter has been found.)

In many areas the incidence of goiter is low and iodine intakes so marginal that any abnormal stress placed upon the thyroid gland, such as the nutritional privations of war or famine, can result in outbreaks of the disease. Endemic goiter may occur, as has been suggested for parts of Australia (47) and Finland (172), from the consumption of milk containing a goitrogen due to the consumption of cruciferous fodder by the cow. The ingestion of inadequately fermented cassava with presumed goitrogenic properties, coupled with low-iodine intakes, has been tentatively incriminated as causal factors in outbreaks of goiter in parts of Nigeria (166). The significance of naturally occurring goitrogens to the incidence of goiter in man and livestock, especially where iodine intakes are low or marginal, requires much further study. Following the original demonstration by Chesney and co-workers (44) of cabbage goiter in rabbits in 1928, many goitrogenic substances were identified and prepared. These proved to be valuable aids in studies of thyroid function and in the treatment of thyrotoxicosis.

Concurrently with the studies just outlined, intensive investigations were undertaken of the metabolic movements of iodine within the body and of the relationship of these movements to hypo- and hyperthyroidism. These investigations were facilitated by the advent of suitable radioactive isotopes of iodine, especially $^{131}I$, and by the development of sensitive catalytic, colorimetric methods capable of measuring the minute amounts of iodine present in the tissues. Neutron activation methods have also proved highly satisfactory for this purpose (35). Chromatographic and radioautographic techniques were also applied, so that a range of iodine-containing organic compounds could be separated and identified in the tissues. By such means the potent thyroactive substance, triiodothyronine, was demonstrated in human thyroid and plasma (86). More recently, methods for the direct estimation of serum thyroxine have been developed (176,234). These possess certain advantages over protein-bound iodine determinations because problems of iodine contamination during processing or from pretreatment of the patient are minimized.

During the period covered by the foregoing studies, the thyroid hormone and, hence, iodine emerged as important factors in the growth, health, and reproduction of farm animals. Goiter was shown to occur in these animals in most areas where human goiter is endemic. In fact, iodine deficiency has been stated to be the most widespread of all mineral deficiencies in grazing stock (5,229). Attention was also devoted to the relation of such productive functions as milk yield, egg yield, wool growth, reproductive performance, and live-weight gains to variations in the level of thyroid activity in farm stock, where these were limited by iodine supply or by some other environmental factor. The use of thyroactive iodinated proteins as an economic means of controlling thyroid activity and, therefore, these productive functions also became a rewarding field of study.

Ever since Chatin's original work over a century ago, investigations have been carried out on the iodine content of plants, mainly as a means of assessing the sources of iodine to man and animals and the factors that influence their magnitude. Unequivocal evidence that iodine is essential for microorganisms or the higher plants has not been produced (113), although there are indications that, under some conditions, iodine applications can stimulate crop growth (134), including yields of potatoes (231) and of sugar beet (41). The application of iodine-containing fertilizers or iodine-rich materials, such as seaweed, has also been shown to produce remarkable increases in the iodine content of vegetables and grasses in areas naturally low in environmental iodine (91).

## II. Iodine in the Animal Body

### 1. *Distribution*

The healthy human adult probably contains a total of 10 to 20 mg iodine, of which a high proportion, amounting to 70–80%, is concentrated in the thyroid gland. This represents a unique degree of concentration for any trace element, since the mass of the thyroid is normally only about 0.05% of the whole body. The concentration of iodine in the skeletal muscles is less than one-thousandth of that of the thyroid, but because of their large mass they contain the next largest proportion of total-body iodine. The level of iodine in the ovaries is 3–4 times that of the muscles and is normally higher than that of most other extrathyroid tissues, other than the bile, hair, pituitary gland (152,219), and the salivary glands (37). The ability of the ova to concentrate iodine at high dietary intakes is discussed later. The iodine-concentrating mechanisms of the salivary gland appear to be similar to those of the thyroid (238). Significant concentrations of iodine also occur in certain parts of the eye, notably the orbitary fat and the orbicular muscle. In one study of a small range of samples, this muscle averaged close to 26 and 23 $\mu$g/100 g wet weight (80).

Iodine occurs in the tissues as organically bound iodine and as inorganic iodide. The latter is normally present in most tissues in extremely low concentrations, of the order of 1 to 2 $\mu$g/100 g (194). In the saliva the iodine is almost entirely in the inorganic form, even in conditions where organic iodine compounds are secreted in the urine (2,171). The salivary iodine concentration is proportional to the plasma inorganic iodide concentration at physiological levels (99) and at plasma concentrations of up to 100 μg/100 ml or about 500 times normal (98). Such increases can be achieved by the administration of iodide in the prophylaxis of simple goiter and particularly at the much higher levels used in the therapy of exophthalmic goiter (34).

In tissues other than the thyroid, the concentrations of organically bound iodine are small. A level of about 5 μg/100 g appears to be normal for muscle, with higher levels in the ovaries. Most of the organic iodine of the tissues consists of thyroxine bound to protein, together with widely distributed minute concentrations of other compounds, including triiodothyronine. The solubility of muscle iodine differs from that of thyroxine added to tissue extracts and its distribution is not uniform between myosin and actin. However, muscle iodine concentrations decrease in hypothyroidism and increase in hyperthyroidism (195).

### 2. *Iodine in the Thyroid Gland*

The total concentration of iodine in the thyroid varies with the iodine intake and age of the animal and with the activity of the gland. Variation among species is small, except that the thyroids of sea fish are richer and those of rats slightly poorer than those of most mammalian species. The normal healthy thyroid of mammals contains 0.2-0.5% I on dry basis, giving an average total of 8 to 12 mg in the adult human gland. This amount may be reduced to as low as 1 mg or less in endemic goiter, and the concentration is reduced even more significantly because of the hyperplastic changes that characterize the disease. Over 60 years ago, Marine (151) showed that hyperplastic changes are regularly found when the iodine concentration falls below 0.1%. This has been substantially confirmed by later studies with several species, using more refined analytical procedures. Thus sheep's thyroids with marked follicular hyperplasia contained 0.01% I, those with moderate hyperplasia 0.04%, and pigs' thyroids showing only very slight hyperplasia, 0.11% on dry basis (10). Eliminating the hyperplasia by iodine feeding increased the iodine content of the glands so that few contained less than 0.2% and none less than 0.12%. Investigations carried out subsequently in Australia (202) and New Zealand (9,207) of neonatal mortality in lambs associated with goiter also indicate that a thyroid iodine concentration of 0.1%, or slightly higher, can be regarded as a critical level below which the gland cannot function properly.

Iodine exists in the thyroid as inorganic iodide, mono- and diiodotyrosine, thyroxine, triiodothyronine, polypeptides containing thyroxine, thyroglobulin, and probably other iodinated compounds (190). Calcitonin or thyrocalcitonin, synthesized by the parafollicular or C cells of the thyroid, is also present (40,53), but this compound apparently does not contain iodine (29). The iodinated amino acids are bound with other amino acids in peptide linkage to form thyroglobulin, the unique, iodinated protein of the thyroid gland. Thyroglobulin, the chief constituent of the colloid filling the follicular lumen, is a glycoprotein with a molecular weight of 650,000. It constitutes the storage form of the thyroid hormone and normally represents some 90% of the total iodine of the gland. The amounts and proportions of the various iodine-containing components of the thyroid vary with the supply of iodine to the gland, with the presence of goitrogenic substances which can inhibit the iodine-trapping mechanism of the gland or the process of hormonogenesis, and with the existence of certain disease states and metabolic defects of genetic origin.

### 3. *Iodine in Blood*

Iodine exists in normal blood in both inorganic and organic forms. Inorganic iodide concentrations are too low for direct determination, except possibly by neutron activation or when dietary iodide intakes are very high. However, plasma inorganic iodide can be estimated indirectly with the use of radioiodine from the specific activity of the urinary (213) or the salivary (100) iodide. In normal subjects the urine and saliva techniques give similar results. In thyrotoxicosis and in dyshormonogenesis, where organic iodine occurs in the urine, falsely high values may be obtained from the use of urine (2,171). The normal range of plasma inorganic iodide in human subjects is stated by Wayne *et al.* (233) in their book, "Clinical Aspects of Iodine Metabolism," to be 0.08-0.60 μg/100 ml, with values below 0.08 suggesting iodine deficiency and values above 1 μg/100 ml, pointing to exogenous iodine administration.

The organic iodine of the blood, which does not occur in the erythrocytes, is present mainly as thyroxine bound to the plasma proteins. Only a very small proportion, normally about 0.05%, is free in human serum (215). Up to 10% of the organic iodine of the plasma is made up of several iodinated substances, including tri- and diiodothyronine (86,188) and minute concentrations of other compounds that are probably tissue metabolites of thyroxine. Thyroglobulin occurs only in pathological states involving damage to the thyroid gland and the iodotyrosines do not normally appear in the peripheral circulation at all, or even in the thyroid vein when the thyroid contains 70% or more of its $^{131}$I in the form of iodotyrosines (224). On the other hand, triiodothyronine is secreted into the blood by the thyroid because $^{131}$I-triiodothyronine has been demonstrated in the thyroid vein to the extent of 10 to 20% of the

$^{131}$I-thyroxine concentrations – a level much higher than in the circulating plasma.

The protein-bound iodine of the serum (serum PBI) or the butanol-extractable iodine of the serum (serum BEI) corresponds well with the circulating thyroid hormone level and a number of methods has been developed to measure serum PBI (24, 65, 196, 209) or serum BEI (147). The PBI levels in the serum vary significantly with species, age, pregnancy, and level of thyroid activity. Individual variability is high in all species, although some of the reported variability probably reflects contamination or insufficient care with the method. Thus in one study with dairy cattle involving 106 animals in 9 herds the individual values ranged from 2.1 to 18.3 μg/100 ml (123) and in another study, involving 120 samples from cows at pasture, values ranged from 1.4 to 6.8 (means 3.4) μg/100 ml (227). In adult man the limits of normality have been placed at 4 to 8 or 3 to 7.5 μg/100 ml, with a "mean" close to 5 to 6 μg (87, 92,183,233). Slightly lower PBI norms (3-4 μg/100 ml) have been reported for mice, rats, and dogs (222), for adult sheep (233a), and for beef cattle (130). Still lower mean serum PBI levels have been recorded in other studies with the domestic fowl (230), beef cattle (142), and horses (114). In the study carried out by Irvine (114) involving 84 racehorses maintained under different conditions, the mean serum PBI value of 48 untrained horses was 1.86 μg/100 ml. Training was associated with a decrease of 40%, whereas exposure of unadapted horses caused a marked increase. Stallions and colts had a slightly but significantly lower PBI level than mares and fillies. This appears to be the only recorded incidence of a sex difference in PBI levels in animals.

Serum PBI levels rise significantly during human pregnancy. As early as the third week the levels rise to the upper part of the normal range or even to levels which, outside of pregnancy, are characteristic of hyperthyroidism (59,92,148). This rise is not attended by any clinical evidence of excessive thyroid activity and occurs well before the increase in basal metabolic rate (BMR) of pregnancy. Limited data suggest a rise in pregnancy in cows (123) and in ewes (132a) but not in mares (114). Serum PBI levels in newborn babies (175) and calves (140), before the ingestion of colostrum, are similar to those of their mothers. After the ingestion of colostrum they rise sharply to significantly higher levels and then fall again to those of adults after some weeks or months. Foals and yearlings also show significantly higher PBI levels than mature horses (114).

Increased serum PBI values are highly characteristic of hyperthyroidism and decreased values of hypothyroidism in man. There is so little overlap that PBI determinations represent a convenient and generally satisfactory means of assessing variations in thyroid secretion rate

(183,194,233). The magnitude of the changes in man is apparent from Fig. 17. Caution in the interpretation of results is necessary because misleadingly high PBI values can occur when pharmaceutical iodine preparations have previously been administered. If thyroidal radioiodine uptake is also high, hyperthyroidism can be diagnosed with more confidence. Iodide can be removed by the use of a resin column as proposed by Farrell and Richmond (65) or the problem of iodine contamination can be solved by direct assays of serum thyroid hormone levels (176,234).

Serum PBI levels also provide a useful index of thyroid activity in farm animals. Extremely low or negligible PBI values have been observed in thyroidectomized bulls (139) and horses (114) and a strong relation ($r = 0.70$–$0.78$) between PBI levels and thyroid secretion rate has been demonstrated in cattle (179). However, attempts to demonstrate a relationship between PBI levels and growth rate and other productive processes in farm animals have met with varying success. Kunkel *et al.* (131) found that plasma PBI was related to growth rates of beef cattle in a curvilinear manner, so that correlations with gain were positive below an optimal level and were negative above this level. Post (178) obtained a linear relationship between PBI and weight gain of cattle over the whole range of PBI observed, indicating that optimal levels of PBI had not been exceeded. He suggested that PBI values might be useful as selection criteria in beef cattle breeding in the subtropical environment where the study was carried out. Significant nega-

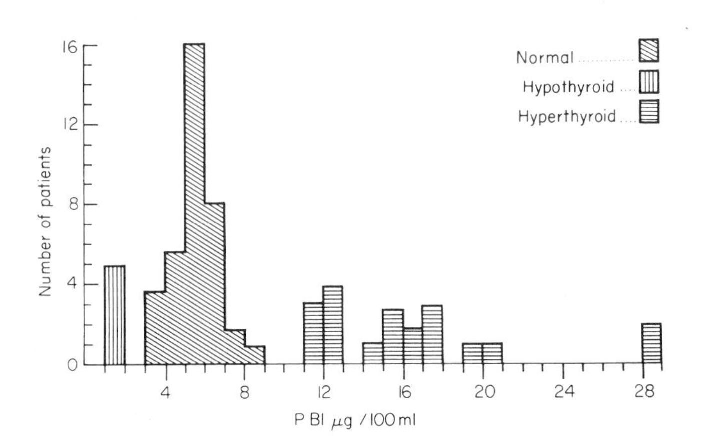


**Fig. 17.** Distribution of protein-bound iodine (PBI) in normal subjects and in patients whose thyroid disease was confirmed by a satisfactory response to treatment. [From Hallman *et al.* (92).]

tive correlations have been reported between PBI levels and daily rates of gain in pigs (72) and between these levels and interval from first breeding to conception and services per conception, in dairy cows. In this latter study, Lennon and Mixner (136) could find no relationship between either PBI levels or thyroid secretion rate and lactational performance in Holstein and Jersey cows.

4. *Iodine in Milk*

The concentration of iodine in milk is influenced by stage of lactation and dietary iodine intakes. Individual variability is so large that it is not known if real species differences exist. Iodine is unique among the trace elements for the ease with which it passes the mammary barrier and in the extent to which the level in the milk is influenced by variations in dietary iodine levels. In an early study, Orr and Leitch (167) raised the iodine content of the milk of cows from 40 to 70 μg/liter to 330 μg/liter by adding 180 mg of iodine as KI daily to the diet. Blom (32) was able to increase the iodine content of cow's milk even further from a "normal" of 20 to 70 μg/liter to 510–1070 μg/liter by feeding a supplement of only 100 mg KI daily. Subsequently Kirchgessner (125) demonstrated rising milk iodine concentrations from graded increments of dietary iodine. At an intake of 1.6 mg I/cow/day the iodine level in the milk was 28 ± 6 μg/liter, at 12.7 mg it was 78 ± 18, and at an intake of 20 mg/cow/day it was 267 ± 55 μg/liter. Such treatment has been suggested as an effective means of raising the iodine intakes of women and children in goitrous areas.

In a converse manner, the iodine content of milk is reduced below normal in goitrous areas and the determination of iodine in milk has been proposed as a convenient means of establishing the iodine status of an area and of providing an index of endemic goiter incidence (28,168). Binnerts (28) found the following mean differences between goitrous and nongoitrous regions in the Netherlands: summer milk, 9.7 and 21.1 μg/liter and winter milk, 20.6 and 83.5 μg/liter, respectively.

Cow's colostrum is much higher in iodine than true milk and there is a fall in concentration in late lactation. Kirchgessner (125) reports a mean value of 264 ± 100 μg/liter for colostrum, compared with 98 ± 82 for milk. Lewis and Ralston (140) found the colostrum of 5 cows to range from 200 to 350 μg/liter, compared with 72 to 136 μg/liter in the later milk of the same animals. Salter (194) quotes values of 50 to 240 μg/liter for the iodine content of human colostrum and of 40 to 80 μg/liter for human milk once lactation is established.

In ruminants, iodine appears to be present in the milk entirely in the form of iodide, since only iodide has been detected after $^{131}$I administration, and no thyroactive compounds can be found by chromatographic procedures (74,241). This is in agreement with biological tests of bovine milk (30) and with the finding that the normal bovine mammary gland is impervious to thyroxine (185). In the milk of the rat and the rabbit (38) and the dog (229a), an iodine-containing protein can be detected after $^{131}I$ administration. On enzymic hydrolysis, monoiodotyrosine but not diiodotyrosine is present, together with at least two other labeled compounds which do not appear to be any of the known naturally occurring amino acids or their derivatives (38).

## III. Iodine Metabolism

Iodine metabolism and thyroid function are closely linked, since the only known role of iodine is the synthesis of the thyroid hormones—thyroxine and triiodothyronine. The activity of the thyroid is regulated by a negative feedback mechanism involving the adenohypophysis and the hypothalamus. The hypothalamus secretes the thyrotropin-releasing factor (TRF) (89), a peptide which reaches the adenohypophysis via the portal vessels of the pituitary stalk and provokes the secretion of the thyroid-stimulating hormone (TSH) by the $b_2$ cells. Thyroid-stimulating hormone is a glycoprotein with a molecular weight of about 25,000, which stimulates the gland to release its hormones and to trap iodide. The thyroid hormones, in turn, inhibit the release of both TRF by the hypothalamus and TSH by the pituitary, in this way keeping the plasma level of the thyroid hormones normal. Triiodothyronine, which has some 4 times the potency of thyroxine, is also stronger in inhibiting TSH secretion. Iodine metabolism thus consists essentially of the synthesis and degradation of the thyroid hormones and of the re-utilization or excretion of the iodine so released.

### 1. *Absorption and Excretion*

Iodine occurs in foods largely as inorganic iodide and is absorbed in this form from all levels of the gastrointestinal tract. Other forms of inorganic iodine are reduced to iodide prior to absorption (50). Iodide administered orally is rapidly and almost completely absorbed from the tract, with very little appearing in the feces (133,232). Iodinated amino acids are well absorbed as such, although more slowly and less completely than iodide. A proportion of their iodine may be lost in the feces in organic combination. The remainder is broken down and absorbed as iodide (117).

Iodine is excreted mainly in the urine, with smaller amounts appearing in the feces and the sweat. In those tropical areas which have a low dietary iodine status, losses in the sweat could impose a significant drain upon limited iodine supplies. Vought and co-workers (232) found fecal iodine excretion in normal adults to range from 6.7 to 42.1 $\mu$g/day. Koutras (126) gives 5-20 $\mu$g/day as the normal range of fecal iodine excretion in man. The dominant role of the kidney in iodine excretion has led to numerous efforts to relate stable iodine excretion, and radioiodide urinary excretion after a standard interval of time, to thyroid states. The level of urinary iodine excretion correlates well with the plasma iodide concentration (233) and with $^{131}$I thyroid uptakes (69). In a study of patterns of urinary iodine excretion in goitrous and nongoitrous areas, Follis (69) has shown that these patterns provide valuable information of definitely adequate and definitely inadequate iodine nutriture but are less satisfactory indicators of marginal intakes. This worker has set a urinary level of 50 $\mu$g I/g of creatinine as the "tentative lower limit of normal" for adolescents, with 32.5 $\mu$g/g as the corresponding figure for children aged 5-10 and 75 $\mu$g/g for adult men. Koutras (126) considers that a urinary iodine excretion below 40 $\mu$g/day is suggestive of iodine deficiency in man, if renal iodide clearance is normal. The results of studies of iodine excretion per 24 hr in various goitrous areas, brought together by Follis (69) and presented in Table 35, lend support to this claim.

The extrarenal disposal rate index or T index, expressed

$$\frac{0\text{-}8 \text{ hr } \% \times 100}{(8\text{-}24 \text{ hr } \%) \times (0\text{-}48 \text{ hr } \%)}$$

is of particular value because it is virtually independent of the renal condition. This index has been closely correlated with early thyroid clearance tests in individuals with established diagnoses ranging from slight myxedema, normal, and definite thyrotoxicosis (70).

At high dose levels, absorbed labeled thyroxine is partially deiodinated and the iodine excreted in the urine as iodide. Part of such thyroxine is also taken up by the liver and excreted through the bile into the feces, either unchanged or in conjugated form. Small amounts of thyroxine also pass directly into the stomach, jejunum, and colon but the bile accounts for the major portion of intestinal thyroxine in the rat and the dog (85,221). At physiological doses of thyroxine or following biosynthetic labeling of plasma hormone, the liver plays a less prominent part in excretion and conjugation. A higher proportion is deiodinated so that only the resulting iodide which escapes the efficient thyroid trap reaches the urine and very little appears in the feces. Berson and Yalow (26) found only 10-15% of circulating organically bound $^{131}$I to be ex-

TABLE 35
*Human Urinary Iodine Excretion per 24 Hours in Goiter Areas*[a]

| Area | No. of observations | Mean iodine ($\mu$g/24 hr) | Range or S.D. |
|---|---|---|---|
| Mendoza, Argentina | 25 | 23.6 | 9.0-40.0 |
| Alto Ventuari, Venezuela | 12 | 22.5 | 9.0-40.0 |
| Aland Is., Finland | 102 | 41.2 | 6.0-119 |
| Krk Is., Yugoslavia | | 26.7 | 12.9-44.6 |
| Holland | 14 | 30.1 | 14-52 |
| Uele Region, Congo | 14 | 18.7 | ±6.9 |
| Punjab, Bihar States, India | 11 | 10.0 | |
| Prae Prov., Thailand | 18 | 8.6 | 3-33 |

[a]Adapted from Follis (69).

creted in the feces in man, the remainder appearing in the urine primarily after degradation to a nonorganic form. The daily fecal clearance of plasma hormone was less than 5% of the total extrathyroidal organic iodine pool, from which they concluded that most of the hormonal iodine in man is fated for metabolic degradation and return to the body iodide pool.

## 2. *Intermediary Metabolism*

Iodide ions resemble chloride ions in that they permeate practically all tissues and are distributed in a similar manner in the extracellular fluids. The total iodide pool thus consists of the iodide present in the whole extracellular space, together with the red blood cells and certain areas of selective concentration, including particularly the thyroid and the salivary glands and, to a lesser extent, the gastric glands and dense connective tissue. Equilibrium within the total iodide pool is reached extremely rapidly (25). In the rat about 52% of either absorbed or intraperitoneally injected $^{131}$I is excreted, mainly in the urine, with a biological half-life of 6 to 7 hr (103). This half-life is considered to represent the turnover of the inorganic iodide or the inorganic iodide phase. Metabolic equilibrium between the retained $^{131}$I and the whole-body iodine pool was obtained within 4 days and the biological half-life of this iodine was shown to reach 9.5 days and to remain constant at that level. This is considered to represent the turnover of the organic iodine.

Despite its high-iodine content and the efficiency with which it traps iodine, the thyroid gland contributes little to the iodide pool. This is because the binding into organic form is so rapid, except when this process is blocked by antithyroid agents. Significant quantities of iodide are also trapped by the salivary glands (37), apparently by mechanisms similar to those of the thyroid (95,96,238). Since salivary iodide is not converted into organic form and is normally all reabsorbed (94), this process represents little net loss to the iodide pool.

The iodide pool is replenished continuously from three sources—exogenously from the diet and endogenously from the saliva and from the breakdown of iodine hormones produced by the thyroid. Iodide is continuously lost from this pool by the activities of the thyroid, the kidneys, and the salivary glands, which compete for the available iodine. Koutras (126) has represented iodine metabolism as a metabolic cycle consisting of three main pools: the plasma inorganic iodide, the intrathyroidal iodine pool, and the pool comprising the hormonal or protein-bound iodine of the plasma and the tissues. The rates of removal of iodide from the first of these pools by the thyroid and the kidneys are expressed as thyroidal and renal clearances, calculated as organ accumulation of iodide per unit time divided by plasma iodide concentration. In healthy humans, total clearance from the iodide pool occurs at the rate of about 50 ml/min or 3 liters/hr, to which the kidneys contribute about two-thirds and the thyroid, one-third. If the volume of the total iodide pool is taken as 25 liters the rate of turnover is about 12%/hr (118). In normal man the renal iodide clearance is very constant at about 35 ml/min over all ranges of plasma iodide examined (45). In fact there is no renal homeostatic mechanism to keep the plasma iodide level constant (186). Thyroid clearance, by contrast, is sensitive to changes in plasma iodide concentration and varies greatly with the activity of the gland. In normal individuals the thyroid clears an average of 10 to 20 ml/min, whereas in exopthalmic goiter or Grave's disease, a clearance of 100 ml/min is usual (194) and over 1000 ml/min is possible (25). The measurement of thyroid $^{131}I$ clearance is a valuable diagnostic aid in hyperthyroidism but is much less sensitive to myxedema (170). For this purpose the extrarenal disposal rate, described earlier, is more satisfactory. In goitrous cretins with congenital inability of the thyroid to trap iodide, the salivary glands are similarly affected and the saliva/plasma iodide ratio is less than unity (239). This ratio may, therefore, be used diagnostically where this type of dyshormonogenesis is suspected but has no such value in patients with iodine deficiency goiter nor in patients with goiter and a normal plasma iodide level (97). The saliva $^{131}I$/plasma protein-bound $^{131}I$ ratio has also been suggested as a test of thyroid function (144), but as Harden and Alexander (94) have pointed out, this ratio has the disadvantage that it will be altered by changes in salivary flow rate and in salivary iodide trapping, and it is too indirect for routine clinical diagnosis.

The intrathyroidal iodine pool involves a series of consecutive steps in the synthesis of the thyroid hormones. First, the iodide brought to the gland in the plasma is trapped by an energy-requiring active mechanism which can be blocked by antithyroid agents of the perchlorate and thiocyanate type. This iodide is oxidized to elemental iodine, or to some similar reactive form, as a preliminary to its incorporation into organic combination by a peroxidase (1). The more reactive iodine combines with the tyrosine residues and thyroglobulin to form 3-monoiodotyrosine and 3,5-diiodotyrosine. This reaction takes place near the boundary between follicular cells and follicular lumen, with the same enzyme system probably acting as iodide-peroxidase and as tyrosine-iodinase (58). It can be blocked by a great number of antithyroid substances of the thiouracil type and even by iodide itself in high concentrations. The ability of the gland to trap iodide is not reduced by such agents. In fact the thiouracil-blocked gland can maintain an iodide concentration several hundred times that in the plasma (223). Two diiodotyrosine molecules combine to form 1 molecule of thyroxine, or 1 mono- and 1 diiodotyrosine combine to form 1 molecule of triiodothyronine. These are stored in the colloid bound to thyroglobulin. The normal human gland contains some 8-12 mg of iodine, mainly in the form of 2 iodotyrosine and 2 iodothyronine molecules bound in this way (186). In a recent study of pancreatin hydrolyzates of normal human thyroids the distribution of the iodine was found to be 16.1% as iodide, 32.7% as monoiodotyrosine, 33.4% as diiodotyrosine, 16.2% as thyroxine, and 7.6% as triiodothyronine (62).

The thyroid hormones are released through proteolysis of thyroglobulin by a protease system which yields both iodotyrosines and thyroactive thyronines (154). The normal human thyroid has been estimated to secrete daily 51.6 $\mu g$ of iodine as thyroxine and 11.9 $\mu g$ as triiodothyronine (182). The iodotyrosines liberated from the proteolysis of thyroglobulin, unlike the iodothyronines, are not secreted into the circulation but are deiodinated by an enzyme called "deiodinase" or "dehalogenase" (187). The iodine so released is not lost from the gland but is reutilized for hormone synthesis. This leads to an economical use of iodine and ensures that virtually all the iodine entering the normal thyroid as iodide leaves it in the form of hormones after one or more entries into thyroglobulin (156).

In order to ensure an adequate supply of hormones, the human thyroid must trap about 60 $\mu g$ of iodide daily. This is primarily achieved, irrespective of the plasma iodide level, by adjustment of the thyroidal iodide clearance rate, so that when the plasma iodide decreases the thyroidal clearance increases, with the actual iodide uptake remaining more or less constant. Adaptation to iodine deficiency thus occurs by increasing the thyroidal iodide clearance rate. Such functional overactivity of the iodide-trapping mechanism is usually associated with an increase in the gland mass, or goiter, but in mild iodine deficiency the biochemical manifestations of the deficiency, namely low plasma iodide and urinary iodine and high thyroidal iodide clearance and radioiodine uptake, have been demonstrated without any obvious goiter (233).

The circulating thyroid hormones, comprising the third metabolic iodine pool as stated by Koutras (126), occur mostly bound to a thyroxine-binding globulin, to prealbumin, and to albumin itself (79, 111), with only about 0.05% of the thyroxine normally present in the free state (215). It is for this reason that the serum PBI corresponds so well with the circulating thyroid hormone level. According to Rall and co-workers (182), some 10% of the circulating thyroxine and 56% of the circulating triiodothyronine is metabolised daily in man. Once these hormones enter the tissues, about 80% is broken down by several deiodinating enzymes, with the iodine so liberated returning to the iodide pool, thus completing the iodide cycle. The small remainder, which is proportionately larger in the rat than in man (26), enters the enterohepatic cycle and is lost to the body either through excretion into the bile unchanged, or through excretion by this route in conjugated form following detoxication in the liver (221).

Disturbances in the peripheral metabolism of thyroxine occur in some forms of thyroid disease in man (135, 164). Nicoloff and Dowling (164) were able to demonstrate increased hepatic incorporation of labeled thyroxine, with an increased rate of deiodination in Grave's disease, whereas in patients with primary hypothyroidism hepatic incorporation and rate of deiodination were reduced. The pathogenetic basis of these changes is unknown, although it seems that the amount of labeled thyroxine incorporated in the liver is directly related to the deiodination rate of that organ.

## IV. Iodine Deficiency and Thyroid Function

Since the entire functional significance of iodine can be accounted for through its presence in the thyroid hormones, the manifestations of iodine deficiency are those of a deficient supply of these hormones to the organism. However, the reverse is not necessarily true. Many factors can upset the capacity of the thyroid gland to accumulate iodine and to convert it into the required thyroactive compounds. These factors can act independently of iodine supply or may only become apparent in circumstances of borderline iodine deficiency. An enlargement of the thyroid, or goiter, must be regarded as a final common expression of a number of separate and distinguishable disease processes. An absolute dietary deficiency of iodine is only one of these processes. A deficiency of iodine "conditioned" by the presence of goitrogens is another. A metabolic defect in thyroid hormonogenesis, due to the presence of certain types of goitrogens or to a constitutional disability, represents a third disease process resulting in the development of goiter.

The thyroid hormones and, therefore, iodine are involved in a variety of metabolic activities, described below, most of which derive from their fundamental effect on the rate of cellular oxidation.

1. *Metabolic Processes*

The rate of energy exchange and the quantity of heat liberated by an organism at relative rest is elevated in hyperthyroidism and reduced below normal in hypothyroidism. After total thyroidectomy the BMR gradually falls to about one-half normal and can be raised again by the administration of thyroactive substances. Where thyroid output is limited by lack of dietary iodine the BMR is similarly lowered and can be restored to normal by supplemental iodine or by thyroid hormone therapy. The calorigenic action of the thyroid hormones can be demonstrated by measuring the oxygen consumption of excised tissues. Liver, kidney, muscle, and other tissues from iodine-deficient or thyroidless animals consume less oxygen than normal, whereas these tissues from hyperthyroid animals consume more oxygen than normal (42). Tissues excised and cultured *in vitro* are not affected by thyroxine, suggesting that they are deprived of some essential organismic factor.

The changes in cellular oxidation are accompanied by disturbances in the metabolism of water, salts, proteins, carbohydrates, and lipids. In hypothyroidism the retention of water and salt is increased and plasma volume is considerably reduced. The administration of thyroid hormone, or of iodine where the hypofunctioning thyroid results from iodine deficiency, induces a pronounced water and salt diuresis and an increased plasma volume. Hypothyroid states are accompanied by above-normal, and hyperthyroid states are accompanied by subnormal serum cholesterol levels in man. Important evidence has recently been obtained that the oxygen-releasing factor present in blood (157) may be deficient or absent in hypothyroidism. Grosz and Farmer (88) found that the blood redox potentials of hypothyroid patients declined significantly faster in the first 60 min after deoxygenation than did the blood redox potentials of hyperthyroid patients. By direct measurements it was shown that this results from the more rapid decline of the partial pressures of oxygen in the blood of the former patients. As pointed out by these workers, this means that hemoglobin will not release oxygen sufficiently into solution at low-oxygen tensions to maintain an adequate partial pressure of oxygen in the blood perfusing tissue capillaries and will, incidentally, affect the redox state of the tissues.

### 2. *Growth and Differentiation*

The thyroid hormone is essential for growth during early life in all mammals and birds [see Pitt-Rivers and Tata (177)]. Total thyroidectomy induces severe dwarfism in rats or birds, whereas athyreosis in human infants leads to the type of dwarfism known as cretinism which occurs in severely goitrous areas. In those areas iodine administration during adolescence increases the rate of growth as well as reducing the incidence of goiter. In an early Swiss study assessed by Stocks (217), in which 2 mg sodium iodide tablets were administered weekly to schoolgirls for a 3-year period, large increases in growth rates above the mean growth curves were observed for those with pronounced goiters and smaller but significant increases for those with moderate to small goiters. Stunted growth in domestic animals in iodine-deficient areas, responsive to iodine and not always associated with visible goiter, has also been reported (63, 116), and numerous studies have correlated growth rates with thyroid secretion rates in chicks (75, 180), lambs (208), and calves (179).

The retardation of early growth brought about by a severely hypofunctioning thyroid is accompanied by a delay in almost all developmental processes. This effect is strikingly evident in the metamorphosis of amphibian larvae. The hastening of metamorphosis of frog tadpoles brought about by thyroactive compounds is apparently not due to a mere stimulation of metabolic rate because the dinitrophenols elevate metabolism in tadpoles but do not stimulate their development into frogs.

### 3. *Neuromuscular Functioning*

Endemic goiter is frequently associated with an incidence of feeblemindedness and deaf-mutism in children. Whether there is a direct causal relationship or whether goiter directly diminishes intelligence or learning capacity is less certain. Thus Shee (205), working in a goiter area in Ireland, found a significant inverse correlation between established goiter and the intelligence of schoolchildren assessed by a standard intelligence test. On the other hand, Stocks (217) in his assessment of the Swiss study mentioned earlier, could find no evidence that proficiency in school work is lowered in adolescents with enlarged thyroids. Human hypothyroidism is, nevertheless, characterized by mental sluggishness and apathy, and hyperthyroidism by emotional instability, nervousness, and irritability, accompanied usually by muscle tremors and hyperactivity of the sweat glands. These are regarded as evidence that the thyroid affects both the central nervous system and the autonomic system. Hypothyroids have generally also a lowered resistance to narcotics and heightened thresholds to light and sound stimuli.

In myxedema and other thyroid deficiencies the cerebral cortex originates fewer alpha waves per second than does the normal brain, the electrical activity of the brain can be returned to normal by the administration of thyroxine and a good correlation between the BMR and the rate of emanation of alpha waves from the nervous system has been observed (193). Furthermore, the respiratory centers in the medulla are less sensitive to $CO_2$, and the rate of breathing is diminished in the hypothyroid state, whatever its cause. Treatment with thyroid substance, or with iodine if the thyroid insufficiency arises from a lack of this element, results in a rapid restoration of the sensitivity of these centers, with an increased respiratory rate. The mechanism involved in this sensitizing of the nervous system by the thyroid secretion is possibly connected with changes in the rate of oxidation of this tissue, as was demonstrated many years ago by Cohen and Gerard (49), for the brain in different thyroid states.

### 4. *Reproduction and Interactions with the Gonads and Other Glands*

The delicate balance and interaction between the thyroid and the adenohypophysis and the hypothalamus and their hormones which control the activity of the thyroid have already been described. It can be stated further that thyrotropin can produce hypertrophy of the thyroid cells when these are grown in blood serum (8) and that the atrophy of the thyroid which follows hypophysectomy can be prevented by the administration of this hormone. The quantitative stimulation of thyroid function brought about by thyrotropin takes place without qualitatively influencing its metabolism (214). Furthermore, thyroidectomy is followed by a reduction in the size of the adrenals in rats, and administration of thyroxine induces adrenal cortical hypertrophy (109). Conversely, hyperfunction of the adrenal cortex may induce decreased thyroid activity, and administration of adrenaline can induce hyperplastic changes in the thyroid (210).

A relationship between the thyroid and the gonads is apparent in all male and female mammals and birds [see Maqsood (149)]. In man, colloid goiter often develops at puberty and hyperthyroidism is sometimes precipitated at the menopause. Goitrous cretins are usually sterile and invariably fail to develop normal sexual vigor, with a delayed maturation of the genitalia. Thyroidectomy at an early age is followed, in all species, by a long period in which the gonads and the secondary sex organs remain in an infantile condition.

In some types of birds thyroid-gonadal interrelationships can be very conspicuous due to plumage changes. In the brown leghorn male, thyroidectomy is followed by a long period in which the testes remain small and free from spermatozoa. The comb decreases in size, molting is inhibited, and the characteristic male plumage is lost. Administration of estrogen to such birds does not induce the female plumage pattern as it does in normal males, which suggests that this pattern results from a synergistic action of the thyroid and ovarian hormones (27). Thyroidectomy reduces egg production in hens (237), and there is some evidence that the seasonal cycle of egg production in poultry is related in part to seasonal variation in thyroid activity. Turner and co-workers (228) have achieved some success in preventing this seasonal decline, associated with high summer temperatures, as well as the decline in productivity with advancing age, by feeding thyroactive iodinated proteins but this effect is not related to a lack of dietary iodine. In fact, poultry can withstand considerable iodine deficiency without any marked lowering of yield or hatchability of eggs. Thus Rogler (191) maintained hens for 35 weeks on an iodine-deficient diet without affecting hatchability or embryo weight, whereas hens maintained on such a diet for 2 years revealed decreased hatchability, prolongation of hatching time, and retarded embryo development.

Reproductive failure is often the outstanding manifestation of iodine deficiency and consequent impairment of thyroid activity in farm animals. The birth of weak, dead, or hairless young in breeding stock has long been recognized in goitrous areas (63). Fetal development may be arrested at any stage, leading either to early death and resorption, to abortion and stillbirth, or to the live birth of weak young, often associated with prolonged gestation and parturition and retention of fetal membranes (4, 105, 161). Allcroft *et al.* (4) demonstrated subnormal serum PBI levels in herds showing a high incidence of aborted, stillborn, and weakly calves. Thyroidectomy is not completely incompatible with the production of apparently normal offspring (211), presumably due to the capacity of the fetal thyroid to maintain pregnancy. However, Falconer (64) has shown that thyroidectomy of ewes some months before conception severely reduced both the prenatal and the postnatal viability of the lambs, despite the presence of an apparently adequate thyroid in the lamb itself. Neonatal mortality in lambs from ewes fed goitrogenic kale, responsive to iodine administration during pregnancy, has frequently been observed (9, 203).

In addition to the reproductive disturbances just described, irregular or suppressed estrus in dairy cattle causing infertility has been associated with goiter or iodine deficiency and shown to respond to iodine therapy (105, 116, 153a, 162). Moberg (162), working with 190 herds totalling 1572 cows in goitrous areas in Finland, obtained a significant improvement from iodine therapy in first service conception rate and in the number of cows with irregular estrous incidence. McDonald *et al.* (153a), working in an iodine-deficient area in Canada, also obtained a marked improvement in first service conception rate by feeding an organic iodine preparation, ethylenediamine dehydroiodide, beginning 8-10 days before the cows came into estrus.

The thyroid gland plays an equally important part in the maintenance of male fertility (149). A decline in libido and a deterioration in semen quality have been associated with iodine deficiency in bulls and stallions (105, 116), and a seasonal decline in semen quality in rams has been related to a mild hypothyroid state. This condition is responsive in part to doses of thyroactive proteins, but not to iodine except in known iodine-deficient areas and can be simulated by thiouracil administration (189).

### 5. *Condition of the Integument*

Changes in the skin and its appendages, hair, wool, fur, and feathers, are among the most constant features of iodine deficiency. Cretins exhibit a pale, gray skin lacking in mobility and their hair tends to be dry and scanty. In human myxedema the skin is dry, rough, and thickened, and shedding of the hair is a characteristic symptom. In birds the relation of different levels of thyroid activity to the molting process and to the form, structure, and pigmentation of the feathers can be particularly striking, as pointed out earlier. Calves and pigs born to iodine-deficient mothers are often hairless and have thick pulpy skins, with less severe deficiencies manifested in milder disorders of the pelt, such as rough, dry skin, harsh coat, scanty wool, and hairiness of the fleece (10, 116). The normal development of the wool-producing follicles, particularly the secondary follicles, requires thyroid activity in excess of that needed for general body growth, and a deficiency in the growing lamb may permanently reduce the quality of the adult fleece (67).

### 6. *Nature and Mode of Action of Thyroid Hormones*

Many different compounds possess thyroidal activity equal to or in excess of that of thyroxine. Thus, in comparison with thyroxine (3,5,3′,5′-tetraiodothyronine), 3,5,3′-triiodothyronine is 3-5 times as active, depending upon the dose used; 3,3′-diiodothyronine is similar in potency; 3,5-diiodo-3′,5′-dibromothyronine is nearly as active; and

3,5,3′-triiodothyropropionic acid is 300 times more active in accelerating amphibian metamorphosis (156, 163, 189). On the other hand, 3,3′,5′-triiodothyronine has only one-twentieth of the activity of thyroxine or one-hundredth of that of its isomer (189). Studies of various thyroxine analogs have revealed positive correlations between their thyromimetic activities and the ability of various substituent groups to attract or release electrons and to form hydrogen bonds (39). Although these findings have limited physiological meaning, they permit the following generalizations: the thyronine nucleus (see Fig. 18) is essential to any activity;

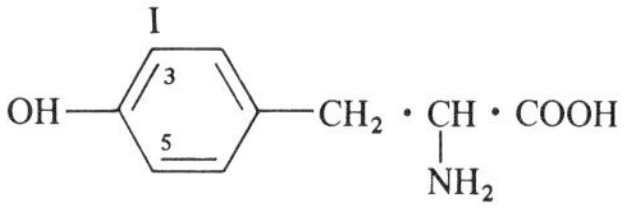

I
OH— [ring 3, 5] —$CH_2 \cdot CH(NH_2) \cdot COOH$

3-Monoiodotyrosine

OH— [ring 3′, 5′] —O— [ring 3, 4] —$CH_2 \cdot CH(NH_2) \cdot COOH$

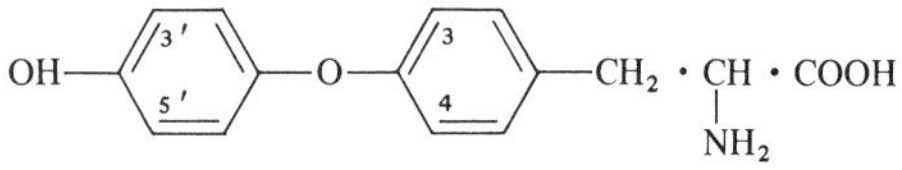

Thyronine
[4-(4′-Hydroxyphenoxy)phenylalanine]

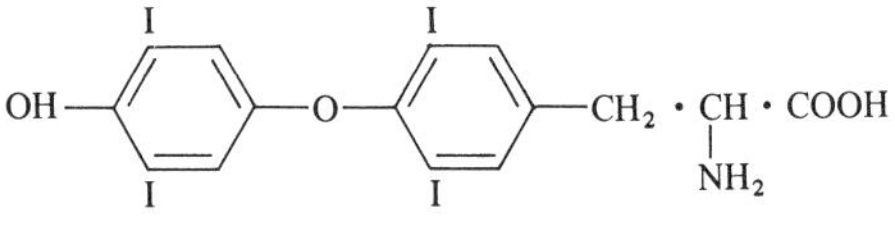

I I
OH— [ring] —O— [ring] —$CH_2 \cdot CH(NH_2) \cdot COOH$
I I

Thyroxine
(3,3′,5,5′-Tetraiodothyronine)

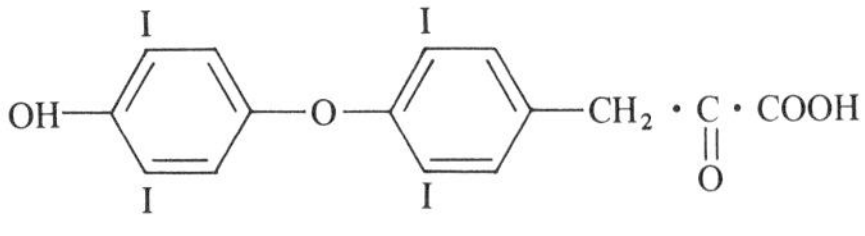

I I
OH— [ring] —O— [ring] —$CH_2 \cdot C(=O) \cdot COOH$
I I

3,5,3′-Triiodothyropyuvic acid

**Fig. 18.** Structural formulas of monoiodotyrosine, the thyronine nucleus, thyroxine, and triiodothyropyruvic acid.

iodine substitution in the inner aromatic ring is required for substantial activity; partial replacement of iodine by bromine results in little loss of activity; an alanine side chain is not essential; and, whatever the aliphatic side chain on the nucleus, a 3,5,3′-substitution in the rings ensures maximal activity.

The mechanism of action of thyroid hormone at tissue level is still obscure. Many enzymes in various tissues are affected by variations in thyroid activity, or by treatment of the animal with thyroactive substances (22, 93), but direct involvement of thyroxine or triiodothyronine in an enzyme system needed for the energy-transforming processes of the cells has not been demonstrated. These hormones have been shown to increase the initial rate of oxidation of succinate by submitochondrial particles of rat liver, associated with increased efficiency of phosphorylation (36). The role of mitochondria in thyroid hormone action is emphasized further by Hoch (107). The active forms of the hormones at the cellular level may actually be elaborated in the cells. Thus 3,5,3′-triiodothyropyruvic acid and 3,5,3′-triiodothyroacetic acid, both of which occur in the tissues, are as effective as thyroxine in promoting the metamorphosis of tadpoles and the latter has an immediate effect in raising the BMR of rats. However, this compound quickly loses its effect and does not lead to a rise in the BMR of myxedema patients (137).

## V. Iodine Requirements

It is difficult to assess the minimum dietary requirements of man or other species because of considerable individual variability and of the effects of environment, including the nature of the rest of the diet. Calculations based on average daily urine losses give an adult human requirement of 100 to 200 μg/day (55), and the results of several balance studies indicate that equilibrium or positive balance can be achieved at iodine intakes ranging from 44 to 162 μg/day (52, 197). This wide range no doubt reflects the wide normal range in both the renal clearance of iodide and the fecal excretion of iodine in healthy individuals (233). From an assessment of many investigations, Elmer (61) placed the optimum iodine requirements of man at 100 to 200 μg/day. Wayne and associates (233) arrived at a figure in the region of 160 μg/day "as the minimum certainly safe amount of iodine which must be available in the individual's diet if iodine-deficiency goiter is to be avoided." They suggested that it might be advisable to raise this figure to 200 μg/day for children and for pregnant women.

Some years ago, Levine and associates (138) set the minimum iodine requirements of the rat at 0.9 μg/day to maintain a concentration of

0.1% I in the moisture-free thyroid gland and at 1 to 2 μg/day to prevent any significant enlargement of the gland. More recently, Heinrich *et al.* (103) placed the nutritional iodine requirement of this species at 0.7 to 0.9 μg/day, on the basis of whole-body iodine turnover rates. If these measures are taken as limits, it can be calculated that the rat requires 20-40 μg I per 1000 food Calories (kcal) consumed. If this reasoning is applied to human adults consuming 3000 kcal daily, the iodine requirement would be 60-120 μg/day.

The minimum iodine requirements of farm stock can be given with even less precision than those of man. Mitchell and McClure (159) have pointed out that iodine requirements are more properly related to heat production than to energy intake. They calculated the minimum iodine requirements of different classes of animals on this basis as shown in the following tabulation:

| Animal | Body weight (lb) | Heat production (cal) | Iodine requirement (μg/day) |
|---|---|---|---|
| Poultry | 5 | 225 | 5-9 |
| Sheep | 110 | 2500 | 50-100 |
| Pigs | 150 | 4000 | 80-160 |
| Cow in milk (40 lb/day) | 1000 | 20,000 | 400-800 |

These estimates, except for the cow where they are much lower, compare well with the minimum intake figures given by Orr and Leitch (168) for these species in nongoitrous areas. Wilgus *et al.* (235) place the minimum iodine needs of growing and breeding poultry at close to 1 ppm and those of laying hens at 0.2 to 1.0 ppm of the dry diet. Other workers have obtained lower figures, namely 0.03 and 0.15 ppm (77) and 0.07 for normal growth in chickens and 0.3 ppm for completely normal thyroid structure (54). The dietary iodine requirements of pheasants and quail have been shown to be no greater than 0.3 ppm, either for growth or for the production of normal thyroids (199).

The lack of complete conformity between a low iodine level in the environment and the incidence of nontoxic goiter in some areas is probably explained by environmental, particularly dietary, variations in the levels of other minerals and of goitrogens which influence iodine uptake by the thyroid and, therefore, iodine requirements. Thiocyanates and perchlorates (7, 242) and rubidium salts (20) are known to interfere with iodine uptake and high levels of arsenic can induce goiter in rats (204).

Reports of an arsenic-thyroid antagonism in man have appeared, and it is perhaps significant that a high incidence of goiter occurs in the Styrian Alps, the home of the arsenic eaters, and in the Cordoba Province of Argentina, where chronic arsenic poisoning is endemic (198). High fluoride intakes have also been proposed as contributing factors in the incidence of goiter in parts of South Africa (216) and in the Punjab of India (236). It is difficult to assess the significance of these claims, especially as experimental evidence on the effects of fluoride on the thyroid is rather confusing. Thus the administration of NaF to normal rats and rabbits lowered basal metabolism and reduced the rate of thyroid hormonogenesis in some experiments (78, 184, 218) and had no such effect in others (73, 174). Fluoride feeding in concentrations up to 100 ppm for 7 years also had no gross, histological or functional effects upon the thyroid gland of dairy cows (108).

The incidence of goiter in various areas has been associated with the presence of limestone formations yielding hard waters but, as with fluoride, experimental evidence of a calcium-thyroid interaction is confusing. An enhancement by calcium of goiter in rats on low-iodine diets, together with increased radioiodine uptake, was reported by Hellwig (104) and by Taylor (225). No such effect was observed in mice (19) and as Scrimshaw (200) has pointed out, the differences in thyroid weights obtained by Taylor become much less striking when calculated per unit of body weight. Moreover, Malamos and Koutras (145) were unable to demonstrate an increase in renal iodide clearance by oral or intravenous administration of calcium to man.

Interactions between iodine and cobalt and effects of cobalt deficiency and excess upon goiter incidence are more convincingly documented than those postulated for arsenic, fluorine, and calcium. Novikova (165) showed with rats that the absolute content and concentration of iodine in the thyroid is decreased when the cobalt intake is deficient and when it is excessive. The greatest decrease in the iodine concentrating power of the gland was observed when the intake of both iodine and cobalt is low. Blokhina (31) reported that cobalt, and also manganese, are necessary for the synthesis of the thyroid hormones. Cobalt increased the height of the follicular cells and the hormonogenetic capacity of the gland in iodine deficiency. Several investigators in the Soviet Union have demonstrated an inverse correlation between the cobalt content of the soils, foods, and water supplies in certain areas and the incidence of goiter in man and farm animals (127, 128). A depressing effect on the thyroid from large doses of bivalent cobalt has been demonstrated by others in rats (129) and in man (170).

## VI. Sources of Iodine

### 1. *Iodine in Water*

The iodine content of the drinking water generally reflects the iodine content of the rocks and soils of a region and hence of the locally grown foods. It has been correlated with the incidence of goiter in many areas ever since the pioneering studies of Chatin more than a century ago. Three examples from widely separated areas can be cited to illustrate this relationship. Kupzis (132) found the water supplies in goitrous areas in Latvia to range from 0.1 to 2.0 μg I/liter, compared with 2 to 15 μg/liter for nongoitrous areas. Young and co-workers (243) reported that the drinking water in English villages, where the incidence of goiter in the human population was assessed at 56%, averaged 2.9 μg/liter, compared with 8.2 μg/liter in other villages where the goiter incidence was only 3%. More recently, Coble *et al.* (48), in a study of goiter incidence in Egyptian oases, found the iodine content of well and spring waters in goitrous villages to range from 7 to 18 μg/liter, compared with the very high levels of 44 and 100 μg/liter in two samples of water from oases where there was no goiter. These latter figures indicate that in some circumstances the water supply can contribute substantial quantities of iodine to the daily diet. In most areas, the proportion of the total intake obtained from this source is of minor importance. In fact, the classical early studies of von Fellenberg (66) carried out in Switzerland, indicate that the drinking water supplies less than 10% of the total daily intake, in goitrous and nongoitrous areas alike.

### 2. *Iodine in Human Foods*

Reported values for the iodine content of foods are characterized by high variability from sample to sample (112). Some of this variability can be ascribed to analytical errors, especially with data obtained before 1930, at which time more reliable methods became available. The iodine concentration in plants is very susceptible to differences in the content and availability of iodine in the soil and to the amount and nature of the fertilizers applied. Chilean nitrate of soda, the only mineral fertilizer naturally rich in iodine, can double or treble the iodine content of food crops when applied in the amounts required to supply their nitrogen needs. Gurevich (91) has shown that the iodine concentration in a range of vegetable and cereal crops can be increased by 10 to 100 times or more, by applications of seaweed and of by-products of the fish-, crab-, and whale-processing industries.

Animal products, such as milk or eggs, can be very much richer in iodine when they come from animals consuming iodine-enriched diets. Examples of this effect upon cow's milk were given earlier when considering the iodine content of milk. On average diets, hen's eggs contain some 4-10 $\mu g$ of iodine, most of which is located in the yolk (150, 192). Feeding the hens sufficiently large amounts of iodine, either as iodized salt, as sodium iodide, or as seaweed, can raise this amount as much as 100-fold (90, 115) or 1000-fold (150).

The great variability in the iodine concentration in foods, as just described, makes it difficult to classify human foods into groups in accordance with their "normal" iodine contents. In fact, the iodine content of foods of marine origin is so much higher than that of any other class of foodstuff that differences among the latter become relatively insignificant. Sea fish and shellfish are particularly rich in iodine and their oils are even higher in this element (112). The edible flesh of sea fish and shellfish may contain 300-3000 ppb on fresh basis, compared with 20-40 ppb for freshwater fish. The order of magnitude of iodine concentrations of other classes of foods, in parts per billion on fresh edible basis, are as follows: fresh fruits, 20; leafy and root vegetables (excluding spinach and watercress which are usually higher), 30; cereal grains, 40; meats, 40; milk and milk products, 40; eggs, 90. On dry matter basis the cereal grains and their mill products are the lowest in iodine concentration of all common foods, other than such highly refined products as white sugar.

Overall iodine intakes are determined more by the source of foods composing a given dietary than by the choice or proportions of different foods, except for those of marine origin. The residents of an endemic goiter area cannot, therefore, normally obtain sufficient iodine for their full requirements merely by a more judicious choice of foods grown or produced within that area. This can only be achieved by the consumption of a substantial proportion of their total dietary in the form of marine foods, or of foods imported from iodine-rich regions elsewhere, or by the use of iodized salt in some form. The adoption of the first of these procedures, by consuming large amounts of imported seaweeds, is believed to be one reason for the low incidence of goiter in Japan, although this country is geologically low in iodine. The concentration of iodine in seaweed varies with the species and season of the year. Values as high as 0.4-0.6% on dry basis are common (113). The people of Taiwan, among whom the incidence of goiter was high until iodized salt was introduced, do not consume seaweed in the same quantities as do the Japanese.

### 3. *Iodine in Animal Feeds*

The iodine content of land plants varies greatly with the species, the soil type, the fertilizer treatment given and, to some extent, with the climatic and seasonal conditions. The iodine levels in common pasture species usually range from 300 to 1500 ppb on dry basis (115, 168). Roughages (hays and straws) generally contain 300–500 ppb and are substantially higher in iodine than the cereal grains or the oilseed meals commonly employed as protein supplements in farm rations (113). Protein concentrates of animal origin, other than fish meal, cannot be relied upon as significant sources of dietary iodine, unless the animals from which these products are obtained have been ingesting exceptionally large amounts of this element.

### 4. *Iodine Supplementation*

The principal methods that have been adopted for the control of endemic goiter in man are (*a*) the use of iodized table or domestic cooking salt or both, (*b*) the administration of iodized tablets or confections to individuals, (*c*) the use of iodized salt in bread, and (*d*) the addition of iodine to municipal water supplies. The last of these procedures has mostly been abandoned because of the number of rural people who cannot participate and the small proportion of the iodized water actually used as drinking water. The use of medicated tablets or confections has the advantage of providing known doses of iodine but requires the continuous cooperation of many people and authorities. It is, therefore, rarely completely satisfactory after the initial enthusiasm has waned. The incorporation of iodized salt into all factory-baked bread is a convenient and effective means of iodine supplementation in goitrous areas where bread is a staple foodstuff and home baking is not a common practice (106). Iodine added as potassium iodide suffers no loss during the baking process (106). Compulsory iodization of domestic salt is the most economical, convenient, and effective means of mass prophylaxis in most goitrous areas and is now carried out as a public health measure in many countries.

Satisfactory processes are available for producing and distributing fortified salt and of maintaining the stability of the added iodine. The more stable iodate is steadily replacing iodide for this purpose. The optimum level of iodization depends on the average consumption of salt per head and the average daily dose of iodine needed to prevent goiter in a community. In the United States and New Zealand the average consumption of salt is 6 g daily, in European countries it is somewhat higher and in hot, humid countries such as southern India it may approach 30 g daily.

In endemic goiter areas, each person should receive a supplement of not less than 100 μg of iodine daily, with supplements as high as 300–400 μg daily in some areas, presumably to counteract the effect of goitrogenic substances in the diet. In the United States, Canada, and most countries of Latin America the salt is iodized at a level of 1 in 10,000. This would provide a supplement of about 500 μg I/head/day from the table salt alone. In Switzerland, where all food salt is iodized, the level is much lower namely 1:100,000, and in Britain where only table salt is iodized, a level of 1:40,000 is employed. According to Holman and McCartney (107a) a level of 1:20,000 is probably the most satisfactory for countries with a moderate goiter rate.

With farm animals, the best iodine supplementation method to adopt depends upon the conditions of husbandry. Stall-fed animals are usually provided with iodized salt licks or pellets, or the iodine is incorporated into the mineral mixtures or concentrates provided to supply other nutritional needs. Salt licks containing potassium iodide lose iodine readily from volatilization or leaching if exposed for any length of time to hot or humid conditions (57). Stabilization procedures have been developed (119) or potassium iodate, which is more stable than iodide and is nontoxic at the levels required, may be used (207, 240). Inclusion of this compound into salt licks or mineral mixtures for stock at levels of 0.01% (143) or 0.05% (56) iodine is recommended to ensure adequate intakes. In a comparison of potassium iodide, calcium iodate, and 3,5-diiodosalicylic acid as iodine sources for livestock, Shuman and Townsend (206) found the first two of these to be rapidly lost from the surface layer of salt blocks when exposed to outdoor weather conditions, whereas the diiodosalicylic acid remained constantly on the surface and was obtained by the animals in a normal manner. This latter source of iodine had been earlier shown to be well utilized by rats (160), but Aschbacher and coworkers (14, 16) found it to be an unsatisfactory source of supplemental iodine for calves and diary cattle. Compared with rats, cows have much less ability to remove iodine from this molecule before it is excreted (15). Subsequently Aschbacher (13) showed further that allowing ewes free access to salt containing 0.007% I as diiodosalicylic acid did not prevent signs of iodine deficiency in their newborn lambs, whereas access to salt containing the same concentration of iodine as KI successfully achieved this aim. These workers (158) then compared the nutritional availability of iodine from calcium iodate, pentacalcium-orthoperiodate (PCOP), and sodium iodide to pregnant cows. All three forms were found to supply the fetal thyroid with equal efficiency. Since PCOP has much greater physical stability than iodide and appreciably greater physical stability than calcium iodate under field conditions (155), this compound is clearly a valuable form of supplemental iodine for use in livestock salt blocks or mineral mixes.

With sheep and cattle under permanent grazing in goitrous areas the problem of providing supplemental iodine requires different forms of treatment. Iodized fertilizers cannot be relied upon to maintain satisfactory iodine levels in the herbage for long periods after application, and the provision of iodized salt licks is subject to the hazard, as with all trace elements, of spasmodic and uncertain consumption by the grazing animal. Unless stable and insoluble compounds such as PCOP are used, there is also the problem of physical loss by volatilization and leaching. Regular dosing or "drenching" with inorganic iodine solutions is effective but is costly and time-consuming, except where the animals are otherwise frequently handled. This form of treatment, consisting of two oral doses of 280 mg KI or 360 mg $KIO_3$, given at the beginning of the fourth and the fifth months of pregnancy, is satisfactory for the prevention of the neonatal mortality and associated goiter in lambs which arise when ewes are grazed on goitrogenic kale (*Brassica oleracea*) (207). This condition can also be controlled by intramuscular injections of an organic iodized poppyseed oil preparation (Neohydriol), containing some 40% by weight of bound iodine. A single 1-ml injection of this preparation into ewes 2 months before lambing raised the iodine concentrations in the thyroid glands of lambs from kale-fed ewes to normal levels and prevented marked goiter and high neonatal mortality. Similar injections given 1 month later only partially prevented the thyroid enlargement, although the iodine concentrations in these glands were normal and the death rate was reduced (9, 207) (see Table 36).

TABLE 36
*Effects of Organic Iodine (Neohydriol) on the Thyroid Glands and Neonatal Mortality of Lambs from Kale-fed Ewes*[a]

| Ewe treatments | Thyroid glands: Mean wt. (g) | Thyroid glands: Mean iodine (% dry wt) | Mortality (%) |
|---|---|---|---|
| Controls | 10.3 | 0.01 | 59 |
| Neohydriol: 2 months before lambing | 1.9 | 0.40 | 10 |
| Neohydriol: 1 month before lambing | 5.5 | 0.45 | 14 |

[a]From Andrews and Sinclair (9).

## VII. Goitrogenic Substances

Goitrogens are substances capable of producing thyroid enlargement by interfering with thyroid hormone synthesis. The pituitary responds by increasing its output of thyroid-stimulating hormone which induces hypertrophy in the gland in an effort to increase thyroid hormone production, in the manner described earlier. The extent to which the increased thyroid tissue mass compensates for the inhibition or blocking of thyroid hormonogenesis will depend upon the dose of goitrogen and, in some circumstances, upon the level of iodine intake by the animal.

The first clear evidence of a goitrogen in food was obtained by Chesney and co-workers in 1928 (44). Rabbits fed a diet consisting mainly of fresh cabbage developed goiters which could be prevented by a supplement of 7.5 mg iodine per rabbit per week. "Cabbage goiter" was subsequently demonstrated in other animal species (33), and goitrogenic activity was reported for a wide range of vegetable foods, including virtually all cruciferous plants (17, 33, 82, 153, 212). During this time Kennedy, Purves, and others in New Zealand showed that rapeseed goiter arises from interference with the process of thyroid hormonogenesis and, unlike cabbage goiter, is only partially controlled by supplemental iodine (84). Kennedy (122) found allyl thiourea to be a potent goitrogen and suggested that it might be the active agent of *Brassica* seeds. This proved to be incorrect, and within a few years Astwood and co-workers (18) isolated and identified a new compound named "goitrin" (L-5-Vinyl-2-thiooxalidone) from rutabaga and showed that the goitrogenic activity of *Brassica* seeds could largely be accounted for by the presence of this compound in combined form (progoitrin).

Goitrin occurs in the edible portions of most *Brassica* species but is not responsible for the whole goitrogenicity of all foods. Other thioglycosides with antithyroid activity have been found in cruciferous plants (21), and similar activity has been associated with the presence of cyanogenetic glucosides in white clover (*Trifolium repens*) (68). The latter owe their potency to the conversion of the HCN into thiocyanate in the tissues. Thiocyanate is a goitrogenic agent which acts by inhibiting the selective concentration of iodine by the thyroid (237). Its action is reversible by iodine, whereas that of goitrin which acts by limiting hormonogenesis in the gland, is either not reversible by such means or is only partly so. The number of goitrogens of this latter type is large. Most of them possess the thiouryelene radical —NH—C=(S)—NH—. All of them act by inhibiting the iodination of tyrosine, presumably through inhibition of the thyroidal peroxidase, and appear to affect the conversion of mono- to diiodotyrosine (DIT) to a greater extent than the iodination of tyrosine to form monoiodotyrosine (MIT) (110), thus leading to an increased MIT/DIT ratio in the thyroid.

Goitrogens are widely employed in the treatment of thyrotoxicosis in man and have found some use in livestock and poultry husbandry through their favorable influence on the fattening process. Those in clinical use include methyl and propyl thiouracil, carbimazole, and methimazole. The derivatives of aminobenzene, phenylbutazone and paraaminosalicylates, are also used. Iodide itself, in large doses, can act as an antithyroid agent in thyrotoxicosis, and this form of therapy was common before goitrogenic drugs became available. Goitrogens are also of considerable interest as eitiological factors in the production of nontoxic goiter in man and animals. Where the diet is composed wholly or largely of goitrogenic foods such as cabbage or kale, especially when consumed raw, the amounts of antithyroid material ingested can be sufficient to induce goiter, even when iodine intakes are normal. Examples of such diets, forced upon some communities during World War II and accompanied by an increased incidence of goiter, have been described (44, 81). Reference has been made earlier to the frequent reports of goiter in farm stock intensively fed on kale. Where such foods are consumed by man in normal quantities and are cooked, goiter is much less likely to develop because the cooking of vegetables destroys the enzyme which liberates this substance from its inactive precursor (18). However, Greer and associates (83) have shown that fresh human and rat feces can hydrolyze pure progoitrin into goitrin, which suggests that goitrin may be formed in the gastrointestinal tract even from cooked vegetables.

The goitrogenicity of plants varies with the species and with the conditions under which they have grown, including the fertilizer treatment (3). This could be due to an effect upon the level of iodine in the plant, as well as upon the levels and types of goitrogen present, because the goitrogenicity even of *Brassica* species is due in part to the presence of thiocyanate and other goitrogens, the effects of which can be overcome by a sufficient increase in iodine intake. In fact, there is some evidence, at least with kale, that such goitrogens can contribute a major proportion of the goitrogenicity. These findings emphasize the necessity of maintaining adequate iodine intakes in all animal and human populations exposed to the hazards of goitrogenic foods, especially where environmental supplies of this element are otherwise low or marginal.

The actual significance of natural goitrogens in the production of human goiter is far from clear. Clements and co-workers (46, 47, 76) working in Australia, have shown that milk from cows consuming certain cruciferous plants in the grazing, contains a potent goitrogen, the effect of which on children could not be overcome by feeding 10 mg KI weekly. This suggests that the goitrogen is of the goitrin or thiouracil type. Goitrogens have also been found in the milk of cows fed cruciferous fodder in England (124) and in Finland (172). These interesting and important observations deserve a great deal further study and clinical evaluation.

## VIII. Iodine Toxicity

Very high intakes of iodine when ingested over prolonged periods are of interest because of their effects on the thyroid itself and on the clearance of radioiodine and in relation to their more general pathological effects, particularly upon the reproductive processes in the female. Prolonged administration of large doses of iodine to normal individuals markedly reduces thyroidal iodine uptake. This is the antithyroidal or goitrogenic effect, referred to earlier, which can be used in the treatment of thyrotoxicosis. Iodide goiter and hypothyroidism rarely occur from this cause, although there is an area in Japan in which the consumption of large quantities of iodine with the diet is reported to be the cause of endemic goiter (220). Several reports of a similar effect upon individuals, under exceptional circumstances, have appeared [see Wayne *et al.* (233)]. Large doses of stable iodine will similarly reduce radioiodine uptake by the thyroid and decrease retention of this radionuclide, an effect of particular interest in relation to protection against radioactive fallout involving $^{131}$I. Thus Driever and co-workers (60) found that previous administration to calves of stable KI decreased whole-body retention of $^{131}$I, given subsequently, by approximately one-half in 23 days.

The effects of iodine doses many times greater than those likely to be obtained from ordinary diets have been examined in a series of experiments with several species carried out by Arrington and his co-workers. Significant species differences in tolerance to high intakes of iodide were shown to exist. In all species studied the tolerance was high, i.e., relative to normal dietary iodine intakes, pointing to an extremely wide margin of safety for this element. Thus adult female rats fed 500, 1000, 1500, and 2000 ppm of iodine, as KI, from zero to 35 days prepartum, revealed increasing neonatal mortality of the young with increasing levels of iodine, but the effects of the lowest level of supplemental iodine fed (500 ppm) were slight when compared with those receiving no supplemental iodine (6). Histological examination of mammary gland tissue from females fed iodine indicated that milk secretion was absent or markedly reduced. The fertility of male rats fed 2500 ppm iodine from birth to 200 days of age appeared to be unimpaired (6). In subsequent studies with rats, rabbits, hamsters, and pigs (11), rabbits fed 250 ppm. of iodine or more for 2 to 5 days in late gestation showed significantly higher mortality of the young than controls receiving no supplemental iodine. No such effects were apparent in swine even at an intake of 2500 ppm iodine (see Table 37). Hamsters were similarly unaffected except for a slightly reduced feed intake and a decreased weaning weight of the young.

In a comparable series of experiments with poultry, Arrington *et al.* (12, 150, 173) have demonstrated profound effects upon egg production and hatchability. When laying hens were fed 312 to 5000 ppm iodine as KI in a practical laying ration, egg production ceased within the first week at the highest level and was reduced at the lower levels (173). The fertility of the eggs produced was not affected but early embryonic death, reduced hatchability, and delayed hatching resulted. Within 7 days after cessation of iodine feeding the hens resumed egg production, indicating that the adverse effects of the excess iodine are only temporary. The effects upon lactation and neonatal mortality in the young in rats and rabbits, described in the preceding paragraph, are also only temporary. Similar results were obtained in a subsequent experiment with pullets and hens fed supplementary iodine in amounts ranging from 625 to 5000 ppm for 6 weeks, but the effects were much smaller for sexually mature pullets than for the mature hens (12). Furthermore, molting accompanied the cessation of lay in some of the mature hens but was not observed in the pullets. Mature ova were present in the hens not laying but ovulation did not occur.

The mechanism by which the excess iodine affects egg production and embryonic mortality, or reproduction in female rats and rabbits, is not understood. Preliminary experiments conducted by Marcilese *et al.* (150) indicate that thyroxine production is not impaired in hens fed high

TABLE 37

*Tolerance of Pregnant Sows to High Dietary Intakes of Iodine as KI*[a,b]

| Dietary iodine | No. sows | Av. pigs per litter | % Surviving | | Av. Body Weight | |
|---|---|---|---|---|---|---|
| | | | 3 days | 14 days | Birth | 14 days |
| Control | 3 | 11.6 | 87 | 84 | 1.45 | 3.45 |
| 1500 ppm[b] | 2 | 14.5 | 79 | 79 | 1.18 | 2.91 |
| 2500 ppm[b] | 3 | 9.7 | 82 | 79 | 1.36 | 3.45 |

[a]From Arrington *et al.*, (11).

[b]Fed for 30 days before farrowing.

levels of iodine. However, the growing ova were shown to have a marked ability to concentrate iodine from the high doses administered—a finding in keeping with the earlier demonstration of the specific incorporation of orally administered radioiodine into hens' eggs and follicles (226). Thus iodine in the eggs of hens given 100 mg iodine daily, as NaI, increased linearly for 10 days and reached a plateau of 3 mg/egg at that time. When hens were given 500 mg iodine daily the level of iodine in the eggs increased rapidly to an average of 7 mg/egg by 8 days, at which time most hens ceased production. Ova continued to develop in hens not laying and many ova were found to be regressing (150). It was suggested that when a threshold amount of iodine reaches the ova, development ceases and regression takes place.

## REFERENCES


1. N. M. Alexander, *J. Biol. Chem.* **234**, 1530 (1959); *Endocrinology* **68**, 671 (1961).
2. W. D. Alexander, S. Papadopoulos, R. McG. Harden, S. MacFarlane, D. K. Mason, and E. Wayne, *J. Lab. Clin. Med.* **67**, 808 (1966).
3. R. Allcroft and F. J. Salt, *Advan. Thyroid Res., Trans. Int. Goitre Conf., 4th 1960* p. 4 (1961).
4. R. Allcroft, J. Scarnell, and S. L. Hignett, *Vet. Rec.* **66**, 367 (1954).
5. R. T. Allman and T. S. Hamilton, *FAO Agr. Stud.* No. 5 (1948).
6. C. B. Ammerman, L. R. Arrington, A. C. Warnick, J. L. Edwards, R. L. Shirley, and G. K. Davis, *J. Nutr.* **84**, 107 (1964).
7. M. Anbar, S. Guttman, and Z. Lewitus, *Nature* (*London*) **183**, 1517 (1959).
8. R. K. Anderson and H. L. Alt, *Amer. J. Physiol.* **119**, 67 (1937).
9. E. D. Andrews and D. P. Sinclair, *Proc. N. Z. Soc. Anim. Prod.* **22**, 123 (1962).
10. F. N. Andrews, C. L. Shrewsbury, C. Harper, C. M. Vestal, and L. P. Doyle, *J. Anim. Sci.* **7**, 298 (1948).
11. L. R. Arrington, R. N. Taylor, C. B. Ammerman, and R. L. Shirley, *J. Nutr.* **87**, 394 (1965).
12. L. R. Arrington, R. A. Santa Cruz, R. H. Harms, and H. R. Wilson, *J. Nutr.* **92**, 325 (1967).
13. P. W. Aschbacher, *J. Anim. Sci.* **27**, 127 (1968).
14. P. W. Aschbacher, R. G. Cragle, E. W. Swanson, and J. K. Miller, *J. Dairy Sci.* **49**, 1042 (1966).
15. P. W. Aschbacher and V. J. Feil, *J. Dairy Sci.* **51**, 762 (1968).
16. P. W. Aschbacher, J. K. Miller, and R. G. Cragle, *J. Dairy Sci.* **46**, 1114 (1963).
17. E. B. Astwood, *Ann. Intern. Med.* **30**, 1087 (1949).
18. E. B. Astwood, M. A. Greer, and M. G. Ettlinger, *J. Biol. Chem,* **181**, 121 (1949); M. A. Greer, M. G. Ettlinger, and E. B. Astwood, *J. Clin. Endocrinol.* **9**, 1069 (1949).
19. A. A. Axelrad, C. P. Leblond, and J. Isler, *Endocrinology* **56**, 387 (1955).
20. I. Bach, S. Braun, T. Gati, P. Kertai, J. Sos, and A. Udvardy, *Advan. Thyroid Res., Trans. Int. Goitre Conf., 4th, 1960* p. 505 (1961).
21. H. S. Bachelard and V. M. Trikojus, *Nature* (*London*) **185**, 80 (1960).
22. S. B. Barker, *Physiol. Rev.* **31**, 205 (1951).

23. E. J. Baumann, *J. Physiol. Chem.* **21**, 319 (1896); **22**, 1 (1896).
24. J. Benotti and N. Benotti, *Clin. Chem.* **9**, 408 (1963).
25. S. A. Berson, *Amer. J. Med.* **20**, 653 (1956).
26. S. A. Berson and R. S. Yalow, *J. Clin. Invest.* **33**, 1533 (1954).
27. B. B. Bilvais, *Physiol. Zool.* **20**, 67 (1947).
28. W. T. Binnerts, *Nature (London)* **174**, 973 (1954).
29. P. Blanquet, M. Croizet, M. Moura, and F. Dumara, *Ann. Endocrinol.* **28**, 293 (1967).
30. K. L. Blaxter, *Vitam. Horm. (New York)* **10**, 217 (1952).
31. R. I. Blokhima, *Proc. 1st Int. Symp. Trace Element Metab. Anim., 1969* p. 426, (C. F. Mills, ed.), Livingstone, Edinburgh, 1970.
32. I. J. B. Blom, *Onderstepoort J. Vet. Sci. Anim. Ind.* **2**, 139 (1934).
33. F. Blum, *Schweiz. Med. Wochenschr.* **70**, 1301 (1942); **72**, 1046 (1943).
34. R. Bogard and D. T. Mayer, *Amer. J. Physiol.* **147**, 320 (1946).
35. H. J. M. Bowen, *Biochem. J.* **73**, 381 (1959).
36. J. R. Bronk, *Ann. N.Y. Acad. Sci.* **86**, 494 (1960).
37. K. Brown-Grant, *Physiol. Rev.* **41**, 189 (1961).
38. K. Brown-Grant and V. A. Galton, *Biochim. Biophys. Acta* **27**, 423 (1958).
39. T. C. Bruice, N. Kharasch, and R. J. Winzler, *Arch. Biochem. Biophys.* **62**, 305 (1956).
40. G. Bussolati and A. G. E. Pearse, *J. Endocrinol.* **37**, 205 (1967).
41. N. F. Buzanov and A. A. Popova, *in* "Problems in Agricultural Engineering and the Breeding of Sugar Beets," p. 45, Moscow, 1958.
42. A. Canzanelli, R. Guild, and D. Rapport, *Endocrinology* **25**, 707 (1939).
43. A. Chatin, *C. R. Acad. Sci.* **30-39**, (1850-1854).
44. A. M. Chesney, T. A. Clawson, and B. Webster, *Bull. Johns Hopkins Hosp.* **43**, 261 (1928).
45. D. S. Childs, F. R. Keating, J. E. Rall, M. M. Williams, and M. H. Power, *J. Clin. Invest.* **29**, 726 (1950).
46. F. W. Clements, *Brit. Med. Bull.* **16**, 133 (1960).
47. F. W. Clements and J. W. Wishart, *Metab., Clin. Exp.* **5**, 623 (1956); F. W. Clements, *Med. J. Aust.* **2**, 645 (1957).
48. Y. Coble, J. Davis, A. Schulert, F. Heta, and A. Y. Awad, *Amer. J. Clin. Nutr.* **21**, 277 (1968).
49. R. A. Cohen and R. W. Gerard, *J. Cell. Comp. Physiol.* **10**, 223 (1937).
50. B. N. Cohn, *Arch. Intern. Med.* **49**, 950 (1932).
51. J. F. Coindet, *Ann. Chim. Phys.* **15**, 49 (1820).
52. V. V. Cole and G. M. Curtis, *J. Nutr.* **10**, 493 (1935).
53. C. W. Cooper and A. H. Tasjian, *Endocrinology* **79**, 819 (1966).
54. R. D. Creek, H. E. Parker, S. M. Hauge, E. N. Andrews, and C. W. Carrick, *Poultry Sci.* **33**, 1052 (1954).
55. G. M. Curtis, I. D. Puppel, V. V. Cole, and N. L. Matthews, *J. Lab. Clin. Med.* **22**, 1014 (1937).
56. D. P. Cuthbertson, *in* "Progress in the Physiology of Farm Animals," Vol. I, p. 56. Butterworth, London and Washington, D.C. 1954.
57. W. M. Davidson and C. J. Watson, *Sci. Agr.* **28**, 1 (1948).
58. L. J. De Groot and A. M. Davis, *Endocrinology* **70**, 492 (1962).
59. J. T. Dowling, N. Freinkel, and S. H. Ingbar, *J. Clin. Endocrinol. Metab.* **16**, 280 (1956).

60. C. W. Driever, J. E. Christian, W. F. Bousquet, M. P. Plumlee, and F. N. Andrews, *J. Dairy Sci.* **48**, 1088 (1965).
61. A. W. Elmer, "Iodine Metabolism and Thyroid Function," Oxford Univ. Press, London and New York, 1938.
62. A. M. Ermans, J. Kinthaert, C. Delcroix, and J. Collard, *J. Clin. Endocrinol. Metab.* **28**, 169 (1968).
63. J. M. Evvard, *Endocrinology* **12**, 539 (1928).
64. I. R. Falconer, *Nature (London)* **205**, 703 (1965).
65. L. P. Farrell and M. H. Richmond, *Clin. Chim. Acta* **6**, 620 (1961).
66. T. von Fellenberg, *Ergeb. Physiol.* **25**, 176 and 284 (1926).
67. K. A. Ferguson, P. G. Schinckel, H. B. Carter, and W. H. Clarke, *Aust. J. Biol. Sci.* **9**, 575 (1956).
68. D. S. Flux, G. W. Butler, J. M. Johnson, A. C. Glenday, and G. B. Peterson, *N. Z. J. Sci. Technol., Sect. A* **38**, 88 (1956).
69. R. H. Follis, Jr., *Amer. J. Clin. Nutr.* **14**, 253 (1964).
70. R. Fraser, Q. J. G. Hobson, D. G. Arnott, and E. W. Emery, *Quart. J. Med.* **22**, 99 (1953).
71. A. Fyfe, *Edinburgh Phil. J.* **1**, 254 (1819).
72. A. M. Gawienowski, D. T. Mayer, and J. F. Lasley, *J. Anim. Sci.* **14**, 3 (1955).
73. L. Gedalia, J. Gross, S. Guttmann, J. E. Steiner, F. G. Sulman, and M. M. Weinreb, *Arch. Int. Pharmacodyn. Ther.* **129**, 116 (1960).
74. R. F. Glascock, *J. Dairy Res.* **21**, 318 (1954).
75. E. Glazner and C. S. Shaffner, *Poultry Sci.* **28**, 834 (1949).
76. H. B. Gibson, J. F. Howeler, and F. W. Clements, *Med. J. Aust.* **5**, 875 (1960).
77. P. R. Godfrey, C. W. Carrick, and F. W. Quackenbush, *Poultry Sci.* **32**, 394 (1953).
78. L. Goldemberg, *J. Physiol. Pathol. gen.* **28**, 566 (1930).
79. A. H. Gordon, J. Gross, D. O'Connor, and R. Pitt-Rivers, *Nature (London)* **169**, 19 (1952).
80. F. B. de Gorge and N. K. José, *Nature (London)* **214**, 491 (1967).
81. M. A. Greer, *Physiol. Rev.* **30**, 513 (1950).
82. M. A. Greer and E. B. Astwood, *Endocrinology* **43**, 105 (1948).
83. M. A. Greer, S. Iiono, S. Barr, and H. Whallon, *Advan. Thyroid Res., Trans. Int. Goitre Conf., 4th, 1960* p. 1 (1961).
84. W. E. Griesbach, T. H. Kennedy, and H. D. Purves, *Brit. J. Exp. Pathol.* **22**, 249 (1941); W. E. Griesbach and H. D. Purves, *ibid.* **24**, 174 (1943); H. D. Purves, *ibid.* p. 171.
85. J. Gross and C. P. Leblond, *J. Biol. Chem.* **171**, 309 (1947).
86. J. Gross and R. Pitt-Rivers, *Lancet* **1**, 439 (1952).
87. A. Grossmann and G. F. Grossmann, *J. Clin. Endocrinol. Metab.* **15**, 354 (1955).
88. H. J. Grosz and B. B. Farmer, *Nature (London)* **222**, 875 (1969).
89. R. Guillemin, E. Yamazaki, D. A. Gard, M. Jutisz, and E. Sakiz, *Endocrinology* **73**, 564 (1963).
90. G. P. Gurevich, *Vop. Pitan.* **18**, 65 (1959); *Nutr. Abstr. Rev.* **30**, 697 (1960).
91. G. P. Gurevich, *Izv. Akad. Nauk SSSR, Ser. Biol.* No. 5, 791 (1962); *Fed. Proc., Fed. Amer. Soc. Exp. Biol.* **23**, Trans. Supp. 511 (1964).
92. B. L. Hallman, P. K. Bondy, and M. A. Hagewood, *Arch. Intern. Med.* **87**, 817 (1951).
93. M. W. Hamolsky and A. S. Freedberg, *N. Engl. J. Med.* **262**, 23, 70 and 129 (1960).
94. R. McG. Harden and W. D. Alexander, *Proc. Roy. Soc. Med.* **61**, 647 (1968).

95. R. McG. Harden, W. D. Alexander, J. Shimmins, H. Kostalas, and D. K. Mason, *J. Lab. Clin. Med.* **71**, 92 (1968).
96. R. McG. Harden, W. D. Alexander, J. Shimmins, and J. Robertson, *Quart. J. Exp. Physiol.* **53**, 227 (1968).
97. R. McG. Harden, J. S. Chisholm, J. Shimmins, and W. D. Alexander, *J. Clin. Endocrinol. Metab.* **28**, 117 (1968).
98. R. McG. Harden, T. Hilditch, I. Kennedy, D. K. Mason, S. Papadopoulos, and W. D. Alexander, *Clin. Sci.* **32**, 49 (1967).
99. R. McG. Harden, D. K. Mason, and W. D. Alexander, *Quart. J. Exp. Physiol.* **51**, 130 (1966).
100. R. McG. Harden, D. K. Mason, and W. W. Buchanan, *J. Clin. Endocrinol. Metab.* **25**, 957 (1965); *J. Lab. Clin. Med.* **65**, 500 (1965).
101. C. R. Harington, "The Thyroid Gland, Its Chemistry and Physiology," Oxford Univ. Press, London and New York, 1933.
102. C. R. Harington and G. Barger, *Biochem. J.* **21**, 169 (1927).
103. H. C. Heinrich, E. E. Gabbe, and D. H. Whang, *Atomkernenergie* **9**, 279 (1964).
104. C. A. Hellwig, *Arch. Surg. (Chicago)* **40**, 98 (1940).
105. S. L. Hignett, *J. Brit. Dairy Farmers' Ass.* **56**, 5 (1952).
106. E. H. Hipsley, *Med. J. Aust.* **1**, 532 (1956).
107. F. L. Hoch, *N. Engl. J. Med.* **266**, 446 and 498 (1962).
107a. J. C. M. Holman and W. McCartney, *World Health Organ., Mon. Ser.* **44**, (1960).
108. B. H. Hoogstratten, N. C. Leone, J. L. Shupe, D. A. Greenwood, and J. Lieberman, *J. Amer. Med. Ass.* **192**, 26 (1965).
109. R. G. Hoskins, *J. Amer. Med. Ass.* **55**, 1724 (1910).
110. S. Iino, *Acta Endocrinol. (Copenhagen)* **36**, 212 (1961)., quoted by Wayne *et al.* (233).
111. S. H. Ingbar, *Endocrinology* **63**, 256 (1958).
112. "Iodine Content of Foods," Chilean Iodine Educ. Bur., London, 1952.
113. "Iodine and Plant Life," Chilean Iodine Educ. Bur., London, 1950.
114. C. H. G. Irvine, *Amer. J. Vet. Res.* **28**, 1687 (1967).
115. J. M. Johnson and G. W. Butler, *Physiol. Plant.* **10**, 100 (1957).
116. M. Jovanovic, V. Pantic, and B. Markovic, *Acta Vet. (Belgrade)* **3**, 31 (1953).
117. F. R. Keating and A. Albert, *Recent Progr. Horm. Res.* **4**, 429 (1949).
118. F. R. Keating, M. H. Power, J. Berkson, and S. F. Haines, *J. Clin. Invest.* **26**, 1138 (1947).
119. F. C. Kelly, *Bull. W. H. O.* **9**, 217 (1953).
120. F. C. Kelly and W. W. Snedden, *World Health Organ., Mon. Ser.* **44**, (1960).
121. E. C. Kendall, *J. Biol. Chem.* **39**, 125 (1919).
122. T. H. Kennedy, *Nature (London)* **150**, 233 (1942).
123. G. K. Kiesel and M. J. Burns, *Amer. J. Vet. Res.* **21**, 226 (1960).
124. R. Kilpatrick, G. D. Broadhead, C. J. Edmonds, D. S. Munro, and G. M. Wilson, *Advan. Thyroid Res., Trans. In. Goitre Conf., 4th, 1960* p. 273 (1961).
125. M. Kirchgessner, *Z. Tierphysiol., Tierernaehr. Futtermittelk.* **14**, 270 and 278 (1959).
126. D. A. Koutras, *in* "Activation Analysis in the Study of Mineral Metabolism in Man," I.A.E.A., Teheran, 1968.
127. M. M. Kovalev, *Collect. Sci. Pap. Chernovtsy Med. Inst.* No. 10, p. 327 (1959)., cited by Novikova (165).
128. V. V. Kovalsky, N. P. Shergin, and A. D. Gololobov, *Proc. 8th All-Union Congr. Physiol. Biochem. & Pharmacol.*, p. 306 (1955).
129. J. P. Kriss, W. H. Carnes, and R. T. Gross, *J. Amer. Med. Ass.* **157**, 117 (1955).

130. H. O. Kunkel, R. W. Colby, and C. M. Lyman, *J. Anim. Sci.* **12**, 3 (1953).
131. H. O. Kunkel, G. G. Green, J. K. Riggs, R. L. Smith, and M. C. Shrode, *J. Anim. Sci.* **16**, 1030 (1957).
132. J. Kupzis, *Z. Hyg. Infektionskr.* **113**, 551 (1932).
132a. A. K. Lascelles and B. P. Setchell, *Aust. J. Biol. Sci.* **12**, 455 (1959).
133. C. P. Leblond, *Rev. Can. Biol.* **1**, 402 (1942).
134. J. J. Lehr, J. M. Wybenga, and M. Rosanov, *Plant Physiol.* **33**, 421 (1958).
135. E. J. Lennon, N. H. Engbring, and W. W. Engstrom, *J. Clin. Invest.* **40**, 996 (1961).
136. H. D. Lennon and J. P. Mixner, J. Dairy Sci. **40**, 357 and 541 (1957).
137. J. Lerman and R. Pitt-Rivers, *J. Clin. Endocrinol. Metab.* **15**, 653 (1955).
138. H. Levine, R. E. Remington, and H. von Kolnitz, *J. Nutr.* **6**, 347 (1933).
139. R. C. Lewis, *J. Dairy Sci.* **44**, 2265 (1961).
140. R. C. Lewis and N. P. Ralston, *J. Dairy Sci.* **36**, 33 and 363 (1951).
141. W. T. London, D. A. Koutras, A. Pressman, and R. L. Vought, *J. Clin. Endocrinol. Metab.* **25**, 1091 (1965).
142. J. F. Long, L. O. Gilmore, G. M. Curtis, and D. C. Rife, *J. Dairy Sci.* **35**, 603 (1952).
143. J. K. Loosli, R. B. Becker, C. F. Huffman, P. H. Phillips, and J. C. Shaw, "Nutrient Requirements of Dairy Cattle." Natl. Res. Council, Washington, D.C., 1956.
144. A. A. Maglione, C. J. Collica, and S. Rubenfeld, *Amer. J. Roentgenol., Radium Ther. Nucl. Med.* [N.S.] **97**, 896 (1966).
145. B. Malamos and D. A. Koutras, *Proc. 7th Int. Congr. Intern. Med.*, Vol. 2, p. 678 (1962).
146. B. Malamos, D. A. Koutras, P. Kostamis, A. C. Kralios, G. Rigopoulos, and N. Zerefos, *J. Clin. Endocrinol. Metab.* **26**, 688 (1966).
147. E. B. Man and P. K. Bondy, *J. Clin. Endocrinol. Metab.* **17**, 1373 (1957).
148. E. B. Man, M. Heinemann, C. E. Johnson, D. C. Leary, and J. P. Peters, *J. Clin. Invest.* **30**, 157 (1951).
149. M. Maqsood, *Biol. Rev.* **27**, 281 (1952).
150. N. A. Marcilese, R. H. Harms, R. M. Valsecchi, and L. R. Arrington, *J. Nutr.* **94**, 117 (1968).
151. D. Marine and W. Williams, *Arch. Intern. Med.* **1**, 349 (1908); D. Marine and C. H. Lenhart, *ibid.* **3**, 66 (1909).
152. E. Maurer and H. Dugrue, *Biochem. Z.* **193**, 356 (1928).
153. R. McCarrison, *Indian J. Med. Res.* **20**, 957 (1933).
153a. R. J. McDonald, G. W. McKay, and J. D. Thomson, *Proc. 4th Int. Congr. Anim. Reprod., Artif. Insem., 1961* Vol. 3, p. 679 (1962).
154. M. T.McQuillan, P. G. Stanley, and V. M. Trikojus, *Aust. J. Exp. Biol. Med. Sci.* **6**, 617 (1953).
155. R. J. Meyer, intern. rep., Morton Salt Co. (unpublished) (1968).
156. R. Michel, *Amer. J. Med.* **20**, 670 (1956).
157. J. E. Miles, J. N. Miles, C. Walters, and J. Metcalfe, *J. Clin. Invest.* **47**, 1851 (1968).
158. J. K. Miller, B. R. Moss, E. W. Swanson, P. W. Aschbacher, and R. G. Cragle, *J. Dairy Sci.* **51**, 1831 (1967).
159. H. H. Mitchell and F. J. McClure, *Bull. Nat. Res. Counc. (U.S.)* **99**, (1937).
160. S. Mittler and G. H. Benham, *J. Nutr.* **53**, 53 (1954).
161. R. Moberg, *Proc. 3rd World Congr. Fert. Steril.*, p. 71 (1959).
162. R. Moberg, *Proc. 4th Int. Congr. Anim. Reprod., Artif. Insem., 1961* Vol. 3, p. 682 (1962).
163. M. V. Musset and R. Pitt-Rivers, *Lancet* **2**, 1212 (1954).
164. J. T. Nicoloff and J. T. Dowling, *J. Clin. Invest.* **47**, 2000 (1968).

165. E. P. Novikova, *Vop. Pitan.* **22**, 45 (1963); *Fed. Proc., Fed. Amer. Soc. Exp. Biol.,* **23,** Transl. Suppl., 459 (1964).
166. *Nutr. Rev.* **26**, 77 (1968).
167. J. B. Orr and I. Leitch, *J. Roy. Agr. Soc. Engl.* **87**, 43 (1926).
168. J. B. Orr and I. Leitch, *Med. Res. Counc.* (*Gt. Brit.*), *Spec. Rep. Ser.* **123**, (1929).
169. A. Oswald, *Z. Physiol. Chem.* **27**, 14 (1899).
170. K. R. Paley, E. S. Sobel, and R. S. Yalow, *J. Clin. Endocrinol. Metab.* **15**, 995 (1955); **18**, 850 (1958).
171. S. Papadopoulos, S. MacFarlane, R. McG. Harden, D. K. Mason, and W. D. Alexander, *J. Endocrinol.* **36**, 341 (1966).
172. P. Peltola, *Advan. Thyroid Res., Trans. Int. Goitre Conf., 4th, 1960* p. 10 (1961).
173. J. T. Perdomo, R. H. Harms, and L. R. Arrington, *Proc. Soc. Exp. Biol. Med.* **122**, 758 (1966).
174. P. H. Phillips, H. E. English, and E. B. Hart, *Amer. J. Physiol.* **113**, 441 (1935); *J. Nutr.* **10**, 399 (1935).
175. D. E. Pickering, N. E. Kontaxis, R. C. Benson, and R. J. Meecham, *Amer. J. Dis. Child.* **95**, 616 (1958).
176. V. J. Pileggi, N. D. Lee, O. J. Golub, and R. J. Henry, *J. Clin. Endocrinol. Metab.* **21**, 1272 (1961).
177. R. Pitt-Rivers and J. R. Tata, "The Thyroid Hormones." Pergamon Press, Oxford, 1959.
178. T. B. Post, *Aust. J. Agr. Res.* **14**, 572 (1963).
179. T. B. Post and J. P. Mixner, *J. Dairy Sci.* **44**, 2265 (1961).
180. B. N. Premachandra, G. W. Pipes, and C. W. Turner, *J. Anim. Sci.* **17**, 1237 (1958).
181. J. Prevost (1830); cited by B. T. Bowery, *Bull. W.H.O.* **9**, 175 (1953).
182. J. E. Rall, J. Robbins, and C. G. Lewallen, *in* "The Hormones" ( G. Pincus, K. V. Thimann, and E. B. Astwood, eds.), Vol. 5, p. 159. Academic Press, New York, 1964.
183. R. L. Rapport and G. M. Curtis, *J. Clin. Endocrinol.* **10**, 735 (1950).
184. W. S. Raveno, *J. Mich. State Med. Soc.* **33**, 359 (1944).
185. R. P. Reineke and C. W. Turner, *J. Dairy Sci.* **27**, 793 (1944).
186. D. S. Riggs, *Pharmacol. Rev.* **4**, 282 (1952).
187. J. Roche, R. Michel, O. Michel, and S. Lissitzky, *Biochim. Biophys. Acta* **9**, 161 (1952).
188. J. Roche, R. Michel, J. Nunez, and W. Wolf, *C. R. Soc. Biol.* **149**, 884 (1955).
189. J. Roche, R. Michel, R. Truchot, and W. Wolf, *C. R. Soc. Biol.* **149**, 1219 (1955).
190. J. Roche, R. Michel, and W. Wolf, *C. R. Acad. Sci.* **240**, 251 and 921 (1955).
191. J. C. Rogler, *Diss. Abstr.* **18**, 1925 (1958).
192. A. L. Romanoff and A. J. Romanoff, "The Avian Egg." Wiley, New York, 1949.
193. D. A. Ross and R. S. Schwab, *Endocrinology* **25**, 75 (1939).
194. W. T. Salter, *in* "The Hormones" (G. Pincus and K. V. Thimann, eds.), Vol. 2, p. 181. Academic Press, New York, 1950.
195. W. T. Salter and McA. W. Johnson, *J. Clin. Endocrinol.* **8**, 924 (1948).
196. W. T. Salter and I. Rosenblum, *J. Endocrinol.* **7**, 180 (1950).
197. L. Sceffer, *Biochem. Z.* **259**, 11 (1933).
198. M. Scott, *Trans. 3rd Int. Goitre Conf.*, p. 34 (1938).
199. M. L. Scott, A. van Tienhoven, E. R. Holm, and R. E. Reynolds, *J. Nutr.* **71**, 282 (1960).
200. N. S. Scrimshaw, *Fed. Proc., Fed. Amer. Soc. Exp. Biol.* **17**, Suppl. 2, 59 (1958).
201. N. S. Scrimshaw, *Pub. Health Rep.* **75**, 731 (1960).

202. B. P. Setchell, D. A. Dickinson, A. K. Lascelles, and R. B. Bonner, *Aust. Vet. J.* **36**, 159 (1960).
203. A. Shand, *Brit. Vet. Ass. Publ.* No. 23 (1952).
204. G. R. Sharpless and M. Metzger, *J. Nutr.* **21**, 341 (1941).
205. J. C. Shee, *Ir. J. Med. Sci.* p. 752 (1939).
206. A. C. Shuman and D. P. Townsend, *J. Anim. Sci.* **22**, 72 (1963).
207. D. P. Sinclair and E. D. Andrews, *N. Z. Vet. J.* **6**, 87 (1958); **7**, 39 (1959); **9**, 96 (1961).
208. O. N. Singh, H. A. Henneman, and E. P. Reineke, *J. Anim. Sci.* **15**, 625 (1956).
209. E. M. Smith, H. N. Wagner, J. M. Mozley, and R. Reba, *Clin. Res.* **10**, 403 (1962).
210. L. J. Soffer, J. L. Gabrilove, and J. W. Jailer, *Proc. Soc. Exp. Biol. Med.* **71**, 117 (1949).
211. A. A. Spielman, W. E. Petersen, J. B. Fitch, and B. S. Pomeroy, *J. Dairy Sci.* **28**, 329 (1945).
212. V. Srinovasan, N. R. Moudgal, and P. S. Sarma, *J. Nutr.* **61**, 87 (1957).
213. M. M. Stanley, *J. Clin. Endocrinol.* **9**, 941 (1949).
214. M. M. Stanley and E. B. Astwood, *Endocrinology* **44**, 49 (1949).
215. K. Sterling and M. A. Bremner, *J. Clin. Invest.* **45**, 153 (1966).
216. D. G. Steyn, *Rep. S. Afr. Med. Ass., 1938.*
217. P. Stocks, *Biometrika* **19**, 272 (1927); *Ann. Eugen.* **2**, 382 (1927).
218. V. Stolc and J. Podoba, *Nature (London)* **188**, 855 (1960).
219. A. Sturm and B. Buchholz, *Arch. Klin. Med.* **161**, 227 (1928).
220. H. Suzuki, T. Higuchi, K. Sawa, S. Ohtaki, and Y. Horiuchi, *Jap. Acta Endocrinol. (Copenhagen)* **50**, 161 (1965).
221. A. Taurog, F. N. Briggs, and I. L. Chaikoff, *J. Biol. Chem.* **191**, 29 (1951).
222. A. Taurog and I. L. Chaikoff, *J. Biol. Chem.* **163**, 313 (1946).
223. A. Taurog, I. L. Chaikoff, and D. D. Feller, *J. Biol. Chem.* **171**, 189 (1947).
224. A. Taurog, J. D. Wheat, and I. L. Chaikoff, *Endocrinology* **58**, 121 (1956).
225. S. Taylor, *J. Clin. Endocrinol.* **14**, 1412 (1954).
226. C. B. Thorell, *Acta Vet. Scand.* **5**, 224 (1964).
227. J. R. Todd and R. H. Thompson, *Res. Vet. Sci.* **3**, 449 (1962).
228. C. W. Turner, M. R. Irwin, and E. P. Reineke, *Poultry Sci.* **24**, 171 (1945); C. W. Turner and H. L. Kempster, *ibid.* **27**, 453 (1948).
229. E. J. Underwood, "The Mineral Nutrition of Livestock." FAO/CAB Publ. Central Press, Aberdeen, 1966.
229a. L. Van Middlesworth, *J. Clin. Endocrinol. Metab.* **16**, 989 (1956).
230. A. Van Zyl and J. E. Kerrich, *S. Afr. J. Med. Sci.* **20**, 9 (1955).
231. I. G. Vazhenin and V. I. Belyakova, *in* "Trace Elements in Plant and Animal Life," p. 121. Moscow, 1952.
232. R. L. Vought, W. T. London, L. Lutwak, and T. D. Dublin, *J. Clin. Endocrinol. Metab.* **23**, 1218 (1963).
233. E. J. Wayne, D. A. Koutras, and W. D. Alexander, "Clinical Aspects of Iodine Metabolism." Blackwell, Oxford, 1964.
233a. M. H. Weeks, J. Katz, and N. G. Farnham, *Endocrinology* **50**, 511 (1952).
234. C. D. West, V. J. Chavré, and M. Wolfe, *J. Clin. Endocrinol. Metab.* **25**, 1189 (1965).
235. H. S. Wilgus, Jr., F. X. Gassner, A. R. Patton, and G. S. Harshfield, *Colo., Agr. Exp. Sta., Bull.* **49** (1953).
236. D. C. Wilson, *Lancet* **1**, 211 (1941).
237. J. Wolff, I. L. Chaikoff, A. Taurog, and L. Rubin, *Endocrinology* **39**, 140 (1946).

238. J. Wolff and J. R. Maury, *Biochim. Biophys. Acta* **47**, 467 (1961).
239. J. Wolff, R. H. Thompson, and J. Robbins, *J. Clin. Endocrinol. Metab.* **24**, 699 (1964).
240. E. Wright and E. D. Andrews, *N. Z. J. Sci. Technol., Sect. A* **37**, 83 (1955).
241. W. E. Wright, J. E. Christian, and F. N. Andrews, *J. Dairy Sci.* **38**, 31 (1955).
242. J. B. Wyngaarden, B. M. Wright, and P. Ways, *Endocrinology* **50**, 1537 (1952).
243. M. Young, M. G. Crabtree, and I. M. Mason, *Med. Res. Counc. (Gt. Brit.), Spec. Rep. Ser.* **217**, (1936).

# 12

# SELENIUM

## I. Introduction

Interest in the biological significance of selenium was initially confined to its toxic effects upon animals. The toxicity of this element was first demonstrated in 1842 (99), but nearly a century passed before the existence of naturally occurring selenosis was established. Two diseases of livestock, known as "blind staggers" and "alkali disease," occurring in parts of the Great Plains of North America, were then identified as manifestations of acute and chronic selenium poisoning, respectively (12, 166). These discoveries stimulated investigations of the distribution of selenium in rocks, soils, plants, and animal tissues at toxic levels, with a view to determining minimum harmful levels and of developing practical means of prevention and control.

Attention became focused upon the physiological rather than the toxicological role of selenium in 1957, when Schwarz and Foltz (180) showed that the liver necrosis that develops in rats fed certain diets can be prevented by selenium supplements, and two groups of workers demonstrated that this element also prevents exudative diathesis in chicks fed similar diets (155, 179). Very soon after, Muth and co-workers (141) in Oregon, and McClean and associates (129) in New Zealand, discovered that the muscular dystrophy, which occurs naturally in lambs and calves in those areas, is a manifestation of selenium deficiency and can be prevented by selenium therapy. Subsequently, naturally occurring selenium deficiency areas, in which the growth, health, and fertility of animals were impaired, were demonstrated in many countries, especially in New Zealand (89) and in the Scandinavian countries (9, 67). It soon became apparent that the total areas of the world affected by selenium deficiency are far greater than those afflicted by selenium excess.

These important discoveries were followed by studies of selenium metabolism in animals in physiological quantities and of the functional relationship of this element to vitamin E. Highly sensitive analytical methods were developed for the estimation of the minute amounts of selenium involved (1, 17, 79, 115, 203) and the metabolic movements of the element were investigated further with the aid of radioactive $^{75}$Se. Many of the disorders of animals responsive to selenium were found also to be responsive to vitamin E, so that the interrelationships between the two nutrients became a fertile field for study. Conclusive evidence that selenium is a nutritionally essential element, with a role beyond that of a substitute for a normal intake of vitamin E, finally emerged from the critical work of Thompson and Scott (196) with chicks.

Neither selenium deficiency nor selenium poisoning has been clearly established in human populations, even in areas where the deficiency or where selenosis occurs naturally in livestock, although some evidence has been produced that selenium deficiency can be a complicating factor in certain types of kwashiorkor (21, 119, 176). An association between above normal levels of selenium and the incidence of human dental caries has also been advanced (74).

Many years ago it was reported that certain Se accumulator plants may require this element for healthy growth (197). Convincing evidence that selenium is essential for other plants, or for microorganisms has not yet appeared. However, selenite was reported to be essential for the production of formic dehydrogenase in *Escherichia coli* (159) and may, therefore, be concerned in protein synthesis.

## II. Selenium in Animal Tissues and Fluids

### 1. *General Distribution*

Selenium occurs in all the cells and tissues of the body in concentrations that vary with the tissue and with the level of selenium in the diet. The kidney and particularly the kidney cortex, is by far the highest in selenium concentration, followed by the glandular tissues, especially the pancreas, and the pituitary, and by the liver. The muscles, bones, and blood are normally relatively low and adipose tissue very low in selenium. Cardiac muscle is consistently higher in selenium than skeletal muscle (47, 82a). Burk and co-workers (22) observed a decline in whole carcass selenium, from 0.2 ppm to approximately half this level, in rats fed a torula yeast diet for a period of 4 weeks. The mean levels in the kidneys of these animals fell from 1.0 to 0.3 ppm Se, and in the liver from 0.7 to 0.1 ppm Se (fresh basis). Similar levels occur in these tissues in normal and Se-deficient sheep (23, 32, 88, 92, 157, 158) and pigs (115) (see Table 38). The kidney and the liver are the most sensitive indicators of the selenium status of the animal, and the Se concentrations in these organs can provide valuable diagnostic criteria. Andrews *et al.* (10) suggest that selenium levels of less than 0.25 ppm (fresh basis) in the kidney cortex, and of 0.02 ppm in the liver, are indicative of marked selenium deficiency in the sheep. Concentrations greater than 1.0 ppm Se in the kidney cortex and 0.1 ppm in the liver are considered normal, and one-half the quantity of these levels indicates a marginal degree of Se-responsive unthriftiness.

Few data on the selenium levels in normal human tissues, other than blood, have appeared. Dickson and Tomlinson (38) examined autopsy specimens of the liver, skin, and muscles of 10 adults with the following results: liver, range 0.18-0.66, mean 0.44; skin, range 0.12-0.62, mean 0.27; muscle, range 0.26-0.59, mean 0.37 $\mu$g Se/g of whole tissue. Analysis of a wider range of tissues for two individuals, one infant and one adult, revealed the highest selenium concentrations in the kidney and the thyroid and the lowest in the fat.

At toxic intakes of selenium, i.e., 10-100 times or more greater than those normally ingested, tissue selenium concentrations rise steadily until levels as high as 5-7 ppm in liver and kidneys and 1-2 ppm in the muscles are reached. Beyond these tissue levels, excretion begins to keep pace with absorption (32, 131, 158, 193). Selenium is, therefore, not continuously cumulative in the tissues. Even higher levels may be attained in the hair and hooves of severely affected animals. Olson *et al.* (152) found the hair of yearling cattle on seleniferous range to average over 10 ppm Se, with individual values as high as 30 ppm, compared with 1 to 4 ppm Se for the hair of cattle from normal areas. They concluded that hair values consistently below 5 ppm indicate that the diet is unlikely to contain sufficient selenium to induce clinical signs of selenosis. The tissue selenium levels just cited for selenotic animals are well beyond those that arise in animals treated at recommended rates to prevent selenium deficiency. Food products from such treated animals are unlikely to contain undesirably or dangerously high levels of selenium (10, 32, 82a, 109, 153).

TABLE 38

*Mean Selenium Concentrations in the Tissues of Normal Pigs and of Pigs with Nutritional Muscular Dystrophy*[a,b]

| Pigs (No.) | Kidney | Liver | Skel. muscle | Heart | Pancreas | Spleen | Lung |
|---|---|---|---|---|---|---|---|
| Healthy pigs (6)[c] | 11.47 ± 1.18 | 1.82 ± 0.16 | 0.52 ± 0.06 | 1.05 ± 0.10 | 1.42 ± 0.14 | 1.26 ± 0.09 | 1.13 ± 0.13 |
| NMD pigs (6)[d] | 2.48 ± 0.29 | 0.20 ± 0.05 | 0.16 ± 0.08 | 0.19 ± 0.09 | 0.24 ± 0.13 | 0.40 ± 0.10 | 0.25 ± 0.11 |

[a] From Lindberg (115).
[b] Concentrations in parts per million (dry basis).
[c] Food averaged 0.126 ppm Se.
[d] Food averaged 0.021 ppm Se.

## 2. *Selenium in Blood*

The concentration of selenium in the blood is so responsive to changes in the selenium level of the diet that it is difficult to determine if significant differences due to the species or age of the animal exist. Burk *et al.* (22) found the level in the whole blood of rats fed a torula yeast diet to fall rapidly from a mean of 0.3 μg/ml to approximately 0.05 μg/ml in 4 weeks. The latter value is identical with that considered by Hartley (88) to be "satisfactory" for sheep. In sheep suffering from various Se-responsive diseases, much lower levels, of the order of 0.01 to 0.02 μg Se/ml were reported. Similar values for the blood of normal and of Se-deficient sheep have been reported in other studies (32, 92, 108, 115, 157). In two experiments reported by Kuchel and Buckley (108), the concentration of selenium in the whole blood of sheep grazing pastures of normal selenium status ranged from 0.06–0.20 (mean 0.10) μg/ml and to 0.04–0.08 (mean 0.06) μg/ml. The administration of selenium pellets induced a rapid rise in blood selenium to levels as high as 0.15–0.25 μg/ml, depending on the amount of selenium in the pellets (see Fig. 19). The blood of sheep dying from experimentally induced Se toxicity was reported by Rosenfeld and Beath (168) to contain 1.34–3.1 ppm Se, an order of magnitude higher than the levels just cited. Comparable high

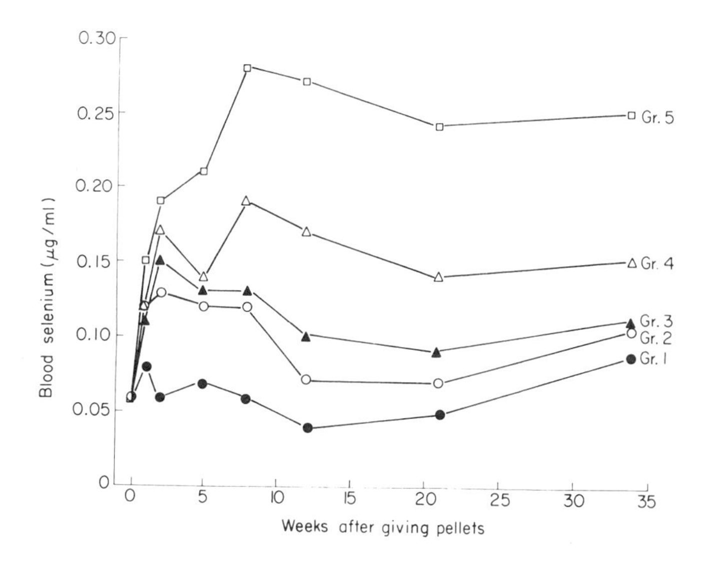


**Fig. 19.** Blood selenium levels in sheep following administration of heavy selenium pellets. (●) Fe only; (○) 1.25% Se; (▲) 2.5% Se; (△) 5% Se; (□) 10% Se. [From Kuchel and Buckley (108).]

blood selenium levels have been observed in sheep fed toxic amounts of selenium (158).

The normal levels of selenium in human blood have been studied in several countries. Allaway and co-workers (6) examined samples from 210 male donors in 19 sites in U.S.A. The Se concentrations ranged from 0.10 to 0.34 μg/ml, with a mean of 0.21 μg/ml. The mean Se concentration varied among different locations, with some evidence of a geographic pattern reflecting established regional differences in the Se levels in crops. These values are similar to those of Burk *et al.* (21) for U.S. residents. Those from northeastern United States are similar to the levels reported by Dickson and Tomlinson (38) for an adjacent area in Canada. Bowen and Cawse (17) obtained somewhat higher blood Se levels in England, whereas Brune and co-workers (19) reported a lower mean level (0.12 μg/g) for a small group of normal Swedish individuals. Whether these small variations represent real regional differences or are a reflection of differences in analytical procedures is unknown.

In a study of whole-blood Se levels in kwashiorkor, Burk *et al.* (21) obtained mean levels of 0.08 and 0.11 μg/ml in two groups of affected children, compared with 0.14 and 0.22 μg/ml, respectively, in control children or recovered patients. The most likely explanation of the low blood Se levels in kwashiorkor children is that they are anemic and hypoproteinemic and, therefore, circulate reduced levels of Se-containing proteins.

Little information is available on the distribution of selenium between cells and plasma in different selenium states. In a study of the distribution of selenium in the tissues of normal and dystrophic chicks, Taussky *et al.* (195) reported levels of 0.1 to 0.2 μg Se/g in the serum and of 0.25 to 0.35 μg/g in the red cells. These concentrations are similar to those found by Burk and co-workers (21) in control children and in children recovered from kwashiorkor. Their mean values were 0.15 and 0.10 μg Se/ml for plasma and 0.36 and 0.23 μg/ml for red blood cells, respectively. Dickson and Tomlinson (38) also found human red blood cells to be higher in selenium than plasma but the difference was not nearly as large.

### 3. *Selenium in Milk*

The concentration of selenium in milk varies greatly with the selenium intake of the animal. Hadjimarkos and Bonhorst (78) found significant differences between locations in cow's milk collected from farms in three counties of Oregon. In two of these counties the mean levels in the milk were 0.049 and 0.067 ppm; in the third, a known low-Se area, the mean level was only 0.005 ppm. Similarly, Allaway *et al.* (6) reported that cow's milk from a low-Se area in Oregon contained less than 0.02 μg Se/ml, compared with 0.05 μg/ml for milk from a high-Se area in South Dakota. Grant and Wilson (66) demonstrated substantial increases, over a period of 3 to 4 weeks, in the selenium level of milk from cows in New Zealand receiving a single oral or subcutaneous dose of 50 mg Se as sodium selenate. The levels in the milk from untreated cows remained steady at 0.003 to 0.004 μg Se/ml. Gardner and Hogue (59a) trebled the selenium concentration of the milk of ewes fed a low-Se diet by supplementing this diet with 2.25 mg Se/day as sodium selenite. Comparable increases in the selenium content of milk were achieved by Paulson *et al.* (157) from selenium supplementation of the diet. A considerable proportion of the total selenium in milk, probably 70-75%, occurs in the protein fractions, with the remainder presumably present as selenite, selenate, or free selenoamino acids (103, 120, 122).

The only study of selenium in human milk appears to be that of Hadjimarkos (72). Neutron activation analysis of samples from 15 healthy mothers revealed selenium concentrations ranging from 0.013 to 0.053 ppm, with a mean of 0.021 ppm.

### 4. *Selenium in the Avian Egg*

The data of Taussky *et al.* (195) indicate that a normal hen's egg averages a total of 10 to 12 μg of selenium, most of which is present in the yolk. Hadjimarkos and Bonhorst (78) also found a high proportion of the total selenium to be present in the yolk. The yolk-white Se concentration ratio averaged 6.3:1. These workers obtained indications that the amount present varies markedly with the Se intake of the hen. Thus the mean whole-egg Se concentration was close to 0.4 and 0.5 ppm in samples from two counties in Oregon of normal Se status, compared with only 0.056 ppm in eggs from a known Se-low area.

At toxic intakes of selenium by hens, extremely high levels can occur in the eggs. In these circumstances, higher levels are present in the white than in the yolk. For instance, Moxon and Poley (134) found that increasing the Se content of the ration from 2.5 to 10.0 ppm raised the amount of selenium in the yolks from 3.6 to 8.4 ppm and that in the whites from 11.3 to 41.3 ppm (dry basis).

## III. Selenium Metabolism

The extent to which selenium is absorbed from the gastrointestinal tract and its retention and distribution within the body vary with the species and with the chemical form and amount of the element ingested. The amounts, forms, and routes of excretion of selenium are also affected by these factors and can be greatly influenced by the presence of other elements, notably arsenic. The level of tocopherol in the diet does not appear to affect significantly the pattern of selenium absorption or retention (22, 47, 95).

At toxic or near-toxic levels, selenium is rapidly and efficiently absorbed from naturally seleniferous diets and from soluble salts of the element added to normal diets. Rats consuming a seleniferous wheat diet, containing 18 ppm Se, retained 63% of the ingested selenium within the first week and a similar proportion from the same level of selenium intake as sodium selenite (8, 131). Evidence from toxicity studies with rats suggests a somewhat higher absorption from seleniferous grains than from selenites and selenates, and a very low absorption from selenides and elemental selenium (53, 193). Some organic compounds, including selenodiacetic and selenopropionic acids, are distinctly less toxic to rats, per unit of selenium, than is selenite, presumably as a consequence of lower absorption (133). Different chemical forms of selenium also vary considerably in their capacity to prevent liver necrosis in rats and exudative diathesis in chicks, but it cannot be assumed that these variations merely reflect differences in absorbability.

Studies with $^{75}Se$ at physiological levels indicate that the duodenum is the main site of absorption of selenium and that there is no absorption from the rumen or abomasum of sheep, or the stomach of pigs (214). In the studies of Wright and Bell (214), total net absorption represented approximately 35% of the ingested isotope in sheep and 85% in pigs, when rations containing 0.35 and 0.50 ppm Se, respectively, were consumed. Evidence that monogastric animals have a higher intestinal absorption of selenium than ruminants has been obtained in other studies (24, 32). Since fecal selenium in sheep is mostly present in insoluble forms, it seems that selenite is reduced to insoluble compounds in the rumen (24, 32). Ehlig and co-workers (47) showed that Se retention in lambs is greater for selenomethionine than for $Na_2SeO_3$. This was due to a higher rate of urinary excretion from the selenite than from the selenomethionine treatments rather than to differences in apparent rates of absorption from the two selenium sources.

Radioselenium is more efficiently retained from Se-deficient than from Se-supplemented diets by chicks (101), rats (22, 95), and sheep (116, 213). This increased retention no doubt reflects increased intestinal absorption in response to greater tissue demands for selenium, but the pattern of excretion is also affected by the level of selenium intake. Lopez *et al.* (116) measured the whole-body retention of selenium, urine and fecal losses, and retention in selected tissues following administration of $^{75}Se$ to lambs fed varying levels of dietary selenium. Whole-body loss of $^{75}Se$, 48-336 hr after administration of the isotope, could be described by a first-order rate constant which was inversely proportional to the dietary level of selenium. The concentration of $^{75}Se$ in various tissues was also inversely related to the dietary selenium level. Hopkins and co-workers (95) demonstrated a similar inverse relationship in whole-carcass retention of injected $^{75}Se$ in rats which had previously been fed Se-low and Se-high diets.

Absorbed selenium is at first carried mainly in the plasma (20), apparently in association with the plasma proteins (124, 193), from which it enters all the tissues, including the bones, the hair, and the erythrocytes (20, 32, 135, 193). The intracellular distribution of radioselenium varies with the tissue and with the level of selenium. Thus Wright and Bell (213) found $^{75}Se$ to be distributed rather uniformly among the particulates and soluble fraction in liver, whereas nearly 75% of the activity in kidney cortex was found in the nuclear fraction. Alterations in intracellular particulate matter distribution of $^{75}Se$ occur with changing levels of dietary selenium from low to high, and from high to low (213, 215). McConnell and Roth (125) maintain that, of the subcellular fractions, the microsomes appear to be the initial site for incorporating selenium into protein.

It has long been accepted that selenium replaces sulfur in sulfur-containing compounds in the tissues and occurs predominantly as selenocystine and selenomethionine in both protein-bound and non-protein-bound forms (123). McConnell and Wabnitz (127) first reported that these two selenium amino acids were present in dog liver protein, and Rosenfeld (167) reported their presence in wool hydrolyzates. Selenium-75 is also incorporated into the $\alpha$- and $\beta$-lipoproteins of rat and dog serum, with the greater proportion in the $\alpha$-lipoprotein (124). Cummins and Martin (33) have produced new evidence which challenges the whole concept of sulfur replacement by selenium. Their findings suggest that in the rabbit there is no pathway for the *in vivo* synthesis of selenocystine or selenomethionine from selenite and that the selenium-containing compounds associated with fractions containing cystine or methionine merely consist of selenite bound to the sulfur compounds.

Selenium is incorporated into leukocytes (121) as well as erythrocytes. Enhanced *in vitro* uptake of $^{75}Se$ by the red blood cells has been observed in Se-deficient sheep (116, 212) and in children suffering from kwashiorkor (21). McConnell and Roth (125) found that the rate of disappearance of $^{75}Se$ from whole blood, plasma, and red cells in dogs conformed to a multiple-component rate function. Following the administration of subtoxic amounts of $^{75}Se$, the isotope was detected in various blood proteins for as long as 310 days. The greatest rate of disappearance of radioactivity from the red cells was at 100 to 120 days after the initial injection. This suggests that once selenium enters these cells it remains there throughout their life-span. Selenium is also incorporated into myoglobin, cytochrome c, the muscle enzymes, myosin and aldolase, and nucleoproteins (123).

Most of the selenium deposited in the tissues is highly labile. Following transfer of animals from seleniferous to nonseleniferous diets (8, 193), or following injections of stable or radioactive selenium (15, 116, 219), the retained selenium is lost from the tissues, at first rapidly and then more slowly. Selenium is excreted in the feces, the urine, and the expired air, the amounts and proportions depending upon the level and form of the intake, the nature of the rest of the diet, and the species. Exhalation of selenium is an important route of excretion at high dietary intakes of the element (57, 122) but is much less so at low intakes (57, 82a). Increasing the protein and methionine contents of the diet increases the pulmonary excretion of injected $^{75}Se$ (57). Injections of arsenic, mercury, and thallium (105, 111, 154a), and of cadmium (56), also increase such excretion. Lead (111) and zinc (154a) have no such effect on selenium volatilization.

Fecal excretion of ingested selenium is generally greater than urinary excretion in ruminants (24, 32, 156) but not in monogastric species (122, 193). With injected selenium, the urine is a major pathway of excretion in sheep (116, 211), as well as in nonruminant species. The effect of route of administration and of increasing dietary levels of selenium upon the pattern of $^{75}Se$ excretion in lambs is shown in Fig. 20. Most of the selenium in the feces consists of selenium which has not been absorbed from the diet, together with small amounts excreted into the bowel with the biliary, pancreatic, and intestinal secretions (112, 171). Levander and Baumann (112) showed that excretion of selenium into the gastrointestinal tract via the bile fluid is markedly increased, and retention in the carcass, liver, and blood greatly decreased, when subacute injections of arsenic are given with the selenium. Injections of mercury, thallium, and lead had no such effect on the biliary excretion of selenium (111). In these studies the As injections did not alter urinary selenium excretion. Palmer *et al.* (154) have reported that As injections, prior to the selenium, reduce the amount of selenium excreted in the urine of rats. A new selenium metabolite identified as trimethylsenelonium ion was isolated from the urine of rats injected with $^{75}Se$ selenite. Substantial quantities of this substance, amounting to 30 to 50% of urinary Se/24 hr and considered to be a normal excretory product of selenite Se, were excreted after a single dose of selenite (154).

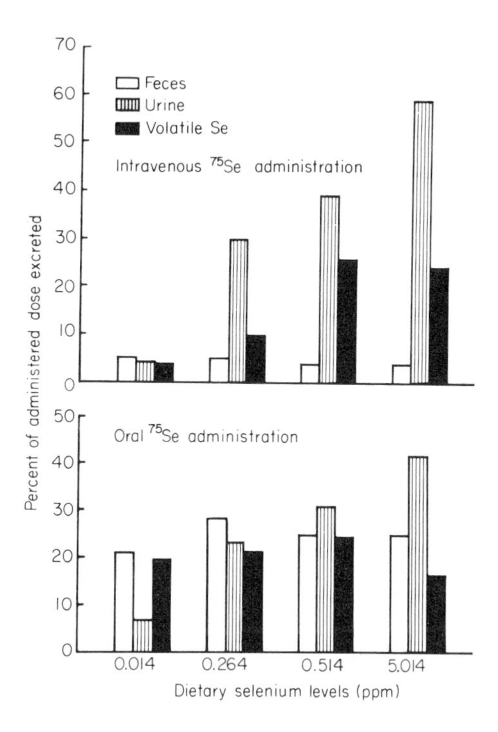


**Fig. 20** Excretion of $^{75}$Se by lambs fed varying selenium intakes after administration of the isotope during a 12-day period. [From Lopez, Preston, and Pfander (116).]

Selenium metabolism can be altered appreciably by the presence of sulfate. This is apparent from experiments with $^{75}$Se with laboratory animals, and from studies demonstrating an ameliorating effect of sulfate upon selenium toxicity and an exacerbating effect upon selenium deficiency. The urinary excretion of selenium following a parenteral dose of sodium selenate was increased nearly threefold in rats given sulfate parenterally and in the diet (56). Similarly, 2% sodium sulfate in a diet containing 5 ppm Se as selenate increased the urinary excretion of selenium over a 16-day period to 68% of that ingested, compared with 51% in the absence of added sulfate (80). Sulfate has only a slight effect on the urinary excretion of selenium administered in the form of selenite (56). The specificity of the sulfate effect is also apparent in chronic selenium toxicity, where sulfate is very much more effective against selenate than against selenite or seleniferous grain (80, 81). The addition of sulfate to the diet decreases the effectiveness of selenium as selenite in preventing white muscle disease (WMD) in lambs and calves (93, 142), whereas the addition of cystine and methionine has no such effect (93, 206). In a recent study (206), sulfate had no influence upon the incidence of WMD in lambs, although it significantly increased the numbers with degenerative lesions of the heart. These results can probably be partially explained by differences in the forms of selenium ingested. When the selenium is mostly in organic combination, sulfate would not be expected to affect the metabolism of this selenium to the extent that it would affect its selenium analog, selenate, or even selenite. As Whanger *et al.* (206) have suggested, the forms of selenium and sulfur used must be taken into account in interpreting the results of sulfur interference with selenium metabolism.

Selenium is transmissible through the placenta to the fetus, whether supplied in inorganic or organic forms. This has been demonstrated in the mouse (83), rat (208), dog (126), and sheep (23, 116, 213). It is also evident from reports that selenium administration to the mother during pregnancy prevents WMD in lambs and calves. The placenta, nevertheless, presents something of a barrier to the transfer of selenium in inorganic forms. Several workers have shown that the concentration of $^{75}Se$ in the blood and most organs of the fetus, following injection of the ewe with $^{75}Se$ sodium selenite, is lower than in the mother (23, 97, 98, 213). Jacobson and Oksanen (98) showed further that when ewes are injected with $^{75}Se$-selenomethionine or $^{75}Se$-selenocystine, the $^{75}Se$ concentration in the lambs is higher than when selenite is injected, and is nearly as high as in the mother. Similar results with $^{75}Se$-selenomethionine have been obtained in mice (83), indicating that these forms of selenium more readily pass the placental barrier than do inorganic forms.

## IV. Selenium Deficiency and Functions

### 1. *Discovery of the Essentiality of Selenium*

Selenium is necessary for growth and fertility in animals and for the prevention of various diseases which show variable responses to vitamin E. The discovery of selenium as an essential element arose from the observation of Schwarz that rats fed a torula yeast diet develop a fatal liver necrosis which could be prevented by brewer's yeast, despite the absence of sufficient cystine or vitamin E to account for its protective action. This led Schwarz (175) to postulate the presence of a third anti-liver-necrosis factor, designated "Factor 3." In the course of procedures designed to identify the nature of Factor 3, a correlation between biopotency and selenium content was noticed, and supplements of selenite were shown to protect against liver necrosis in the same manner as Factor 3 itself (180). Scott and associates (185) had previously found that chicks fed torula yeast, vitamin E-deficient diets grew poorly, revealed a high mortality, and developed the condition known as exudative diathesis. Cooperative studies between these workers and Schwarz and his colleagues (179) disclosed a potency and distribution of the antiexudative diathesis factors similar to that of Factor 3. Supplements of selenite were found to be effective in promoting the growth of the chicks and in preventing exudative diathesis, as they prevented liver necrosis in rats. Meanwhile, Patterson *et al.* (155) made the significant independent observation that the ash of alkali-hydrolyzed kidney was potent against exudative diathesis, whereas the ash of acid-hydrolyzed kidney was ineffective. This suggested that the activity was due to an element that forms a volatile inorganic acid, such as arsenic, selenium, or tellurium. Selenite was then found to be highly effective in preventing the growth depression and mortality and the development of exudative diathesis.

Investigations in several parts of the world with farm animals subsequently disclosed a nutritional significance for selenium far beyond its relationship to the disorders that arise in rats and chicks fed specialized diets. Selenium therapy was found to be effective in the treatment of various myopathies, including white muscle disease (WMD) in lambs, calves, and foals (90, 129, 141, 163) and hepatosis dietetica in pigs (46, 67). Selenium was also shown to promote growth, improve fertility, and reduce postnatal losses in sheep in certain areas (89, 129). In all the above conditions, variable but usually less satisfactory responses could also be obtained with tocopherol administration. Neither the resorption sterility in rats (85) nor the encephalomalacia in chicks (35) that arises on vitamin E-deficient diets responds to selenium. Despite the experimental and field evidence of growth and fertility responses to selenium greater than could be achieved by tocopherol, unequivocal evidence that selenium is a dietary essential for growth, independent of, or additional to, its function as a substitute for vitamin E, was not obtained until Thompson and Scott (196) prepared purified diets with crystalline amino acids containing less than 0.005 ppm Se. Chicks consuming this diet had poor growth and high mortality even when 200 ppm of *d*-$\alpha$-tocopherol acetate was added. Higher levels of vitamin E prevented mortality but even with 1000 ppm growth was inferior to that obtained with selenium and no added vitamin E. These findings, together with those of McCoy and Weswig (128) and of Wu *et al.* (216) with rats, indicate clearly that selenium does not function merely as a substitute for vitamin E.

### 2. *Selenium and Muscular Dystrophy*

Nutritional muscular dystrophy is a degenerative disease of the striated muscles which occurs, without neural involvement, in a wide range of animal species. It was described in calves in Europe as early as the 1880s and has been observed as field occurrences in many countries in lambs, calves, foals, rabbits, and even marsupials [see Anderson (9)]. The disease was first produced experimentally by Goettsch and Pappenheimer (64) while studying the requirements of vitamin E for reproduction in rabbits and guinea pigs.

Extensive muscular dystrophy cannot be produced in calves with vitamin E-free diets unless these diets contain highly unsaturated fats (14) and only partial control of congenital muscular dystrophy can be secured with tocopherol administration (89, 129, 141, 163). The occurrence of muscular dystrophy is affected by the extent to which factors influencing the utilization of vitamin E are present in the diet, as well as by the intakes of vitamin E itself and of selenium. Codliver oil was the first of such factors found to aggravate vitamin E deficiency, due to its high content of unsaturated fatty acids (64). Muscular dystrophy has been produced in several species by feeding diets high in fats of this type (14, 47, 48, 205). The presence of nontocopherol antioxidants in the diet or of oxidation products of tocopherol, such as quinone and hydroquinone, may be significant in this respect, since such substances afford some protection against muscular dystrophy (14, 44, 47, 118). Selenium utilization by the animal may also be affected by the presence of succinoxidase inhibitors, first shown to be present in some feedstuffs by Hogue (94) and by Cartan and Swingle (29) and subsequently demonstrated in white clover in New Zealand (172). The extent to which such factors actually influence the incidence of Se-responsive muscular dystrophy (WMD) in particular areas is not yet clear. In some countries the incidence is low and sporadic. Less than 1% of the flock or herd may be affected, and in some seasons or years the disease may not appear at all. In other countries, notably in parts of New Zealand, of Turkey, and of Estonia (104), the incidence of WMD is higher and more consistent, unless appropriate treatment with selenium is undertaken. As many as 20-30% of the lambs born in some flocks can be affected.

White muscle disease rarely occurs in mature animals. In lambs it can occur at birth (congenital muscular dystrophy) or at any age up to 12 months. It is most common between 3 and 6 weeks of age. Lambs affected at birth usually die within a few days. The deep muscles overlying the cervical vertebrae are particularly affected with the typical chalky white striations. Lambs affected later in life (delayed muscular dystrophy) show a stiff and stilted gait and an arched back. They are disinclined to move about, lose condition, become prostrate, and die. Animals with severe heart involvement may die suddenly without showing any such signs. Clinical signs of WMD may not appear until the lambs are driven or moved about. Mildly affected animals may recover spontaneously.

These disabilities are associated with a noninflammatory degeneration, or necrosis, of varying severity, of the skeletal or the cardiac musculature, or of both. A bilaterally symmetrical distribution of the skeletal muscle lesions is characteristic of WMD in lambs. The symmetry frequently extends beyond a simple bilateral involvement of paired muscles to the distribution of lesions within the muscles (217). The lesions are usually most readily discernible in the thigh and shoulder muscles. The lesions in the cardiac muscle are commonly confined to the right ventricle but may occur in other compartments. They are seen either as subendocardial grayish-white plaques or as more diffuse lesions of a similar color extending up to about 1 mm into the myocardium (58). Alterations in the electrophoretic pattern of affected muscles are a consistent feature of the disease (217). Characteristic abnormalities in the electrocardiograms (ECG) have also been demonstrated by Godwin (62) in lambs with WMD. The changes in the ECG pattern develop early and become very marked as death approaches. Similar changes in ECG pattern have been observed in rats fed a torula yeast, low-Se diet (61) and in lambs fed similar diets (63).

White muscle disease is characterized further by subnormal concentrations of selenium in the blood and tissues (see Table 38) and by abnormally high levels of serum glutamic oxalacetic transaminase (SGOT) and of lactic dehydrogenase (LDH) (157a, 206, 207). In normal lambs and calves, SGOT activity rarely exceeds 200 units/ml and is usually only about half that level. However, in animals with pronounced WMD, SGOT concentrations 5-10 times higher (Table 39) are found. Blincoe and Dye (16) showed that the increase above normal is roughly proportional to the amount of muscle damage. Young *et al.* (218) maintain that SGOT determinations are useful aids in the diagnosis of WMD, especially in lambs. However, individual variability in SGOT activity is high

TABLE 39

*Serum Glutamic Oxalacetic Transaminase Activity in Lambs and Calves*[a]

| Parameters | Normal | | | White muscle disease | |
|---|---|---|---|---|---|
| | Lambs | | Calves | Lambs | Calves |
| Mean (units/ml) | 128 | 56 | 57 | 1890 | 1313 |
| SE | ±18 | ±31 | ±17 | ±254 | – |
| Range (units/ml) | 97-191 | 22-160 | 19-99 | 687-3460 | 295-2360 |
| No. of animals | 20 | 21 | 69 | 17 | 4 |

[a]From Blincoe and Dye (16).

in both healthy and diseased animals. Furthermore, as Oksanen (148) has pointed out, "the SGOT value may be only moderately increased in animals with marked clinical symptoms caused by extensive sub-acute or chronic degenerative processes" and "a degeneration in the myocardium, even an acute one, may also produce only a moderate increase in the SGOT value, although this often ends in sudden death." He, nevertheless, considers that SGOT values, properly applied, can be of considerable diagnostic significance, particularly in subclinical cases.

Although WMD in lambs and calves has received most attention, because of its economic importance and occurrence under natural conditions, similar degenerative changes occur in foals (89), in association with hepatosis dietetica in pigs (148), and with exudative diathesis in chicks (145). Scott (183) states that nutritional muscular dystrophy in chicks is characterized by degeneration of the skeletal muscle fibers, especially of the pectoral muscles. In all these species the disease can be prevented by selenium supplementation of the diet at appropriate levels. Muscular dystrophy in rabbits is not similarly responsive to selenium (43, 96).

### 3. *Exudative Diathesis in Chicks*

A disease described as exudative diathesis was first observed in chicks consuming vitamin E-free diets containing codliver oil (34). It first appears as an edema on the breast, wing, and neck and later has the appearance of massive subcutaneous hemorrhages, arising from abnormal permeability of the capillary walls and accumulation of fluid throughout the body. The greatest accumulation occurs under the ventral skin giving it a greenish-blue discoloration. The level of plasma proteins is low in the blood of affected chicks, and an anemia develops which is probably a consequence of the hemorrhages that occur (146). Growth rate is subnormal and mortality can be high. In the outbreaks of the disease that occur commercially, as a result of consuming wheat very low in selenium as the main dietary component, chicks are most commonly affected between 3 and 6 weeks of age. They become dejected, lose condition, show leg weakness, and may become prostrate and die (10, 89).

The addition of $\alpha$-tocopherol and certain antioxidants, such as ethoxyquin, to diets that induce exudative diathesis greatly reduces the incidence and severity of the disease in chicks and poults. The addition of selenium is completely effective in preventing all signs of the disease and promotes better growth than is obtained from tocopherol supplements alone (155, 173, 179). White muscle disease affecting particularly the breast muscles often develops concurrently with exudative diathesis, as mentioned earlier, and a congenital myopathy, characterized by the hatching of dead chicks or chicks dying 3-4 days after hatching, has been reported as a further Se-responsive condition in this species (173). Such chicks show extensive pale areas in the gizzard and sometimes also of the hind limb skeletal musculature. Selenium is thus concerned in the prevention of two of the three well-defined conditions which appear in chicks deprived of vitamin E, namely exudative diathesis and white muscle disease, but not encephalomalacia.

### 4. *Hepatosis Dietetica in Pigs*

Hepatosis dietetica is the name given to a disease that occurs in young pigs on diets low in vitamin E or selenium. The disease has been produced experimentally on vitamin E-free diets based on torula yeast (46) or soybean meal (67) and occurs spontaneously in New Zealand (10, 89) and in Scandinavia (148), when pigs are fed grain rations low in selenium. The disease is most common in fattening pigs at an age of 3 to 15 weeks. Mortality is high and deaths occur suddenly, often without premonitory signs or serious reduction in growth rate. Severe necrotic liver lesions are invariably apparent on postmortem examination, usually with some degree of myocardial and skeletal muscle degeneration. There is also deposition of ceroid pigment in adipose tissue, which gives a yellowish-brown color to the body fat, together with a generalized subcutaneous edema.

The mortality and liver lesions characteristic of hepatosis dietetica are accompanied by a marked depletion of tissue selenium concentrations (see Table 38) and an increase in the level of the liver-specific enzyme, ornithine carbamoyltransferase (OCT) in the blood (148). The degenerative changes that occur in the muscles in this disease do not affect the OCT level, so that determination of blood OCT values can be a useful aid in the diagnosis of hepatosis dietetica. Selenium supplements are completely effective in preventing the mortality and liver lesions. There is some evidence that selenium is not as effective as vitamin E in preventing the skeletal muscle degeneration or pigment deposition which occurs in this disease (67, 148).

### 5. *Selenium-Responsive Unthriftiness in Sheep and Cattle*

Animals affected by the Se-responsive diseases considered above usually grow poorly or fail to maintain satisfactory live weights. In parts of New Zealand a serious condition known as "ill-thrift" occurs in lambs at pasture and can occur in beef and dairy cattle of all ages, particularly in the autumn and winter months (10, 89). This Se-responsive unthriftiness varies from a subclinical growth deficit (89, 129) to clinical unthriftiness characterized by rapid loss of condition and sometimes mortality (42, 89). In affected lambs the fleece is harsh and dry but diarrhea is not a common feature as it is in affected calves. No characteristic microscopic lesions are apparent and there is no increase in SGOT levels. The unthriftiness is often associated with cobalt deficiency and intestinal parasitism and may or may not be associated with WMD and infertility. A further Se-responsive syndrome in lactating ewes, characterized by rapid loss in condition and scouring in association with heavy parasitic infestation has been described (130).

Ill-thrift can be prevented by selenium treatment, with striking increases in growth and wool yield in some instances (10, 42, 89, 103, 209). Significant increases in wool yield from Se supplements have also been observed in Canadian (189) and Scottish experiments (164). In one series of New Zealand trials, lambs 5 months old were divided into two groups, one of which remained as controls and the other received 5 mg Se/lamb, orally as selenite, at commencement and again at 2 and 6 weeks. The mortality was reduced from 27 to 8% by the treatment, and highly significant weight gains were observed in the selenium-treated groups. The remarkable increase in lamb production is illustrated in Fig.

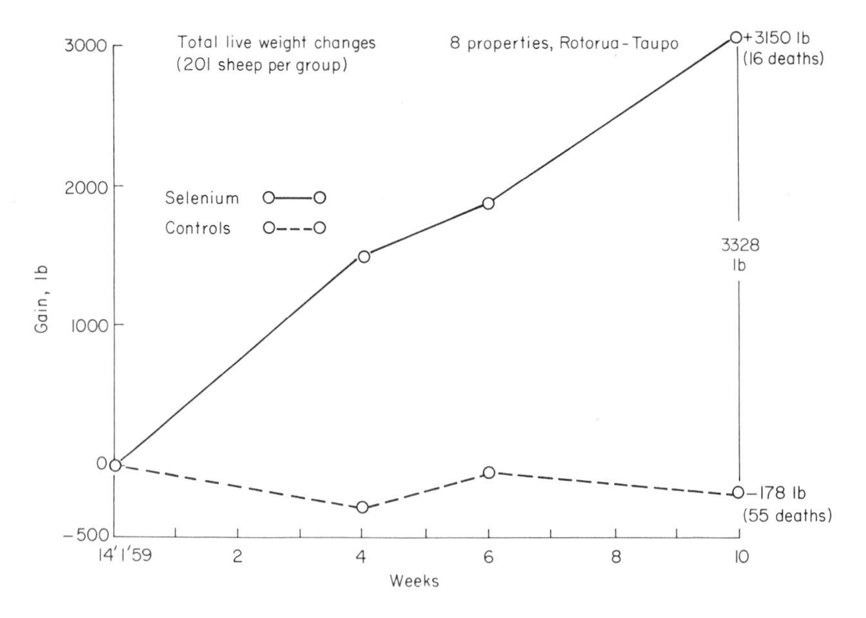


**Fig. 21.** Effect of treatment with selenium on lamb production. (Graph kindly supplied by Dr. W. J. Hartley, New Zealand.)

21, which depicts the change in total live weight by including deaths as total losses. Smaller oral doses, namely 1 mg Se at docking, 1-5 mg at weaning, and 1-5 mg at 3-month intervals thereafter, have since been found adequate in New Zealand. Neither vitamin E nor the antioxidant, ethoxyquin, has any effect on the unthriftiness (88).

Similar effects of selenium upon growth of lambs and calves occur in restricted areas in the western states of the United States (149), but in most parts of the world growth responses to selenium are less spectacular than those reported in New Zealand (13, 158). For instance, in trials carried out by Blaxter (13) on 76 farms involving 4448 lambs in northeast Scotland, 3-mg doses of selenium given at 4-week intervals for several months increased overall weight gain by only 0.4 kg/lamb. On 31 of those farms situated on soils with the lowest presumptive selenium content, the mean response was 0.8 kg/lamb. Variations in growth response from selenium treatment are probably largely accounted for by differences in the Se content of the herbage. Other environmental differences affecting Se utilization, or Se needs, almost certainly exist. The possibility that genetic differences among sheep in Se-responsiveness may significantly affect the position must also be considered (164).

### 6. *Periodontal Disease of Ewes*

This disease was first described in New Zealand (86, 172a) before its relationship to selenium was established. It appears in ewes 3-5 years of age and causes loss of condition resulting from difficulty in mastication. The disease is characterized by loosening and shedding of permanent molars, and sometimes also of incisors, in association with gingival hyperplasia, resorption, and replacement fibrosis of alveolar bone, alveolar infection, and bony exostoses on the adjacent part of the mandible or maxilla (86). Since the disease occurred in areas where various Se-responsive conditions were apparent, and virtually disappeared on affected properties after selenium dosing, effects of selenium administration were observed. It was found that selenium administration greatly reduces the incidence of periodontal disease but does not eliminate it completely. It seems, therefore, that other factors are also involved.

### 7. *Selenium and Fertility in Animals*

Infertility is not a prominent feature of selenium deficiency in farm animals in most areas where WMD occurs spontaneously. In parts of California annual losses occur due to the birth of premature, weak, or dead calves. These losses were shown to be completely prevented by injecting pregnant cows, 50-60 days before calving, with a sodium selenite-vitamin E mixture (117). In parts of New Zealand a high seasonal incidence of barren ewes occurs in association with WMD in lambs or with Se-responsive unthriftiness. In certain of these areas, 30% of ewes may be infertile and losses of lambs are high. Selenium administered to the ewes before mating dramatically increases the fertility, and further treatment with selenium before lambing reduces the lamb losses and incidence of WMD (87). The recommended doses are 5 mg Se (as $Na_2SO_3$ by mouth) 2-4 weeks before mating plus a similar dose 1 month before lambing.

Estrus, ovulation, fertilization, and early embryonic development proceed normally in affected flocks. Between 3 and 4 weeks after conception, i.e., at about the time of implantation, embryonic mortality is high (87). This mortality can be prevented by selenium, administered as just described, but not by either vitamin E or by an antioxidant. These effects are illustrated in Table 40. On the other hand, Buchanan-Smith and co-workers (19a) obtained satisfactory reproductive performance in ewes fed a Se-deficient, purified diet only when both selenium and vitamin E were administered in combination.

Selenium deficiency results in reduced hatchability of fertile eggs and reduced viability of newly hatched Japanese quail (100). Selenium deficiency, imposed over successive generations on rats fed a low-Se, torula yeast diet with adequate vitamin E, also severely affects fertility (128). The animals grew and reproduced normally but their offspring were almost hairless, grew more slowly, and failed to reproduce. Sterility was determined by lack of breeding with fertile rats and, in the males, by examining the male organs for spermatozoa. Immotile sperms, with separation of heads from tails, were found in 5 of the 8 males. No spermatozoa were located in the other untreated males. The condition was uninfluenced by supplements of methionine, sodium sulfate, or vitamins, whereas 0.1 ppm Se as $Na_2SO_3$ restored hair coat, growth, and repro-

TABLE 40

*Effects of Selenium, Vitamin E, and Antioxidant upon Ewe Fertility*[a,b]

| Fertility[b] | Untreated | Selenium[c] | Vitamin E | Antioxidant |
|---|---|---|---|---|
| Barren ewes (%) | 45 | 8 | 50 | 43 |
| Lambs born/100 ewes lambing | 105 | 120 | 109 | 112 |
| Lamb mortality (%) | 26 | 15 | 16 | 14 |
| Lambs marked/100 ewes lambing | 43 | 93 | 46 | 54 |

[a]From Hartley (87).

[b]Two hundred ewes per group.

[c]Treatment was confined to the mating period. Normal practice involves a further dose of Se (1 month before lambing) to reduce lamb mortality.

ductive capabilities. Wu and co-workers (216) showed further that the rats fed Se-deficient diets develop degenerative changes in the epididymis, in addition to impaired testicular growth and function. Indications were obtained that epididymal function, probably related to sperm maturation, was even more sensitive to selenium deficiency than the development of the testis and the functioning of the seminiferous tubules. The motility of spermatozoa was improved almost linearly with increasing amounts of selenium, from 0.01 to 0.08 ppm added to the basal diet. No such drastic effects upon testicular development or semen quality were observed in ram lambs fed a purified diet, very low in selenium, for 140 days (19a).

The remarkable protective action of selenium injections against the testicular necrosis induced by single subcutaneous doses of cadmium chloride is discussed in Chapter 10.

### 8. *Mode of Action of Selenium*

The functions of selenium in cellular metabolism are not yet clear. The demonstration that several selenium compounds can exert antioxidant effects *in vitro* (220), and the many known nutritional interrelationships between selenium and vitamin E, led to the hypothesis that the sole metabolic action of selenium is that of a nonspecific antioxidant providing protection against peroxidation in tissues and membranes (194).

Although selenium may act as an antioxidant in some cases, there is increasing evidence that the antioxidant theory is insufficient to account for the nutritional and metabolic functions of this element. Desai *et al.* (37) performed a time-sequence study of the interrelationships of peroxidation, lysosomal enzymes, and nutritional muscular dystrophy in the chick. They concluded that increased susceptibility of muscle lipids to peroxidation, and increased lysosomal enzyme activity, are not the primary abnormalities responsible for the onset of muscular dystrophy. However, Whanger *et al.* (207) have shown that two of the lysosomal enzymes, aryl sulfatase and $\beta$-glucuronidase, are significantly elevated in the skeletal muscles of myopathic (Se-deficient) lambs and that this elevation is accompanied by a significant reduction in the concentrations of LDH, GOT, and peroxidase in these muscles. They postulate the sequence of biochemical events in WMD in the following terms: "Due to conditions in the cells, possibly peroxidative or other related conditions, the lysosomes rupture releasing their enzymes into the cytoplasm. The hydrolytic nature of these enzymes results in destruction of the tissue cells releasing the cellular enzymes into the blood. This apparently is the reason for the frequently observed increase of plasma activity in myopathic animals and for the reduced enzyme concentration (LDH, GOT and peroxidase) in muscles of WMD lambs."

In a series of studies with rats, Green and associates (41, 68) showed that lipid peroxidation is not a significant process in the living animal and concluded that the biological role of vitamin E and selenium does not involve their prevention of lipid peroxidation *in vivo*. Selenium was shown to be present in rat liver fractions in at least three oxidation states, tentatively designated "selenide," "selenite," and "selenate." The amount of selenide present was dependent on the vitamin E status of the animal. They put forward the hypothesis that a function of vitamin E may be to provide the oxidoreductive environment necessary for the maintenance of the selenium in the reduced state and that the selenide may form a part of the active center of a new class of nonheme iron proteins. This hypothesis is consistent with the finding of Schwarz (177) that the biochemical lesion in vitamin E and Se-deficient rat liver is associated with mitochondrial oxidative changes. As Diplock (41) has pointed out, these changes might be caused by a defect in the nonheme Fe proteins of the electron transfer chain.

Green and co-workers (69) had previously suggested that selenium is concerned in the biosynthesis of ubiquinone (coenzyme Q), a benzoquinone derivative related to $\alpha$-tocopherol and involved in the respiratory chain. Deficiencies of tocopherol and selenium in the rat and the rabbit produced decreased tissue levels of ubiquinone, and oral administration of either substance resulted in marked increases in ubiquinone concentrations in these species and in birds (45, 69, 70). Although either tocopherol or selenium can control ubiquinone levels in most tissues of the rat, only tocopherol is effective in the uterus (71). These data are interpreted as indicating that (*a*) some of the disturbances in respiration, oxidation, and tissue differentiation in vitamin E-deficient states may be explicable in terms of the effects on ubiquinone levels, and (*b*) there are two mechanisms, one involving $\alpha$-tocopherol and one selenium, influencing ubiquinone synthesis. It is suggested that the existence of such alternate biochemical systems, and their unequal significance in different tissues, may account for the observed differences between the physiological effects of selenium and tocopherol in animals, both from species to species and at different periods during the life cycle of the same species.

Schwarz (178) has presented evidence indicating that the ability of selenium, sulfur amino acids, and tocopherol to halt the development of respiratory decline in rat liver slices can be rationalized by considering the effects of these substances on one enzyme system, $\alpha$-ketoglutarate oxidase. He considers that the respiratory defect that develops during the latent phase of the development of hepatic necrosis in rats may be due to a defect in the lipoyl dehydrogenase moiety of the $\alpha$-oxoglutarate dehydrogenase system. Evidence that selenium may act on hydrogen transport along the respiratory chain has come from another source. Giese *et al.* (60) investigated the activity of nicotinamide adenine dinucleotide (NAD), reduced NAD ($NADH+H^+$), NAD phosphate (NADP), and $NADP+H^+$ in the livers of chicks fed a semisynthetic diet, with and without supplements of vitamin E and selenite. Selenium was shown to have an inhibitory effect on NAD and NADP activity compared with all other treatments, thus causing a shift in concentration from the oxidized to the reduced coenzyme.

The recent studies of Scott and his associates (184, 196) with chicks have shed new light on the functions and requirements of selenium and their relationship to vitamin E. Chicks exhibited a requirement for selenium for growth, even when abundant amounts of $\alpha$-tocopherol were supplied. The nutritional role of selenium thus emerges as more than a substitute for a normal intake of vitamin E. In fact, it appears more likely that vitamin E merely reduces the requirement for selenium. The status of vitamin E as an essential nutrient, rather than of selenium, is, therefore, now open to question, at least in respect to growth and the prevention of exudative diathesis in chicks. Scott (184) has since reported that selenium affects the absorption and retention of vitamin E and of triglyceride. Tissue uptake of $^{14}C$-labeled dl-$\alpha$-tocopherol, and of its acetate, was low in chicks receiving an amino acid basal diet containing less than 0.005 ppm Se. In chicks receiving this diet with added selenite, the blood and tissue uptake of vitamin E was as much as 100 times greater. It was shown further that chicks receiving the basal low-Se diet were unable to absorb fat normally, and almost completely lacked the ability to hydrolyze triglyceride. After about 14 days on the deficient diet, neutral fat began to increase in the feces until, at about 20 days of age, most of the dietary fat was recovered, unhydrolyzed. The specific cause of the marked decrease in fat hydrolysis in selenium deficiency remains to be identified, but the decreased absorption and retention of vitamin E is thought to be a secondary phenomenon resulting from the failure of formation of lipid-bile salt micelles in the intestinal tract and possibly from a low production of the particular plasma lipoproteins concerned in vitamin E transport. These revealing studies indicate that selenium is required in metabolic processes that are not protected by vitamin E and that one of the important functions of this vitamin may be to protect the traces of selenium in the animal body. On the other hand, Cheeke and Oldfield (30) found no evidence that selenium influences the absorption or retention of vitamin E in the rat, although the plasma vitamin E level was affected. They suggest that selenium may modify the distribution of this vitamin in the tissues.

## V. Selenium Sources and Requirements

### 1. *Selenium Requirements*

The minimum selenium requirements of animals vary with the form of the selenium ingested and with the nature of the rest of the diet, particularly its content of $\alpha$-tocopherol. A dietary intake of 0.1 ppm Se provides a satisfactory margin of safety against any dietary variables or environmental stresses likely to be encountered by grazing sheep and cattle (5). A minimum requirement of 0.06 ppm Se for the prevention of WMD in lambs is indicated by the experiments of Oldfield *et al.* (149). However, Shirley *et al.* (188) found no evidence of a need for supplementary selenium in cows and calves maintained for 2 years on Florida pastures containing less than 0.03 ppm Se. New Zealand experience indicates that grazing lambs can grow normally and remain free from clinical signs of Se deficiency on pastures containing 0.03-0.04 ppm Se (89). Gardiner and Gorman (59) have shown that WMD can occur in lambs grazing Australian pastures estimated to contain 0.05 ppm Se.

The differences just cited illustrate the difficulty of determining a single, precise, critical selenium level in herbage, if such exists. Apart from analytical uncertainties, including variable losses of selenium in drying and storing samples (60a), and the problem of relating the material analyzed to that actually eaten by animals with selective grazing habits, variations exist in the herbage levels of tocopherol and of other substances affecting selenium utilization and needs. For these reasons attempts to correlate selenium concentrations in herbage with the incidence of WMD, or with responses to selenium treatment, have not always been successful (5). In fact, a direct, rather than an inverse, relationship between the selenium content of alfalfa and the occurrence of WMD in calves has been reported, despite indisputable evidence of the effectiveness of selenium supplementation (174). This can probably be explained by the unusually high levels of sulfate in the alfalfa. High-sulfate intakes reduce the availability of selenium to animals at high-selenium intakes (81). It seems likely, therefore, that selenium requirements are greater when sulfate intakes are high than when they are low. In some areas selenium deficiency can be precipitated in animals, where the intakes are already low or marginal, by the application to the soil of gypsum, elemental sulfur, or heavy dressings of superphosphate. This could be due to a depression in selenium uptake by plants, to a reduction in availability of selenium to the animal due to higher sulfate levels in the herbage, or to both together.

The minimum selenium requirements of poultry have recently been raised to a new plane of precision. Nesheim and Scott (145) found that a torula yeast diet containing 0.056 ppm Se and 100 IU vitamin E/lb could not sustain maximum growth rate in chicks unless it was supplemented with 0.04 ppm as sodium selenite. Subsequently, Scott *et al.* (186), using a practical corn-soy diet, found the selenium requirement in turkey poults to range from 0.18 ppm in the presence of vitamin E to approximately 0.28 ppm in the absence of added vitamin E. The profound importance of the level of vitamin E in the diet to the selenium requirements of chicks is illustrated further by the experiments of Thompson and Scott (196), referred to previously. With purified diets containing 100 ppm vitamin E, the selenium requirement was less than 0.01 ppm. With 10 ppm vitamin E, it was more than 0.02 ppm, and with no added vitamin E, approximately 0.05 ppm Se.

### 2. *Potency of Different Forms of Selenium*

Significant differences in the availability of different chemical forms of selenium, with resultant differences in selenium requirements, have become apparent from a series of studies by Schwarz. Selenium compounds were divided into three categories with respect to their potency against liver necrosis in rats (181). The first category includes elemental selenium and certain compounds which are practically inactive, presumably due to poor absorption. A second group, which includes most inorganic salts such as selenates and selenites and the selenium analogs of cystine and methionine, affords protection at only a few micrograms per 100 g of diet. They appear to be all more or less equally efficiently utilized as sources of selenium to the rat and the chick. Most of the selenium present in the plant materials which comprise the bulk of the diet of farm stock can be accounted for as members of this second category. The extent to which variations in the proportions of these forms of selenium affect dietary requirements of this element by farm species is largely unknown. It is unlikely to be great if the data of Schwarz and Foltz (181) with rats and chicks are taken as a guide.

The third category of selenium compounds consists of organic forms of selenium more active per unit of selenium than those of the second group. Studies of these forms of selenium began with the preparation of an organic component of hydrolyzates of kidney powder and other potent sources which was shown to be 3-4 times as potent against liver necrosis as selenite selenium. This fraction was originally designated Factor 3 and was established as organic in nature, strongly bound to proteins and separable into two factors known as $\alpha$- and $\beta$-factors 3 (176). Efforts to isolate and identify the active compound in these potent preparations were halted because of very low yields and chemical instability of the purified fractions. Attention was then turned to various organic selenium compounds which revealed large differences in activity (176). In a later study of the activity of aliphatic monoseleno- and diselenodicarboxylic acids, Schwarz and Fredga (182) observed remarkable differences in their capacity to prevent dietary liver necrosis in rats. The most active substances studied were monoseleno-11,11′-di-*n*-undecanoic acid, diseleno-4,4′-di-*n*-valeric acid, and diseleno-11,11′-di-*n*-undecanamide. Their potencies were 169, 138, and 180% of that of selenite Se, respectively. In both the mono- and diselenodicarboxylic acid series, a minimum of potency was seen at chain length C-7, and a maximum at chain length C-10 to C-11. Branching of the chain reduced activity and introduction of two methyl groups at the same carbon, forming a quaternary carbon atom, eliminated biological potency almost entirely. In addition, a clear-cut alternating effect, with the even-numbered acids rather inactive and the uneven numbers potent, was observed in the series of symmetrical monoselenodicarboxylic acids. No such alternating effect was apparent in the analogous diselenodicarboxylic acids.

Satisfactory explanations of these differences, and of the critical dependence of potency upon the nature of the organic acid moieties attached to the selenium atom, must await identification of the exact role of selenium (Factor 3) in intermediary metabolism. Schwarz and Fredga (182) state the present position, "It is conceivable that the various compounds used in this study are converted within the organism into a specific selenium-containing cofactor needed in intermediary metabolism. In that case the differences in biological potency could be a reflection of differences in efficiency with which the individual compounds are converted into the biologically active form. An alternate but less likely hypothesis could be based on the assumption that the various selenium compounds fulfil the essential biological function unchanged."

### 3. *Selenium in Human Foods and Animal Feeds*

The level of selenium in individual foods is highly variable. This variability stems largely from soil differences in the areas in which the foods are grown. Thus in early studies of foods from seleniferous areas, carried out by Smith *et al.* (191), values were reported ranging from 0.25 to

1.0 ppm Se for white bread, from 0.16 to 1.27 ppm for milk, and from 1.17 to 8.0 ppm for meat (all on dry basis). These are very much higher than the levels found in similar materials from "normal" areas or from Se-deficient areas. For instance, Hadjimarkos (77) observed a tenfold difference in the selenium concentration in milk and eggs from a known "Se-low" county and from two "Se-normal" counties in Oregon. The mean levels for the former county were 0.005 and 0.056 ppm for milk and eggs, respectively, compared with 0.05–0.07 and 0.4–0.5 ppm for these foods obtained from the latter counties. A similar variation in the selenium content of muscle and organ meats is apparent from studies with sheep and pigs grown under Se-deficient and Se-supplemented conditions. The liver and kidneys are substantially richer in this element than the muscles, which normally contain 0.5–0.8 ppm Se (dry basis). The meat, viscera, and bone of different types of fish have been reported to contain 0.17–0.77 ppm Se (147) and fish flour, made from whole, defatted fish for human consumption, 1.8 ppm (73).

The powerful influence of locality is apparent from the studies of Lindberg (115) with wheat and rye grain from different countries. Samples of Swedish wheat contained 0.007–0.022 ppm Se, which is similar to the concentrations observed in wheat from Se-deficient areas in New Zealand. The corresponding values for samples of wheat from other countries were as follows: Argentina, 0.05; U.S. (hard winter wheat), 0.37; U.S. (durum wheat), 0.70; and Canada (Manitoba), 1.30 ppm. Analysis of a series of samples of rye revealed the following differences: Sweden, 0.006; U.S.A., 0.08; Turkey, 0.06; and Argentina, 0.07 ppm Se.

The effect of cooking and processing on the selenium content of foods presents further difficulties in assessing selenium intakes by man. The effect of cooking on the selenium remaining in meat is unknown, but much of the selenium in vegetables is discarded with the cooking water (11). The selenium content of skimmed milk powder also varies with the method of drying (49). These aspects of human dietary sources of selenium warrant further investigation.

The selenium content of animal feeds, particularly pastures and forages, is much more thoroughly documented. It varies with the species of plants and with the selenium status of the soil in which they have grown. The effect of species is most apparent with the variously called accumulator, converter, or indicator plants which occur in seleniferous areas and which carry Se concentrations frequently lying between 1000 and 3000 ppm. One sample of *Astragalus racemosus* from Wyoming contained no less than 14,920 ppm, and over 4000 ppm Se (dry basis) was observed in the annual legume. *Neptunia amplexicaulis,* growing on a selenized soil in northern Queensland (106). The role of such plants in selenosis is considered later, but they play little part in the more general question of selenium sources to animals. For all ordinary edible grasses and legumes the primary determinant of selenium concentration is the level of available selenium in the soil. This is apparent from the area studies of forages in the United States carried out by Allaway and associates (2,5,140), and from numerous investigations in New Zealand and elsewhere. The Se concentrations of pastures and forages in areas free from Se-responsive diseases in animals generally contain 0.1 ppm Se or more, whereas the levels in areas with a variable incidence of such diseases are most frequently below 0.05 ppm and sometimes as low as 0.02 ppm. Selenium levels as low as 0.006-0.007 ppm occur in cereal grains from Se-deficient areas, compared with values 10-100 times higher in grains from normal, nonseleniferous soils. Values, a further order of magnitude higher, occur in grains from seleniferous soils.

Water supplies do not constitute a significant source of selenium to animals or man, either in Se-deficient, in normal (77), or in seleniferous areas (150).

### 4. *Selenium Supplementation*

The methods available for providing selenium supplements to animals include periodic injections or oral dosing with selenium salts, provision of salt mixtures or licks containing small amounts of selenium salts, use of feeds obtained from selenium-rich areas, treatment of the soil with Se compounds, and, more recently, administration of Se-containing heavy pellets to ruminants. The method of choice depends primarily upon the conditions of husbandry and is influenced by the desirability of minimizing the amounts of selenium introduced into the soil-plant-animal-man selenium cycle. In the U.S., deliberate addition of selenium to animal feeds is not at present permitted, so that this form of supplementation, which is the most common procedure with other minerals, is impossible. For this reason grains and alfalfa grown in areas where the levels of selenium in the soil are naturally high are often obtained and used in low-Se areas. As Allaway *et al.* (3) have pointed out, this system of blending imposes the need for controlled analyses of the feeds because the selenium content of feeds grown in the so-called high-selenium parts of the U.S. varies widely. In New Zealand, selenium-containing supplements are commercially available for preventing Se-responsive conditions in pigs and poultry and are recommended for incorporation into rations to provide 0.15 ppm of added selenium (10).

Direct subcutaneous injections of selenium, usually as sodium selenite, constitute the most economical procedure in terms of the amounts of selenium used and are widely practiced to control a range of Se-responsive diseases, particularly in cattle. New Zealand experience indicates that doses ranging from 10 mg Se for calves to 30 mg for adults, given at 3-month intervals, are satisfactory (10). Excessive concentrations of selenium in edible animal products from the use of selenium injections, or oral doses, in the amounts prescribed are extremely unlikely (32,109,153). With sheep, oral dosing with selenite or selenate, in doses from 1 to 5 mg Se, at intervals as described previously, is the most common means of preventing Se-responsive diseases in this species, because the drenching can often be carried out when the animals are yarded for routine management procedures. Both injections and oral dosing have the advantage of providing known amounts of selenium to individual animals but have the disadvantage of requiring individual handling and movements of stock, with consequent increased costs in time and labor.

The disadvantages attached to selenium dosing or injecting led Kuchel and Buckley (108) to investigate the possibilities of Se-containing heavy pellets for grazing sheep, to be used in a similar manner to the heavy cobalt pellets. Such pellets, in which the source of selenium was calcium selenate, barium selenate, and elemental selenium, were compared when dosed singly to sheep grazing pastures of normal selenium status. Blood and tissue selenium levels were found to be significantly enhanced by all three types of pellets for periods up to 12 months. It was concluded that pellets composed of finely divided metallic iron and elemental selenium were the components of choice, since these pellets were less liable to disintegration, or to coating with calcium phosphate. In an experiment with this type of pellet containing increasing proportions of selenium, namely 1.25, 2.5, 5, and 10%, blood selenium concentrations increased significantly in all cases within 1 week. Within 10 to 12 weeks the extent of the rise was related to the proportion of selenium in the pellets (see Fig. 19). The pellets were well retained in the reticulorumen, and the selenium concentrations in the edible tissues of the treated sheep after 6 to 12 months were no greater than those reported for untreated lambs in the U.S. (157) and in New Zealand (32). Handreck and Godwin (82a) have confirmed these findings with sheep, using dense pellets consisting of elemental iron and selenium (9:1 by weight) labeled with $^{75}$Se. During the experimental period of 1 month the pellets released 0.5–1.3 mg Se/day, of which about 30% was excreted in the urine and about 1% in the expired air. Blood selenium levels levelled out at 0.18 to 0.34 $\mu$g/ml, and there was no evidence of toxicity or excessive accumulation of selenium in the tissues. Heavy selenium pellets must, therefore, be considered a safe, reliable, and economical means of preventing selenium deficiency in grazing ruminants.

Treatment of the soil with various selenium compounds, by spraying of herbage or by application alone or following incorporation of the selenium into normal fertilizers, offers further possibilities for raising the selenium levels in pastures and forages to safe and satisfactory levels. In terms of the amounts of selenium required and introduced into the selenium cycle this is an expensive and hazardous process, because of the inefficiency with which such added selenium is taken up by most plants, especially from acid soils (3,7). Under experimental conditions, Allaway *et al.* (7) showed that the addition of 2 lb Se/acre, as $Na_2SeO_3$, increased the selenium content of alfalfa from very low levels to concentrations that protect lambs from WMD. Furthermore, the effects of one application lasted for at least 3 years. It is apparent from the extensive studies of Davies and Watkinson (36,204), and of Grant (65) and others, that a serious difficulty in supplying animal requirements by application to pasture is the high levels that occur immediately following application, whether these are due to foliar or root uptake, or to initial surface contamination. The problem is to supply sufficient selenium in a form which will maintain adequate concentrations of the element in the plants for a sufficiently long period to avoid the necessity of frequent application, without inducing potentially toxic levels following treatment.

Watkinson and Davies (204) maintain that, "with proper precautions to minimize pasture contamination," 1 oz/acre and possibly 2 oz/acre (as sodium selenite) should present no hazard, "at least for a few years." In an experiment on a pumice soil (pH 5.0), a comparison was made between sodium, ferric, and barium selenites, elemental selenium, and selenite incorporated in a "frit," each at 4 oz Se/acre. Uptake of selenium by white clover was greatest and equal from the three selenites, and there was no effect from the elemental selenium, apart from a passing surface contamination. The selenium frit proved to be the most hopeful of the treatments because the initial selenium levels in the clover were only half those of the selenite treatments and they were similar at the end of 18 months. The application of a selenium frit may enable somewhat higher rates of application than the 1-2 oz/acre recommended above to be used, with longer-lasting effects. A further possibility has been proposed by Allaway *et al.* (4) involving the use of "selenized superphosphate." This material, prepared by incorporating the selenium into the superphosphate during the treatment of the rock phosphate with acid, seems to have promise on some soils since plants containing adequate levels of selenium can be produced at additions equivalent to 16 oz Se/acre and 5 times this amount has not resulted in plants with toxic levels of selenium. On very sandy soils the apparent safe range of application of selenized superphosphate is very much narrower.

## VI. Selenium Poisoning in Animals

### 1. *Discovery of Naturally Occurring Selenosis*

As early as 1856, Madison described the occurrence of a fatal disease in horses in Nebraska, characterized by loss of hair from mane and tail and soreness of the feet [see Moxon (131)]. Similar disorders of grazing stock and of stock fed grain and roughages grown in the area were experienced by the early settlers. The name "alkali disease" was given to this disorder in the erroneous belief that it was caused by the alkali (high-salt) waters and seepages of the area. At about the same time, reports of serious stock losses from an apparently different disease, to which the name "blind staggers" was given, appeared from parts of Wyoming.

In 1929, Franke began active work on the alkali disease problem in South Dakota. The syndrome as it occurred in different farm species was described, and the toxicity of feeds from affected areas was demonstrated in laboratory animals, leading to the hypothesis that the disease was caused by the presence in the feeds of toxic elements. A search for such elements led to Robinson's discovery, in 1933, of the presence of selenium (166). High concentrations of this element in the soils and plants of the affected regions were subsequently demonstrated. Conclusive evidence that selenium is the toxic element responsible for alkali disease came from the experiments of Franke and associates (54), which disclosed signs of toxicity in animals from the administration of selenium salts similar to those that arise from ingestion of the natural toxicant. During this time, Beath and co-workers in Wyoming were devoting particular attention to native species of the genus *Astragalus* which they suspected to be associated with the occurrence of blind staggers. Certain species of this genus were found to contain extraordinarily high concentrations of selenium (12), and blind staggers was later established as an expression of relatively acute selenium poisoning. Rosenfeld and Beath (170) in their outstanding monograph, entitled "Selenium," describe three types of selenium poisoning in animals: (*a*) acute, (*b*) blind staggers, and (*c*) chronic, alkali disease.

Seleniferous soils and vegetation occur over considerable areas of the Great Plains of North America and localized seleniferous areas have been identified in Ireland (51,202), Israel (165), northern Australia (106), the Soviet Union (107), and South Africa (18). In these areas, toxic intakes of selenium by animals arise either (*a*) by the consumption of selenium accumulator plants, with their extremely high-selenium concentrations, or (*b*) from the ingestion of more normal pasture and forage species and grains with high concentrations of selenium due to the presence of above-normal levels of available selenium in the soils of the affected areas, or (*c*) from both together. Analysis of soils for total selenium does not provide a reliable index of the toxicity of the herbage growing upon those soils because of differences in the chemical forms of soil selenium and in the ability of various plant species to absorb and retain selenium. Nevertheless, soils containing more than 0.5 ppm total Se are regarded by Moxon (131) as potentially dangerous to livestock. This means that they can produce herbage containing 4-5 ppm Se or more.

Selenium accumulator plants play an important dual role in the incidence of selenosis in grazing animals. They rarely occur outside seleniferous areas and, therefore, may be used as indicators in the identification and mapping of such areas. For this reason they are sometimes referred to as "indicator" plants. They are also called "converter" plants because they have the ability to absorb selenium from soils in which the selenium is present in insoluble forms, relatively unavailable to other common plant species, and upon the death of the plants to return selenium to the soil in organic forms which are available to these other species. Accumulator or converter plants thus provide a direct source of selenium to animals consuming them, and they convert unavailable forms of selenium into available forms, thus raising the selenium levels in plants that otherwise might contain "safe" levels of the element. In some areas this may not be important, but in others the presence of converter plants can intensify the severity of selenosis and extend its incidence. Native stock usually avoid converter plants, except in times of drought or overgrazing, but stock brought in from other areas are not so discriminating and deaths from blind staggers result.

### 2. *Manifestations of Selenosis*

All degrees of selenium poisoning exist from a mild, chronic condition to an acute form resulting in death of the animal. Chronic selenium poisoning is characterized by dullness and lack of vitality; emaciation and roughness of coat; loss of hair from the mane and tail of horses and body of pigs; soreness and sloughing of the hoofs; stiffness and lameness, due to erosion of the joints of the long bones; atrophy of the heart ("dishrag" heart); cirrhosis of the liver; and anemia. In acute selenium poisoning the animals suffer from blindness, abdominal pain, salivation, grating of the teeth, and some degree of paralysis. Respiration is disturbed and death results from respiration failure. Death also results from starvation and thirst because, in addition to loss of appetite, the lameness and pain from the condition of the hoofs are so severe that the animals are unwilling to move about to secure food and water. This situation is comparable to that of animals in fluorosis areas. With the possible exception of loss of appetite and reduced rate of gain or increased loss of body weight, sheep have not been observed to exhibit signs of selenosis similar to those just described for horses, cattle, and pigs (150,158).

In the rat and the dog, a marked restriction of food intake occurs, together with anemia and severe pathological changes in the liver. This organ becomes necrotic, cyrrhotic, and hemorrhagic to varying degrees. Anemia is a common manifestation of selenosis in all species, but in the rat and the dog a microcytic, hypochromic anemia of progressive severity usually develops, and animals may die with hemoglobin levels as low as 2 g/100 ml (135,192).

Growing chicks exhibit a reduction in food intake and growth rate when consuming seleniferous diets, and there is a fall in egg production in hens. Franke and Tully (55) showed that the eggs produced by these hens have low hatchability at concentrations of selenium in the feed too low to cause manifest signs of poisoning in other animals. The eggs are fertile but a proportion produce grossly deformed embryos, characterized by missing eyes and beaks and distorted wings and feet. Deformed embryos were also produced by injecting selenite in concentrations as low as 0.1 ppm into the air cell of normal fertile eggs (52). Similar effects were later reported in turkeys (28).

Consumption of seleniferous diets interferes with the normal development of the embryo in rats (54,169), pigs (200), sheep (169), and cattle (40). This effect is also apparent from the birth of foals and calves with deformed hoofs in seleniferous areas. According to Olson (150), a reduction in reproductive performance is the most significant economic effect of selenium poisoning of the alkali disease type, and the effect on reproduction can be quite severe without the animals showing other typical lesions of selenosis. Wahlstrom and Olson (200) found that 10 ppm of selenite selenium fed to young sows lowered the conception rate, and increased the number of services per conception, as well as the proportion of piglets small, dead or weak at birth.

### 3. *Factors Affecting Selenium Toxicity*

The toxicity of selenium to animals varies with the amounts and chemical forms of the selenium ingested, with the duration and continuity of intake, and with the nature of the rest of the diet. Species differences appear to be small. Munsell and co-workers (139) place the minimum dietary level at which selenium will accumulate in the tissues and ultimately produce signs of toxicity at 3 to 4 ppm of the dry diet, but this will clearly vary with the nature of the diet, including its protein and sulfate contents. Furthermore, levels of 3 to 4 ppm Se do not adversely affect the growth of chicks or the hatchability of eggs. Diets based on seleniferous grains containing 5 ppm Se, reduce hatchability slightly, and, at 10 ppm, hatchability is reduced to zero (160). Starting rations for chicks should contain less than 5 ppm Se for normal growth rate (161).

Signs of chronic poisoning are common in rats and dogs given diets containing 5-10 ppm Se, and, at 20 ppm, there is complete refusal of food and death in a short time (139). Young pigs fed seleniferous diets containing 10-15 ppm Se develop signs of selenosis within 2 to 3 weeks (169). Diets containing lower levels than these would induce similar effects if fed for a longer period. The minimum toxic levels for grazing stock are more difficult to determine. They probably lie close to 5 ppm Se. Edible herbage in seleniferous areas commonly contain 5-20 ppm Se. At the lower levels, signs of selenium toxicity in cattle would take weeks, or even months, to appear. Mature sheep fed regular oral doses of sodium selenite, up to 600 μg Se/kg body weight per day, for periods as long as 15 months, revealed no pathological changes in the tissues. After 5 to 6 months of treatment, a depression in food consumption and body weight increase was observed at intakes of 200 and 300 μg Se/kg. per day, with some mortality at 400 μg Se/kg per day (158). The 200-μg/kg per day treatment is equivalent to a dietary intake of approximately 8 ppm Se.

The tolerance of animals to different levels of selenium is affected by the chemical forms in which the selenium is presented and by the ratio of the selenium to other components of the diet. High-protein diets afford some protection. Thus, Smith (190) found that 10 ppm Se in a 10% protein diet was highly toxic to rats, whereas an additional 20% protein as casein scarcely produced any adverse effects. Linseed oil meal was later shown to be superior to casein in protecting rats against Se poisoning and some fraction in the meal, other than protein, was found to be responsible for its superior effect (82). A curious feature of this protective action of linseed oil meal is that it is accompanied by higher selenium levels in the liver and kidneys compared with those of rats fed casein-containing seleniferous diets (113). This suggests that the selenium is bound in a less toxic form in this organ. Recent studies by Levander and co-workers (113) showed that chelating agents can increase the amount of selenium dialyzed from liver homogenates of rats fed the casein diets but have no such effect with the linseed oil meal, seleniferous diet.

The sulfur-containing amino acids cannot be implicated with any certainty in the protein-protecting effect (151). Increasing levels of sulfate, up to 0.87% of a sulfate-free diet, progressively relieve the growth inhibition induced in young rats by 10 ppm Se, as selenite or selenate (81). The sulfate supplements are comparatively ineffective against the liver degeneration of selenosis but can alleviate the growth depression by as much as 40%. These findings invite speculation on the possibility that the protective effect of high-protein diets may be related in part to the production of endogenous sulfate. Inorganic sulfate is much less effective against organic forms of selenium than against selenate (80,81).

The most effective dietary factor reducing selenium toxicity so far discovered is arsenic. This was first demonstrated by Moxon (132) who showed that 5 ppm As in the drinking water completely prevented all signs of selenosis in rats. Arsenic has since been used successfully to alleviate Se poisoning in pigs, dogs, chicks, and cattle. Sodium arsenite and arsenate are equally effective, arsenic sulfides are ineffective, and arsenic in the form of various organic compounds, such as arsanilic acid and 3-nitro-4-hydroxyphenylarsonic acid provide a partial protection (91,109,110,210) (see Table 41).

### 4. *Mechanism of Selenium Toxicity*

The precise ways in which selenium at toxic intakes interferes with tissue structure and function are far from clear. Selenium has long been known to affect certain unicellular organisms and enzyme systems (114) and to inhibit alcoholic fermentation by yeast and some of the enzymes concerned with cellular respiration (162). The inhibition of oxygen consumption by tissues, disclosed by these *in vitro* studies, appears to be

TABLE 41
*Arsenicals and Selenium Poisoning in Swine*[a]

| Treatment | Av. daily gain (lb) | No. showing symptoms[b] |
|---|---|---|
| Basal diet | 1.53 | None |
| Basal diet + 0.01% arsanilic acid | 1.59 | None |
| Basal diet + 0.005% 3-nitro-4-hydroxy-phenylarsonic acid | 1.64 | None |
| Basal diet + 7 ppm Se | 1.22 | 3 |
| Basal diet + 7 ppm Se + 0.01% arsanilic acid | 1.36 | None |
| Basal diet + 7 ppm Se + 0.005% 3-nitro-4-hydroxy-phenylarsonic acid | 1.64 | None |

[a]From Wahlstrom *et al.* (201).
[b]Five pigs in each lot.

mediated through a poisoning of succinic dehydrogenase. The liver succinic dehydrogenase levels of rats fed seleniferous diets are reduced below normal and can be maintained at normal levels by appropriate dietary intakes of arsenic (105a). It is unlikely that these effects are sufficient to account for the various manifestations of selenosis or their prevention by arsenic, especially as arsenic is known to enhance the biliary excretion of selenium. No other enzymes appear to be affected in the living tissues of selenized animals, although the *in vitro* experiments of Wright (210) with tissues to which sodium selenite or selenate was added, indicate a general inhibition of dehydrogenating enzymes and of urease, with no impairment of the cytochrome-indophenoloxidase system or of catalase or liver arginase. These findings point to the removal of sulfhydryl groups essential to oxidative processes as possible biochemical sites of the injurious effects of selenium.

### 5. *Prevention and Control of Selenosis*

Three possibilities theoretically exist for the prevention or treatment of selenium poisoning. These are (*a*) treatment of the soil so that selenium uptake by plants is reduced and maintained at nontoxic levels, (*b*) treatment of the animal so that selenium absorption is reduced or excretion increased, thus preventing toxic accumulations in the tissues, and (*c*) modifying the diet of the animal by the inclusion of substances that antagonize or inhibit the toxic effects of selenium within the body tissues and fluids.

The addition of sulfur or gypsum to soils in a toxic area in North America has been unsuccessful in reducing the absorption of selenium by cereals (53), probably because these soils are generally already high in gypsum and carry a high proportion of their selenium in organic combinations that are relatively little affected by changes in the inorganic sulfur-selenium ratio. On the other hand, the addition of sulfur to soils to which selenate has been added can inhibit selenium absorption by plants. Selenium uptake by alfalfa from a seleniferous soil has been strikingly reduced by additions of calcium sulfate and of barium chloride (165). The latter salt reduced the selenium levels in plants by 90 to 100% when applied in quantities that did not affect plant growth nor resulted in significant concentrations of barium in the tissues. The practical possibilities of these interesting findings appear to be limited.

Urinary loss of selenium from the body can be enhanced by the administration of bromobenzene to rats and dogs fed a seleniferous diet and to steers on seleniferous range (137). However, this form of treatment has obvious practical limitations. On the other hand, dietary modifications, including high-protein and -sulfate intakes and the feeding of arsenic at appropriate levels, have a useful potential in alleviating selenium toxicity, where dietary control of the animals can be achieved. The position is much more difficult under range conditions. Early studies by Moxon *et al.* (136) indicated that 25 ppm as sodium arsenite added to the salt of cattle on seleniferous range gave some protection but observations by ranchers and additional studies by Dinkel and co-workers (39) showed that this method of control is ineffective, probably because the arsenic intake is neither sufficiently high nor regular enough. However, various management possibilities exist, following mapping of the land into pastures of high- and low-selenium contents (150). Furthermore, the more highly seleniferous areas can be used for grain production and the grain sold into normal market channels where it would become so diluted that it should not contribute to a public health problem; moreover, it could help in raising selenium intakes by animals in Se-deficient areas.

### 6. *Selenium and Dental Caries*

The results of epidemiological studies with children and experiments with laboratory animals have led Hadjimarkos (74,77) to propose that the consumption of small amounts of selenium during the development period of the teeth increases the incidence of caries. The epidemiological evidence comes from regional studies of caries incidence in children in the states of Oregon and Wyoming in the United States. These studies were prompted by early observations that a high prevalence of caries is a characteristic of human populations in the seleniferous areas (191). In the Oregon study, children born and reared in a county where selenium deficiency in livestock is common, had a lower incidence of caries (9.0 DMF teeth/child) than children from three other counties of known higher-Se status (13.4–14.4 DMF teeth/child). Analyses of 24-hr urine specimens from a selection of children in the former county revealed significantly lower mean Se concentrations than those from comparable children in the other three counties. The level of selenium in the urine was considered a reliable criterion of total selenium ingestion. This was supported by samples of milk and eggs from three of these counties which disclosed significantly lower mean selenium levels in these dietary items from the low-Se status county than from the other two counties (78).

A study was made in Wyoming of 388 children, 10–18 years of age, who were born and had lived either in known seleniferous areas or in areas where no seleniferous soils exist and selenosis in animals had not been reported. These children were separated further into areas where the water supplies were low (0.0–0.5 ppm) and high (0.6–2.6 ppm) in fluoride content. It was found that the children from the seleniferous areas were more susceptible to dental caries than those from nonseleniferous areas. This was true whether the children consumed low- or high-fluoride water, although the normal fluoride depressing effect on caries was clearly apparent. An interesting feature of this study was an indication that the presence of selenium partially inhibited the beneficial influence of fluoride on caries. The number of children in certain of the groups was quite small, and satisfactory studies of urinary selenium excretion do not appear to have been carried out. The actual comparative selenium intakes of the children could, therefore, not be deduced. The results are, nevertheless, in line with those of the more extensive Oregon survey and lend some further support to the contention that selenium increases susceptibility to caries. No such support is apparently provided by a small study of schoolboys in two rural areas of New Zealand (26), in which the incidence of caries did not differ but the mean urinary selenium excretion of the two areas differed significantly. As Hadjimarkos (74,77) has pointed out, the urinary selenium concentrations were considerably lower in both areas than those in the Oregon study. He suggests that selenium intakes were, in fact, below a threshold value below which selenium does not exert any influence on caries susceptibility.

The relation of selenium intake to caries incidence has been studied in several experiments with rats with variable results. Buttner (25) added sodium selenite at two levels, 5 and 10 ppm Se, to the drinking water of rats during pregnancy and lactation and, subsequently, to their offspring for 120 days. The selenium significantly increased caries incidence in both groups (Table 42). In experiments carried out by Navia *et al.* (143), rats fed a cariogenic diet with 4 ppm Se added at the time of birth developed a numerical increase in caries which was not significantly different from the controls. In an earlier study with rats fed a sucrose-skim milk diet, the ingestion of selenium postdevelopmentally actually decreased the incidence of caries (138). Subsequently Hadjimarkos (76) found that the administration of selenium postdevelopmentally had no influence on the incidence of caries in rats fed a corn cariogenic diet. This worker also showed that the ingestion of selenium can alter the pattern of food and water intake by rats, with the consumption of water being reduced on some diets appreciably more than that of food (75). He suggests that the consumption of selenium postdevelopmentally may increase somewhat the incidence of caries through an "increased adhesiveness of cariogenic food particles on the surfaces of the teeth as a result of a reduc-

TABLE 42
*Effect of Prenatal and Postnatal Administration of Sodium Selenite on Dental Caries in the Rat*[a]

| Selenite in water | Sex | No. rats | Wt gain | Mean No. carious lesions[b] | $t^c$ | Extent of carious lesions | $t^c$ |
|---|---|---|---|---|---|---|---|
| Control | M | 17 | 330±9 | 5.3±0.5 | – | 11.6±1.4 | – |
| | F | 14 | 207±7 | | | | |
| 5 ppm | M | 13 | 240±5 | 7.2±0.7 | 2.5 | 18.7±2.2 | 3.0 |
| | F | 7 | 176±7 | | | | |
| 10 ppm | M | 5 | 220±18 | 8.6±1.1 | 2.8 | 25.2±3.2 | 4.0 |
| | F | 2 | 148±3 | | | | |

[a]From Buttner (25).
[b]Sexes combined.
[c]*t*-Values as compared with controls.

tion in water intake which is proportionately greater than that of food." The effect of selenium consumed during the developmental period of the teeth on caries susceptibility is probably explained by changes in the protein components of the enamel (77). No direct evidence of any such changes is available, although there are indications that selenium is associated with the protein of bone matrix and not with the mineral components (27). It is also evident that any action of selenium in enhancing caries or in diminishing the anticaries effect of fluoride cannot be mediated through a decrease in the amount of fluoride incorporated into the teeth during their development (74).

### 7. *Selenium and Cancer*

Selenium has been implicated as a carcinogen in rats and as a possible protective agent against human cancer. In neither case is the evidence entirely convincing. Shamberger and Frost (187) examined the human cancer death rates in ten of the cities, with populations of 40,000 to 70,000, from which Allaway *et al.* (6) had taken blood samples from male donors for the determination of selenium concentrations. They obtained an almost perfect inverse relationship ($r = -0.96$) between blood selenium levels and human cancer death rates. In addition, these workers compared the human cancer death rate for 1966, in certain provinces of Canada where selenium indicator plants have been found extensively, with others where these plants have not been found. The human cancer death rate for the former provinces was 122±7.8 compared with 139.9±4.9 ($P < 0.1$) for the latter. Clayton and Baumann (31) had previously reported that 5 ppm Se appeared to decrease the susceptibility of rats to liver tumors from azo dye.

Selenium acquired its reputation as a carcinogen from the early studies of Nelson *et al.* (144) with rats. Cirrhosis of the liver was seen within 3 months in the rats which died early in the study. No tumors were observed in 73 rats surviving less than 18 months, but 11 of 53 rats surviving more than 18 months developed adenomas or low-grade carcinomas in their cirrhotic livers. Four others were reported to show advanced adenomatoid hyperplasia. Subsequently, Fitzhugh *et al.* (50) noted that the diet used in this study was low in protein which brought about some liver degeneration in both the Se-treated and the control group. They also noted that the hepatic cell carcinomas did not metastasize. Later Tscherkes and co-workers (198,199) studied the effects of selenium with rats on both low- (12%) and high- (30%)protein diets. On the low-protein diet with 4.3 ppm of added selenium as sodium selenite, tumors were observed in 10 of the 23 rats that survived 19-32 months. On the high-protein diet, 3 of 40 rats were reported to have sarcomas by the twenty-second month of the experiment. It is impossible to incriminate selenium as a carcinogen from these latter observations because no controls, without added selenium, were used. Harr and co-workers (84) fed selenium at varying levels to rats and evaluated its carcinogenic potential in comparison with the known hepatocarcinogen, *N*-2-fluorenylacetamide. Hyperplastic liver lesions occurred in the rats fed selenium which did not regress when the added selenium was withdrawn from the diet. Sixty-three neoplasms were observed in the experimental animals but none of these could be attributed to the addition of selenium. It is obvious that the whole question of selenium and cancer requires critical reevaluation and reexamination.

## REFERENCES

1. W. H. Allaway and E. E. Cary, *Anal. Chem.* **36**, 1359 (1964).
2. W. H. Allaway and E. E. Cary, *Feedstuffs* **38**, 62 (1966).
3. W. H. Allaway, E. E. Cary, and C. F. Ehlig, *in* "Selenium in Biomedicine" (O. H. Muth, ed.), p. 273. Avi Publ. Co., Westport, Connecticut, 1967.
4. W. H. Allaway, E. E. Cary, J. Kubota, and C. F Ehlig, *Proc. Cornell Nutr. Conf.* p. 9 (1964).
5. W. H. Allaway and J. F. Hodgson, *J. Anim. Sci.* **23**, 271 (1964).
6. W. H. Allaway, J. Kubota, F. L. Losee, and M. Roth, *Arch. Environ. Health* **16**, 342 (1968).
7. W. H. Allaway, D. P. Moore, J. E. Oldfield, and O. H. Muth, *J. Nutr.* **88**, 411 (1966).
8. H. D. Anderson and A. L. Moxon, *J. Nutr.* **22**, 103 (1941).
9. P. Anderson, *Acta Pathol. Microbiol. Scand.* Suppl. 134 (1960).
10. E. D. Andrews, W. J. Hartley, and A. B. Grant, *N. Z. Vet. J.* **16**, 3 (1968).
11. K. Aterman, *Brit. J. Nutr.* **13**, 38 (1959).

12. O. A. Beath, H. F. Eppson, and C. S. Gilbert, *Wyo., Agr. Exp. Sta., Bull.* **206** (1935).
13. K. L. Blaxter, *Brit. J. Nutr.* **17**, 105 (1963).
14. K. L. Blaxter and R. F. McGill, *Vet. Rev. Annot.* **1**, 91 (1955).
15. C. Blincoe, *Nature* (*London*) **186**, 398 (1960).
16. C. Blincoe and W. B. Dye, *J. Anim. Sci.* **17**, 224 (1958).
17. H. J. M. Bowen and P. A. Cawse, *Analyst* **88**, 721 (1963).
18. J. M. M. Brown and P. J. De Wet, *Onderstepoort J. Vet. Res.* **34**, 161 (1967).
19. D. Brune, K. Samsahl, and P. O. Westov, *Clin. Chim. Acta* **13**, 285 (1966).
19a. J. G. Buchanan-Smith, E. C. Nelson, B. I. Osburn, M. E. Wells, and A. D. Tillman, *J. Anim. Sci.* **29**, 808 (1969).
20. R. G. Buescher, M. C. Bell, and R. K. Berry, *J. Anim. Sci.* **19**, 1251 (1960).
21. R. F. Burk, W. N. Pearson, R. F. Wood, and F. Viteri, *Amer. J. Clin. Nutr.* **20**, 723 (1967).
22. R. F. Burk, R. Whitney, H. Frank, and W. N. Pearson, *J. Nutr.* **95**, 420 (1968).
23. V. Burton, R. F. Keeler, K. F. Swingle, and S. Young, *Amer. J. Vet. Res.* **23**, 962 (1962).
24. G. W. Butler and P. J. Peterson, *N. Z. J. Agr. Res.* **4**, 484 (1961).
25. W. Buttner, *J. Dent. Res.* **42**, 453 (1963).
26. P. D. Cadell and F. B. Cousins, *Nature* (*London*) **185**, 863 (1960).
27. R. D. Campo, C. D. Tourtellotte, and J. W. Ledrick, *Proc. Soc. Exp. Biol. Med.* **125**, 512 (1967).
28. C. W. Carlson, W. Kohlmeyer, and A. L. Moxon, *S. Dak. Farm Home Res.* **3**, 20 (1951).
29. G. H. Cartan and K. F. Swingle, *Amer. J. Vet. Res.* **20**, 235 (1959).
30. P. R. Cheeke and J. E. Oldfield, *Can. J. Anim. Sci.* **49**, 169 (1969).
31. C. C. Clayton and C. A. Baumann, *Cancer Res.* **9**, 575 (1949).
32. F. B. Cousins and I. M. Cairney, *Aust. J. Agr. Res.* **12**, 927 (1961).
33. L. M. Cummins and J. L. Martin, *Biochemistry* **6**, 3162 (1967).
34. H. Dam, *J. Nutr.* **27**, 193 (1944).
35. H. Dam, G. K. Nielsen, I. Prange, and E. Sondergaard, *Nature* (*London*) **182**, 802 (1958).
36. E. B. Davies and J. H. Watkinson, *N. Z. J. Agr. Res.* **9**, 317 and 641 (1966).
37. I. D. Desai, C. C. Calvert, M. L. Scott, and A. L. Tappel, *Proc. Soc. Exp. Biol. Med.* **115**, 462 (1964).
38. R. C. Dickson and R. H. Tomlinson, *Clin. Chim. Acta* **16**, 311 (1967).
39. C. A. Dinkel, J. A. Minyard and O. E. Olson, *S. Dak., Agr. Exp. Sta., Circ.* **135**, (1957); cited by Olson (150).
40. C. A. Dinkel, J. A. Minyard, and D. E. Ray, *J. Anim. Sci.* **22**, 1043 (1963).
41. A. T. Diplock, *Proc. 1st Int. Symp. Trace Element Metab. Anim., 1969* p. 190, (C. F. Mills, ed.) Livingstone, Edinburgh, 1970.
42. C. Drake, A. B. Grant, and W. J. Hartley, *N. Z. Vet. J.* **8**, 4 and 7 (1960).
43. H. Draper, *Nature* (*London*) **180**, 1419 (1957).
44. H. Draper and B. C. Johnson, *J. Anim. Sci.* **15**, 1154 (1956).
45. E. E. Edwin, A. T. Diplock, J. Bunyan, and J. Green, *Biochem. J.* **79**, 108 (1961).
46. R. G. Eggert, E. L. Patterson, W. T. Akers, and E. L. R. Stokstad, *J. Anim. Sci.* **16**, 1037 (1957).
47. C. F. Ehlig, D. E. Hogue, W. H. Allaway, and D. J. Hamm, *J. Nutr.* **92**, 121 (1967).
48. E. S. Erwin, W. Sterner, R. S. Gordon, L. J. Machlin, and L. Tureen, *J. Anim. Sci.* **19**, 1320 (1960).
49. H. Fink, *Abstr. 5th Int. Congr. Nutr., 1960* p. 19 (1960).

50. O. G. Fitzhugh, A. A. Nelson, and C. I. Bliss, *J. Pharmacol. Exp. Ther.* **80**, 289 (1944).
51. G. A. Fleming and T. Walsh, *Proc. Roy. Irish Acad., Sect. B* **58**, 151 (1957).
52. K. W. Franke, A. L. Moxon, W. E. Poley, and W. C. Tully, *Anat. Rec.* **65**, 15 (1936).
53. K. W. Franke and E. P. Painter, *Cereal Chem.* **15**, 1 (1938).
54. K. W. Franke and V. R. Potter, *J. Nutr.* **10**, 213 (1935); K. W. Franke and E. P. Painter, *ibid.* p. 599.
55. K. W. Franke and W. C. Tully, *Poultry Sci.* **14**, 273 (1935); **15**, 316 (1936).
56. H. E. Ganther and C. A. Baumann, *J. Nutr.* **77**, 208 and 408 (1962).
57. H. E. Ganther, O. A. Levander, and C. A. Baumann, *J. Nutr.* **88**, 55 (1966).
58. M. R. Gardiner, *Aust. Vet. J.* **38**, 387 (1962).
59. M. R. Gardiner and P. C. Gorman, *Aust. J. Exp. Agr. Anim. Husb.* **3**, 284 (1966).
59a. R. W. Gardner and D. E. Hogue, *J. Nutr.* **93**, 418 (1967).
60. W. Giese, W. Cappell, and H. Hill, *Proc. FAO/IAEA Panel, Trace Min. Stud. Isotopes Domestic Animals,* Vienna, *1968.*
60a. G. Gissel-Nielsen, *Plant Soil.* **32**, 242 (1970).
61. K. O. Godwin, *Quart. J. Exp. Physiol.* **51**, 94 (1966).
62. K. O. Godwin, *Nature* (*London*) **217**, 1275 (1968).
63. K. O. Godwin and F. J. Frazer, *Quart. J. Exp. Physiol.* **51**, 94 (1966).
64. M. Goettsch and A. M. Pappenheimer, *J. Exp. Med.* **54**, 154 (1931).
65. A. B. Grant, *N. Z. J. Agr. Res.* **8**, 681 (1965).
66. A. B. Grant and G. F. Wilson, *N. Z. J. Agr. Res.* **11**, 733 (1968).
67. C. A. Grant and B. Thafvelin, *Nord. Veterinuer Med.* **10**, 657 (1958).
68. J. Green and J. Bunyan, *Nutr. Abstr. Rev.* **39**, 321 (1969).
69. J. Green, A. T. Diplock, J. Bunyan, E. E. Edwin, and D. McHale, *Nature* (*London*) **190**, 318 (1961).
70. J. Green, A. T. Diplock, J. Bunyan, and E. E. Edwin, *Biochem. J.* **79**, 108 (1961).
71. J. Green, E. E. Edwin, A. T. Diplock, and J. Bunyan, *Nature* (*London*) **189**, 748 (1961).
72. D. M. Hadjimarkos, *J. Pediat.* **63**, 273 (1963).
73. D. M. Hadjimarkos, *Lancet* **1**, 605 (1965).
74. D. M. Hadjimarkos, *Arch. Environ. Health* **10**, 893 (1965); **14**, 881 (1967).
75. D. M. Hadjimarkos, *Experientia* **22**, 117 (1966); **23**, 930 (1967).
76. D. M. Hadjimarkos, *Advan. Oral Biol.* **3**, p 253 (1968).
77. D. M. Hadjimarkos, *Caries Res.* **3**, 14 (1969).
78. D. M. Hadjimarkos and C. W. Bonhorst, *J. Pediat.* **59**, 261 (1961).
79. R. J. Hall and P. L. Gupta, *Analyst* **94**, 292 (1969).
80. A. W. Halverson, P. L. Guss, and O. E. Olson, *J. Nutr.* **77**, 459 (1962).
81. A. W. Halverson and K. J. Monty, *J. Nutr.* **70**, 100 (1960).
82. A. W. Halverson, C. M. Hendrick, and O. E. Olson, *J. Nutr.* **56**, 51 (1955).
82a. K. A. Handreck, K. O. Godwin, *Aust. J. Agr. Res.* **21**, 71 (1970).
83. E. Hansson and S. O. Jacobson, *Biochim. Biophys. Acta* **115**, 285 (1966).
84. J. R. Harr, J. F. Bone, I. J. Tinsley, P. H. Weswig, and R. S. Yamamoto, *in* "Selenium in Biomedicine" (O. H. Muth, ed.), p. 153. Avi. Publ. Co., Westport, Connecticut, 1967.
85. P. L. Harris, M. I. Ludwig, and K. Schwarz, *Proc. Soc. Exp. Biol. Med.* **97**, 686 (1958).
86. K. E. Hart and M. M. Mackinnon, *N. Z. Vet. J.* **6**, 118 (1958); M. M. Mackinnon, *ibid.* **7**, 18 (1959).
87. W. J. Hartley, *Proc. N. Z. Soc. Anim. Prod.* **23**, 20 (1963).

88. W. J. Hartley, *in* "Selenium in Biomedicine" (O. H. Muth, ed.), p. 79. Avi Publ. Co., Westport, Connecticut, 1967.
89. W. J. Hartley and A. B. Grant, *Fed. Proc., Fed. Amer. Soc. Exp. Biol.* **20**, 679 (1961).
90. W. J. Hartley, C. Drake, and A. B. Grant, *N. Z. J. Agr.* **99,** 259 (1959).
91. C. M. Hendrick, H. L. Klug, and O. E. Olson, *J. Nutr.* **51,** 131 (1953).
92. M. Hidiroglou, R. J. Jenkins, R. B. Carson, and R. R. Mackay, *Can. J. Anim. Sci.* **48,** 335 (1968).
93. H. F. Hintz and D. E. Hogue, *J. Nutr.* **82,** 495 (1964).
94. D. E. Hogue, *Proc. Cornell Nutr. Conf.* p. 32 (1958).
95. L. L. Hopkins, Jr., A. L. Pope, and C. A. Baumann, *J. Nutr.* **88,** 61 (1966).
96. E. L. Hove, G. S. Fry, and K. Schwarz, *Proc. Soc. Exp. Biol. Med.* **98,** 1 and 27 (1958).
97. S. O. Jacobson and E. Hansson, *Biochim. Biophys. Acta* **6,** 287 (1965).
98. S. O. Jacobson and H. E. Oksanen, *Acta Vet. Scand.* **7,** 66 (1966).
99. A. Japha, Dissertation Halle, 1842; quoted by Moxon and Rhian (135).
100. L. S. Jensen, *Proc. Soc. Exp. Biol. Med.* **128,** 970 (1968).
101. L. S. Jensen, E. D. Walter, and J. S. Dunlap, *Proc. Soc. Exp. Biol. Med.* **112,** 899 (1963).
102. R. D. Jolly, *N. Z. Vet. J.* **8,** 13 (1960).
103. G. B. Jones and K. O. Godwin, *Aust. J. Agr. Res.* **14,** 716 (1963).
104. J. Kaarde, *Wien. Tierarztl. Monatsschr.* **52,** 391 (1965); *Vet. Bull. (London)* **36,** 234 (1966)(abstr.).
105. L. D. Kamstra and C. W. Bonhorst, *Proc. S. Dak. Acad. Sci.* **32,** 72 (1953).
105a. H. L. Klug, A. L. Moxon, D. F. Peterson, and V. R. Potter, *Arch. Biochem.* **28,** 253 (1950).
106. S. G. Knott, C. W. R. McCray, and W. T. K. Hall, *Queensl. J. Agr. Sci.* **15,** 43 (1958); S. G. Knott and C. W. R. McCray, *Aust. Vet. J.* **35,** 161 (1959).
107. V. V. Kovalsky, *Priroda (Moscow)* **4,** 11 (1954); *Nutr. Abstr. Rev.* **25,** 544 (1955).
108. R. E. Kuchel and R. A. Buckley, *Aust. J. Agr. Res.* **20,** 1099 (1969).
109. K. L. Kuttler and D. W. Marble, *Amer. J. Vet. Res.* **22,** 422 (1961).
110. E. Leitis, I. S. Palmer, and O. E. Olson, *Proc. S. Dak. Acad. Sci.* **35,** 189 (1956).
111. O. A. Levander and L. C. Argrett, *Toxicol. Appl. Pharmacol.* **14,** 308 (1969).
112. O. A. Levander and C. A. Baumann, *Toxicol. Appl. Pharmacol.* **9,** 98 and 106 (1966).
113. O. A. Levander, M. L. Young, and S. A. Meeks, *Toxicol. Appl. Pharmacol.* **16,** 79 (1970).
114. V. E. Levine, *J. Bacteriol.* **10,** 217 (1925).
115. P. Lindberg, *Acta Vet. Scand.* Suppl. 23 (1968).
116. P. L. Lopez, R. L. Preston, and W. H. Pfander, *J. Nutr.* **94,** 219 (1968); **97,** 123 (1968).
117. D. L. Mace, J. A. Tucker, C. B. Bills, and C. J. Ferreira, *Calif., Dep. Agr., Bull.* No. 1, p. 63 (1963).
118. C. G. Mackenzie, *Proc. Soc. Exp. Biol. Med.* **84,** 388 (1953).
119. A. S. Majaj and L. L. Hopkins, Jr., *Lancet* **2,** 592 (1966).
120. M. M. Mathias, D. E. Hogue, and J. K. Loosli, *J. Nutr.* **93,** 14 (1967).
121. K. P. McConnell, *Tex. Rep. Biol. Med.* **17,** 120 (1959).
122. K. P. McConnell, *J. Biol. Chem.* **141,** 427 (1941); **145,** 55 (1942); **173,** 653 (1948).
123. K. P. McConnell, *J. Agr. Food Chem.* **11,** 385 (1963).
124. K. P. McConnell and R. S. Levy, *Nature (London)* **195,** 774 (1962).

125. K. P. McConnell and D. M. Roth, *Biochim. Biophys. Acta* **62**, 503 (1962).
126. K. P. McConnell and D. M. Roth, *J. Nutr.* **84**, 340 (1964).
127. K. P. McConnell and C. H. Wabnitz, *J. Biol. Chem.* **226**, 765 (1957).
128. K. E. M. McCoy and P. H. Weswig, *J. Nutr.* **98**, 383 (1967).
129. J. W. McLean, G. G. Thompson, and J. H. Claxton, *Nature (London)* **184**, 251 (1959); *N. Z. Vet. J.* **7**, 47 (1959).
130. J. W. McLean, G. G. Thompson, and B. M. Lawson, *N. Z. Vet. J.* **11**, 59 (1963).
131. A. L. Moxon, *S. Dak., Agr. Exp. Sta., Bull.* **311** (1937).
132. A. L. Moxon, *Science* **88**, 81 (1938).
133. A. L. Moxon, K. P. Dubois, and R. L. Potter, *J. Pharmacol. Exp. Ther.* **72**, 184 (1941).
134. A. L. Moxon and W. E. Poley, *Poultry Sci.* **17**, 77 (1938).
135. A. L. Moxon and M. Rhian, *Physiol. Rev.* **23**, 305 (1943).
136. A. L. Moxon, M. Rhian, H. D. Anderson, and O. E. Olson, *J. Anim. Sci.* **3**, 299 (1944).
137. A. L. Moxon, A. E. Schaefer, H. A. Lardy, K. P. Dubois, and O. E. Olson, *J. Biol. Chem.* **132**, 785 (1940).
138. H. R. Mühlemann and K. G. König, *Helv. Odontol. Acta* **8**, 79 (1964); quoted by Hadjimarkos (77).
139. H. E. Munsell, G. M. Devaney, and M. H. Kennedy, *U.S., Dep. Agr., Tech. Bull.* **534** (1936).
140. O. H. Muth and W. H. Allaway, *J. Amer. Vet. Med. Ass.* **142**, 1379 (1963).
141. O. H. Muth, J. E. Oldfield, L. F. Remmert, and J. R. Schubert, *Science* **28**, 1090 (1958).
142. O. H. Muth, J. R. Schubert, and J. E. Oldfield, *Amer. J. Vet. Res.* **22**, 466 (1961).
143. J. M. Navia, L. Menaker, J. Seltzer, and R. S. Harris, *Fed. Proc., Fed. Amer. Soc. Exp. Biol.* **27**, 676 (1968) (abstr.).
144. A. A. Nelson, O. G. Fitzhugh, and H. O. Calvery, *Cancer Res.* **3**, 230 (1943).
145. M. C. Nesheim and M. L. Scott, *J. Nutr.* **65**, 601 (1958).
146. M. C. Nesheim and M. L. Scott, *Fed. Proc., Fed. Amer. Soc. Exp. Biol.* **20**, 674 (1961).
147. N. Ohta, N. Onuma, and K. Kawasaki, *Nippon Kagaku Zasshi* **81**, 920 (1960); cited by Hadjimarkos (73).
148. H. E. Oksanen, *in* "Selenium in Biomedicine" (O. H. Muth, ed.), p. 215. Avi Publ. Co., Westport, Connecticut, 1967.
149. J. E. Oldfield, J. R. Schubert, and O. H. Muth, *J. Agr. Food Chem.* **11**, 388 (1963).
150. O. E. Olson, *Proc. Ga. Conf. Feed Mfrs., 1969.*
151. O. E. Olson, C. W. Carlson, and E. Leitis, *S. Dak., Agr. Exp. Sta., Tech. Bull.* **20** (1958).
152. O. E. Olson, C. A. Denkel, and L. D. Kamstra, *S. Dak. Farm Home Res.* **6**, 12 (1954).
153. K. Ostadius and B. Aberg, *Acta Vet. Scand.* **2**, 60 (1961).
154. I. S. Palmer, D. D. Fischer, A. W. Halverson, and O. E. Olson, *Biochim. Biophys. Acta* **177**, 336 (1969).
154a. J. Parizek, I. Benes, A. Babicky, J. Benes, V. Proschazkova, and J. Lener, *Physiol. Bohemslov.* **18**, 105 (1969).
155. E. L. Patterson, R. Milstrey, and E. L. R. Stokstad, *Proc. Soc. Exp. Biol. Med.* **95**, 621 (1957).
156. G. D. Paulson, C. A. Baumann, and A. L. Pope, *J. Anim. Sci.* **25**, 1054 (1966).
157. G. D. Paulson, G. A. Broderick, C. A. Baumann, and A. L. Pope, *J. Anim. Sci.* **27**, 195 (1968).

157a. G. D. Paulson, A. L. Pope, and C. A. Baumann, *Proc. Soc. Exp. Biol. Med.* **122**, 321 (1966).
158. A. W. Peirce and G. B. Jones, *Aust. J. Exp. Agr. Anim. Husb.* **8**, 277 (1968).
159. J. Pinsent, *Biochem. J.* **57**, 10 (1954).
160. W. E. Poley and A. L. Moxon, *Poultry Sci.* **17**, 72 (1938); A. L. Moxon and W. E. Poley, *ibid.* p. 77.
161. W. E. Poley, W. O. Wilson, A. L. Moxon, and J. B. Taylor, *Poultry Sci.* **20**, 171 (1941).
162. V. R. Potter and C. A. Elvehjem, *Biochem. J.* **30**, 189 (1936); *J. Biol. Chem.* **117**, 341 (1937).
163. J. F. Proctor, D. E. Hogue, and R. G. Warner, *J. Anim. Sci.* **17**, 1183 (1958).
164. J. Quarterman, C. F. Mills, and A. C. Dalgarno, *Proc. Nutr. Soc.* **25, xxiii** (1966).
165. S. Ravikovitch and M. Margolin, *Ktavim* **7**, 41 (1957); *Emp. J. Exp. Agr.* **27**, 235 (1959).
166. W. O. Robinson, *J. Ass. Off. Agr. Chemists* **16**, 423 (1933).
167. I. Rosenfeld, *Proc. Soc. Exp. Biol. Med.* **111**, 670 (1962).
168. I. Rosenfeld and O. A. Beath, *J. Nutr.* **30**, 443 (1945).
169. I. Rosenfeld and O. A. Beath, *J. Agr. Res.* **75**, 93 (1947); *Proc. Soc. Exp. Biol. Med.* **87**, 295 (1954).
170. I. Rosenfeld and O. A. Beath, "Selenium." Academic Press, New York, 1964.
171. I. Rosenfeld and H. F. Eppson, *Wyo., Agr. Exp. Sta., Bull.* **414** (1964).
172. P. G. Roughan, *N. Z. J. Agr. Res.* **8**, 607 (1965).
172a. R. M. Salisbury, M. C. Armstrong, and K. G. Gray, *N. Z. Vet. J.* **1**, 51 (1953).
173. R. M. Salisbury, J. Edmondson, W. S. H. Poole, F. C. Bobby, and H. Birnie, *World's Poultry Congr., Proc., 12th, 1962*, p. 379 (1962).
174. J. R. Schubert, O. H. Muth, J. E. Oldfield, and L. F. Remmert, *Fed. Proc., Fed. Amer. Soc. Exp. Biol.* **20**, 689 (1961).
175. K. Schwarz, *Proc. Soc. Exp. Biol. Med.* **77**, 818 (1951); **78**, 852 (1951).
176. K. Schwarz, *Fed. Proc., Fed. Amer. Soc. Exp. Biol.* **20**, 666 (1961).
177. K. Schwarz, *Vitam. Hormo. (New York)* **20**, 463 (1962).
178. K. Schwarz, *Fed. Proc., Fed. Amer. Soc. Exp. Biol.* **24**, 58 (1965).
179. K. Schwarz, J. G. Bieri, G. M. Briggs, and M. L. Scott, *Proc. Soc. Exp. Biol. Med.* **95**, 621 (1957).
180. K. Schwarz and C. M. Foltz, *J. Amer. Chem. Soc.* **79**, 3293 (1957).
181. K. Schwarz and C. M. Foltz, *J. Biol. Chem.* **233**, 245 (1958).
182. K. Schwarz and A. Fredga, *J. Biol. Chem.* **244**, 2103 (1969).
183. M. L. Scott, *in* "Selenium in Biomedicine" (O. H. Muth, ed.), p. 231. Avi Publ. Co., Westport, Connecticut, 1967.
184. M. L. Scott, *Symp. Fat-Soluble Vitam., 1969*. Univ. Wisconsin Press, 1970.
185. M. L. Scott, F. W. Hill, L. C. Norris, D. C. Dobson, and T. Nelson, *J. Nutr.* **56**, 387 (1957).
186. M. L. Scott, G. Olson, L. Krook, and W. R. Brown, *J. Nutr.* **91**, 573 (1967).
187. R. J. Shamberger and D. V. Frost, *Can. Med. Ass. J.* **100**, 682 (1969).
188. R. L. Shirley, M. Koger, H. L. Chapman, P. E. Loggins, R. W. Kidder, and J. E. Easley, *J. Anim. Sci.* **25**, 648 (1966).
189. S. B. Slen, A. S. Demiruren, and A. D. Smith, *Can. J. Anim. Sci.* **41**, 263 (1961).
190. M. I. Smith, *Pub. Health Rep.* **54**, 1441 (1939).
191. M. I. Smith, K. W. Franke, and B. B. Westfall, *Pub. Health Rep.* **51**, 1496 (1936); M. I. Smith and B. B. Westfall, *ibid.* **52**, 1375 (1937).
192. M. I. Smith, E. F. Stohlman, and R. D. Lillie, *J. Pharmacol. Exp. Ther.* **60**, 449 (1937).

193. M. I. Smith, B. B. Westfall, and E. F. Stohlman, *Pub. Health Rep.* **52**, 1171 (1937); **53**, 1199 (1938).
194. A. L. Tappel, *Fed. Proc., Fed. Amer. Soc. Exp. Biol.* **24**, 73 (1965); A. L. Tappel and K. A. Caldwell, *in* "Selenium in Biomedicine" (O. H. Muth, ed.), p. 345. Avi Publ. Co., Westport, Connecticut, 1967.
195. H. H. Taussky, A. Washington, E. Zubillaga, and A. T. Milhorat, *Nature* (*London*) **200**, 1211 (1963); **206**, 509 (1965).
196. J. N. Thompson and M. L. Scott, *J. Nutr.* **97**, 335 (1969); **100**, 797 (1970).
197. S. F. Trelease and H. M. Trelease, *Amer. J. Bot.* **26**, 530 (1939).
198. L. A. Tscherkes, S. G. Aptekar, and M. N. Volgarev, *Byull. Eksp. Biol. Med.* **53**, 78 (1961).
199. L. A. Tscherkes, M. N. Volgarev, and S. G. Aptekar, *Acta Unio Int. Contra Cancrum* **19**, 632 (1963).
200. R. C. Wahlstrom and O. E. Olson, *J. Anim. Sci.* **18**, 141 (1959).
201. R. C. Wahlstrom, L. D. Kamstra, and O. E. Olson, *J. Anim. Sci.* **14**, 105 (1955).
202. T. Walsh, G. A. Fleming, R. O'Connor, and A. Sweeney, *Nature* (*London*) **168**, 881 (1951).
203. J. H. Watkinson, *Anal. Chem.* **38**, 92 (1966).
204. J. H. Watkinson and E. B. Davies, *N. Z. J. Agr. Res.* **10**, 116 and 122 (1967).
205. J. G. Welch, W. G. Hoekstra, A. L. Pope, and P. H. Phillips, *J. Anim. Sci.* **19**, 620 (1960).
206. P. D. Whanger, O. H. Muth, J. E. Oldfield, and P. H. Weswig, *J. Nutr.* **97**, 553 (1969).
207. P. D. Whanger, P. H. Weswig, O. H. Muth, and J. E. Oldfield, *J. Nutr.* **99**, 331 (1969).
208. B. B. Westfall, E. F. Stohlman, and M. I. Smith, *J. Pharmacol. Exp. Ther.* **64**, 55 (1938).
209. G. F. Wilson, *N. Z. J. Agr. Res.* **7**, 432 (1964).
210. C. I. Wright, *J. Pharmacol. Exp. Ther.* **68**, 220 (1940); *Pub. Health Rep.* **53**, 1825 (1940).
211. E. Wright, *N. Z. J. Agr. Res.* **8**, 284 (1965).
212. P. L. Wright and M. C. Bell, *Proc. Soc. Exp. Biol. Med.* **114**, 379 (1963).
213. P. L. Wright and M. C. Bell, *J. Nutr.* **84**, 49 (1964).
214. P. L. Wright and M. C. Bell, *Amer. J. Physiol.* **211**, 6 (1966).
215. P. L. Wright and F. R. Mraz, *Poultry Sci.* **43**, 947 (1964).
216. S. H. Wu, J. E. Oldfield, O. H. Muth, P. D. Whanger, and P. H. Weswig, *Proc. West. Sect. Amer. Soc. Anim. Sci.* **20**, 85 (1969).
217. S. Young and R. F. Keeler, *Amer. J. Vet. Res.* **23**, 955 and 966 (1962).
218. S. Young, W. W. Hawkins, and K. F. Swingle, *Amer. J. Vet. Res.* **22**, 416 (1961).
219. M. K. Yousef, W. J. Coffman, and H. D. Johnson, *Nature* (*London*) **219**, 1173 (1968).
220. H. Zalkin, A. L. Tappel, and J. P. Jordan, *Arch. Biochem. Biophys.* **91**, 117 (1960).

# 13

# FLUORINE

## I. Introduction

An association between high-fluoride intakes and dental defects was first demonstrated experimentally in rats in 1925 (99). By 1931, chronic endemic fluorosis in man and farm stock was discovered in several parts of the world (27,147,165). Fluorine-bearing fumes and dusts from various industrial plants processing fluoride-containing raw materials, such as aluminum and superphosphate works, steel mills, and enamel factories, were also found to constitute a health hazard to man and animals living in nearby areas (3,129). In some circumstances the use of rock phosphates as mineral supplements resulted in a further type of fluorine hazard to livestock. These discoveries combined to stimulate studies in many countries of the distribution of fluorine in rocks, soils, water, plants, animal tissues, and the atmosphere, and of the maximum safe levels of dietary fluoride in different forms.

Studies of the physiology and toxicology of fluorine received an additional impetus in the late 1930's when it was discovered that the fluoride ion can play a significant role in the prevention of human dental caries. The margin between beneficial and toxic intakes of fluoride appeared so small that long-term investigations were initiated with man and animals to assess the benefits and the hazards of artificially fluorided water supplies. Recent evidence pointing to the benefits of certain levels of fluoride ingestion upon the maintenance of a normal skeleton in the human adult has opened up a new area of interest in fluorine as an element of importance in human health and nutrition.

## II. Fluorine as an Essential Element

The classification of fluorine as an essential or as a nonessential element depends upon the criteria employed in determining essentiality. If an essential element is considered as one that must be provided in the diet to permit survival, then fluorine cannot yet be regarded as essential for plants, microorganisms, or animals. Fluoride has been known to be a constant constituent of bones and teeth since 1805, and efforts to produce animals without fluoride in the bone have so far failed. Attempts to demonstrate a vital function for this element by the use of low-fluorine diets have also failed.

Sharpless and McCollum (136) fed rats a highly purified low-fluorine diet and compared their performance with that of other rats fed the same diet to which 10 ppm F had been added. Growth and reproduction were equally satisfactory on the two diets, and there were no abnormalities in tooth or bone structure. Evans and Phillips (38) found that a mineralized milk diet containing 1.6 ppm F (dry basis) was adequate for the growth, health, and fertility of rats through five generations. The bones and teeth of the animals remained strong, smooth, and well-calcified, and there was no depletion of body stores of fluoride. No improvement from graded doses of supplementary fluoride was observed. Lawrenz (76) fed rats on a basal diet containing even lower-fluoride levels (0.47 ppm) for a period of 207 days. Evidence of fluorine depletion was obtained, but the growth rate and the weight of the skeleton and teeth were similar to those of comparable animals receiving a diet containing 2.5 ppm F. No signs of malnutrition or dental defects were apparent. Essentially similar results were obtained in later studies with rats fed a low-fluorine diet through three generations (90a) and a "minimal" fluoride diet containing less than 0.005 ppm F for 10 weeks (33). Significantly lower bone fluoride levels than those of control animals given the same diet plus 2 ppm F were demonstrated, but no significant treatment differences in weight gains, health, feed utilization, or in serum phosphatase activities could be detected. Serum and liver levels of the glutamic oxalacetic and glutamic pyruvic transaminases and of isocitric dehydrogenase were also investigated (33). The only significant differences noted were an increase in the serum isocitric dehydrogenase and a decrease in the activity of this enzyme in the liver of the rats on the minimal fluoride diet.

By contrast, McClendon and Gershon-Cohen (92) fed weanling rats for 66 days upon a basal diet prepared from materials grown in water culture and claimed to be "fluorine-free." The rats averaged only 51 g live weight and had 10 carious molars per animal, compared with an average of 128 g live weight and 0.5 carious molar per animal in rats fed the same diet for the same period but receiving supplementary fluorine. Although these differences are highly significant, they are difficult to evaluate in the absence of any data on the actual fluorine levels in the diets or in the tissues of the animals.

The results of the experiments just cited do not permit the conclusion that fluorine is either an essential or a nonessential component of the diet of rats, since different results could conceivably be obtained with diets still lower in fluorine content. Indisputable evidence is available that fluoride, additional to that normally consumed by man in most areas, is required to confer maximal resistance to human dental caries. In addition, appropriate levels of dietary fluoride assist in the maintenance of a normal skeleton and in reducing the incidence of osteoporosis in the mature, adult population. If an essential element is defined as one which is ordinarily required for health and well-being, under the usual conditions in which individuals live, then in the light of the above evidence, fluorine must be considered as an essential element in human nutrition (58).

## III. Fluorosis in Farm Animals

### 1. *Sources of Fluorine*

The pastures, fodders, and grains commonly consumed by animals can rarely be incriminated as a major factor in the incidence of chronic fluorosis in animals or man, unless these materials have been subjected to soil or surface contamination from fluoride-bearing dusts, fumes, or waters. Most plant species have a limited capacity to absorb fluorine from the soil, even when fluoride-containing fertilizers are applied (56). Certain plant species, notably the tea plant and the camellia, are exceptional in this respect. Fluorine concentrations of 100 ppm or more are common in these species (50,55,125), and levels as high as 1700–1900 ppm in the dry matter have been reported (90,127). Highly toxic concentrations, mainly as fluoracetate, also occur naturally in the South African plants, *Dichapetulam cymosum* and *Dichapetulam toxicaria* (88). Uncontaminated pasture plants, forages, and grains, by contrast, are much lower in fluorine content. In two studies of English mixed pastures, the fluorine concentrations ranged from 2 to 12 ppm (5) and from 2 to 16 (mean 5.3) ppm (dry basis) (6). Most of the Australian grasses examined by Harvey (57) contained only 1-2 ppm F, with some samples that had been irrigated with fluorided water reaching 9-13 ppm. Cereal and other grains and their by-products usually contain 1-3 ppm F (73,94).

Feeds of animal origin, if free from bone, are similarly low in fluorine since the dry matter of the soft tissues and fluids of the body rarely contain more than 2-4 ppm F. Milk and milk products are even lower in fluorine. Normal cow's milk has been reported to contain 1-2 ppm F (dry basis) (94).

The fluorine levels of common, uncontaminated feed materials are too low to constitute a fluorine hazard to livestock. The ingestion of toxic or potentially toxic amounts of fluorine by farm animals occurs most commonly under the following circumstances: (*a*) in areas where the drinking water is naturally high in fluoride—this is known as endemic fluorosis and usually results in chronic rather than acute fluorine intoxication, (*b*) when the animal's diet is injudiciously supplemented with fluoride-bearing minerals as a source of extra calcium and phosphorus, and (*c*) in restricted areas adjacent to industrial plants or mines which result in the emission of fluorine fumes or fluoride dusts which contaminate the soil and the herbage and water consumed by the animals. Industrial fluorine intoxication of this type has been extensively studied and thoroughly considered in the classic monograph of Roholm (129) and in several more recent reviews (3,6,23,65,153).

A disease of sheep, cattle, horses, and man, known locally as "darmous," has long been recognized in parts of North Africa. It was characterized by Velu (165) in 1931 as a chronic fluorosis caused by contamination of the herbage and water supplies of the affected areas with dusts blown from rock phosphate deposits and mines. An overwhelming proportion of the fluoride ingested in these areas comes from this surface contamination of the herbage and the water supplies (45).

The chronic endemic fluorosis of sheep and cattle that occurs over extensive areas in parts of the United States of America, Australia, India, and South Africa results primarily from the consumption of the naturally highly fluorided waters of those areas. Such waters are invariable derived from deep wells or bores in which the fluoride comes from deep-seated rock formations rather than from surface leachings. The surface water supplies in these areas are not unduly high in fluoride. They usually contain less than 1 ppm F, or even only 0.1 ppm F or less, as in nonfluorosis areas (26). In the endemic fluorosis areas, by contrast, the deep well or bore waters upon which the animals are largely dependent commonly contain 3-5 ppm F and not infrequently 10-15 ppm F (26,57). Where evaporation has taken place in troughs and bore drains, levels as high as 40 ppm have been observed (57). The consumption of fluorided waters of this nature will inevitably result in fluorosis in animals if long continued.

The consumption of mineral phosphates as dietary supplements presents a further fluorine hazard to livestock, depending upon the duration and level of intake, the source of the phosphate, and the treatment it has received during processing. North African and North American rock phosphates are mostly higher in fluoride content (3-4% F) than those from Pacific and Indian Ocean island deposits which usually contain only about half this concentration or less. The former sources can supply harmful amounts of fluoride to livestock when used in the proportions ordinarily employed to supplement farm rations (121); the latter sources apparently do not constitute a similar fluorine hazard (149). During the manufacture of superphosphate and dicalcium phosphate about 25-50% of the original fluoride may be lost. The fluoride content of mineral phosphates can be reduced further by means of special defluorinating processes.

### 2. *Clinical and Pathological Signs of Fluorosis*

Lethal doses of fluorine produce a rapid loss of appetite and body weight, accompanied by gastroenteritis, muscular weakness, clonic convulsions, pulmonary congestion, and respiratory and cardiac failure (151). The ingestion of smaller amounts at first produce no observable ill-effects, although those amounts may eventually prove toxic. During this latent period the animal is protected by two physiological mechanisms acting concurrently: (*1*) a rise in urinary excretion of fluorine, which follows directly upon increased intakes but soon reaches a maximum that cannot easily be surpassed, and (*2*) deposition of retained fluorine in the skeleton—a process that initially proceeds rapidly and then more slowly as the fluorine levels in the bones rise until a stage of saturation is reached. Beyond this point, which approximates 30-40 times the fluorine level of normal bone, "flooding" of the soft tissues with fluorine occurs, plasma fluoride levels rise, and metabolic breakdown occurs. When such excretion and retention ceilings are reached, there is a voluntary refusal of food, so that typical starvation phenomena are imposed upon the signs of fluorine toxicosis (122). The consequent reduction in fluorine intake may be considered the last physiological mechanism available to the animal for protecting the vulnerable soft tissues and fluids against excessive fluorine concentrations.

Clinical signs of fluorosis, notably dental defects and some loss of appetite, may appear well before bone saturation is reached, if the fluoride intake is sufficient to produce a significant rise in plasma fluoride levels. Such changes have been observed in animals fed various levels of dietary fluoride, without any discernible effects upon productive functions. Tooth formation and appetite are extremely sensitive to a rise in plasma fluoride concentrations (20,142). Anemia has also been reported in fluorosis but this was mostly associated with very high levels of intake (23,114,133). More recent studies of chronic fluorosis in cattle have detected no significant changes in blood hemoglobin concentrations (20, 61,66).

Animals exposed to excess fluorine prior to the eruption of the permanent teeth develop dental defects which are the most sensitive visible indicators of elevated fluorine intakes and of elevated plasma fluoride levels (20). The teeth become modified in shape, size, color, orientation, and structure. The incisors become pitted and the molars abraded. There may also be exposure of the pulp cavities due to fracture or wear. Once the permanent teeth are fully formed and erupted, their architecture is no longer susceptible to high-fluoride intakes (44,65,156). However, the dentin of mature teeth, exposed after maturity, increases in fluorine content with time and with dose, as do bones (161).

Osseous lesions are similarly characteristic of fluorosis and can occur in animals exposed at any age. Exostoses develop, particularly of the jaw and long bones, usually accompanied by a general thickening and change in shape of the bones. Such bones appear chalky, rough, and porous compared with normal bones. Mineralization of the tendons at the point of attachment to the long bones may also occur so that the joints become thickened and ankylosed. When fluorosis develops to this stage the animal becomes stiff and lame, making movement difficult and painful.

The histopathology of skeletal fluorosis has been extensively studied, sometimes with divergent conclusions (30,39,129,170). From the detailed studies of skeletal fluorosis in the sheep and the rabbit carried out by Weatherall and Weidmann (170), it appears that (*a*) bone resorption is a marked but not invariable feature of the condition, (*b*) the newly formed bone of exostoses is histologically similar to normal bone found in the fetus, (*c*) the exostoses of fluorotic bone are less calcified than nonfluorotic bone and contain wide areas of osteoid, and (*d*) no area of hypermineralization occurs in any part of fluorotic bone. Overproduction of osteoid has been consistently observed in skeletal fluorosis (30, 116, 170), and there is evidence of a disturbance of mucopolysaccharide production in the bones and teeth of fluorosed pigs (12). These findings point clearly to fluorine-induced defects in the mechanism of osteogenesis and chondrogenesis in the functioning bone cells.

There is no evidence that fluoride ingestion directly affects reproduction or lactation or that it significantly depresses the digestibility or the utilization of the energy and protein of the diet (139). Nor is inappetence necessarily a serious feature of fluorosis during the latent period. Appetite and the productive processes are impaired during the more advanced stages of the disease, or earlier if plasma fluoride levels are high enough. These effects are secondary to the inappetence and to the dental lesions and bone abnormalities, which can seriously limit the willingness and ability of the animal to gather and masticate fodder. Thus the reduced milk production of cows receiving toxic levels of fluorine results from, and is in proportion to, the reduction in food consumption that occurs (156). Similarly, the poor lamb and calf crops of flocks and herds in fluorosis areas result from mortality of the newborn, due to the impoverished condition of the mother rather than to a failure of the reproductive process itself (57). Such secondary effects are likely to be more severe under harsh grazing conditions because the reduced food consumption brought about by the lesions of the teeth and joints makes movement and mastication difficult and painful, whereas under pen-feeding conditions, the feed is more accessible and usually more digestible. In fact, reproductive performance is extremely resistant to toxic levels of fluorine as long as appetite is not seriously depressed (115). Signs of fluorosis rarely appear in newborn or suckling animals, except under conditions of acute fluorine intoxication, because both placental and mammary transfer of fluorine is limited.

### 3. *Fluorine Retention in Bones and Teeth*

The levels of fluorine in the bones and teeth increase in proportion to the amount, form, duration, and continuity of the fluorine intake, and with the age of the animal (Tables 43 and 45). The relationship is logarithmic rather than linear, so that the rate of deposition falls with increasing fluoride content of the bone (48,91). The rate also falls with increasing age of the animal, since older animals retain less and deposit less in their bones than do younger animals (103,157). In normal adult farm animals, not unduly exposed to fluorine, the concentrations in whole, dry, fat-free bones rarely exceed 1200 ppm F and more usually lie within the range 300-600 ppm (52,57,161,170). The fluorine concentrations of normal teeth parallel those of the long bones but usually at lower levels. Thus normal enamel is reported to contain 100-270 ppm, normal dentin 240-625 ppm, and normal molar teeth 200-537 ppm F (dry, fat-free basis) (25,95,114,121).

TABLE 43
*Effect of Sodium Fluoride in the Diet on Skeletal Fluoride*[a]

| F added (ppm) | F (ppm) dry fat-free weight | | | |
|---|---|---|---|---|
| | Metacarpus | Metatarsus | Frontal | Rib |
| 0 | 645 | 522 | 645 | 677 |
| 20 | 2580 | 2810 | 3270 | 4000 |
| 30 | 4030 | 3870 | 4750 | 5380 |
| 40 | 4900 | 4640 | 5640 | 5770 |
| 50 | 5710 | 5690 | 7250 | 7990 |

[a]From Suttie (153).

In dairy cattle, fluorine toxicosis is associated with levels in excess of 5500 ppm in compact and 7000 ppm F in cancellous bone, with a "saturation" point of the order of 15,000 to 20,000 ppm F (122,161). Concentrations between 4500 and 5500 ppm indicate a marginal zone, and those below 4500 ppm F are considered to be innocuous (161). The toxic thresholds for fluorine in the bones of sheep have been placed somewhat lower, namely 2000–3000 ppm in bulk cortical and 4000–6000 ppm in bulk cancellous bone (68). Apparently normal swine bones have been reported to contain upwards of 3000 to 4000 ppm F (72).

The remarkable capacity of the skeleton to immobilize safely and retain fluorine, so acting as a buffer against systemic effects, varies with the age of the animal and its rate of skeletal growth. It also varies in different parts of the skeleton (see Table 43) and in different parts of the same bone. During the period of most rapid skeletal development the percentage and the total fluorine content of teeth and bones increase rapidly. In the fully mature bone the rate of deposition and retention is greatly reduced (119).

Bones that exhibit a marked tendency to exostosis formation, such as the mandibles and shaft bones, have a particularly high affinity for fluorine. The exostoses are invariably higher in fluorine and can be as much as 20 times higher than the other parts of the bones. Fluoride is also incorporated more rapidly into active areas of bone growth than into static regions, into cancellous parts faster than cortical (compact) parts, and into surface (both periosteal and endosteal) regions of the shaft of adults faster than into the middle regions (174). In a study of the metatarsal bones of adult cows variously exposed to excess fluoride intakes, Suttie and Phillips (160) found a gradation of fluoride concentration from a maximum at the periosteal surface to a minimal value in the interior of the shaft. They also found higher-fluoride concentrations at the ends of the shaft than at the center. These findings indicate the importance of bone sampling procedures when bone fluoride levels are used as diagnostic criteria of fluorosis.

Studies with radioactive fluorine indicate that uptake by bone depends upon its vascularity and growth activity (169). Hodge (64) postulates that fluorine deposition in bone occurs by two processes. The first involves a rapid exchange of F ions in the tissue fluids with OH or $CO_3$ ions on the mineral crystal surface. This hypothesis receives support from a study with poultry (162). The second process involves slow bone formation leading to storage, in which the fluorine is incorporated into the hydroxylapatite lattice, i.e., the bone salt itself. The latter process suggests that fluorine uptake by bone depends upon the same factors, namely circulation of tissue fluids, available surface area, and exchangeable fraction of bone, which influence normal bone-salt constituents. Since there is no change in the Ca:P ratio of fluorotic bone, it seems that the phosphate group of bone salt is not replaced by fluoride ion. However, the carbonate content of fluorotic bone is decreased and the magnesium content is increased (72,91,162,175). This suggests a replacement of the carbonate group by fluoride ion and possibly a precipitation of some fluorine as $MgF_2$. Acceptance of a two-phase deposition of fluorine in bone implies that surface concentration may depend mainly on dose but the amount of fluorine in the bulk of the bone will depend upon dose and upon time.

### 4. *Fluorine in Soft Tissues, Blood, and Urine*

a. *Soft Tissues.* The concentrations of fluorine in the soft tissues of normal animals are generally low and do not increase with age. Fluorine does not concentrate in any tissues other than the bones and teeth, although the placenta and the aorta sometimes contain elevated levels of fluorine, probably as a result of calcification which secondarily holds fluoride (65). The kidneys are usually richer in fluorine than other organs but this may be due in part to retained urine (Table 44). Fluorine does not accumulate significantly in the thyroid gland. Investigations with stable (127) and radioactive fluorine (59,117,169) show clearly that the thyroid has no special ability to concentrate this element, even when intakes are high. Thyroid fluorine levels are normally no higher than those of many other soft tissues.

Increases in soft tissue concentrations are small in chronic fluorosis (see Table 44). Dairy cows receiving a ration supplemented with 50 ppm F as NaF for 5 1/2 years, only increased the fluorine levels in their heart, liver, thyroid, and pancreas two- to threefold above the 2-3 ppm F (dry basis) found for these tissues in control cows (161). Even smaller

TABLE 44
*Fluorine Levels in Soft Tissues of Animals*[a]

| Tissue | Rats[b] | | Sheep[c] | | Cows[d] | |
|---|---|---|---|---|---|---|
| | Normal | 2 mg F/day for 76 days | Normal | 10 ppm F in water for 2 years | Normal | 50 ppm F in ration for 5 1/2 years |
| Liver | 0.21 | 0.28 | 3.5 | 2.4 | 2.3 | 3.6 |
| Kidney | 0.62 | 1.50 | 4.2 | 20.0 | 3.5 | 19.3 |
| Thyroid | – | – | 3.0 | 7.6 | 2.1 | 7.3 |
| Heart | 2.6 | 5.4 | 3.0 | 2.3 | 2.3 | 4.6 |
| Pancreas | – | – | 2.8 | 3.2 | 2.8 | 4.2 |
| Muscle | 0.53 | 1.60 | – | – | – | – |

[a]Expressed as parts per million (dry basis).
[b]From Venkateswardu and Narayanarao (167).
[c]From Harvey (57).
[d]From Suttie *et al.* (161).

increases were observed in the same tissues of sheep consuming water containing 5 or 10 ppm F for a period of 2 years (57). Similar results were reported for rats receiving 2 mg F/day for 62 days (167). Schroeder and co-workers (134) were unable to detect any soft tissue accumulation in mice given 10 ppm F in the drinking water for a period of 2 years.

The tissues of newborn and suckling animals are usually lower in fluorine than the corresponding tissues of their mothers. In one study, rats and rabbits were fed varying levels of added fluorine, up to 300 ppm of the diet, for several weeks before and during pregnancy, and the fluorine accumulated by the dam and offspring was measured. On the basal, low-F diet the total fluorine in the fetuses at term was negligible. Some increase occurred at the higher intakes but the amounts accumulated were extremely small compared with those in the mothers (87). Suckling rats and puppies subsisting on the milk of fluoride-fed mothers accumulated more fluorine in their bodies than similar animals consuming the milk of mothers on normal diets (87,109,116). However, the former animals may have acquired some fluorine directly from the diet or water of their mothers. The bulk of evidence indicates that the transfer of fluorine across the mammary barrier into the milk is small, except at very high fluorine intakes.

By contrast, birds consuming high-fluoride diets can readily transfer fluorine to the egg, especially into the yolk. The yolk of eggs from hens on normal, low-F diets contains only 0.8-0.9 ppm F (120). This concentration can be raised as high as 3 ppm by supplementing the hen's diet with 2% rock phosphate (120) or by injection of the hens with NaF (130).

b. *Blood.* The concentration of fluorine in plasma is so low in normal animals (0.1-0.2 ppm) that, until recently, its precise estimation was extremely difficult. The development of the micro methods of Singer and Armstrong (145) has improved the position greatly and a clearer picture of the relation of plasma fluoride levels to fluoride intakes by the animal has now emerged. The initial studies by these workers emphasized the capacity of the body to maintain low plasma fluorine concentrations, in the face of elevated intakes of the element, by means of skeletal deposition and rapid urinary excretion. Human plasma fluoride levels remained constant under varying concentrations of fluoride in the communal water supplies (143). The plasma fluoride levels of rats also remained unaltered by fluoride intakes sufficient to produce a threefold increase in skeletal fluoride load (144). Apparently the levels of ingested fluoride in those experiments were not high enough, because Armstrong *et al.* (8) subsequently reported significant increases in rat plasma fluoride concentrations following large increments in dietary fluoride. Levels as high as 1.0 ppm F were obtained. Even higher plasma fluoride concentrations up to 3.3 ppm were reported by Simon and Suttie (142) in rats receiving as much as 600 ppm of dietary fluoride as NaF.

Elevated plasma fluoride levels during periods of high-fluoride intakes are well documented in cattle and sheep. Dairy cows, ingesting 109 ppm F in the diet for 7 years, averaged 0.45 ppm F in the blood, compared with 0.07 ppm in control animals (140). In a long-term study of dairy cows, continuously and periodically exposed to high-fluoride intakes, plasma fluoride concentrations were found to be related to the current level of fluoride ingestion (20). The plasma fluoride levels of controls were consistently below 0.1 ppm, whereas those of the fluoride-treated animals were generally significantly higher. A level of 1.0 ppm F in the plasma represented a high level, observed only after extended periods of ingestion. A feature of this experiment was the rapidity with which plasma fluoride levels changed in response to changes in fluoride intake. A further finding of great importance was the extreme sensitivity of the teeth, during the period of formation, to small changes in plasma fluoride concentration. Severe dental lesions were obtained when plasma fluoride levels approached 0.5 ppm or more. Significant but less severe damage occurred with plasma fluoride above 0.2 ppm, whereas few adverse effects could be attributed to levels below 0.2 ppm F in the plasma (20).

Marked diurnal variation in plasma fluoride levels has been observed in the rat (137), the sheep (142), and in dogs and humans following the administration of a single oral dose (108). As Shearer and Suttie (137) have pointed out, plasma samples must be taken very soon after the actual ingestion of the fluoride if they are to reflect the total daily intakes. Apparently the loss of fluoride to the urine and the skeleton is so rapid that control concentrations are approached within hours of the completion of intake. The magnitude and rapidity of the short-term changes that can take place in plasma fluoride levels, compared with those in the liver and femur, are illustrated in Table 45.

c. *Urine.* A positive correlation between urinary fluoride levels and the levels of fluoride ingested has been demonstrated in several animal species. Data illustrating this relationship in cattle are given in Fig. 22. In sheep and cattle not exposed to excess fluorine the urinary concentration rarely exceeds 10 ppm and is more usually close to 5 ppm. With elevated fluoride intakes the urinary fluoride levels rise quickly to 15 to 30 ppm and may reach upper limits of 70 to 80 ppm. Higher values are occasionally observed and great variation can exist among animals con-

TABLE 45

*Tissue Fluorine Concentration after Ingestion by Rats of Diets Supplemented with Fluoride*[a]

| Tissue | Control | 200 ppm F in diet | | 450 ppm F in diet | |
|---|---|---|---|---|---|
| | | 7 am[b] | 3 pm | 7 am[b] | 3 pm |
| Plasma (ppm F) | 0.09±0.06 | 0.32±0.09 | 0.09±0.05 | 1.31±0.22 | 0.24±0.16 |
| Liver (ppm wet wt) | 0.41±0.05 | 0.52±0.05 | 0.48±0.06 | 0.90±0.05 | 0.54±0.05 |
| Femur (ppm in ash) | 332±19 | 810±29 | 809±29 | 1144±40 | 1192±43 |

[a]From Shearer and Suttie (137).
[b]End of normal nocturnal feeding period.

suming the same amounts of fluorine, as well as among samples from the same animal taken on different days or at different times on the same day (138,139,153,155). In the experiments of Suttie and co-workers (155), normal cows excreted urine with a concentration below 5 ppm F; cows that were on the border line of fluorine toxicity, as judged by other criteria, excreted urine with 20-30 ppm F; and those with systemic signs of toxicity excreted urine containing over 35 ppm F. Similar results were obtained by Shupe *et al.* (138,139) with heifers fed different amounts and forms of fluoride for prolonged periods.

Urinary fluorine does not arise only from the fluorine in the feed and the water. It may come also from the release of the element from skeletal sources. Mobilization and excretion of a portion of the fluoride

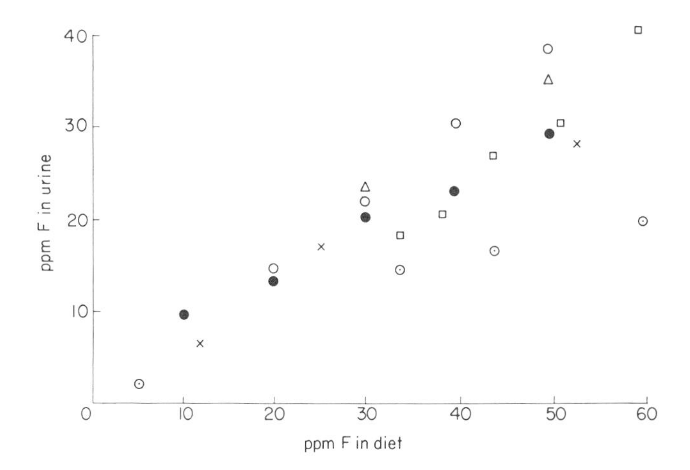


**Fig. 22.** The general relationship between urinary fluoride concentration of cattle and the fluoride content of the ration. [Taken from data assembled by Suttie (153).]

present in the skeleton of animals removed from a high-fluoride diet has been observed in rats (48,103,132), humans (81), and cattle (20,123). High urinary fluoride levels can, therefore, reflect either current ingestion or previous exposure to high intakes. In some circumstances cattle can maintain above-normal urinary fluoride levels for long periods after high intakes have ceased (123). A rapid return to normal concentrations in the urine, and in the blood, can also occur following transfer of cattle from high- to low-fluoride intakes (20).

### 5. *Enzyme Changes in Chronic Fluorosis*

Fatty acid oxidase activity is inhibited by 0.01 *M* $F^-$ (70), and the related acetate-activating system by 0.001 to 0.005 *M* $F^-$ *in vitro* (4). Following their discovery that high dietary fat enhances fluorine toxicity in rats (103), Sievert and Phillips (141) observed a marked decline in the fatty acid oxidase activity of the kidney and a decrease in the nitrogen and fat contents of that organ in fluorosis. They were unable to demonstrate any such effects upon the liver. Fecal lipid levels were then shown to be greatly increased in the fluorotic rat, without any increase in metabolic fat or impairment of efficiency in utilizing dietary free fatty acids (159). It was concluded that there is a partial inhibition of lipase activity in the intestine of the fluorotic rat. Subsequently it was found that a larger portion of a palmitate-$^{14}C$ dose remains in the serum and liver lipids of fluoride-fed rats than occurs in normal animals (181), suggesting a blockage of fatty acid utilization. An impaired ability to metabolize fat must, therefore, be included among the biochemical lesions of fluorosis.

Disturbances in carbohydrate metabolism also occur in fluorosis. Glycogen turnover is depressed in fluoride-fed rats and in liver slices prepared from those animals (180). The level of liver glucose-6-phosphate dehydrogenase decreases significantly in fluorotic rats (21) but the decrease is an indirect effect brought about by changes in the pattern of food consumption. In these experiments fluoride ingestion not only reduced food consumption, it resulted in a "continual nibbling" pattern of food intake, which, in turn, decreased the glucose-6-phosphate dehydrogenase activity. Similarly, the inability of the fluoride-fed rat to metabolize glycogen in a normal manner appears to be indirectly mediated. Rats consuming as much as 450 ppm F in the diet catabolized glucose at a normal rate and maintained a normal level of liver glycogen, but they were unable to metabolize glycogen normally, due to some effect at the liver enzyme level (179). Since the level of the adaptive hepatic enzymes regulating glycogen turnover can be varied by regulating food intake (152), it seems likely that the level of those enzymes is reduced in fluoride-fed animals as a consequence of the changed pattern of food intake. Such a reduced enzyme level could account for the impaired glycogen turnover. The above observations, as Zebrowski and Suttie (179) have stated, "point out a defect in carbohydrate metabolism and add to the list of metabolic changes observed in animals fed toxic amounts of fluoride. The evidence also indicates that this defect is probably secondary to a more direct effect of fluoride on some other area of metabolism."

In an attempt to throw some light on these "other areas" and to identify specific sites of fluoride action on cellular metabolism, Carlson and Suttie (22) examined the effects of fluoride upon HeLa cells. Limited fluoride inhibition of HeLa cell growth was found to be associated with a decreased efficiency of glucose utilization. The per cent conversion of glucose to lactic acid and $CO_2$ production from glucose were unaffected. It was concluded that fluoride (30 ppm) did not directly and specifically inhibit glycolysis under the conditions of the experiment and that the alterations in carbohydrate metabolism may have been related to the slower growth rate brought about by the fluoride. However, fluoride produced a specific metabolic alteration in rapidly growing HeLa cells by decreasing the cellular adenosine triphosphate more rapidly and to a lower level than other metabolic inhibitors.

Many attempts have been made to correlate the level of alkaline phosphatase in the blood and tissues with the level of fluoride ingestion by animals. The results have not been consistent, either within or between species. Motzok and Branion (105) observed significant increases in alkaline phosphatase levels in the serum and bones of fluorotic chicks but not in the liver, kidneys, or intestinal mucosa. In one study with rabbits (175), serum alkaline phosphatase levels were elevated in fluorosis, whereas in another study with this species very little change could be detected when high-fluorine diets were fed (87). Many years ago, Phillips (118) observed a marked stimulation of alkaline phosphatase activity in the blood of dairy cows affected with fluorosis. The increase was dependent upon the amount of fluoride ingested and was correlated with the severity of clinical signs in the cattle. Later experiments by Olson and co-workers (113) revealed little correlation between the level of fluoride in the ration and the phosphatase activity of the blood of dairy cattle, mainly because of high individual variability independent of treatment.

Alkaline phosphatase activity in bone appears to be more securely related to fluoride intakes and can be used as a valuable diagnostic criterion of fluorosis. In cows and dairy heifers, Miller and Shupe (102) observed a close correlation between fluorine ingested and the fluorine content, osseous abnormalities, and alkaline phosphatase activity of bones. In an experiment lasting over 7 years, dairy heifers 6 months old at the commencement were fed on rations containing 12, 27, 49, and 93 ppm fluorine as NaF. Osseous abnormalities, excessive accumulation of fluorine in bone, and significant increases in alkaline phosphatase activity in bone occurred at the higher intakes of 49 and 93 ppm; no such changes were apparent at intakes of 12 or 27 ppm F. Selected data from this experiment are presented in Table 46.

### 6. *Diagnosis of Fluorosis*

There is no single completely satisfactory criterion of fluorosis in domestic animals, but when evidence from several criteria is combined a more certain diagnosis is possible. Phillips *et al.* (119) place the criteria that are important in recognizing the developing syndrome of fluorine toxicosis in the following order of reliability: (*1*) chemical analyses which indicate an increased amount of fluorine in the diet, the urine, the bones, and the teeth; (*2*) tooth effects such as chalkiness or mottling, erosion of enamel, enamel hypoplasia, and excessive wear; (*3*) systemic evidence as reflected by anorexia, inanition and cachexia, exostoses, and bone changes. Plasma fluoride levels and bone alkaline phosphatase activity may now be added to the first of these criteria, as just discussed. The dental defects, included above as the second criterion, apply only to animals exposed to high-fluoride intakes during the period of tooth formation. Tooth damage is, therefore, not necessarily related to current levels of fluoride intake.

According to Suttie (153), "analysis of bone fluoride content, and an estimation of the age of the animal, would perhaps give the best indication of potential damage to an animal. This measurement would be independent of day to day variations in intake, and would stress the accumulative nature of the disease process." Fluoride determination on tail bones obtained by a simple and safe biopsy technique provides a useful and reasonably accurate means of measuring bone fluorine accumulations in cattle, for either diagnostic or experimental purposes (16). The value of vertebral biopsies in the diagnosis of bovine fluoride toxicosis has been studied further by Suttie (154). As the fluoride content of the large metacarpal bone is well related to other more subjective signs of fluorosis, this worker compared the fluoride concentration of the coccygeal vertebrae and the large metacarpal of 114 cattle that had been subjected to a wide range of fluoride intakes. An extremely high correlation was found between the values for the two bones, the fluoride concentration of the dry fat-free metacarpus being approximately 50% of the ver-

TABLE 46
*Fluorine Content and Alkaline Phosphatase Activity of Bones from Cows Fed on Different Levels of Fluoride for 7 Years, 108 Days*[a]

| | Metatarsal bone[b] | | Ribs[c] | |
|---|---|---|---|---|
| Dietary F level (ppm) | F Content (ppm dry basis) | Phosphatase activity[d] | F Content (ppm dry basis) | Phosphatase activity[d] |
| 12 | 551 | 37 | 895 | 55 |
| 27 | 1312 | 81 | 2417 | 97 |
| 49 | 2545 | 123 | 5323 | 154 |
| 93 | 5290 | 134 | 7798 | 165 |

[a]From Miller and Shupe (102).
[b]Midshaft.
[c]Eleventh, twelfth, and thirteenth rib.
[d]Milligrams P hydrolyzed in 15 min per gram bone.

tebrae ash. These data confirm the value of fluoride determinations of vertebrae biopsy specimens as a diagnostic aid in bovine fluorosis.

## IV. Fluorine Tolerance

Tolerance levels of dietary fluorine are influenced by the age and species of the animal, the chemical form, duration, and continuity of fluorine ingested, the nature of the diet, and the conditions of grazing. Tolerance is also influenced by the criteria employed in deciding when a condition becomes intolerable. At massive intakes, resulting in immediate flooding of the soft tissues with fluorine, the importance of these interacting factors is diminished. At lower intakes, such as generally occur in chronic fluorosis, tolerance levels are highly dependent upon the extent to which one or more of the factors are operating. Under practical conditions of livestock husbandry the problem of fluorine tolerance is complicated by the fact that mild dental and skeletal changes and elevated bone, blood, and urine levels are not necessarily accompaned, for prolonged periods, by any measurable decline in health, growth, or productive performance, such as fertility or milk yield. The problem is complicated further by great variations in grazing, other environmental conditions, and in the continuity of intake of fluorine.

### 1. *Species Differences*

Poultry exhibit a greater tolerance to high-fluorine intakes than other livestock. Maximum safe dietary levels of 300 to 400 ppm F, as rock phosphate, have been reported for growing chicks and of 500 to 700 ppm for laying hens (47,49). Similar tolerance levels to those stated for chicks have been given for growing female turkeys, but 200 ppm F as sodium fluoride was found to result in decreased weight gains in young male turkeys (7). Mice seem to be particularly tolerant of high-fluoride intakes. Weber and Reid (171) found that a dietary level of 1500 ppm F (as NaF) was required to induce a significant growth reduction in young mice over a 3-week period, although a small reduction in feed consumption occurred at a level of 1000 ppm F. Rats are usually affected by about 500 ppm F in the diet.

With sheep, cattle, and pigs, 80-100 ppm F of the dry ration is on the border line of toxicity when fed as rock phosphate for long periods (104,121). When sodium fluoride is administered to dairy cows, for 5 1/2 years from 2 years of age, 40 ppm F is considered to be near the margin of tolerance (156). At 50 ppm F from this source, signs of fluorosis appeared within 3 to 5 years (156). When the cows were first exposed to 50 ppm of dietary fluorine as NaF at 4 to 6 years of age there were no adverse effects through three lactations, other than mild exostosis of the long bones and increased fluorine concentrations in the body (158). From the results of another experiment, beginning with young calves and lasting for 7 years, it was concluded that the tolerance for soluble fluoride is not more than 30 ppm F of the total dry diet (140). Suggested safe ration levels for different farm species, taken from the publication of Phillips *et al.* (119), are given in Table 47. Practical permissible limits for cattle dependent upon forages contaminated with inorganic fluorides from industrial sources have been proposed by Suttie (153) as follows: a yearly average of 40 ppm F (dry basis) in the herbage, based on monthly sampling; 60 ppm F for more than 2 consecutive months; or 80 ppm F

TABLE 47

*Safe Levels of Fluorine in the Total Ration of Livestock*[a]

| Species | NaF or other soluble fluoride (ppm F) | Rock phosphates or phosphatic limestones (ppm F) |
|---|---|---|
| Dairy cow | 30-50 | 60-100 |
| Beef cow | 40-50 | 65-100 |
| Sheep | 70-100 | 100-200 |
| Swine | 70-100 | 100-200 |
| Chickens | 150-300 | 300-400 |
| Turkeys | 300-400 | - |

[a]From Phillips *et al.* (119).

for more than 1 month, even though the yearly average does not exceed 40 ppm.

The minimum toxic levels of water-borne fluoride appear to be quite variable under field conditions, although the reported variations probably reflect differences in the volume and the continuity of the fluoridated water consumed. Water containing 7-8 ppm F induced no visible effect upon the teeth of sheep, cows, goats, and donkeys in South Africa and Morocco (164,165), whereas artesian bore water containing 5 ppm F induced severe dental abnormalities and other signs of fluorosis in sheep in Queensland (57). In other areas, mottled enamel has been observed in lambs consuming water containing 2 ppm F (106) and in cows dependent upon drinking water containing 4-5 ppm F (31). In the experiments of Peirce (115) with sheep given drinking water containing various concentrations of sodium fluoride up to 20 ppm F, remarkable tolerance to the fluoridated water was observed. With mature sheep, initially aged 2.5-3.5 years, no adverse effects on health, food consumption, or wool production were apparent and no dental defects developed, even with water containing 20 ppm F. When fluoridated water was given to ewes during pregnancy and lactation and was continued with their lambs for 7 years, the reproductive performance of these lambs was unaffected over six successive seasons even by the 20-ppm F water. However, consumption of water containing 10 and 20 ppm F induced changes characteristic of fluorosis in the teeth and reduced wool yields. When fluoridated water was given for 3.5 years to younger sheep, aged 10-11 months at the beginning of the experiment, mottling of the teeth was apparent in the group drinking the water containing 5 ppm F or more. Again no adverse effects upon general health or food consumption were observed even in the group consuming the 20 ppm water.

These experiments strikingly demonstrate that tolerance to fluoride is vitally affected by the age of exposure of the animals and by the continuity of the intake. For approximately half of each year, during the cool, wet months of winter and spring, the actual consumption of the fluorided water was low. Total annual fluoride intake by the sheep would, therefore, be smaller than from waters of comparable fluoride concentrations in the sub-tropical endemic fluorosis areas of Queensland (57). Moreover, the low stocking rates and harsher grazings of the latter areas would demand more movement by the sheep to obtain food and would impose more wear and tear on the teeth than under the experimental conditions of Peirce. Environmental differences of this nature can clearly influence greatly the tolerance of animals to fluorided waters under field conditions.

### 2. *Relative Toxicity of Different Forms of Fluorine*

Under most conditions the more soluble forms of fluorine, such as sodium fluoride, are more toxic per unit of fluorine than are the highly insoluble compounds, such as calcium fluoride. The fluorine of cryolite and of rock phosphate appears to be intermediate in toxicity. In the rat, Mitchell and Edman (104) place the relative toxicity of these substances, per unit of fluorine, as follows: sodium fluoride, 110-190; cryolite, 150-230; rock phosphate, 120-260; and calcium fluoride 230-2300 ppm of the dry diet. In cattle, the fluorine in sodium fluoride is approximately twice as toxic as that in rock phosphate (121,156). In a direct comparison with dairy cattle of the toxicity of NaF, $CaF_2$ and the fluoride residue on industrially contaminated hay, in which the fluorine from each source was fed at a level of about 65 ppm in the ration, little difference in the toxicity of the sodium fluoride and the contaminated hay was observed. The calcium fluoride was only about half as available as the other two sources as judged by fluorine retention in the skeleton and by blood and urine levels (139).

At a level of 14 ppm F, or similar low levels, it is difficult to distinguish between soluble and insoluble forms of fluorine, judging by their effects upon the teeth of rats (77). Apparently under these conditions the volume of digestive fluids is sufficient to dissolve even the more insoluble compounds. At relatively high levels of intake, the characteristic differences among compounds become manifest (104).

Particle size may also have a bearing on toxicity. The finer the particle size the more nearly the toxicity of cryolite approaches that of sodium fluoride (77,129). Sodium fluoride or cryolite provided in the drinking water promotes greater fluorine retention in the rat than a dry diet supplying the same intakes of those compounds (78, 172). Further, NaF given in milk to young rats in relatively low concentrations is less available than when similarly administered in water (107). However, the depressing effect of milk decreased to insignificance when the fluoride level was increased from 2 to 5 or 10 ppm (107).

### 3. *Duration and Continuity of Intake*

Tolerance to a particular level of fluoride depends greatly upon the duration of intake. Tolerance is also affected by the continuity of ingestion, because a portion of the fluoride deposited in bones is in dynamic equilibrium with the body tissues and fluids. During periods of low intake the exchangeable skeletal stores of fluoride are depleted and excreted in the urine. Opportunity is thus provided for further skeletal immobilization of fluorine during any subsequent periods of high fluoride intake. The effect of these processes has been demonstrated in experiments comparing continuous and intermittent fluoride dosage in the rat (78), the sheep (57), and the cow (20). The influence of intermittent intake is also apparent in the experiments of Peirce (115), considered above, because the sheep were subjected to alternating periods of high- and low-fluoride intake as a result of seasonal periods of high- and low-water consumption.

In those experiments and in those of Harvey (57), the benefits of alternating periods of exposure arise largely from the consequent reduction in total fluoride intake over the whole period. When compared on the basis of similar total yearly intake, as in the experiments of Carlson (20) with cattle, skeletal storage of fluoride, as measured by vertebral biopsy, was similar for continuous and periodic exposure. Animals obtaining elevated levels of dietary fluoride (1.5 mg F/kg body weight per day or 40-50 ppm F in the dry ration), for 6 months of the year, stored only half as much fluoride as animals on this intake continuously for the whole year. Animals receiving fluoride in periods of high and then low exposure stored the same amount of fluoride as those on a continuous intake of the same yearly average. Periodic exposure was, nevertheless, found to result in a more dynamic metabolism of fluoride in skeletal tissue. Rapid increases in fluoride content during periods of ingestion and rapid loss during periods of low-fluoride intake can be important in determining the overall storage of fluoride by the skeleton. They are also important in relation to systemic reactions to fluoride which may arise with short-term ingestion of high levels. Such reactions, namely weight loss due to decreased appetite and unthriftiness, were apparent in the experiments of Carlson (20) during periods of high-fluoride ingestion (3.0 mg F/kg body weight per day or 90 ppm F in the dry ration).

### 4. *Influence of Other Dietary Components*

Several cations influence the toxicity of fluorine. Aluminum salts exert a protective effect against fluorine in rats (11,135), sheep (11), and cattle (86) and the incorporation of aluminum sulfate into feeding "cakes" has been proposed as a means of rendering fluorine nontoxic to dairy cows (163). From the results of mineral balance studies with sheep it appears that aluminum salts function by reducing the absorption of fluorine from the intestinal tract (11). Calcium salts apparently function similarly. Thus 3 mg fluorine was found in the whole carcass of rats at the end of a 2-week period when 1% $CaCl_2$ was administered with NaF in the drinking water, compared with 6 and 11 mg, respectively, when this calcium salt was added at levels of 0.1 and 0.01% (172). When $MgCl_2$ and $AlCl_3$ were administered at the same levels, a smaller inhibition of fluorine retention was observed.

The level of fat in the diet is a further factor influencing fluorine toxicity. Raising the level of dietary fat from 5 to 15%, or to 20%, enhances the growth-retarding effect of high-fluorine intakes in rats (17,103) and in chicks (15). This effect is unrelated to the chain length of the fat (103) and has been attributed, in part, to increased retention of fluorine in the heart, kidneys, and skeletal tissues (15,17). The drastic increase in fecal fat in the fluorotic rat, arising in part from an inhibition by fluorine of lipase activity in the intestine, as demonstrated by Phillips and coworkers (141,159), is probably of additional significance. It seems logical to assume that this phenomenon becomes increasingly deleterious to the fluorotic animal as its dependence upon dietary fat as a source of energy is raised.

## V. Fluorine Metabolism

### 1. *Absorption*

Soluble fluorides are rapidly and almost completely absorbed from the gastrointestinal tract. Radiofluoride ($^{18}F$) has been detected in the blood of sheep 5 min after a dose was placed in the stomach, the peak being reached in 3 hr (117). In humans given small oral doses of soluble fluoride, blood maxima were observed within 1 hr and 20-30% appeared in the urine in the succeeding 3-4 hr (18). Rats administered small doses of $Na^{18}F$ absorbed 75% in 1 hr (169) and 80-90% within 8 hr (36). When a dilute solution of NaF was given orally to rats and the animals killed immediately, 5, 15, 30, 60, and 90 min later, 12, 22, 36, 50, 72, and 86%, respectively, of the dose was absorbed (181).

When ingested in small amounts in solution the relatively insoluble forms of fluorine are almost as well absorbed as the more soluble compounds. Thus the fluorine of calcium fluoride was 96% absorbed and that of cryolite 93% absorbed when administered in this manner (75). When added to the diet in solid form the percentage absorption was only 60-70% for calcium fluoride and 60-77% for cryolite (83). Even poorer absorption, ranging from 37 to 54%, has been reported for the fluoride in bone (75,83). Approximately 50% of rock phosphate fluoride is absorbed, judging by the toxicity of this form of fluoride relative to that of sodium fluoride.

In addition to the chemical form, the absorbability of fluoride is affected by the method of administration and the nature of the rest of the diet. Reference has already been made to the depressing effect upon fluoride absorption of calcium and aluminum (11,86,135,163,166) and of high fat diets (17,103) and to the evidence of lower retention of fluoride administered in milk than in water (107). It seems that the latter effect is related to a slower *rate* of absorption from milk than from water rather than to any reduction in ultimate amount or percentage absorbed (36). The complexity of the process of fluoride absorption is illustrated further by the experiments of Ericsson (37). This worker investigated the effect of sodium chloride, wheat flour, and various other food components on the intestinal absorption and skeletal storage of $^{18}F$-labeled fluoride in the rat. The most notable finding was that sodium chloride significantly depressed skeletal F uptake from both water and flour.

2. *Distribution*

Absorbed fluorine is absorbed and distributed rapidly throughout the body as the fluoride ion, in a pattern similar to that of chloride, i.e., in the chloride space. It readily crosses cell membranes including that of the red cell. The fluoride concentration of the plasma is normally about twice that of the red cells, and about 75% of the total blood fluoride is present in the plasma, with less than 5% bound to plasma solutes under physiological conditions (18). The speed with which fluoride leaves the blood is illustrated by several experiments with radioactive fluorine. Thus Perkinson *et al.* (117) calculated that 40% of the blood fluoride, following intravenous injection, left the blood per minute in lambs and 32% in cows, whereas Bell *et al.* (13) found that after 2 min only 53% of the $^{18}F$ tracer dose administered to cattle remained in the blood. The disappearance curve of $^{18}F$ from the blood is triphasic in nature. In lambs, the first and most rapid phase has a half-time of only 3 to 4 min, the second phase has a half-time of about 1 hr, and the third a half-time of about 3.3 hr (117). The first phase presumably represents mixing of the fluoride with the body water; the second, uptake by the skeleton; and the third, excretion in the urine.

The fluoride concentrations of the soft tissues tend to reflect the level in the plasma (19). Placental and mammary transfer of fluoride has been demonstrated in several species, although this is limited except at relatively high intakes [see Hodge and Smith (65)]. Fetal blood and maternal blood contain roughly similar fluoride concentrations. Salivary $^{18}F$ concentrations follow blood concentrations in man, but at lower levels (18). Muscle fluoride concentration does not vary directly with plasma fluoride levels (8). In studies with parenterally injected $^{18}F$ in the rat, Armstrong and co-workers (8) obtained indications of a fluorine distribution in muscle in a volume greater than extracellular fluid but less than total tissue water. The rates of removal of $^{18}F$ from muscle (and liver) pointed to the presence in muscle tissue of a form of bound fluorine that does not readily reach equilibrium with ionic fluoride in extracellular fluid.

### 3. *Excretion*

The relative proportions of fluoride excreted in the urine and the feces are highly variable and depend upon the amount, the solubility, and the method of ingestion. Most of the absorbed fluoride which escapes retention by the bones and teeth is excreted rapidly in the urine. Much smaller amounts are excreted in the perspiration, except in excessive sweating (96), and into the gastrointestinal tract (60).

With soluble fluorides the urine is the main route of excretion and the level in the urine rises with increasing intakes. With the less soluble or more slowly soluble forms, such as solid calcium fluoride or cryolite added to the diet, a lower percentage appears in the urine, as a consequence of lower absorption. Similarly, lower percentages of fluoride were found in the urine of young men given fluoride in the form of bone meal or of solid calcium fluoride then when fluoridated water or sodium fluoride was administered (96). Average individuals on normal diets and not exposed to unusual fluoride intakes from the food, water, or atmosphere excrete 80% or more of their ingested fluorine in the urine (84). Sheep and cows normally excrete 50-90% of their dietary fluoride in the urine. The proportion excreted in this way can be decreased by increasing the level of fluoride in the diet (62) and by the addition of aluminum salts (11). Adding calcium to the diet increased fecal fluoride excretion in rats (174) but not in sheep (57). The method of administration may also influence the route of excretion. Thus, Simon and Suttie (142) found that sheep continuously infused with sodium fluoride solution into the rumen excreted more in the urine and less in the feces than comparable animals receiving a similar single daily dose. No significant difference between treatments in fluoride retention was observed.

## VI. Fluorine in Human Health and Nutrition

### 1. *Sources of Fluorine*

Human populations obtain widely varying quantities of fluorine from the food, the water supply, and the atmosphere, depending upon location. Except in areas in direct association with sources of pollution, the concentrations of fluorine in the air are very low and do not constitute a significant source of this element to man. Cholak (26) collated data on the occurrence of fluoride in the atmosphere from a number of communities in North America. Mean levels ranging from 0.02 to 2 ppb were reported, but the higher values were from single determinations and are probably not widely representative. Coal smoke is generally the chief contributor of fluorine to the atmosphere.

Surveys of the occurrence of fluoride in drinking water have been made in many countries (26). Fluoride occurs in rainwater in amounts that vary with the distance from urban centers and industrial installations. Surface waters, such as are used for drinking and cooking in most communities, generally contain less than 1 ppm F and many of these contain only 0.1 ppm or less. Fluoride intake from such sources would clearly be very low. In some parts of the world the population is dependent upon the waters of deep wells and artesian bores which are naturally high in fluoride. Fluoride concentrations of 4 to 8 ppm are common in such areas. In certain endemic fluorosis areas in southern India (113a) and South Africa (112), concentrations as high as 20 or even 50 ppm F occur.

Food is the major source of fluorine to individuals not exposed to industrial contamination or to naturally or artificially fluoridated drinking water. Normal North American diets, which include little tea, contribute about 0.3-0.5 mg F daily (85,93), approximately 80% of which is absorbed (26,84). Where artificially fluoridated water is consumed the amount obtained from the food would be somewhat higher because of the use of such water in cooking. The consumption of 1200 to 1500 ml/day of water containing 1 ppm F provides 1.2-1.5 mg F, which is more than is normally ingested in the food.

Very few foods contain more than 1-2 ppm F and most of them contain less than 0.5 ppm (dry basis) (28,82,94). In a study of raw, unpeeled potatoes the fluorine concentration ranged from 1 to 3.5 ppm of the dry matter. More than 75% of this was present in the peelings and would usually be discarded (125). Sea fish and fish products are much higher in fluorine than most other foods. Concentrations between 5 and 10 ppm F have been reported (94). The very high fish diet consumed by the inhabitants of Tristan da Cunha, where the drinking water contains only 0.2 ppm F, has been proposed as a cause of the relatively high incidence of mottled teeth found in that population (150).

The consumption of tea can be an important determinant of total dietary fluoride intake. Fluorine concentrations of 100 ppm in tea are common, two-thirds or more of which passes into the infusion, so that one cup of tea can increase the fluorine of the diet by 0.1 to 0.2 mg. As much as 1 mg F can be ingested daily by adults in some communities from this source alone (55). The importance of tea as a source of fluorine is illustrated by balance studies carried out by Ham and Smith (51) with young women. The average fluorine intakes rose from 0.4-0.8 mg/day, when a tea-free diet was consumed, to 1.2-1.4 mg/day when this diet included 1360-1815 ml of tea infusion daily. In a further study, these workers (51) supplemented a normal adult diet with 70 g/day of a baby food containing powdered beef bone and carrying 11-12 ppm F. In these circumstances the fluorine intake of the women rose from the "normal" 0.4-0.8 mg/day to 0.9-1.4 mg/day.

## 2. *Fluorine Retention in Man*

Fluorine normally accumulates with age in human bones, even where intakes from the food and water supply are very low. In an English survey (68), a rise in bone fluoride levels with age was observed at all levels of intake, with a plateau at about 55 years of age. The height of the plateau was greater with individuals who had consumed throughout life water containing 0.8-1.2 or 1.9 ppm F than with those who had consumed water containing virtually no fluorine. Smith *et al.* (146) concluded from their studies in the United States that "there is a striking approximation to a linear relationship for the average bone fluoride concentration when 10-year age groups are plotted against the logarithm of the age."

The level of fluoride intake below which all ingested fluoride is excreted and, therefore, none retained in the bones, must be very low indeed, if any such level exists. Metabolic studies with rats, using $^{18}F$, suggest that there is no level at which complete elimination occurs (169), but long-term studies with this species indicate continuous retention in the bones at the lowest dietary levels of fluorine employed (126, 146, 178). Machle and co-workers (84) found that the input and output of fluorine was approximately equal over many weeks when the daily intake was as low as 0.5 mg, whereas definite retention occurred in all instances when the intake ranged from 3 to 36 mg daily.

Although skeletal retention of fluoride is normal and continuous in man, the capacity of the skeleton to sequester fluoride without serious pathological change is so great that such retention is not necessarily harmful and in some circumstances can be beneficial. Hodge and Smith (65) distinguish three stages in this process, namely bone fluoridation or chemical fluorosis; bone mottling; and abnormal bone. Bone fluoridation alone is characteristic of bones generally containing less than 2500 ppm F and is unaccompanied by detectable gross or microscopic abnormalities or effect on cell activity. Bone mottling occurs in bones containing 2500-5000 ppm, although such bones are normal to both gross and roentgenographic examination. Bone mottling, like enamel and dentine mottling, is a result of the direct action of fluoride upon the osteoblasts or bone-forming cells. Abnormal bones, with gross and roentgenographic abnormalities, an enlarged, heavy chalky-white appearance, and a very irregular surface are characteristic of bones with fluoride levels above 5000-6000 ppm.

The critical levels in bones associated with the changes just described, and the critical levels of long-term fluoride intake related to them, are not so well established for man as they are for cattle and other farm animals. A daily intake of 2 ppm in the diet for 50 years results in chemical fluorosis only, i.e., with no bone mottling or other abnormalities. McClure *et al.* (97) have estimated that a continuous intake of 8 ppm F in the diet for 35 years is necessary before the critical levels of 6000 ppm are attained. Prolonged intake of 10 ppm of dietary fluorine for many years can result in symptomatic skeletal fluorosis in man (98).

The remarkable ability of the skeletal tissues to tolerate accumulations of fluoride is further apparent from the study carried out by Zipkin *et al.* (182). These workers determined the ash and fluoride levels of selected bones from 69 individuals in the United States who had drunk, for at least 10 years, water of different fluoride concentrations ranging from 0.1 to 4.0 ppm. The concentration of fluoride in the bones increased linearly with an increase of fluoride in the drinking water. Some of the bones examined from the high-fluoride group contained up to 5480 ppm F (dry, fat-free basis), without any evidence of tissue damage or histological change. Microscopic examination of the bones in this study and of the lumbar vertebral body joints revealed no significant differences among the various fluoride intakes nor between these and the bones from 33 controls who had lived in areas where the water contained less than 0.5 ppm F (46). Finally no significant difference was observed in the incidence of bone cancer between long-term (more than 20 years) and short-term (less than 5 years) residents of an area where the water fluoride level was 2.5 ppm (46).

### 3. *Endemic Dental Fluorosis or Mottled Enamel*

Mottled enamel was first described in man in 1901 (34) and again in 1910 (42). In 1916, McKay (100) established that mottled enamel was a water-borne disease caused by the presence of some toxic element in the drinking water during the period of the calcification of the teeth. It was not until 1931 that Smith and co-workers in Arizona (147), Churchill in Pennsylvania (27), and Velu in Morocco (165) demonstrated independently that the causative agent in mottled enamel is fluoride. Chronic endemic dental fluorosis, or mottled enamel, due to the consumption of naturally fluoridated waters has since been reported from many countries.

Mottled teeth are characterized by chalky-white patches, distributed irregularly over the surface of the tooth, with a secondary infiltration of yellow to brown staining. The enamel is structurally weak, and in severe cases there is a loss of enamel accompanied by "pitting," which gives the tooth surface a corroded appearance. Changes in the size or shape of the affected teeth are rare in mottled enamel areas. The condition may be mild, moderate, or severe, depending upon the level of intake of fluoride and upon individual susceptibility. Mottled enamel is almost entirely confined to the permanent teeth and develops only during their period of formation. The fully formed enamel of adult teeth is unaffected by fluorine, and the deciduous teeth are only affected at high-fluoride intakes where other signs of fluorosis are evident.

In its mild forms, mottled enamel has little public health significance. In the more severe forms, involving enamel hypoplasia and pitting, the unsightly appearance of the mouth may be accentuated by excessive wear on the teeth and mastication can be affected. At the levels of fluoride common in most mottled enamel areas, the dental lesions are the only fluoride-induced defects apparent. The growth and health of the individual are not affected, and disturbances seriously affecting the skeletal bones or joints do not develop. In areas with highly fluorided waters (over 8 ppm F), not only is there a high incidence of severe mottled enamel, but disturbances in ossification and systemic signs of fluorosis can develop, similar to those in individuals suffering from severe industrial skeletal fluorosis.

Numerous surveys in several countries have established a relationship between the levels of fluoride in the drinking water and the incidence and severity of mottled enamel in children (65). Dean (32) studied 5824 children aged 12–14 years, none of whom had been away from their water supply for periods longer than 1 month, in twenty-two cities in ten states of the United States. As the fluoride content of the drinking water increased, the "community index," i.e., the average index of mottling based on the two teeth most severely affected in each child, increased along an S-shaped curve. At about 2 to 3 ppm F in the water the community index corresponded to "very mild" mottling; at about 4 ppm to "mild" mottling; at 5 to 6 ppm to "moderate" mottling; and even at concentrations of 14 ppm the index lay between "moderate" and "severe" mottling. In a later study, Hodge (63) found that the severity of mottling increased in a linear fashion when the community index of dental fluorosis was plotted against the logarithm of the parts per million of fluoride in the drinking water from about 1 to nearly 10 ppm. Below 1 ppm, a second straight, nearly horizontal line appeared, indicating practically no difference in the community index whether the water contains 0.1 or 1.0 ppm F (see Fig. 23).

The problem of classifying waters of differing fluoride concentrations in terms of their mottled enamel potential is complicated by the amounts of water consumed and by the fluoride content of the diet. The former is affected by the climatic conditions, and the latter by the degree of dependence upon seafoods and upon tea as a beverage. The importance of climate in influencing the incidence and severity of mottling is illustrated by a comparison of the situation in two towns in the United States with water supplies of similar fluoride content, one with a mean annual temperature of 70°F and the other with a mean annual temperature of 50°F (43). In the former town, significantly fewer children had teeth classified as normal and the incidence of mottling was strikingly higher.

The primary defect of the enamel in dental fluorosis is permanent. It is possible to temporarily bleach the disfiguring stain, but there is no known cure once defective enamel is formed. Prevention is possible by changing the water supply to low-fluoride sources or by removal of enough fluoride to maintain the concentration below 2 ppm. The great affinity of fluoride for the tertiary phosphates can be exploited as a means of removing fluoride and practical control of excessive levels has been applied successfully to several community water supplies (111).

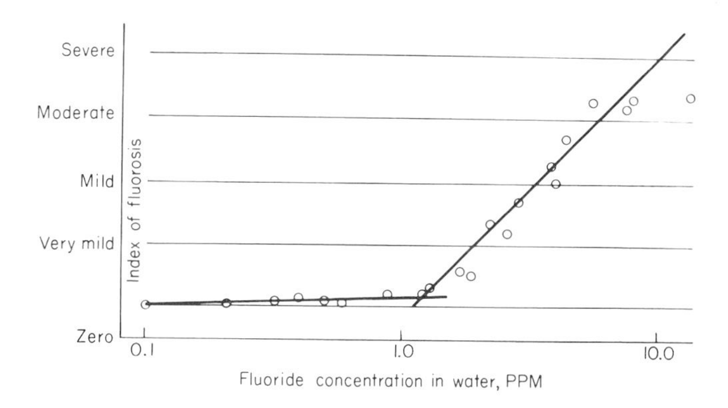


**Fig. 23.** Relationship between the severity of dental mottling and the fluoride concentration of the drinking water. The community index of dental fluorosis is plotted against the logarithm of the parts per million of fluoride in the water. [From Hodge (63).]

### 4. *Dental Caries*

In several of the early studies of mottled enamel, including that of Eager (34) in 1901, it was noted that children with mottled enamel were relatively free from dental decay. Not until the late 1930s was an inverse relationship between mottled enamel and dental caries incidence clearly established. The sequence of events leading to the establishment of this relationship, and the reluctance with which it was accepted, are well described by Hodge and Smith (65).

The quantitative relationship between the decrease in dental caries incidence and increasing fluoride concentration in the drinking water was first clearly shown by Dean (32) in a major survey involving 7257 children, 12-14 years old, living in twenty-one cities in four states in the United States. It was apparent from the data obtained that the dental caries experienced in the permanent teeth decreases as the fluoride concentration of the drinking water increases above about 1.3 ppm. Hodge (63) subsequently showed that the decrease in dental caries is a linear function of the fluoride expressed on a logarithmic scale. When considered in relation to the incidence of mottling, it became evident that the two lines intersect at close to 1 ppm F in the water. This concentration is, therefore, considered to be optimal, or the point of "maximum health with maximum safety." The term "eufluorosis" has been proposed to express optimal protection against dental caries without production of mottling (1).

A close inverse relationship between the incidence of dental caries in man and the fluoride content of the natural water supply has now been established in many countries. The benefit is not confined to children but is continued into adult life, provided that the fluoride exposure has occurred throughout, or for a substantial part of the period of tooth formation (41,101,131). These findings stimulated public health authorities in many parts of the world to fluoridate deliberately communal water supplies, in an attempt to duplicate mass population exposure to natural fluoride waters.

The water supplies of hundreds of communities have been treated with sodium fluoride or fluosilicate to maintain fluoride levels ranging from 0.8 to 1.2 ppm depending upon climatic conditions. Several experiments have been conducted in which a neighboring city or borough was maintained as an unfluoridated control. The populations of the communities were then investigated to determine the relative and actual incidence of dental caries, mottled enamel, and periodontal disease and of any other disabilities or signs of fluorosis, however mild. The results of these studies, after 10 or more years of fluoridation, have been critically assessed (9, 110). They demonstrate with complete certainty the value of artificially fluoridating water supplies as a means of securing better dental health for a community. The very slight increase in the mildest forms of mottling that may occur is insufficient to constitute a significant public health problem (9). Furthermore, where the fluoridation is discontinued for any reason, a clear and definite reversal of trend back to the prefluoridation level of caries experience has been demonstrated, both in respect of mean DMF scores and of children entirely free of decay (110). Fluoridation does not abolish dental decay. Average reductions, ranging from 40 to 70% in the prevalence of caries in the teeth of children born subsequent to the change in water supply, have been reported in different studies, without any evidence of systemic injury or observable effects upon the health of individuals.

The great success of mass water fluoridation stimulated attempts to achieve the same benefits by other routes of administration. Large sections of the community, especially in the developing countries, do not have access to communal water supplies and are, therefore, denied the use of this method of improving their dental health. There has also been considerable emotional opposition to fluoridation of water supplies, based mainly on ignorance and fear, and on the grounds that only a small proportion of the fluorided water is used for human consumption, that individual consumption can be very variable, and, in the erroneous belief, that its benefits are restricted to children. For these reasons, numerous studies have been made of the benefits and practical possibilities of providing supplemental fluoride by means of fluoride tablets, and by fortification of commonly used foods such as salt, milk, flour, or sugar. Topical applications of fluorides in fairly concentrated solutions (1-2%) have also been tried with some success (54,74). In another study, a 20% reduction in caries incidence was achieved as a result of a single application of an 8% solution of stannous fluoride to 12 to 13-year-old children (71).

In Switzerland, salt has shown considerable promise as a fluoride vehicle (89, 176) and fluoride fortification of cereals, possibly in combination with salt, provides further possibilities for caries control, judging by limited European experience (35, 89). Administration of fluoride tablets to children and to women during pregnancy and lactation has shown beneficial results on caries incidence and is quite widely practiced in many communities where the water is low in fluoride. The results of several studies testify to the effectiveness of this form of treatment in children, with a decrease in caries incidence comparable with that found in communities with fluoridated water (10, 40, 177). Unfortunately, conscientious tablet utilization is difficult to maintain widely, and critical large-scale studies of this procedure and of the effectiveness and practicability of fluoridation of salt, milk, flour, or sugar are necessary before these means of community caries control can be recommended with confidence.

An understanding of the mechanisms by which fluoride inhibits dental caries, in the words of Hodge and Smith (65) "has been tantalisingly elusive." It seems that the fluoride produces a more perfect crystal structure and alters the chemical composition of the enamel in a way which confers resistance to caries. It is known that fluoride can replace the hydroxyl ion in the hydroxyapatite lattice of the mineral of tooth enamel and that fluoride can exchange with carbonate ions. It is known also that fluoride is taken up rapidly by the enamel surfaces from very dilute fluoride solutions and that this fluoride is concentrated more in enamel imperfections or in early cavities than in intact surfaces nearby (53, 100). Furthermore, the solution rate and presumably the solubility of the tooth surface are reduced following fluoride treatment (67, 69, 168). Finally, there is the likelihood of a fluoride-induced reduction in the production by bacteria of acids responsible for decalcification (53).

### 5. *Fluoride and Osteoporosis*

Osteoporosis is recognized radiologically as a decrease in bone density and is often accompanied by collapsed vertebrae. The disease is particularly prevalent beyond age 60, females being more often afflicted than males. The seriousness of this condition in the United States is indicated by the work of Smith and Frame (148) who studied 2063 women in Detroit. From the data obtained it was estimated that there are approximately 14 million women in this country with a significant degree of osteoporosis of which they are unaware.

Patients suffering from osteoporosis and other demineralizing diseases have been treated with substantial amounts of sodium fluoride with reported beneficial effects upon back pains, bone density, and calcium balance (2, 24, 29, 124, 128). Fluoride treatment appears to be a useful therapy for these conditions, but more research and experience are required to establish its efficacy and safety. Of wider importance is the finding of Leone and associates (80) that there was substantially less osteoporosis in a high-fluoride area in Texas (8 ppm F in the water) than in a low-fluoride area (0.09 ppm F in the water) in Massachusetts. These workers suggested that fluoride ingestion might be important in the maintenance of a normal skeleton. Subsequently Bernstein *et al* (14) examined approximately 1000 X-rays of the lower lumbar spine of adults over age 45 living in two areas of North Dakota. In one area the water supply provided 0.15-0.3 ppm F and in the other, 4-6 ppm F. A summary of the main findings is presented in Table 48.

TABLE 48
*Percentage of Subjects Showing Reduced Bone Density, Collapsed Vertebrae, and Calcified Aortas*[a]

| | Men | | | Women | | |
|---|---|---|---|---|---|---|
| Age (years) | High-F area (%) | Low-F area (%) | *P* | High-F area (%) | Low-F area (%) | *P* |
| | | | Reduced Bone Density | | | |
| 45 | 11.9 | 21.6 | N.S. | 12.1 | 22.4 | N.S. |
| 55 | 19.4 | 33.7 | <0.05 | 31.0 | 64.2 | <0.01 |
| 65+ | 44.7 | 47.5 | N.S. | 43.6 | 85.2 | <0.01 |
| | | | Collapsed Vertebrae | | | |
| 45 | 20.0 | 18.6 | N.S. | 3.3 | 6.1 | N.S. |
| 55 | 26.5 | 27.6 | N.S. | 3.1 | 20.8 | <0.05 |
| 65+ | 52.7 | 45.1 | N.S. | 9.5 | 35.3 | <0.01 |
| | | | Calcified Aortas | | | |
| 45 | 8.5 | 21.6 | <0.05 | 5.1 | 8.2 | N.S. |
| 55 | 12.1 | 39.5 | <0.01 | 20.7 | 40.8 | <0.05 |
| 65+ | 36.8 | 65.8 | <0.01 | 46.2 | 60.0 | N.S. |

[a]From Bernstein *et al.* (14).

As expected, the number of individuals judged to have decreased bone density increased with age, in both the high- and the low-fluoride areas. At all ages there was substantially less osteoporosis in women in the high-fluoride area; the changes in men were in a similar direction but less obvious. The differences in women showing collapsed or distorted vertebrae were even greater, particularly in those over 55. No such area effect was apparent in the men, who revealed a high incidence irrespective of fluoride intake.

A surprising further fact to emerge from this study was the decreased calcification of the aorta in men in the high-fluoride area. This condition was approximately twice as common in men in the low-fluoride area at all ages. No such differences were found in women. The significance of this in terms of coronary heart disease or other forms of cardiovascular disease is unknown. It is perhaps pertinent that Leone *et al.* (79), in comparing mortality rates in high- and low-fluoride areas in Texas, found the only significant difference to be a somewhat lower incidence of death from heart attacks in the high-fluoride area. It should be appreciated that the reported benefits in respect to the incidence of osteoporosis and collapsed vertebrae in women and of calcification of the aorta in men were obtained at levels of fluoride in the drinking water above those considered safe for children. They would undoubtedly result in a significant incidence of mottled enamel if consumed during early life. Whether a level of 1 ppm F in the water, as recommended for the control of dental caries, is sufficient for the maintenance of a normal skeleton in the adult remains to be determined.

## REFERENCES


1. P. Adler, I. Sarkany, K. Toth, J. Straub, and E. Szerenyi, *J. Dent. Res.* **30**, 368 (1951).
2. M. I. Aeschlimann, J. A. Grant, and J. F. Crigler, Jr., *Metab. Clin. Exp.* **15**, 905 (1966).
3. J. N. Agate, G. H. Bell, G. F. Boddie, R. G. Bowler, M. Buckell, E. A. Cheeseman, T. H. Douglas, H. A. Druett, J. Garrad, D. Hunter, K. Perry, J. D. Richardson, and J. B. Weir, *Med. Res. Counc. (Gt. Brit), Memo.* **22** (1949).
4. A. C. Aisenberg and V. R. Potter, *J. Biol. Chem.* **215**, 737 (1955).
5. R. Allcroft, *Inst. Biol., Symp.* **8**, 95 (1959).
6. R. Allcroft, K. N. Burns, and C. N. Hebert, "Fluorosis in Cattle," Anim. Dis. Surv. Rep. No. 2. H. M. Stationery Office, London, 1965.
7. J. O. Anderson, J. S. Hurst, D. C. Strong, H. M. Nielsen, D. A. Greenwood, W. Robinson, J. L. Shupe, W. Binns, R. A. Bagley, and C. I. Draper, *Poultry Sci.* **34**, 1147 (1955).
8. W. D. Armstrong, L. Singer, and J. J. Vogel, *Fed. Proc., Fed. Amer. Soc. Exp. Biol.* **25**, 696 (1966).
9. F. A. Arnold, Jr., *Amer. J. Pub. Health* **47**, 539 (1957).
10. F. A. Arnold, Jr., F. J. McClure, and C. L. White, *Dent. Progr.* **1**, 8 (1960).
11. D. E. Becker, J. M. Griffith, C. S. Hobbs, and W. M. McIntyre, *J. Anim. Sci.* **9**, 647 (1950).
12. L. F. Belanger, W. J. Visek, W. E. Lotz, and C. L. Comar, *J. Biophys. Biochem. Cytol.* **3**, 559 (1957); *Amer. J. Pathol.* **34**, 25 (1958).
13. M. C. Bell, G. M. Merriman, and D. A. Greenwood, *J. Nutr.* **73**, 379 (1961).
14. D. S. Bernstein, N. Sadowsky, D. M. Hegsted, C. D. Guri, and F. J. Stare, *J. Amer. Med. Assoc.* **198**, 499 (1966).
15. D. Bixler and J. C. Muhler, *J. Nutr.* **70**, 26 (1960).
16. K. N. Burns and R. Allcroft, *Res. Vet. Sci.* **3**, 215 (1962).
17. W. Buttner and J. C. Muhler, *J. Nutr.* **63**, 263 (1957); **65**, 259 (1958).
18. C. H. Carlson, W. D. Armstrong, and L. Singer, *Amer. J. Physiol.* **199**, 187 (1960); *Proc. Soc. Exp. Biol. Med.* **104**, 235 (1960).
19. C. H. Carlson, L. Singer, and W. D. Armstrong, *Proc. Soc. Exp. Biol. Med.* **103**, 418 (1960).
20. J. R. Carlson, Doctoral Thesis, University of Wisconsin, (1966.
21. J. R. Carlson and J. W. Suttie, *Amer. J. Physiol.* **210**, 79 (1966).
22. J. R. Carlson and J. W. Suttie, *Exp. Cell Res.* **45**, 415 and 423 (1967).
23. J. S. Cass, *J. Occup. Med.* **3**, 471 and 527 (1961).
24. R. M. Cass, J. D. Croft, P. Perkins, W. Nye, C. Waterhouse, and R. Terry, *Arch. Intern. Med.* **118**, 111 (1966).
25. C. Y. Chang, P. H. Phillips, and E. B. Hart, *J. Dairy Sci.* **17**, 695 (1934).
26. J. Cholak, *J. Occup. Med.* **1**, 501 (1959).
27. H. N. Churchill, *Ind. Eng. Chem.* **23**, 996 (1931).
28. P. A. Clifford, *J. Assoc. Off. Agr. Chem.* **28**, 277 (1945).
29. P. Cohen, and F. H. Gardner, *J. Amer. Med. Assoc.* **195**, 962 (1966).

30. C. L. Comar, W. J. Visek, W. E. Lotz, and J. H. Rust, *Amer. J. Anat.* **93**, 361 (1953).
31. H. T. Dean, *Pub. Health Rep.* **50**, 206 (1935).
32. H. T. Dean, *in* "Fluorine and Dental Health," pp. 6-11 and 23-31. Am. Assoc. Advance. Sci., Washington, D.C., 1942.
33. A. R. Doberenz, A. A. Kurnick, E. B. Kurtz, A. R. Kemmerer, and B. L. Reid, *Proc. Soc. Exp. Biol. Med.* **117**, 689 (1964).
34. J. M. Eager, *Pub. Health Rep.* **16**, 2576 (1901).
35. R. Ege, *Tandlaegebladet* **65**, 445 (1961); quoted by Ericsson (37).
36. Y. Ericsson, *Acta Odontol. Scand.* **16**, 51 and 127 (1958).
37. Y. Ericsson, *J. Nutr.* **96**, 60 (1968).
38. R. J. Evans and P. H. Phillips, *J. Nutr.* **18**, 353 (1939).
39. J. M. Fargo, G. Bohstedt, P. H. Phillips, and E. B. Hart, *Amer. Soc. Anim. Prod., Rec. Proc. Ann. Meet.* **31**, 122 (1938).
40. R. Feltman and G. Kosel, *J. Dent. Med.* **16**, 190 (1961).
41. J. R. Forrest, G. J. Parfitt, and E. R. Bransby, *Mon. Bull. Min. Health Pub. Health Lab. Serv.* **10**, 104 (1951).
42. H. A. Fynn, *Dent. Items Interest* **32**, 31 (1910); quoted by Harvey (57).
43. D. J. Galagan and G. G. Lamson, *Pub. Health Rep.* **68**, 497 (1953); D. J. Galagan and J. R. Vermillion, *ibid.* **72**, 491 (1957).
44. N. L. Garlick, *Amer. J. Vet. Res.* **16**, 38 (1955).
45. M. Gaud and A. Charnot, *Bull. Mens. Off. Int. Hyg. Pub.* **30**, 1280 (1930).
46. E. F. Geever, N. C. Leone, P. Geiser, and J. Lieberman, *Pub. Health Rep.* **73**, 721 (1958); *J. Amer. Dent. Assoc.* **56**, 499 (1958).
47. R. W. Gerry, C. W. Carrick, R. E. Roberts, and S. M. Hauge, *Poultry Sci.* **26**, 323 (1947); **28**, 19 (1949).
48. G. E. Glock, F. Lowater, and M. M. Murray, *Biochem. J.* **35**, 1235 (1941).
49. J. G. Halpin and A. R. Lamb, *Poultry Sci.* **11**, 5 (1932).
50. M. P. Ham and M. D. Smith, *Can. J. Res., Sect. F* **28**, 227 (1950).
51. M. P. Ham and M. D. Smith, *J. Nutr.* **53**, 215 and 225 (1954).
52. K. Haman, P. H. Phillips, and J. G. Halpin, *Poultry Sci.* **15**, 154 (1936).
53. J. L. Hardwick, J. H. Fremlin, and J. Mathieson, *Brit. Dent. J.* **104**, 47 (1958).
54. R. Harris, *Aust. Dent. J.* **4**, 257 (1959).
55. M. F. Harrison, *Brit. J. Nutr.* **3**, 162 (1949).
56. E. B. Hart, P. H. Phillips, and G. Bohstedt, *Amer. J. Pub. Health* **24**, 936 (1934).
57. J. M. Harvey, *Queensl. J. Agr. Sci.* **9**, 47 (1952); **10**, 127 (1953).
58. D. M. Hegsted, *Proc. 1st Annu. Conf. Trace Substances Environ. Health, 1967* p. 105, Columbia, Missouri, 1968.
59. J. W. Hein, J. Bonner, F. Brudevold, and H. C. Hodge, *J. Dent. Res.* **33**, 661 (1954).
60. J. W. Hein, J. Bonner, F. Brudevold, F. A. Smith, and H. C. Hodge, *Nature (London)* **178**, 1295 (1956).
61. C. S. Hobbs and G. M. Merriman, *Tenn., Agr. Exp. Sta., Bull.* **351** (1962).
62. C. S. Hobbs, R. P. Moorman, J. M. Griffith, G. M. Merriman, S. L. Hansard, and C. C. Chamberlain, *Tenn., Agr. Exp. Sta., Bull.* **235** (1954).
63. H. C. Hodge, *J. Amer. Dent. Assoc.* **40**, 436 (1950).
64. H. C. Hodge, *Trans. 4th Conf. Metab. Interrelations, 1953.*
65. H. C. Hodge and F. A. Smith, *in* "Fluorine Chemistry" (J. H. Simons, ed.), Vol. 4, Academic Press, New York, 1965.
66. B. H. Hoogstratten, N. C. Leone, J. L. Shupe, D. A. Greenwood, and J. Lieberman, *J. Amer. Med. Assoc.* **192**, 26 (1965).
67. S. Isaac, F. Brudevold, F. A. Smith, and D. E. Gardner, *J. Dent. Res.* **37**, 254 (1958).

68. D. Jackson and S. M. Weidmann, *J. Pathol. Bacteriol.* **76**, 451 (1958).
69. G. N. Jenkins, *Arch. Oral Biol.* **1**, 33 (1959).
70. R. B. Johnson and H. A. Lardy, *J. Biol. Chem.* **184**, 235 (1950).
71. W. A. Jordan, J. R. Snyder, and V. Wilson, *Pub. Health Rep.* **73**, 1010 (1958).
72. C. H. Kick, R. M. Bethke, B. H. Edginton, O. H. M. Wilder, R. Record, W. Wilder, T. J. Hill, and S. M. Chase, *Ohio, Agr. Exp. Sta., Bull.* **558** (1935).
73. M. Kirchgessner, U. Weser, H. Friesecke, and W. Oelschläger, *Z. Tierphysiol., Tierernaehr. Futtermittelk.* **18**, 251 (1963).
74. J. W. Knutson and D. W. Armstrong, *Pub. Health Rep.* **58**, 1701 (1943); **60**, 1085 (1945); **61**, 1683 (1946); **62**, 425 (1947).
75. E. J. Largent and F. F. Heyroth, *J. Ind. Hyg. Toxicol.* **31**, 134 (1949).
76. M. Lawrenz, unpublished data cited by Mitchell and Edman (104).
77. M. Lawrenz and H. H. Mitchell, *J. Nutr.* **22**, 451 and 621 (1941).
78. M. Lawrenz, H. H. Mitchell, and W. A. Ruth, *J. Nutr.* **19**, 531 (1940); **20**, 383 (1940).
79. N. C. Leone, M. B. Shimkin, E. A. Arnold, C. A. Stevenson, E. R. Zimmermann, P. P. Geiser, and J. J. Lieberman, *in* "Fluoridation as a Public Health Measure," Am. Assoc. Advance. Sci., Washington, D.C., 1964.
80. N. C. Leone, C. A. Stevenson, B. Besse, L. E. Hawes, and T. R. Dawber, *AMA Arch. Ind. Health* **21**, 326 (1960).
81. R. C. Likins, F. J. McClure, and A. C. Steere, *Pub. Health Rep.* **71**, 217 (1956).
82. H. C. Lockwood, *Analyst* **62**, 775 (1937).
83. W. Machle and E. J. Largent, *J. Ind. Hyg. Toxicol.* **31**, 134 (1949).
84. W. Machle, E. W. Scott, and E. J. Largent, *J. Ind. Hyg. Toxicol.* **24**, 199 (1942).
85. W. Machle, E. W. Scott, and J. Treon, *Amer. J. Hyg.* **29**, 139 (1939).
86. B. N. Majumdar and S. N. Ray, *Indian J. Vet. Sci.* **16**, 107 (1946).
87. D. C. Maplesden, I. Motzok, W. T. Oliver, and H. D. Branion, *J. Nutr.* **71**, 70 (1960).
88. J. Marais, *Onderstepoort J. Vet. Sci. Anim. Ind.* **18**, 203 (1943).
89. M. Marthaler, *Schweiz. Bull. Eidg. Gesundheitsamtes, Part B* No. 2. (1962).
90. S. Matuura, N. Kokubu, S. Wakimoto and M. Tokimasa, *Mem. Fac. Sci., Kyushu Univ., Ser. C* **2**, 37 (1954); cited by Cholak (26).
90a. R. L. Maurer and H. G. Day, *J. Nutr.* **62**, 561 (1957).
91. H. G. McCann and F. A. Bullock, *J. Dent. Res.* **36**, 391 (1957).
92. F. J. McClendon and J. Gershon-Cohen, *J. Agr. Food Chem.* **1**, 464 (1953).
93. F. J. McClure, *Amer. J. Dis. Child.* **66**, 362 (1943).
94. F. J. McClure, *Pub. Health Rep.* **64**, 1061 (1949).
95. F. J. McClure and R. C. Likins, *J. Dent. Res.* **30**, 172 (1951).
96. F. J. McClure, H. H. Mitchell, T. S. Hamilton, and C. A. Kinser, *J. Ind. Hyg. Toxicol.* **27**, 159 (1945).
97. F. J. McClure, H. G. McCann, and N. C. Leone, *Pub. Health Rep.* **73**, 741 (1958).
98. F. J. McClure and I. Zipkin, *Dent. Clin. N. Amer.* p. 441 (1958); quoted by Hodge and Smith (65).
99. E. V. McCollum, N. Simmonds, J. E. Becker, and R. W. Bunting, *J. Biol. Chem.* **63**, 553 (1935).
100. F. McKay, *Dent. Cosmos* **58**, 447 (1916).
101. F. S. McKay, *Amer. J. Pub. Health* **38**, 828 (1948).
102. G. W. Miller and J. L. Shupe, *Amer. J. Vet. Res.* **23**, 24 (1962).
103. R. F. Miller and P. H. Phillips, *J. Nutr.* **51**, 273 (1953); **56**, 447 (1955); **59**, 425 (1956).
104. H. H. Mitchell and M. Edman, *Nutr. Abstr. Rev.* **21**, 787 (1952).

105. I. Motzok and H. D. Branion, *Poultry Sci.* **37**, 1469 (1958).
106. G. R. Moule, *Queensl. Agr. J.* **61**, 352 (1945).
107. J. C. Muhler and D. A. Weddle, *J. Nutr.* **55**, 347 (1955).
108. J. C. Muhler, G. R. Stookey, L. B. Spear, and D. Bixler, *J. Oral Ther. Pharmacol.* **2**, 241 (1966).
109. M. M. Murray, *J. Physiol.* (*London*) **87**, 338 (1936).
110. M. N. Naylor, *Health Educ. J.* **28**, 136 (1969).
111. M. S. Nichols, *in* "Fluoridation as a Public Health Measure," Am. Assoc. Advance. Sci., Washington, D. C., 1954.
112. T. Ockerse, *in* "Dental Caries and Fluorine," Am. Assoc. Advance. Sci., Washington, D. C., 1946.
113. L. E. Olson, H. M. Nielsen, J. L. Shupe, and D. A. Greenwood, *J. Pharmacol. Exp. Ther.* **122**, 7871 (1958).
113a. C. G. Pandit, T. N. Raghavacheri, D. S. Rao, and V. Krishnamurti, *Indian J. Med. Res.* **28**, 533 (1940).
114. A. W. Peirce, *Nutr. Abstr. Rev.* **9**, 253 (1939-1940).
115. A. W. Peirce, *Aust. J. Agr. Res.* **3**, 326 (1952); **5**, 545 (1954); **10**, 186 (1959).
116. J. J. Pendborg and C. M. Plum, *Acta Pharmacol. Toxicol.* **2**, 294 (1946).
117. J. D. Perkinson, I. B. Whitney, R. A. Monroe, W. E. Lotz, and C. L. Comar, *Amer. J. Physiol.* **182**, 383 (1955).
118. P. H. Phillips, *Science* **76**, 239 (1932).
119. P. H. Phillips, D. A. Greenwood, C. S. Hobbs, C. F. Huffman, and G. R. Spencer, *Nat. Acad. Sci.—Nat. Res. Counc., Publ.* **824** (1960).
120. P. H. Phillips, J. G. Halpin, and E. B. Hart, *J. Nutr.* **10**, 93 (1935).
121. P. H. Phillips, E. B. Hart, and G. Bohstedt, *Wis. Agr. Exp. Sta., Bull.* **123** (1934).
122. P. H. Phillips and J. W. Suttie, *Arch. Ind. Health* **21**, 343 (1960).
123. P. H. Phillips, J. W. Suttie, and E. J. Zebrowski, *J. Dairy Sci.* **46**, 513 (1963).
124. M. J. Purves, *Lancet* **2**, 1188 (1962).
125. K. E. Quentin, S. W. Souci, and J. Indinger, *Z. Lebensm.-Unters.-Forsch.* **111**, 173 (1960).
126. W. F. Ramseyer, C. Smith, and C. M. McCay, *J. Gerontol.* **12**, 14 (1957).
127. E. Reid, *Chin. J. Physiol.* **10**, 259 (1936).
128. C. Rich, J. Ensinck, and P. Ivanovitch, *J. Clin. Invest.* **43**, 545 (1964).
129. K. Roholm, "Fluorine Intoxication," Lewis, London, 1937.
130. A. L. Romanoff and A. J. Romanoff, "The Avian Egg," Wiley, New York, 1949.
131. A. L. Russell and E. Elvove, *Pub. Health Rep.* **66**, 1389 (1951).
132. W. B. Savchuk and W. D. Armstrong, *J. Biol. Chem.* **193**, 575 (1951).
133. H. J. Schmidt and W. E. Rand, *Amer. J. Vet. Res.* **13**, 38 (1952).
134. H. A. Schroeder, M. Mitchener, J. J. Balassa, M. Kanisawa, and A. P. Nason, *J. Nutr.* **95**, 95 (1968).
135. G. R. Sharpless, *Proc. Soc. Exp. Biol. Med.* **34**, 562 (1936).
136. G. R. Sharpless and E. V. McCollum, *J. Nutr.* **6**, 163 (1933).
137. T. R. Shearer and J. W. Suttie, *Amer. J. Physiol.* **212**, 1165 (1967).
138. J. L. Shupe, L. E. Harris, D. A. Greenwood, J. E. Butcher, and H. M. Nielsen, *Amer. J. Vet. Res.* **24**, 300 (1963).
139. J. L. Shupe, M. L. Miner, L. E. Harris, and D. A. Greenwood, *Amer. J. Vet. Res.* **23**, 777 (1962).
140. J. L. Shupe, M. L. Miner, D. A. Greenwood, L. E. Harris, and G. E. Stoddard, *Amer. J. Vet. Res.* **24**, 964 (1963).
141. A. H. Sievert and P. H. Phillips, *J. Nutr.* **72**, 429 (1960).
142. G. Simon and J. W. Suttie, *J. Nutr.* **94**, 511 (1968); **96**, 152 (1968).

143. L. Singer and D. W. Armstrong, *J. Appl. Physiol.* **15**, 508 (1960).
144. L. Singer and D. W. Armstrong, *Proc. Soc. Exp. Biol. Med.* **117**, 686 (1964).
145. L. Singer and D. W. Armstrong, *Anal. Chem.* **31**, 105 (1959); *Anal. Biochem.* **10**, 495 (1965).
146. F. A. Smith, D. E. Gardiner, and H. C. Hodge, *At. Energy Comm. (U.S.), Quart. Tech. Rep.* **UR-200** (1952).
147. M. C. Smith, E. M. Lantz, and H. V. Smith, *Ariz., Univ., Agr. Exp. Sta., Tech. Bull.* **32** (1931).
148. R. W. Smith, Jr. and B. Frame, *N. Engl. J. Med.* **273**, 73 (1965).
149. L. C. Snook, *Aust. Vet. J.* **38**, 42 (1962).
150. R. F. Sognnaes, *Brit. Dent. J.* **70**, 433 (1941); *J. Dent. Res.* **20**, 303 (1941).
151. T. Sollman, "A Manual of Pharmacology," Saunders, Philadelphia, Pennsylvania, 1930.
152. D. F. Steiner, V. Rauda, and R. H. Williams, *J. Biol. Chem.* **236**, 299 (1961).
153. J. W. Suttie, *J. Air Pollut. Contr. Assoc.* **14**, 461 (1964); **19**, 239 (1969).
154. J. W. Suttie, *Amer. J. Vet. Res.* **28**, 709 (1967).
155. J. W. Suttie, R. Gesteland, and P. H. Phillips, *J. Dairy Sci.* **44**, 2250 (1961).
156. J. W. Suttie, R. F. Miller, and P. H. Phillips, *J. Nutr.* **63**, 211 (1957); *J. Dairy Sci.* **40**, 1485 (1957).
157. J. W. Suttie and P. H. Phillips, *Arch. Biochem. Biophys.* **83**, 355 (1959).
158. J. W. Suttie and P. H. Phillips, *J. Dairy Sci.* **42**, 1063 (1959).
159. J. W. Suttie and P. H. Phillips, *J. Nutr.* **72**, 429 (1960).
160. J. W. Suttie and P. H. Phillips, *Amer. J. Vet. Res.* **23**, 1107 (1962).
161. J. W. Suttie, P. H. Phillips, and R. F. Miller, *J. Nutr.* **65**, 293 (1958).
162. T. G. Taylor and J. Kirkley, *Calcif. Tissue Res.* **1**, 33 (1967).
163. J. Tesink, *Tijdschr. Diergeneesk.* **80**, 230 (1955).
164. P. K. Van der Merwe, *Onderstepoort J. Vet. Sci. Anim. Ind.* **14**, 335 (1940).
165. H. Velu, *C. R. Soc. Biol.* **108**, 750 (1931); **127**, 854 (1938).
166. K. Venkataramanan and N. Krishnaswamy, *Indian J. Med. Res.* **37**, 277 (1949).
167. P. Venkateswardu and D. Narayanarao, *Indian J. Med. Res.* **45**, 387 (1957).
168. J. F. Volker, *J. Dent. Res.* **19**, 35 (1940); *Proc. Soc. Exp. Biol. Med.* **43**, 643 (1940).
169. P. Wallace-Durbin, *J. Dent. Res.* **33**, 789 (1954).
170. J. A. Weatherall and S. M. Weidmann, *J. Pathol. Bacteriol.* **78**, 233 (1959).
171. C. W. Weber and B. L. Reid, *J. Nutr.* **97**, 90 (1969).
172. D. A. Weddle and J. C. Muhler, *J. Nutr.* **54**, 437 (1954).
173. D. A. Weddle and J. C. Muhler, *J. Dent. Res.* **36**, 386 (1957).
174. S. M. Weidmann and J. A. Weatherall, *J. Pathol. Bacteriol.* **78**, 243 (1959).
175. S. M. Weidmann, J. A. Weatherall, and R. G. Whitehead, *J. Pathol. Bacteriol.* **78**, 435 (1959).
176. H. J. Wespi, *Schweiz. Bull. Eidg. Gesundheitsamtes, Part B* No. 2 (1962).
177. G. Wrzodek, *Zahnaerztl. Mitt.* **47**, 258 (1959); quoted by Hodge and Smith (65).
178. R. E. Wuthier and P. H. Phillips, *J. Nutr.* **67**, 581 (1959).
179. E. J. Zebrowski and J. W. Suttie, *J. Nutr.* **88**, 267 (1966).
180. E. J. Zebrowski, J. W. Suttie, and P. H. Phillips, *Fed. Proc., Fed. Amer. Soc. Exp. Biol.* **23**, 184 (1964).
181. I. Zipkin and R. C. Likins, *Amer. J. Physiol.* **191**, 549 (1957).
182. I. Zipkin, F. J. McClure, and W. A. Lee, *Pub. Health Rep.* **73**, 732 (1958).

# 14

# SILICON

## I. Introduction

Silicon is one of the most abundant elements in the biosphere. It represents 27.7% of the lithosphere and is exceeded in abundance only by oxygen (20). The high content of silica in soils, plants, and atmospheric dust maintains a high, although very variable, intake by animals and man, especially grazing ruminants.

Many years ago it was claimed that silicon is essential for reproduction in marine plankton (46). Silica is known to perform an important skeletal role in organisms such as diatoms. Recent studies have revealed the structural relationship of silica to the organic constituents of the cell wall in these organisms (45) and have thrown some light on the mode of uptake and deposition of silica in wall formation. Thus Coombs and coworkers (11) found that when *Navicula* is grown in a medium without silicon, cell division is blocked after mitosis and cytokinesis have taken place. When silicon is added, synchronous silicon uptake, wall formation, and cell separation are induced. It was further shown that energy is used in the biochemical processes of silicon metabolism in wall formation. Uptake of silica by certain bacteria and accumulation of silicate in cells and particulate fractions of cells in the presence of an energy source have also been demonstrated (24). Some of these particulate fractions catalyzed the reduction of silicate. This seems to be a nonspecific process, analogous to the reduction of selenate or tellurite, in which silicate can serve as electron acceptor and react with the ejected protons to form volatile (oxy-)hydrides (1). Whether silicon compounds can actually participate in enzyme-catalyzed reactions, with the formation of silicon-organic compounds, remains to be determined.

Completely convincing evidence that silicon is essential in the nutrition of the higher plants has not yet been produced [see Jones and Handreck (28)]. In some circumstances silica has beneficial effects upon plant growth or crop yield through promoting resistance to insect and fungal attack and by other means (28).

Silicon occurs in plants as monosilicic acid and as solid silica. The amounts and proportions of these two compounds vary with the plant species and its stage of growth, and with the soil conditions under which the plant has grown. Whole grasses and cereals may contain 30-40% of their total ash or 3-4% or more of the whole dry plants as $SiO_2$. Even higher levels occur in some tropical grasses (42). In clovers and herbs the total silica concentration is appreciably lower, with a high proportion of the relatively low amounts present as monosilicic acid. Solid silica is only sparsely deposited in these species. Cereal grains carry much lower concentrations of silica than the leaves or stems of the same species (22). Grains high in fiber, such as oats, are richer in silica than low-fiber grains such as wheat (41). In mature gramineous plants, most of the silica is present in the form of solid mineral particles, known as opal phytoliths ($SiO_2 \cdot n\text{-}H_2O$) (7). The marked species difference in the total amount of silica absorbed and subsequently secreted as phytoliths is illustrated by one study in which rye grass (*Lolium perenne*) contained 23 times as much insoluble ash as alfalfa (*Medicago sativa*). This difference was reflected in the silica and the opal phytolith contents of the two species (7). In another study, prairie grass hay (mainly *Festuca scabrella*) averaged 6.27% total silica (dry basis), compared with only 0.39% for alfalfa hay (4). An inverse relationship between silica content and digestibility by ruminants has been demonstrated (28,48).

Since the opal phytoliths of pasture plants are much harder than the dental tissues of sheep, and the amounts ingested by grazing animals are so large and continuous, it has been suggested that these minerals may be a major cause of wear in sheep's teeth (5). At a level of 4% $SiO_2$ and at a daily dry matter intake of 1 kg, a sheep would ingest 40 g $SiO_2$/day or 14 kg over a period of a year. At the lower limit, with legumes containing only 0.2% $SiO_2$, a sheep's daily intake of silica would be about 2 g/day or 0.73 kg over a year. The former figures represent very large amounts of abrasive material. Moreover, they exclude the considerable quantities of silica that can occur as quartz particles on the surface of plants as a result of contamination with soil. This source of adventitious silica has been implicated as a significant source of wear of teeth in grazing sheep (23).

## II. Silica in Animal Tissues and Fluids

The use of reliable methods of analysis, particularly the silicomolybdate procedure of King and co-workers (35), has revealed surprisingly high levels of silica in human and animal tissues and fluids. In fetal tissues, the normal range has been reported as 40-400 ppm $SiO_2$ (dry basis), compared with 50-1000 ppm for normal adult human tissues (34). In the fetus the lowest levels occur in the lungs and the highest in the muscles. In adult humans the position is reversed, no doubt due to the inhalation of dust. The concentration of silica in the blood of sheep (27), and of man and other species (34), averages close to 5 $\mu g$ Si/ml. This is higher than the average concentration of 2.2 $\mu g$ $SiO_2$/ml reported for bovine blood by Baumann (8). The blood silica occurs mostly, or entirely, in solution as monosilicic acid and varies little in concentration except after massive oral administration of soluble silicates.

The silicon content of cow's milk is little influenced by dietary silica intakes. There is considerable individual variation and a pronounced tendency for the level in the milk to decrease as lactation progresses. Archibald and Fenner (3) found the milk of 6 cows alternately fed a control ration and one containing added sodium silicate at the rate of 1 g/day (230 mg Si) to average 1.4 mg Si/liter, whether the diet was supplemented or not. Kirchgessner (36) reported cow's colostrum to contain nearly 3 times the mean level of silica of later milk.

The highest silicon concentrations normally occur in the skin and its appendages. For human tissues, Fregert (16) reports the following values: muscle, $18\pm1.3$; tendon, $28\pm1.8$; aorta, $41\pm3.3$; kidney, $42\pm3.9$; nails $56\pm2.2$; hair, $90\pm2.0$; epidermis, $106\pm2.7$ ppm $SiO_2$ (dry basis). Sauer *et al.* (47) obtained somewhat higher average levels in guinea pig tissues, namely 213-346 ppm $SiO_2$ on the dry matter of hair, 55-126 ppm for the lungs, and 30-70 ppm for the rest of the tissues examined. Comparable values have been reported for the mouse (26) and for several other species, including the bovine (34). In an earlier study, 346 ppm $SiO_2$ was found in chicken feathers (21).

## III. Silicon Metabolism

Silicon enters the alimentary tract from the food as monosilicic acid and as solid silica. Quartz and other forms of silica may also be ingested in variable amounts from soil and dust contamination. The amounts of monosilicic acid ingested usually comprise a very small proportion of the total silica, most of which occurs in the solid form. Absorption apparently occurs as monosilicic acid, at least in guinea pigs (47). Some of this comes from the solid silica, which is partly dissolved by the fluids of the gastrointestinal tract. The extent of solution and absorption of solid silica as monosilicic acid varies with the silica content of the diet. This has been demonstrated in sheep by Jones and Handreck (27) who fed three diets, with silica contents ranging from 0.10 to 2.84% in the dry matter, and measured the distribution of silica between feces and urine. They also determined the concentrations of silica in the blood and the rumen liquor. The amounts excreted in the urine increased with increasing silica content of the diet but reached no more than 205 mg $SiO_2$/day. This amount represented less than 4% of the total intake.

Increased urinary output of silica with increasing intake, up to fairly well-defined limits, has been demonstrated in humans (26), rats (31), guinea pigs (47), and cows (4). In experiments with sheep, Nottle (39) found urinary excretion of silica to rise with rising dietary silica intakes up to an intake of 8 g $SiO_2$/day. Thereafter, urinary excretion levelled off at 200 to 250 mg/day, a maximum close to that of Jones and Handreck (27). The upper limits of urinary excretion after oral administration do not seem to be set by the ability of the kidney to excrete more silica, because much greater urinary excretion can occur after intraperitoneal injections of comparable silica doses (47). Those limits are determined by the rate and extent of silica absorption from the gastrointestinal tract into the blood. This is influenced in the ruminant by the solubility of silica in the rumen fluid (27). Once silica has entered the bloodstream, it passes rapidly into the urine because, even at widely divergent levels of silica intake, the concentration of silica in the blood remains practically constant (27,34).

Several of the earlier workers recovered in the feces as little as 85% of the ingested silica. They concluded that the balance was either stored in the body or excreted in the urine. Later studies, employing more reliable methods of analysis, do not support such a conclusion. The proportions of the ingested silica which are stored in the body and excreted in the urine are usually much lower. In the experiments of Jones and Handreck (27), cited earlier, the sum of the amounts excreted in the feces and urine was within 1% of the amounts ingested. Body storage of silica is clearly exceedingly small under such conditions. Urinary excretion is also very low. In three separate experiments with sheep the proportion excreted in the urine decreased progressively from 3.3 to 0.55% of the intake as this intake increased from 0.8 to 31 g $SiO_2$/day (13,27,39). The position is similar in cattle (14,30), although less information is available. These findings suggest that silica may be a useful internal indigestible reference material for determining the fate of the digestible constituents of plant feeds (28,44). However, other workers have claimed that its recovery in feces is too variable for this purpose (17,18).

Microscopic particles of solid silica from plants have been found in the lymph nodes and urinary calculi of sheep and are, to a small extent, absorbed as such from the alimentary tract (6). The permeability of the gut wall to particles of the size of diatoms has also been demonstrated in man. Volkenheimer (51) showed that diatomaceous earth particles were absorbed through the intact intestinal mucosa and passed through the lymphatic and circulatory systems. These particles also reach other tissues in arterial blood via the alveolar region of the lung. Microscopic examination of human organs has revealed the presence of siliceous diatoms in lungs, liver, and kidney, as a consequence of their presence in atmospheric dust and their movement from the respiratory tract (19). The capacity of these particles to travel in the blood and to penetrate body membranes, including the placenta, is illustrated further by their presence in the organs of stillborn and premature infants (19). The discovery of diatoms in human circulatory organs has previously been considered as an important indication of death by drowning. The ubiquitous occurrence of these forms of silica in human organs casts serious doubt on the general validity of such a conclusion.

Recent studies of the processes involved in silica toxicity in man have disclosed further interesting facets of the metabolism of this element. Particles such as silica and asbestos (fibrous silicates of complex composition) have long been known to stimulate a severe fibrogenic reaction in the lungs and elsewhere in the body, as occurs in silicosis and asbestosis in miners. This reaction arises initially from phagocytosis of silica particles by alveolar macrophages. Collagen synthesis by neighboring fibroblasts is stimulated by the continued death of these macrophages. Allison and co-workers (2,38) have now shown that the particular toxicity of silica to macrophages derives from the fact that the particles are taken up into lysosomes and readily damage lysosomal membranes through hydrogen-bonding interactions. The exact relation of macrophage death to fibrogenesis is more obscure. Heppleston and Styles (25) have provided evidence that the macrophage-silica interaction results in the release of a factor, of unknown nature, which stimulates collagen formation. The development of malignant tumors of the pleura and peritoneum constitute a further manifestation of the toxicity of silica and asbestos to man and experimental animals (52). Allison (1) has suggested that lysosomes may be involved in the malignant transformations brought about by silica and that enzymes released from lysosomes damage chromosomes, with a chromosome mutation leading to malignancy.

## IV. Silicon and Bone Calcification

Evidence that silicon is associated with calcium in an early stage of bone calcification has recently been produced by Carlisle (9a). This worker carried out quantitative, electron probe microanalyses of normal tibia from mice and rats 0–28 days old. Silicon was found to be uniquely localized in active calcification sites in young bone, to increase directly with calcium at relatively low calcium concentrations, and to fall below the detection limit at compositions approaching hydroxyapatite. Silicon appeared where a low-to-moderate and varying calcium content suggests active calcification, and was concentrated in sites containing 0.08–1.0% or more silicon and 0.1–2.0% calcium on the edge of trabeculae, or from 0.5 to 15% calcium in the periosteal areas. It was further demonstrated that the amount of bone ash in weanling rats on a low-Ca diet is influenced by the silicon content of the diet and that the amount of silicon in the bones is influenced by the calcium as well as by the silicon concentration of the diet. The latter findings lend additional support to the concept that silicon is associated with calcium in the mineralization process.

## V. Silica Urolithiasis

Normally the urinary silica is readily eliminated. Under some conditions a part of it is deposited in the kidney, bladder, or urethra to form calculi or uroliths. Small calculi may be excreted harmlessly, but sometimes they become so large that they block the passage of urine and cause death of the animal. Urinary calculi can be composed of various minerals, especially calcium, magnesium, and phosphorus, but siliceous stones or uroliths are more common in grazing sheep and cattle than was previously suspected. Silica urolithiasis is a serious problem in grazing wethers in Western Australia (40) and in grazing steers in the western regions of Canada (10,53) and the northwestern regions of the United States (43,49).

The silica of ovine and bovine calculi has been specifically identified as amorphous opal (6,15), most of which is derived from the absorbed monosilicic acid. A small proportion of the opal occurs as phytoliths from plants, with occasional fragments of sponge spicules and diatoms embedded in the calculi (6). In addition to hydrated silica, siliceous uroliths usually contain small amounts of accessory elements, notably magnesium, calcium, and phosphorus, and of organic material. The exact nature of the organic material, or matrix, in siliceous calculi has not been determined, but the chemical studies of Keeler (29,32) indicate the presence of a glycoprotein containing a neutral carbohydrate moiety. A similar glycoprotein has been identified in the organic matrices of phosphatic calculi from sheep and cattle (12).

The factors responsible for the formation of siliceous calculi are poorly defined. Attempts to produce them in sheep and cattle by adding silicates to the diet (9,53) or by restricting water consumption (50) have not been successful, even when the urinary excretion level achieved was two- to threefold greater than the 70-80-$\mu$g/ml level at which silicon normally precipitates in bovine urine (33). The concentration in the urine of sheep (39) and cattle (4) usually exceeds that of a saturated solution of amorphous silica and may reach 1000 ppm $SiO_2$ in sheep. It is clear that high dietary intakes and high-silica outputs in the urine, associated with supersaturation of the urine, are insufficient to explain the polymerization of the monosilicic acid, deposition of silica, and the formation of calculi.

The glycoprotein component of the organic matrix has been assigned a critical role in the formation of urinary calculi in man, through acting as a primary matrix which becomes secondarily mineralized. A similar theory has been adopted to explain the formation of siliceous calculi in cattle (30) and of phosphatic calculi in sheep, cattle, and dogs (12). Jones and Handreck (28) have disputed this theory on the grounds that the main mechanism involved is "precipitation of the inorganic components which, in turn, depends upon both supersaturation and nucleation." They tend to favor the theory that foreign particles of silica and other foreign particles, such as have been found in calculi from sheep, act as nuclei for the deposition of silica. Solid (amorphous) silica is known to accelerate the polymerization and deposition of silica from supersaturated solution (37), but whether solid particles of silica are consistently implicated in the formation of siliceous calculi is unknown. In view of the marked dependence of urinary silica concentration on the rate of urine excretion in cattle, it has been suggested that any method which increases urine output could be used also to prevent urolith formation (4).

## REFERENCES


1. A. C. Allison, *Proc. Roy. Soc., Ser. B* **171**, 19 (1968).
2. A. C. Allison, J. S. Harington, and M. Birbeck, *J. Exp. Med.* **124**, 141 (1966).
3. J. G. Archibald and H. Fenner, *J. Dairy Sci.* **40**, 703 (1957).
4. C. H. Bailey, *Amer. J. Vet. Res.* **28**, 1743 (1967).
5. G. Baker, L. H. P. Jones, and I. D. Wardrop, *Nature* (*London*) **184**, 1583 (1959).
6. G. Baker, L. H. P. Jones, and A. A. Milne, *Aust. J. Agr. Res.* **12**, 473 (1961).
7. G. Baker, L. H. P. Jones, and I. D. Wardrop, *Aust. J. Agr. Res.* **12**, 426 (1961).
8. H. Baumann, *Z. Physiol. Chem.* **319**, 38 (1960); **320**, 11 (1960).
9. W. M. Beeson, J. W. Pence, and G. C. Holan, *Amer. J. Vet. Res.* **4**, 120 (1943).

9a. E. M. Carlisle, *Science* **167**, 279 (1970); *Fed. Proc., Fed. Amer. Soc. Exp. Biol.* **28**, 374 (1969).

10. R. Connell, F. Whiting, and S. A. Forman, *Can. J. Comp. Med. Vet. Sci.* **23**, 41 (1959).
11. J. Coombs, P. J. Halicki, O. Holm-Hansen, and B. E. Volcani, *Exp. Cell Res.* **47**, 315 (1967).
12. C. E. Cornelius and J. A. Bishop, *J. Urol.* **85**, 842 (1961).
13. R. J. Emerick, L. B. Embay, and O. E. Olson, *J. Anim. Sci.* **18**, 1025 (1959).
14. E. B. Forbes and F. M. Beegle, *Ohio, Agr. Exp. Sta., Bull.* **295**, (1916).
15. S. A. Forman, F. Whiting, and R. Connell, *Can. J. Comp. Med. Vet. Sci.* **23**, 157 (1959).
16. S. Fregert, *J. Invest. Dermatol.* **31**, 95 (1958).
17. W. D. Gallup and A. H. Kuhlman, *J. Agr. Res.* **42**, 665 (1931); **52**, 889 (1936).
18. W. D. Gallup, C. S. Hobbs, and H. M. Briggs, *J. Anim. Sci.* **4**, 68 (1945).
19. U. Geissler and J. Gerloff, *Nova Hedwigia* **10**, 565 (1965).
20. V. M. Goldschmidt, "Geochemistry." Oxford Univ. Press (*Clarendon*) London and New York, 1954.
21. M. Gonnerman, *Hoppe-Seyler's Z. Physiol. Chem.* **99**, 255 (1917); *Biochem. Z.* **88**, 401 (1918); **94**, 163 (1919).
22. K. A. Handreck and L. H. P. Jones, *Plant Soil* **29**, 449 (1968).
23. W. B. Healey and T. S. Ludwig, *N. Z. J. Agr. Res.* **8**, 737 (1965).
24. W. Heinen, *Arch. Mikrobiol.* **52**, 69 (1965); *Arch. Biochem. Biophys.* **120**, 86 (1967).
25. A. W. Heppleston and J. A. Styles, *Nature* (*London*) **214**, 521 (1967).
26. P. F. Holt, *Brit. J. Ind. Med.* **7**, 12 (1950).
27. L. H. P. Jones and K. A. Handreck, *J. Agr. Sci.* **65**, 129 (1965).
28. L. H. P. Jones and K. A. Handreck, *Advan. Agron.* **19**, 107 (1967).
29. R. F. Keeler, *Amer. J. Vet. Res.* **21**, 428 (1960).
30. R. F. Keeler, *Ann. N.Y. Acad. Sci.* **104**, 592 (1963).
31. R. F. Keeler and S. A. Lovelace, *J. Exp. Med.* **109**, 601 (1959).
32. R. F. Keeler and K. F. Swingle, *Amer. J. Vet. Res.* **20**, 249 (1959).
33. R. F. Keeler and S. A. Lovelace, *Amer. J. Vet. Res.* **22**, 617 (1961).
34. E. J. King and T. H. Belt, *Physiol. Rev.* **18**, 329 (1938).
35. E. J. King, B. D. Stacy, P. F. Holt, D. M. Yates, and D. Pickles, *Analyst* **80**, 441 (1955).
36. M. Kirchgessner, *Z. Tierphysiol., Tierernaehr. Futtermittelk.* **14**, 270 and 278 (1959).
37. K. B. Krauskopf, *Geochim. Cosmochim. Acta* **10**, 1 (1956); quoted by Jones and Handreck (28).
38. T. Nash, A. C. Allison, and J. S. Harington, *Nature* (*London*) **210**, 259 (1966).

39. M. C. Nottle, *Aust. J. Agr. Res.* **17**, 175 (1966).
40. M. C. Nottle and J. M. Armstrong, *Aust. J. Agr. Res.* **17**, 165 (1966).
41. M. C. Nottle, private communication (1962).
42. V. A. Oyenuga, *Nutr. Abstr. Rev.* **28**, 985 (1958).
43. K. G. Parker, *J. Range Manage.* **10**, 105 (1957).
44. K. Pujszo, S. Seidler, A. Ziolecka, and A. Zolkiewski, *Rocz. Nauk Roln. Ser. B* **74**, 591 (1959); quoted by Jones and Handreck (28).
45. B. E. F. Reimann, J. C. Lewin, and B. E. Volcani, *J. Cell Biol.* **24**, 39 (1965); *J. Phycol.* **2**, 74 (1966).
46. O. Richter, *Verh. Ges. Deut. Naturforsch. Aerzte* **2**, 249 (1904).
47. F. Sauer, D. H. Laughland, and W. M. Davidson, *Can. J. Biochem. Physiol.* **37**, 183 and 1173 (1959).
48. P. J. van Soest and L. H. P. Jones, *J. Dairy Sci.* **51**, 1644 (1968).
49. K. F. Swingle, *Amer. J. Vet. Res.* **14**, 493 (1953).
50. K. F. Swingle and H. Marsh, *Amer. J. Vet. Res.* **14**, 16 (1953).
51. G. Volkenheimer, *Z. Gastroenterol.* **2**, 57 (1964).
52. C. Wagner, *Perugia Quad. Int. Conf. Cancer,* Vol. 3, p. 589 (1966).
53. F. Whiting, R. Connell, and S. A. Forman, *Can. J. Comp. Med. Vet. Sci.* **22**, 332 (1958).

# 15

# VANADIUM

## I. Introduction

Vanadium is an essential element for a species of green algae (4) and several reports of growth stimulation of crop plants from vanadium salts have appeared [see Nason (43)]. Vanadium can partially replace molybdenum for nitrogen fixation by *Azotobacter* but is not required for this process in the presence of molybdenum (33). Nor can it substitute for the molybdenum requirement of the blue-green algae *Anabaena* (3). Clear proof that vanadium is required by the higher plants, including legumes, or by the higher animals,* has not yet been produced. However, there is evidence that this element is involved in lipid metabolism, particularly cholesterol biosynthesis. It may also play some part in the calcification of bones and teeth.

As long ago as 1911, Henze (30) discovered high-vanadium concentrations in the blood of an ascidian worm, and variable but usually high concentrations, ranging from 3 to 1900 ppm, have since been demonstrated in various ascidians (61). A high proportion of this vanadium exists in the blood as the vanadium-protein compound, hemovanadin (14). This compound cannot act as an oxygen carrier, and it is not known if the vanadium performs any vital function in these species (13, 31). The

*Author's Note: Recent evidence suggests that vanadium should now be considered an essential element for the nutrition of chicks. In chicks consuming a diet containing 10 ppb. V, the growth of wing and tail feathers was significantly slower than in V-supplemented controls. (L. L. Hopkins, Jr. and H. E. Mohr, private communication, 1970.)

centrifuged blood cells of *Ascidia nigra* contain the remarkably high concentration of 1.45% V (15), and the vanadium in *Ascidia aspersa* is in a dynamic equilibrium of $V^{(III)}$ and $V^{(IV)}$ in the blood cells (47). This suggests a role for this element in an oxidation-reduction reaction in these cells. Vanadium also occurs in large amounts in certain holothurians (9) and up to 150 ppm (dry weight basis) has been reported in the mollusk *Pleurobranchus plumula* (60). An exceptionally high-vanadium content of 100 ppm has been observed in toadstools (59). These concentrations are several orders of magnitude higher than those found for the tissues of the higher plants and animals.

## II. Vanadium in Plant and Animal Tissues

The body of a normal adult man has been estimated to contain 10–25 mg of vanadium, most of which is present in the bones, teeth, and fat (20). This figure seems unduly high in the light of the analytical data of Söremark (52), mentioned later. Bertrand (9) reported a range of 0.02 (the limit of his method) to 0.3 ppm, with a mean of 0.1 ppm V, for various vertebrate tissues. Tipton and Cook (58) were unable to detect vanadium in some tissues by the spectrographic method employed but reported average concentrations of the order of 0.02 to 0.03 ppm (dry basis) for normal adult liver, spleen, pancreas, and prostate gland. The only organ carrying consistently and substantially higher levels was the lung, which averaged close to 0.6 ppm. Schroeder *et al.* (50) also found significantly higher-vanadium concentrations in adult human lung than in other tissues. Since no such accumulations were observed in the lungs of the newborn and since significant geographic variations were apparent in adults, it was suggested that these moderate accumulations arise from inhalation of atmospheric dusts (50). However, pronounced accumulation in the lungs has been demonstrated in rats injected with $^{48}V$ (54), and carrier-free $^{48}V$ accumulates in the fetuses, especially in the fetal skeleton, and in mammary glands of lactating mice (54).

In a study of blood from male donors from nineteen U.S. cities, Allaway and co-workers (2) found over 90% of the samples to contain less than 1 μg V/100 ml of whole blood. The highest concentration was 2 μg/100 ml. These concentrations are lower than those reported by Butt *et al.* (11) and very much lower than those of Schroeder *et al.* (50). The latter workers reported 23 μg V/100 ml blood in elderly controls and 47.5 μg/100 ml in elderly patients receiving 4.5 mg V/day. This vanadium was present in the serum and not in the erythrocytes. Söremark (52), using a neutron activation method, has obtained generally lower-vanadium concentrations for most biological materials than those reported by others. For instance, calf livers from two sources averaged 0.0024 and 0.01 ppm (fresh basis) and fresh cow's milk from five locations ranged from the extremely low level of 0.00007 to 0.00011 ppm V.

Examination of a variety of fruits and vegetables by Söremark (52) revealed extreme variability in vanadium concentration, both within and among species. Parsley was reported to average 0.79 ppm V and dill 0.14 ppm (fresh basis). Lettuce and radishes were lower, and all other fruits and vegetables much lower still. Data for the vanadium content of pasture grasses were reported by Mitchell (38). This worker obtained concentrations ranging from less than 0.03 to 0.16 ppm for red clover and from less than 0.03 to 0.11 ppm for rye grass. More than half the samples contained 0.03-0.07 ppm V (dry basis).

Estimates of the normal intakes of vanadium by man and farm animals are made difficult by doubts about the validity of many of the reported values for vanadium, as well as by the great variability in the vanadium content of different dietary items. Schroeder *et al.* (50) reported that an institutional diet supplied approximately 1.2 mg/day and that a good, well-balanced diet should provide about 2 mg V/day, with a range of 1 to 4 mg. These workers contend that the major determinant of vanadium intake is the type of fat in the diet. A diet high in unsaturated fatty acids from vegetable sources results in higher-vanadium intakes than one containing saturated fats from animal sources. Large concentrations of vanadium, up to 43 ppm, were found in soybean oil and in corn, olive, linseed, and peanut oils. No such high concentrations were found in castor and codliver oil or in lard or butter. Beef, pork, deer, and chicken fats were relatively rich in this element. Studies with $^{48}V$ in mice (54) and rats (32) do not indicate any particular accumulation in the body fat of these species. It is clear, therefore, that the whole question of the distribution of vanadium in animal tissues and in foods requires more extensive and critical investigation.

## III. Vanadium Metabolism

At intakes well beyond physiological levels, vanadium salts are poorly absorbed from the gastrointestinal tract and appear mostly in the feces. In one study, between 0.1 and 1.0% of the vanadium in 100 mg of the soluble diammonium oxytartarovanadate was absorbed from the human gut (19). In another study, 60% of absorbed vanadium was excreted in the urine in the first 24 hr following dosage, the remainder being retained in the liver and bone (55). The vanadium present in the bone was mobilized and excreted much more slowly than that in the liver. Very small amounts of vanadium (0–8 μg/day) are normally excreted in the urine of man (45, 50). These amounts are increased greatly when vanadium salts are administered orally at subtoxic or toxic levels (22, 50).

Tissue distribution, retention, and excretion of intravenously injected $^{48}V$ have been studied in mice, rats, and chicks. Söremark and Ullberg (54) found the highest amounts of radiovanadium in the bones and teeth of mice, and Hathcock *et al.* (28) by far the greatest retention in the bones and kidneys of chicks. Hopkins and Tilton (32) observed no significant differences in the rate or amount of uptake of the three oxidation states of vanadium tested. The liver, kidneys, spleen, and testes accumulated $^{48}V$ up to 4 hr after injection and retained most of this activity up to 96 hr. By this time, 46% of the isotope had been excreted in the urine and 9% in the feces. During this period the amount of $^{48}V$ in the liver subcellular supernatant fraction decreased from 57 to 11% of the liver radioactivity, whereas the mitochondrial and nuclear fractions increased from 14 to 40%. It was suggested that the marked liver retention of $^{48}V$ was due to its movement into the mitochondrial and nuclear fractions.

## IV. Vanadium and Dental Caries

Vanadium occurs in human enamel and dentin, and the suggestion has been made that it may exchange with phosphorus in the apatite tooth substance (36, 53). Söremark and Ullberg (54) showed by microradioautography that $^{48}V_2O_5$ injected subcutaneously into mice is highly concentrated in the areas of rapid mineralization in tooth dentin and bone. It is also incorporated into the tooth structure of rats and retained in the molars up to 90 days after injection (57). Other evidence implicating vanadium in the mineralization of bones and teeth has been presented by Rygh (49) and by Geyer (25). The former worker reported that the addition of vanadium and of strontium to specially purified diets promoted mineralization of the bones and teeth and reduced the numbers of carious teeth in rats and guinea pigs. The latter worker observed a high degree of protection against caries in hamsters fed a cariogenic diet, when vanadium was administered as $V_2O_5$, either orally or parenterally. Kruger (34) similarly reported that vanadium, administered intraperitoneally to rats during the time of tooth development, is effective in reducing the incidence of caries.

No such beneficial effects of vanadium upon dental caries in animals are apparent in other studies. When young hamsters were given a cariogenic diet and drinking water containing 10 ppm of vanadium the incidence of caries was actually increased (29). In three other independent experiments with rats, administration of varying levels of vanadium in the drinking water was unsuccessful in decreasing caries incidence (12,41,42) (see Table 49).

An inverse association between the vanadium levels in the water supplies and caries incidence in children born and reared in various localities in Wyoming was reported by Tank and Storvick (56). The numbers of children included in some cases were quite small and, as Hadjimarkos (26,27) has pointed out, the drinking water does not constitute an important source of vanadium to man. Its concentration cannot, therefore, serve as a reliable criterion of the total amount of vanadium ingested daily. Conclusive evidence that vanadium exerts a beneficial effect on dental caries incidence, either in animals or man, is thus not available. Evaluation of a possible role for this element in modifying caries and an explanation of the markedly divergent results obtained by different investigators with animals clearly invite further critical research.

## V. Vanadium and Cholesterol Synthesis

More than 30 years ago, vanadium was found to increase markedly the oxidation of phospholipids by rodent liver cells; manganese was shown to inhibit this vanadium effect (8). Later, Curran (18) demonstrated that vanadium, in 0.001 *M* concentration, depresses the synthesis of cholesterol from labeled acetate by surviving rat liver. It was further shown that manganese has an antithetic action with vanadium in this process, as had earlier been found in phospholipid oxidation. These *in vitro* findings have been confirmed by several groups of workers, and, in addition, inhibition of cholesterol synthesis by vanadium has been

TABLE 49

*Effect of Postdevelopmental Vanadium on Caries in Rats*[a,b]

| V in water (ppm) | No. rats surviving | Wt gain (g) | Mean No. carious lesions |
|---|---|---|---|
| 40 | All died | — | — |
| 20 | 27[c] | 156[c] | 9.2[c] |
| 10 | 30 | 163 | 8.6 |
| Control | 39 | 202 | 8.3 |

[a]From Muhler (42).
[b]Groups of 50 animals treated for 140 days.
[c]Sexes combined.

observed *in vivo* in human and animal tissues (19,21,23,35,40). This inhibition was accompanied by decreased plasma phospholipid and cholesterol concentrations and by reduced aortic cholesterol accumulations. Furthermore, the mean serum cholesterol levels in workmen exposed to industrial sources of vanadium was found to be slightly but significantly lower (205 mg/100 ml) than that of controls (228 mg/100 ml). In older human patients and in patients with hypercholesterolemia or ischemic heart disease, vanadium does not lower serum cholesterol in this way (20,22,50,51), whereas in older rats the inhibition can be demonstrated *in vitro* but not *in vivo*. In fact, Curran and Burch (20) observed an actual stimulation of cholesterol biosynthesis by vanadium in 500-600-g rats. These apparently discordant findings give little encouragement to the use of vanadium as an anticholesteremic agent in human medicine but they have stimulated several revealing investigations by Curran and others of its site of action in cholesterol biosynthesis.

Vanadium inhibits synthesis of cholesterol from acetate as well as from mevalonic acid but not from squalene (6). The site of inhibition was thus localized between these two points in the synthetic sequence. Subsequently vanadium was shown to inhibit the conversion of labeled farnesyl pyrophosphate to squalene (5). This indicates that the microsomal enzyme system referred to as squalene synthetase is a site of vanadium inhibition. Later Burch and Curran (10) demonstrated a stimulatory effect of vanadium on acetoacetyl-CoA deacylase in rat liver mitochondria and obtained indications that this is an age-related phenomenon. Their results imply further that vanadium probably exerts an effect on mitochondrial membrane permeability. A possible explanation of the age-related effect of vanadium on cholesterol biosynthesis in the rat has been put forward by Curran and Burch (20) in the following terms: "in the young animal vanadium may decrease cholesterol synthesis both by enhancing acetoacetyl-CoA deacylase, thereby reducing the acetoacetyl-CoA pool and by its inhibition of squalene synthesis. In the older animal there is a lesser effect on squalene synthetase and no effect or an actual increase in the acetoacetyl-CoA pool due to an inhibition of acetoacetyl-CoA deacylase with consequent increased synthesis of cholesterol."

As stated earlier, vanadium counteracts the stimulation of cholesterol synthesis induced by manganese, and manganese nullifies both the stimulatory effect of vanadium on hepatic phospholipid oxidation and its depressant action on cholesterol synthesis (8,18). These antithetic actions of vanadium and manganese are not understood but are further apparent from studies with *Mycobacterium tuberculosis*. The complete inhibition of growth of this organism that can be achieved by including 5 μg of *m*-

vanadate ion in the media is completely reversible by manganous (and by chromate) ions (16). Furthermore, dietary administration of small quantities of vanadium markedly reduces pulmonary tuberculosis lesions in rabbits and produces a minimal but significant depression of disease in mice infected with a mouse-adapted strain of the human tubercle bacillus (17).

## VI. Vanadium Toxicity

Franke and Moxon (24) found that vanadium fed as sodium vanadate is definitely toxic to rats at dietary concentrations as low as 25 ppm. At intakes of 50 ppm the animals exhibited diarrhea and mortality. The relative toxicity at 25 ppm of five different elements, in increasing order, was As, Mo, Te, V, and Se. The toxicity of ingested vanadium is of a similar magnitude in chicks. Thus Romoser *et al.* (48) showed that the addition to practical rations of 30 ppm V as the calcium salt depressed rate of gain, whereas 200 ppm resulted in high mortality. Nelson and coworkers (44) reported that chicks tolerated vanadium intakes of 20 to 35 ppm and that amounts in excess of these resulted in a significant growth depression. In studies carried out by Berg (7) the addition of 10, 15, and 20 ppm V as ammonium *m*-vanadate progressively depressed the growth of chicks in one experiment, and in another experiment a total vanadium intake of 13 ppm also resulted in growth depression. Finally, Hathcock and associates (28) demonstrated toxicity in chicks, evidenced by both growth depression and mortality, from 25 ppm V fed either as ammonium *m*-vanadate or as vanadyl sulfate. The physically and chemically related elements, scandium, titanium, and niobium, were not toxic even when fed at 200 ppm. Radioisotope studies revealed that scandium and niobium were poorly absorbed or poorly retained compared with vanadium. Ethylenediaminetetraacetate completely prevented the toxicity of vanadium, apparently by inhibiting its absorption from the intestinal tract.

Vanadium is not a particularly toxic metal to man. Dimond *et al.* (22) gave vanadium orally, as ammonium vanadyl tartrate, to 6 subjects for 6 to 10 weeks in amounts ranging from 4.5 to 18 mg V/day with no toxic effects other than some cramps and diarrhea at the larger dose levels. Schroeder *et al.* (50) fed patients 4.5 mg/day as the oxytartarovanadate for 16 months with no signs of intolerance appearing. In each case increased amounts of vanadium were present in the urine.

The mechanism of vanadium toxicity in animals is poorly understood. A reduction in the cystine content of the hair of rats fed diets containing

100 ppm V has been reported and the suggestion made that one of the primary modes of action of vanadium resides in its ability to affect the reactions of sulfur-containing compounds (39). An effect on various mammalian enzyme systems has also been observed. Thus vanadium reduces coenzyme A (37) and coenzyme Q (1) levels in rats and stimulates monoamine oxidase activity (46). The significance of these changes in the manifestations of vanadium toxicity is unknown.

## REFERENCES


1. A. S. Aiyar and A. Sreenivasan, *Proc. Soc. Exp. Biol. Med.* **107**, 914 (1961).
2. W. H. Allaway, J. Kubota, F. L. Losee, and M. Roth, *Arch. Environ. Health* **16**, 342 (1968).
3. M. B. Allen, *Sci. Mon.* **83**, 100 (1956).
4. D. I. Arnon and G. Wessel, *Nature (London)* **172**, 1039 (1953).
5. D. L. Azarnoff, F. E. Brock, and G. L. Curran, *Biochim. Biophys. Acta* **51**, 397 (1961).
6. D. L. Azarnoff and G. L. Curran, *J. Amer. Chem. Soc.* **79**, 2968 (1957).
7. L. R. Berg, *Poultry Sci.* **42**, 766 (1963).
8. F. Bernheim and M. L. C. Bernheim, *J. Biol. Chem.* **127**, 353 (1939).
9. D. Bertrand, *Bull. Soc. Chim. Biol.* **25**, 36 (1943); *Bull. Amer. Mus. Natur. Hist.* **94**, 403 (1956).
10. R. E. Burch and G. L. Curran, private communication (1969).
11. E. M. Butt, R. E. Nusbaum, T. C. Gelmour, S. L. Didio, and Sister Mariano, *Arch. Environ. Health* **8**, 52 (1964).
12. W. Buttner, *J. Dent. Res.* **42**, 453 (1963).
13. L. Califano and E. Boeri, *J. Exp. Biol.* **27**, 253 (1950).
14. L. Califano and P. Caselli, *Pubbl. Sta. Zool. Napoli* **21**, 261 (1948).
15. L. S. Ciereszko, E. M. Ciereszko, E. R. Harris, and C. A. Lane, *Comp. Biochem. Physiol.* **8**, 137 (1963).
16. R. L. Costello and L. W. Hedgecock, *J. Bacteriol.* **77**, 794 (1959).
17. R. L. Costello and L. W. Hedgecock, quoted by G. L. Curran, *in* "Metal-Binding in Medicine" (M. J. Seven and L. A. Johnson, eds.), p. 216. Lippincott, Philadelphia, Pennsylvania, 1960.
18. G. L. Curran, *J. Biol. Chem.* **210**, 765 (1954).
19. G. L. Curran, D. L. Azarnoff, and R. E. Bolinger, *J. Clin. Invest.* **38**, 1251 (1959).
20. G. L. Curran and R. E. Burch, *Proc. 1st Annu. Conf. Trace Substances Environ. Health, 1967* p. 96 (1967).
21. G. L. Curran and R. L. Costello, *J. Exp. Med.* **103**, 49 (1956).
22. E. G. Dimond, J. Caravaca, and A. Benchimol, *Amer. J. Clin. Nutr.* **12**, 49 (1963).
23. C. H. Eades and D. G. Callo, *Fed. Proc., Fed. Amer. Soc. Exp. Biol.* **16**, 176 (1957).
24. K. W. Franke and A. L. Moxon, *J. Pharmacol. Exp. Ther.* **61**, 89 (1937).
25. C. F. Geyer, *J. Dent. Res.* **32**, 590 (1953).
26. D. M. Hadjimarkos, *Nature (London)* **209**, 1137 (1966).
27. D. M. Hadjimarkos, *Advan. Oral Biol.* **3**, 253 (1968).
28. J. N. Hathcock, C. H. Hill, and G. Matrone, *J. Nutr.* **82**, 106 (1964).

29. J. W. Hein and J. Wisotzky, *J. Dent. Res.* **34**, 756 (1955).
30. M. Henze, *Hoppe-Seyler's Z. Physiol. Chem.* **72**, 494 (1911); **83**, 340 (1913).
31. M. Henze, R. Stohr, and R. Muller, *Hoppe-Seyler's Z. Physiol. Chem.* **213**, 125 (1932).
32. L. L. Hopkins and B. E. Tilton, *Amer. J. Physiol.* **211**, 169 (1966).
33. C. K. Horner, D. Burk, F. E. Allison, and M. S. Sherman, *J. Agr. Res.* **65**, 173 (1942).
34. B. J. Kruger, *J. Aust. Dent. Assoc.* **3**, (1958).
35. C. E. Lewis, *AMA Arch. Ind. Health* **19**, 419 (1959).
36. F. Lowater and M. M. Murray, *Biochem. J.* **31**, 837 (1937).
37. E. Mascitelli-Coriandoli and C. Citterio, *Nature (London)* **183**, 1527 (1959).
38. R. L. Mitchell, *Research (London)* **10**, 357 (1957).
39. J. T. Mountain, L. L. Delker, and H. E. Stokinger, *Arch. Ind. Hyg. Occup. Med.* **8**, 406 (1953).
40. J. T. Mountain, F. R. Stockwell, and H. E. Stokinger, *Proc. Soc. Exp. Biol. Med.* **92**, 582 (1956).
41. H. R. Mühlemann and K. G. König, *Helv. Odontol. Acta* **8**, 79 (1964); cited by Hadjimarkos (27).
42. J. C. Muhler, *J. Dent. Res.* **36**, 787 (1957).
43. A. Nason, *in* "Trace Elements" (C. A. Lamb, O. G. Bentley, and J. M. Beattie, eds.), p. 269. Academic Press, New York, 1958.
44. T. S. Nelson, M. B. Gillis, and H. T. Peeler, *Poultry Sci.* **41**, 519 (1962).
45. H. M. Perry, Jr. and E. F. Perry, *J. Clin. Invest.* **38**, 1452 (1959).
46. H. M. Perry, Jr., S. Teitelbaum, and P. L. Schwartz, *Fed. Proc., Fed Amer. Soc. Exp. Biol.* **14**, 113 (1955).
47. L. T. Rezayera, *Zh. Obshch. Biol.* **25**, No. 5, 347 (1964); Cited by Söremark (52).
48. G. L. Romoser, W. A. Dudley, L. J. Machlin, and L. Loveless, *Poultry Sci.* **40**, 1171 (1961).
49. O. Rygh, *Bull. Soc. Chim. Biol.* **31**, 1052, 1403, and 1408 (1949); **33**, 133 (1953); *Research (London)* **2**, 340 (1949).
50. H. A. Schroeder, J. J. Balassa, and I. H. Tipton, *J. Chronic Dis.* **16**, 1047 (1963).
51. J. Somerville and B. Davies, *Amer. Heart J.* **64**, 54 (1962).
52. R. Söremark, *J. Nutr.* **92**, 183 (1967).
53. R. Söremark and N. Anderson, *Acta Ondontol. Scand.* **20**, 81 (1962).
54. R. Söremark and S. Ullberg, *in* "Use of Radioisotopes in Animal Biology and the Medical Sciences" (N. Fried, ed.), Vol. 1, p. 103. Academic Press, New York, 1962.
55. N. A. Talvitie and W. D. Wagner, *AMA Arch. Ind. Hyg. Occup. Med.* **9**, 414 (1954).
56. G. Tank and C. A. Storvick, *J. Dent. Res.* **39**, 473 (1960).
57. P. R. Thomassen and H. M. Leicester, *J. Dent. Res.* **43**, 346 (1964).
58. I. H. Tipton and M. J. Cook, *Health Phys.* **9**, 103 (1963).
59. J. H. Watkinson, *Nature (London)* **202**, 1239 (1964).
60. D. A. Webb, *Proc. Roy. Soc. Dublin* **21**, 505 (1937).
61. D. A. Webb, *Pubbl. Sta. Zool. Napoli* **28**, 273 (1956).

# 16

# OTHER ELEMENTS

## I. Aluminum

Despite the abundance of aluminum in the lithosphere, it occurs in most plant and animal tissues in relatively low concentrations and has little biological significance. Concentrations as high as 3000-4000 ppm have been reported in certain trees and ferns (95,206) and 10-50 ppm Al (dry basis) in grasses and clovers (158,178a). These levels appear high compared with the 0.5-5.0 ppm reported for a range of vegetables for human consumption (86). In a later study of the mineral content of fruits the aluminum content was more variable from sample to sample than from one producing area to another. The edible fresh portions of citrus and pome fruits were very low, usually less than 0.1-0.2 ppm Al, and those of berries and stone fruits were mostly much higher [2-4 ppm Al (fresh basis)]. A comparison of peeled and unpeeled apples and pears revealed significantly higher Al concentrations in the peel than in the flesh (215).

Published data for the aluminum content of animal tissues and fluids are extremely discordant. The mean levels for normal human tissues obtained spectrographically by Kehoe *et al.* (104) agree reasonably well with the more recent figures of Tipton and Cook (197) and with those reported earlier for the dog (137,213). A high proportion of their values lies between 0.2 and 0.6 ppm (fresh basis), except for the lungs in which 20-60 ppm, or higher, were observed. The levels of aluminum in this organ increase with age, apparently by accumulation from inhaled atmospheric dust. Several groups of workers (110,162,189) have reported much higher concentrations of aluminum in human tissues. Whether this arises from contamination or analytical errors, or is an expression of variation of environmental origin, is unknown. The aluminum concentration in human hair from individuals in three locations ranged from 1.2-9.2 ppm, with a mean close to 4 ppm. No significant differences due to location were apparent (9). These values are of a similar order to the range of 3 to 11 ppm, with a mean of 7 ppm Al, reported for sheep's wool in two New Zealand locations (79).

Seibold (175) reported a mean level of 0.17 $\mu$g Al/ml (range 0.05-0.50) for 536 samples of human blood serum. This conforms well with the mean level of 0.13 $\mu$g/ml obtained by Kehoe *et al.* (104) for whole human blood. The concentration of aluminum in the blood of farm animals and other species does not appear to have been studied. Considerable individual and month-to-month variation occurs in the aluminum content of cow's milk. Thus, Kirchgessner (108) obtained a mean of 0. 7±0.9 $\mu$g/ml for the milk of 18 cows under normal dietary conditions, and Archibald (5) obtained a mean of 0.5 $\mu$g/ml, with a range from 0.15 to 0.97 $\mu$g/ml. When the cow's ration was supplemented with alum at the rate of 2 g/day, or 114 mg Al, the mean of these cows was raised to 0.8 $\mu$g/ml and the range from 0.4 to 1.4 $\mu$g/ml.

There is no conclusive evidence that aluminum performs any essential function in plants, animals, or microorganisms. The reliability of earlier reports of beneficial effects of aluminum upon plant growth were questioned by Hutchinson (96) in his 1945 review. Subsequently Hackett (73) reported stimulative effects of aluminum on the growth of several pasture plant species, and of oats and rye but not barley, in water-culture experiments. A single attempt to produce a purified Al-deficient diet for rats was unsuccessful. Hove *et al.* (88) concluded from this experiment that "if aluminum is required by the rat the requirement can be met by as little as 1 $\mu$g daily." Aluminum may be involved in the succinic dehydrogenase-cytochrome c system (87,189), and this metal is known to promote the reaction between cytochrome c and succinic dehydrogenase *in vitro*. Whether aluminum is involved in this reaction or in any other enzyme system in the living body is unknown.

Balance experiments in man give no indication that aluminum is essential in the human diet. In fact, negative balances have been reported in children (87) and in adults (104,198). Kehoe *et al.* (104) found the mean intake of aluminum from the food and beverages of a normal adult

North American diet to be 36.4 mg/day over 28 days, and the mean daily fecal excretion to be 41.9 mg over the same period. The mean urinary excretion was only 0.05 mg/liter. Tipton *et al.* (198) studied two adults, one male and one female, over a 30-day period. The mean daily intake of the former was 22 mg Al, with a mean fecal excretion of no less than 45 mg/day and only 1.0 mg/day in the urine. The female subject ingested 18 mg Al/day and excreted 17 mg/day in the feces and 1 mg/day in the urine. In a study of total diet composites, calculated to be nutritionally adequate and to supply the high intake of 4200 kcal/day, Zook and Lehmann (216) found the overall average daily intake of aluminum to be 24.6 mg, but the range was very large. In fact a fourteen-fold range, from as low as 3.8 to as high as 51.6 mg Al/day, was observed in composites made up of items from different geographic locations, in which the overall food group proportions remained the same in all diets. It is impossible to say how far this wide variation in aluminum intakes reflects comparable variations in the original foods, or how far it results from contamination with aluminum from preparation, cooking, or the water supply. The amounts of aluminum in human dietaries are increased by contamination from aluminum vessels used domestically and in processing plants and by the use of aluminum sulfate baking powders, but these amounts are usually small, are very poorly absorbed, and do not constitute an appreciable health hazard (13). Ten times the amounts likely to be ingested in this way can be tolerated by man without harmful effects. Still larger intakes may produce gastrointestinal irritation and produce rickets by interfering with phosphate absorption (40).

## II. Arsenic

Arsenic is widely distributed in the biosphere. It occurs in the air of areas where coal is burned, particularly near smelters and refineries, in seawater to the extent of 2 to 5 ppb and in public water supplies in concentrations which may exceed that of seawater (163). Arsenic occurs in normal soils at levels ranging from less than 1 to as high as 40 ppm, although much higher levels can result from the continued use of arsenical sprays. The amounts of arsenic absorbed by plants from these soils are extremely small (68). Surface contamination of herbage, fruits, and vegetables with spray residues can raise their arsenic concentrations well above the 0.5 ppm common in such materials when uncontaminated. The replacement of arsenical pesticides and weedicides by various organic compounds has reduced this hazard in recent years.

Most human foods contain less than 0.5 ppm and rarely exceed 1 ppm As (fresh basis) (163). This applies to fruits, vegetables, cereal products, meats, and dairy products. By contrast, foods of marine origin are very much richer in arsenic. Many species of bony fish contain 2-8 ppm, oysters 3-10 ppm, and mussels as high as 120 ppm (28, 35, 163). Up to 174 ppm As has been found in prawns from the coastal waters of Britain (28), and 42 ppm in shrimp from the southeastern coastal waters of the United States (34). The arsenic content of commercial fishmeals used in animal feeding ranged from 2.6 to 19.1 ppm with a mean of 6 ppm (120). Fish and Crustacea from freshwater usually contain much lower arsenic concentrations than those just cited. The amounts of arsenic ingested daily in the food are greatly influenced by the amounts and proportions of seafoods included in the diet. An institutional diet which contained no such foods was reported by Schroeder and Balassa (163) to supply 0.4 mg As/day, and an average U.S. diet to supply 0.9 mg/day. These calculated daily intakes are substantially higher than the 0.07 to 0.17 mg/day reported for Japanese individuals (138).

The total amount of arsenic in the normal adult human body has been estimated at 15 to 20 mg or 0.2-0.3 ppm (7, 134). It is difficult to reconcile such large amounts with the more recent studies of human tissues carried out by Smith (182), using neutron activation analysis. This worker examined a wide range of tissues from accident victims in Glasgow with no known abnormal exposure to arsenic. The mean concentration of arsenic in most of the tissues examined, including the muscles, fell within the range 0.04-0.09 ppm (dry basis). Individual variability was very high, but only the skin (mean, 0.12), the nails (0.36), and the hair (0.65 ppm) were substantially and relatively consistently higher in arsenic. The levels found in the muscles, bones, and blood, which together make up the bulk of the weight of the body, suggest an average total-body burden of arsenic for these individuals well below 15-20 mg.

The arsenic content of human hair has excited considerable interest because of its value in the diagnosis of arsenic poisoning. Normal hair always contains arsenic in small amounts, which are greatly increased by excessive intakes of arsenic in certain forms. The levels remain high for some days following cessation of intake, but the return to normal is much more rapid than was formerly claimed (41, 182). Normal human hair was reported 30 years ago to contain 0.3-0.7 ppm As (194). Accurate neutron activation methods have since been devised which allow reliable determination of arsenic on a single hair (41) or on parts of a hair weighing as little as 0.3 $\mu$g (181). By using this method on over 1000 samples chosen at random from living subjects, Smith (182) obtained the values set out in Table 50. The highest value of 74 ppm was found to be due to exposure to arsenic. Smith contends that hair samples showing an arsenic concentration greater than 3 ppm should always be suspect, those with 2-3 ppm require further examination, and those with less than 2 ppm should not be dismissed where As poisoning is suspected. A further point of interest arising from this investigation is the significant difference between the arsenic content of male and female hair. The median value for males was 0.62 and that of females 0.37 ppm. Examination of two samples of the hair of the emperor Napoleon I, probably taken immediately after his death, revealed As concentrations of 10 and of 3-4 ppm. This finding led to the conclusion that Napoleon suffered from arsenic poisoning during his last days on St. Helena (184). This has been denied by Brock (22) and by Cawadias (27). The latter worker proposed that Napoleon actually died as a result of an hepatic, amoebic abscess rupturing into the stomach. Antimony, but apparently not arsenic, was prescribed for this condition. The source of the arsenic in Napoleon's hair is, therefore, not clear.

The levels of arsenic reported for normal human blood vary widely, both among individuals where the same method of analysis has been employed and among different investigations using different analytical procedures. Smith (182) found 12 samples of whole blood to range from 0.001 to 0.92 ppm (dry basis), with a mean of 0.147. Vallee *et al.* (201) quote values ranging from 0.01 to 0.64 ppm of whole fresh blood. Iwataki and Horiuchi (98), using a modified polarographic method, reported As values ranging from 0 to 0.37 μg/g whole blood. Brune *et al.* (23), in a comparison of the concentration of various trace elements in normal and uremic blood by neutron activation analysis, obtained a mean value for 8 samples of normal blood of only 0.004 μg As/g, and for 8 samples of uremic blood of 0.035 μg/g. In the blood of leukemic patients the arsenic is concentrated in the white cells, with less in the red cells and plasma (94). In the rat, about 80% of the blood arsenic is concentrated

TABLE 50
*Arsenic Content of Human Hair*[a]

| Description | As ppm |
|---|---|
| Mean | 0.81 |
| Median | 0.51 |
| Range | 0.03-74 |
| 95% samples | $<2$ |
| 99% samples | $<4.5$ |
| 99.9% samples | $<10$ |

[a]From Smith (182).

in the red cells (89). The affinity of the erythrocytes for arsenic in this species is also apparent from studies with radioactive arsenic (94).

Normal cow's milk has been reported to contain 0.03 to 0.06 (89) and 0.05 ppm As (161). Values ranging from 0.07 to 1.5 ppm have been observed in the milk of cows grazing As-contaminated areas in New Zealand (71). High levels have also been found in the milk of women receiving arsenic therapy for syphilis (52).

The extent to which arsenic is absorbed and retained in the body, and the route of its excretion, vary with the level and chemical form of the arsenic ingested (34, 45, 58, 76, 94, 135, 143). It appears that the more toxic arsenic compounds are retained in the tissues in greater amounts and are excreted more slowly than the less toxic and that toxicity is partly related to this fact (58). Arsenic in the forms in which it ordinarily occurs in foods, including the organically bound arsenic of shrimp, is well absorbed and rapidly eliminated, mainly in the urine (34). Less than 10% of the usual soluble forms of arsenic appears in the feces. Inorganic arsenic ingested as $As_2O_3$ is also well absorbed, but it is retained in greater quantities and for longer periods in the tissues and is excreted almost equally via the urine and feces in man and in the rat (34, 142). The arsenic of organic compounds such as arsanilic acid is similarly well absorbed and is deposited in the tissues of pigs and chicks in amounts proportional to the level fed. It disappears rapidly from the tissues and is excreted mostly in the feces (58, 76, 142, 143). It seems that this form of arsenic is partly catabolized because, as Overby and Frost (143) have shown with pigs, arsanilic acid itself cannot be detected in the urine and only a very small proportion of the total intake appears in the feces.

The word arsenic has become so identified with "poison" in the public mind that the more valuable aspects of this element have tended to be obscured. Its medicinal virtues were acclaimed by the ancient Greeks and solutions containing arsenic (e.g., Fowler's solution) have long been used as tonics for men and animals and for the treatment of human anemias. Reports on the improving effects of arsenic on the appearance of skin and hair of mice, rats, and horses have also appeared (163, 186). In addition, arsenic has been reported to stimulate the growth of tissue cultures (188) as well as the rates of growth and metamorphosis of tadpoles (64). However, arsenic deficiency in animals has not yet been clearly demonstrated. Hove *et al.* (89) observed no improvement in the growth, hemoglobin levels, or red cell numbers of rats consuming a diet supplying 2 μg As daily, when this diet was supplemented with arsenic. Slight indications of a favorable effect of arsenic on hemoglobin production were obtained by these workers and by Skinner and McHargue (179). A diet containing even smaller amounts of arsenic (0.053 ppm) was prepared and fed to mice and rats over long periods by Schroeder and Balassa (163). More than 3000 animals fed such a diet grew and developed normally. It was concluded that "if arsenic is an essential trace element for these mammals, requirements are of the order of 1.0 μg per rat per day or less."

The beneficial effects of various organic arsenicals on the growth, health, and feed efficiency of poultry and pigs have been thoroughly established. These effects have been reviewed by Frost and his associates (55-58). Four organic arsenic compounds have been found of particular value in animal production. These are arsanilic acid, 4-nitrophenylarsonic acid, 3-nitro-4-hydroxyphenylarsonic acid, and arsenobenzene (phenylarsenoxide). No clear relation between structure and growth-promoting effect is discernible. The phenylarsenoxides are more potent than the arsonic acids as coccidiostats, but only the arsonic acids are recognized as growth stimulants for pigs and poultry. The precise mechanism of action is unknown, although the action of arsonic acids closely resembles that of antibiotics and is to some extent complementary to it.

The use and the mode of action of arsenic as a selenium antagonist in the prevention of selenosis are discussed in Chapter 12, and the possible role of arsenic as an antithyroid agent is considered in Chapter 11.

Toxic aspects of arsenic lie largely outside the scope of this text, and no attempt will be made to assess the vast literature on this topic. This has been recently reviewed by Frost (56) and by Schroeder and Balassa (163). In these reviews the remarkable freedom of arsenicals from carcinogenicity and the wide differences in toxicity of different chemical forms of arsenic are emphasized. Schroeder and co-workers (165, 170) fed sodium arsenite, at a level of 5 μg/ml of the element in the drinking water, to mice and rats from weaning to natural death. No evidence of tumorigenicity or carcinogenicity was obtained and no effects on growth, health, or longevity were observed, despite considerable accumulation of arsenic in the tissues, especially in the aorta and the red blood cells.

## III. Barium

Barium occurs widely in soils, plants, and animal tissues in highly variable concentrations. A neutron activation technique, with limits of detection approximating 0.1 μg Ba and 0.04 μg Sr (77) was used by Bowen and Dymond (19) to determine the barium content of plants growing on different soils. Different plant species ranged from 0.5 to 40 ppm, with a mean value of 10 ppm Ba. These levels compare well with those obtained in various spectrographic studies. Thus 10-90 ppm Ba (dry basis) has been reported for Kentucky hay, and 1-20 ppm for grain (85). In a comparison of red clover and rye grass grown together on different soils in Scotland, Mitchell (128) found the former species to contain 12-134 ppm (mean 42) and the latter 8-35 (mean 18) ppm Ba. Certain plant species can accumulate high concentrations of barium from Ba-rich soils. For instance, *Juglans regia* and *Fraxinus pennsylvanica* were reported to contain 2600 and 1700 ppm, respectively (154). Even higher concentrations can occur in Brazil nuts. They vary with the locality but levels of 3000 to 4000 ppm Ba are common and are not accompanied by unusual concentrations of strontium (153, 174).

Barium is poorly absorbed from most diets, with generally little retention in the tissues (47). Tipton and co-workers (198) determined the 30-day mean daily intake of barium by an adult male and an adult female consuming normal diets. The former ingested 1.54 mg Ba/day and was in strong negative Ba balance. The latter ingested 1.77 mg Ba/day and was in strong positive balance. Both subjects excreted an appreciable proportion of the total intake in the urine. The following mean barium concentrations were reported by Tipton and Cook (197) for normal adult human tissues: adrenals, 0.02; brain, 0.04; heart, 0.05; kidney, 0.10; liver, 0.03; lung, 0.10; muscle, 0.05; and spleen, 0.08 ppm of dry tissue. The strontium levels in these tissues were mostly 3-4 times higher, with the exception of the muscles in which the concentrations of the two elements were nearly identical. Sowden and Stitch (188a) examined 35 samples of bone from normal men and women by activation analysis. Barium and strontium were detected in all specimens, with a mean of 7 ppm for Ba and of 100 ppm for Sr in the ashed tissue. Indications were obtained of an increase in Sr content but not in Ba content with age.

Conclusive evidence that barium performs any essential function in plants or animals has not appeared. It has been reported to act as a plant growth stimulant (160), and Rygh's work with rats and guinea pigs fed specially purified diets suggests that this element may be essential to those species (159). When such diets were supplemented with a "complete" mineral mixture, growth and bone development were satisfactory. The omission of either barium or strontium from the mineral supplement resulted in depressed growth. Although these findings were reported over 20 years ago, they do not appear to have been either confirmed or invalidated.

## IV. Boron

For nearly half a century boron has been known to be essential for plants (187, 205), and crop responses to borate applications have been demonstrated in many parts of the world (59). The processes of cell maturation and differentiation, which are boron-dependent in the higher plants (180), are apparently not so dependent in animals—no function for this element in the animal body has been established. Several unsuccessful attempts have been made to induce boron deficiency in rats by the use of highly purified diets containing only 0.15-0.16 ppm B (90, 141, 195). The rats on these diets grew and reproduced as well as those receiving additional boron. If this element is required by the rat it must, therefore, be at a dietary level below 0.15 ppm. There is also no evidence that boron is beneficial to potassium-deficient rats (49), as indicated in earlier experiments (179).

Boron occurs in plant tissues in much higher concentrations than animal tissues. Legumes are the plant materials generally richest in this element (25-50 ppm, dry basis), followed by fruits and vegetables (5-20 ppm), with cereal grains and hays the poorest (1-5 ppm) (12). Levels of 4-7 ppm B in the dry matter have been reported for European pasture grasses (109). In a study of the mineral composition of tropical and subtropical fruits, Zook and Lehmann (215) found great variation in boron concentration both within and among different producing areas. Of the fruits examined, avocados were the highest in boron (7-10 ppm of fresh edible portion), followed by stone fruits (1.4-3.5 ppm). The pome fruits, berries, and citrus fruits were among the lowest (0.3-2.4 ppm of fresh edible portion). Muscle, and other soft tissues of the body, mostly contain only 0.5-1.0 ppm B, or less, of dry tissue (1, 50, 90, 131). The bones contain several times these boron concentrations (1, 50). The ingestion of large amounts of boric acid results in a marked increase in the levels of boron in the tissues, with the highest concentrations appearing in the brain (147). Cow's milk normally contains 0.5-1.0 ppm B, with little variation due to breed or stage of lactation (90, 144). However, the level in the milk varies with the boron intake by the cow. Thus Owen (144) raised the boron level in the milk from 0.7 to over 3 ppm by adding 20 g of borax daily to the cow's normal ration.

The reported intakes of boron by human adults from ordinary diets are remarkably discordant. Some years ago, Kent and McCance (106) reported daily intakes of 10 to 20 mg, the higher amounts being associated with the consumption of large amounts of fruits and vegetables. It is difficult to see how such large quantities of boron could be obtained if the boron levels in foods mentioned in the preceding paragraph are accepted. The 30-day mean boron intakes of only 0.42 and 0.35 mg/day, reported by Tipton *et al.* (198) for two adults consuming their normal diets, are also hard to reconcile with these levels in foods. Zook and Lehmann (216), in their study of the minerals in total diets made up from various sources to supply 4200 kcal/day, obtained an overall average boron content of 3.1 mg. Individual composites of these diets varied relatively little from 2.1 to 4.3 mg. Daily boron intakes of this magnitude appear to be the most acceptable values available, although the very high kilocalorie content of the diets, designed for 16- to 19-year old boys, should be noted.

Boron intakes by grazing animals must be very variable, depending upon the soil type and on the plant species consumed, because the boron concentrations in plants are influenced by the species and the boron status of the soil. These intakes would invariably be much higher, per unit of body weight, than those of humans consuming mixed diets containing a substantial proportion of foods of animal origin. The pastures of the solonetz and solonchak soils of the Kulindisk steppe are reported to be so high in boron that gastrointestinal and pulmonary disorders occur in lambs (148). The water supplies in this region are also unusually high in boron (0.2-2.2 mg/liter), which probably contributes to the boron toxicity.

The boron in food, and boron added as sodium borate or boric acid, is rapidly and almost completely absorbed and excreted, largely in the urine (106, 144, 198). Where high intakes occur either accidentally or from the treatment of large burns with boric acid, similar high absorption and urinary excretion take place, but sufficient boron may be temporarily retained in the tissues, especially in the brain, to produce serious toxic effects (147).

## V. Bromine

Marine plants are usually much richer in bromine than land plants (139), as they are in iodine. Algae are also known to concentrate bromine (176). Some years ago, whole cereal grains were reported to contain 1-11 ppm, white flour 5-8 ppm, and white bread 1-6 ppm Br (38, 51, 211). More recently, Lynn and co-workers (121) investigated the bromine content of dairy feeds from various locations in the United States, where no known bromine fumigant had been used. Most of the green forages contained from 3 to 11 ppm Br, and the grain mixtures varied from 2 to 18 ppm. Heywood (83) reported an even wider range of 0.5 to 25 ppm of natural bromine in untreated grain samples.

All animal tissues other than the thyroid, where the position is reversed, contain 50-100 times more bromine than iodine. Species differences in tissue bromine concentrations appear to be small and the element does not accumulate in any particular organ or tissue (31, 80).

Claims that bromine is concentrated in the thyroid and pituitary glands have not been confirmed (10, 43). The levels in all tissues and fluids can be substantially raised by increasing dietary Br intakes (121, 211). Human dietary bromine intakes have probably increased in recent years in areas where organic bromides are used as fumigants for soils and stored grains, and in motor fuels (46, 83). Bromide concentrations as high as 53-220 ppm in grain (oats and corn) fumigated with methyl bromide and marked increases in the bromine content of milk from cows fed such grains have been observed. (121).

Studies with $^{82}Br$ indicate that this element is retained for only short periods in the tissues and is excreted mostly in the urine (31, 80). Bromide and chloride readily interchange to some degree in the body tissues, so that the administration of bromide results in some displacement of body chloride and vice versa (123). This occurs as a consequence of feeding large amounts of bromide (211) and of injecting physiological quantities (80). Rabbit thyroid glands rendered hyperplastic by lack of iodine are richer in bromine than the blood (10). This suggests that the thyroid distinguishes imperfectly between bromine and iodine and seizes some bromine in the absence of sufficient iodine. The accumulated bromine in the thyroid is quickly lost when iodine is supplied and cannot be used for hormone synthesis (151). However, there is some evidence that injected bromide reduces $^{131}I$ uptake by the rat thyroid and that goiter can occur in rats fed bromide during their first year of life (29).

The levels of bromine in human blood obtained by modern methods of analysis are surprisingly uniform. A mean concentration of 3.7 $\mu$g/ml has been reported by a delicate micromethod (33) and 2.9 $\mu$g/ml by activation analysis (18). Brune *et al.* (23) obtained a mean value of 3.9 $\mu$g/g of whole blood in 8 healthy subjects and of 3.7 $\mu$g/g for 8 euremic patients. These levels are appreciably lower than those reported earlier for 10 normal and 10 manic-depressive psychotics—mean concentrations 7.3 and 7.7 $\mu$g Br/ml, respectively. The latter findings provide no support for the claim that the bromine content of the blood is lower than normal in manic-depressive states.

Extremely variable values have been reported for the bromine content of cow's milk. The magnitude of this variability is illustrated by the following published figures: 1.2-2.6 (51); 0.18-0.24 (26); 0.06-0.20 (37); and 0.005-0.025 ppm (72). Analytical errors probably play a large part in this variability, but the level of bromine in milk is extremely sensitive to differences in bromine intakes by the cow. Lynn *et al.* (121) increased the bromine content in milk from a pretreatment level of 10 ppm to as high as 60 ppm by feeding 12.5 g NaBr/day to cows for 5 days. Smaller increases in milk bromine levels were demonstrated when cows were fed a methyl bromide fumigated grain ration. A high correlation between milk and blood bromide levels was observed.

Bromine has not been shown to perform any vital function in the higher plants or animals. Bromide can completely replace chloride in the growth medium of several halophytic algal species (124) and can substitute for a part of the chloride requirements of chicks (114). A small but significant growth response to trace additions of bromine was reported by Huff and associates (93) with chicks fed a semisynthetic diet. A similar response to bromine was observed by these workers with mice fed a similar diet containing iodinated casein to produce a hyperthyroid-induced growth retardation (17). The bromine content of the basal diet was not given and these indications of a growth requirement for bromine do not appear to have been investigated further. Winnek and Smith (211) were earlier unable to demonstrate any adverse effects on the growth, health, or reproductive performance of rats fed a purified diet containing less than 0.5 ppm Br over a period of 11 weeks, or any improvement when this diet was supplemented with 20 ppm Br as KBr. Bromine is so ubiquitous in nature, and the opportunities for contamination of the animals' dietary and physical environment with bromide are so great that the special closed environment technique devised by Smith and Schwarz (185) appears to be desirable for further nutritional studies with this element.

## VI. Germanium

The biology of germanium has excited little interest, despite its relative abundance in the lithosphere, its chemical properties, and its position in the Periodic Table within the range of the biologically active trace elements with atomic numbers 22 to 53. Germanium has not been included in the numerous studies of the distribution of trace elements in plant and animal tissues by emission spectrography, presumably because of difficulties in detection at low concentrations by these means. Schroeder and Balassa (166) used the photometric method of Luke and Campbell (119), following a special very low-temperature ashing technique, to examine the germanium concentrations of a range of biological materials. This method was reported to have a high repeatability, to give satisfactory recoveries of added germanium, and to have a limit of detection of 0.5 μg/ml. Of 125 samples of foods and beverages analyzed, almost all revealed detectable germanium. Only 4 of these samples contained more than 2 μg/g and 15 others more than 1 μg/g. Germanium was not detected in refined white flour, although it was present in whole wheat and was concentrated in bran. The mean levels in groups of vegetables and leguminous seeds were 0.15 to 0.45 $\mu$g/g (fresh basis). Similar concentrations were observed in meat and in dairy products.

There appear to be no data on the germanium content of normal human organs. The liver, kidneys, heart, lungs and spleen of laboratory mice and rats fed normal diets were shown to contain germanium in concentrations ranging from 0.10 to 2.79 $\mu$g/g (wet weight). When mice were given 5 ppm Ge in the drinking water for their lifetime, higher concentrations were found in these organs, especially in the spleen, but rat tissues accumulated very little germanium under these conditions (165, 170). The germanium in the drinking water at this level gave slight evidence of toxicity in both species, as indicated by reduced life-span and increased incidence of fatty degeneration of the liver. The element was neither tumorigenic nor carcinogenic.

Little is known of the metabolism of the germanium ingested from ordinary diets. Rosenfeld (156) found that oral doses of sodium germanate were rapidly and almost completely absorbed from the gastrointestinal tract within a few hours and were excreted largely in the urine during 4 to 7 days. The data obtained by Schroeder and Balassa (166) for the germanium content of the urine of 4 individuals (mean 1.26 $\mu$g/ml) suggest that dietary germanium is also well absorbed and excreted largely via the kidneys in man. These workers calculated that adults ingest approximately 1.5 mg Ge in the daily diet, of which 1.4 mg appears in the urine and 0.1 mg in the feces.

Germanates have a low order of toxicity in mice and rats (157, 165, 170). Earlier indications that this element has a stimulatory effect on erythropoiesis have not been confirmed (92, 209). The limited evidence so far available suggests that germanium is not an essential element in mammalian nutrition and has little biological significance.

## VII. Lead

Biological interest in lead has been traditionally focused upon its toxic properties as an industrial hazard to man and animals. The amounts required to produce frank lead intoxication are fairly well established, and control procedures designed to reduce lead exposure to "safe" levels have been developed in most countries. Attention is here confined to the biological consequences of the lesser amounts present in modern environments and to the factors that influence these amounts. Over the last half century, human exposures to lead have changed in origin, but have probably not changed significantly in amount. Thus the decrease in exposure over this period from lead in water pipes, food containers, paints, and insecticides has probably been compensated, or more than compensated, by increased exposure to lead from cigarette smoking and cosmetics, and above all from motor vehicle exhausts, especially in highly motorized, urban communities. The latter source of lead arises from direct inhalation and indirectly through deposition on soils and plants and inhalation by domestic animals. Deposition of airborne lead on soils and plants along highways and in urban areas has been clearly demonstrated (24, 149). Thomas *et al.* (196) have obtained data consistent with the existence of coastal-inland, atmospheric and blood lead gradients within the Los Angeles area.

Lead occurs naturally and widely in soils and plants in variable concentrations (21). Mitchell and Reith (130) found the lead content of pastures from areas remote from industrial or auto-exhaust contamination to range from 0.3 to 1.5 ppm in the dry matter during the period of active growth. When mature or senescent, these plants contained lead concentrations from 10 to as high as 30–40 ppm in late winter. On soils low in lead, pasture plants rarely contain more than a few parts per million of the element, unless contaminated by dust (4). Concentrations of several hundred parts per million have been reported from lead-rich areas, the highest values being associated with surface contamination by dust from lead mines (4, 44, 177).

As Patterson (145a) has pointed out, the levels of lead in the blood and tissues of man and domestic animals are not equivalent to those which prevailed during the creation and evolution of physiological responses to lead, i.e., "natural" levels. They represent levels which have been elevated by past and existing activities of man involving the injection of millions of tons of lead into the environment. From geochemical considerations he inferred that the "natural" levels of lead in man are close to 2 mg/70 kg body weight and 0.0025 ppm in blood. The actual total body burden of lead in "normal" adult man is very much higher than this. Amounts ranging from 90 to 400 mg have been reported in separate studies (104, 134, 168). Schroeder and Tipton (168) found the mean total body burden of lead of 150 adult U.S. accident victims to be 121 mg, of which approximately 90% was present in the skeleton. The affinity of bone for lead is apparent from other investigations, in which lead concentrations frequently lie between 5 and 20 ppm of fresh bone (104, 199). The bones of individuals with occupational or accidental exposure to large amounts of lead carry still higher concentrations.

In studies carried out over 30 years ago (104, 199), the soft tissues of the normal adult human body were reported to contain lead concentrations ranging from 0.13 to 0.50 ppm in the brain and from 1.3 to 1.7 ppm (fresh basis) in the liver. The mean levels found in 4 fetuses, of 7 to 8 months gestation, ranged from 0.17 ppm in the brain to 0.68 ppm in the liver (199). Similar relatively high levels of lead in the tissues of stillborn infants were later demonstrated by Schroeder and Tipton (168). These workers also observed accumulations in the tissues with age up to 50-60 years in U.S. residents, particularly in the aorta, bones, kidney, liver, lung, and spleen. The median values for these tissues were generally higher than those from Africa and the Middle East, and there was no comparable increase with age in these latter individuals, except in the aorta. It, therefore, seems reasonable to assume that exposure to lead from all sources was sufficient with the U.S. subjects to cause some tissue accumulation with age, whereas in the "foreign" areas studied it was not. The extent to which this increased exposure comes directly or indirectly from motor vehicle exhausts is difficult to assess (65).

The concentrations of lead in all tissues rises with increased intakes of lead, particularly in the bones, liver, and hair, and much less so in the brain and muscles. Thus Kopito *et al.* (111) found the hair of normal children to average 24 $\mu$g/g (range 2-95), whereas that of patients with chronic plumbism was 282 $\mu$g/g (range 42-975). Schroeder and Nason (167) reported a mean of 17.8±2.17 $\mu$g Pb/g for the hair of 78 normal males and 19.0±2.95 for 47 normal female subjects. The lead in hair does not appear to be a sensitive or reliable index of the lead status of the individual, except where these levels are very high.

The average lead content of the whole blood of groups of healthy U.S. and Mexican individuals was reported by Kehoe and co-workers (104) to be 0.23 and 0.27 $\mu$g/g, respectively. A higher mean of 0.55 $\mu$g/ml, with levels ranging from 0.4 to 0.7 $\mu$g/ml, was obtained at about that time for English subjects (199). A recent study of the levels of lead in the blood of inhabitants of 16 countries is of particular interest because all the analyses were carried out in a central laboratory by a standard method, so that analytical discrepancies were avoided (67). In this study, Goldwater and Hoover (67) reported an overall mean of 0.17 $\mu$g Pb/ml and suggested a "normal" range for healthy humans of 0.15 to 0.40 $\mu$g/ml. It should be noted that these values do not differ greatly from those of earlier investigations. This suggests that total lead intakes have not greatly changed over several decades. The blood lead levels of urban dwellers were slightly but fairly consistently higher than those of rural subjects in most countries. Despite significant mean differences among countries, it was impossible to relate them with much confidence to degree of industrialization. In fact, New Guinea natives, living under conditions completely divorced from industrialization and motorization, revealed average blood lead levels comparable with those of industrialized communities.

The urine analyses from these individuals in different countries point to essentially similar conclusions and further illustrate the difficulties in assuming a simple relationship between location and blood and urine levels of lead, or in defining "natural" levels in the manner proposed by Patterson (145a). The normal urinary concentration of lead centered about a mean of 35 μg/liter, with 95% of the samples containing less than 65 μg/liter. These values compare well with the mean of 27 μg Pb/liter obtained 30 years earlier by Kehoe *et al.* (104). The demonstration of increased urinary δ-aminolevulinic acid (ALA) levels in cases of lead poisoning (36, 74, 132) suggested that estimation of this substance in the urine might provide a means of detecting early and potentially hazardous exposure to lead. Davis and Andelman (39) demonstrated a significant correlation between the levels of urinary ALA and those of urinary lead in random samples collected from children (see Fig. 24). They suggest that the determination of urinary ALA can be

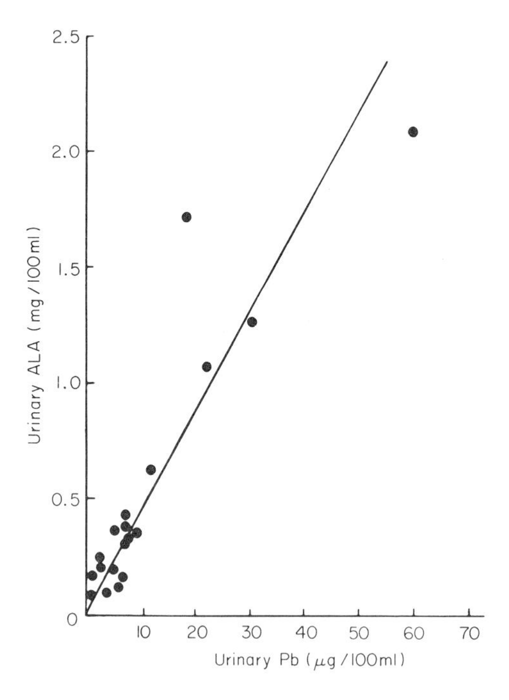


**Fig. 24.** The relation between urinary levels of δ-aminolevulinic acid (ALA) and of lead in children. [From Davis and Andelman (39).]

employed as a large-scale screening procedure for the detection of early lead exposure in asymptomatic children.

The concentrations of lead in the blood, urine, and tissues of normal, healthy laboratory and domestic animals, consuming ordinary rations, are very similar to those given earlier for man. The distribution among the organs and tissues of the body also follows a similar pattern. Thus Ishikawa (97) found variable lead concentrations in the tissues of rats consuming a normal diet, with the highest levels in the bones and the lowest in the brain and muscles. The mean level in the whole blood was given as 0.26 μg Pb/g for rats and as 0.25 μg/g for guinea pigs. The mean blood lead levels for normal sheep and cattle were reported by Allcroft (2) to be $0.14 \pm 0.01$ and $0.13 \pm 0.01$ μg Pb/ml, respectively. The latter value is close to the mean of $0.10 \pm 0.06$ μg/ml obtained by Hammond *et al.* (75) for 16 healthy cows in a quite different environment. The whole fresh livers of 14 healthy calves and 5 cows averaged 0.5 μg/g and ranged from 0.2 to 1.9. These values compare well with the range of 0.3 to 1.5 μg/g reported by Allcroft for healthy heifers and calves. Very much higher levels occur in the tissues of sheep and cattle consuming excessive amounts of lead, especially in the kidney cortex. Allcroft (3) considers that levels higher than 10 ppm in the liver and 40 ppm in the kidney cortex of farm animals are "of definite diagnostic significance where there is collateral evidence of lead poisoning" and are "highly suggestive" where there is no such evidence.

Ingested lead is poorly absorbed and is excreted mainly in the feces. This occurs in man (104, 168, 199), rats (97), guinea pigs (97), dogs (203), rabbits (16), and sheep (16). In three separate studies the mean daily excretion of lead by adult man in the feces and urine was 0.32 and 0.03 mg (104) and 0.22 and 0.05 mg (134, 199), respectively. Alimentary absorption of water-soluble lead is about 10% (103) and food lead is about 5% absorbed in man (112). A lower apparent absorption of $1.3 \pm 0.8\%$ was observed by Blaxter (16) in sheep. This occurred over a wide range of intakes, with very little reduction of true absorption with increasing quantity of lead ingested. Similar results were obtained with rabbits.

The lead which is absorbed enters the blood and reaches the bones and soft tissues of the body, including the liver, from which it is gradually excreted via the bile into the small intestine and from there it is eliminated in the feces. Fecal lead thus consists largely of unabsorbed lead, together with a small proportion which has been absorbed and excreted by this route (3). Up to certain intakes, lead excretion virtually keeps pace with ingestion, so that retention in the tissues is negligible. In the sheep, no lead is retained if less than 3 mg is ingested daily, which is equivalent to approximately 3 ppm of the dry diet (16). Above this quantity, lead is retained in increasing quantities by the tissues. Critical levels of intake appear to be substantially lower in the rat and in man. Cantarrow and Trumper (25) placed the lower limit of safety in man as from 0.2 to 2.0 mg Pb daily. The increase in lead concentrations with age observed in many tissues under U.S. environmental conditions (168) indicates that excretion is not quite keeping pace with total intakes, resulting in a small retention over time. Definite evidence that tissue lead accumulations of this magnitude are either harmful or harmless to man is not available. In the rat, similar concentrations of lead in the tissues, brought about by the consumption of water containing 5 ppm Pb, as lead acetate, were nontoxic as judged by the length of the life-span, although some loss of hair and body weight was evident (165).

Some lead is lost in the hair, sweat, and milk. The level of lead in normal cow's milk has been reported to be 0.02-0.04 mg/kg and that of ewe's milk in early lactation to be 0.11-0.15 mg/kg (16, 25, 104). Dosing with lead salts leads to a marked increase in the amounts appearing in the milk (16, 210). In a recent study of the lead content of U.S. market milk, Murthy *et al.* (136) found the national weighted average to be close to 0.05 mg/kg, with a range from 0.02 to 0.08 mg/kg. No significant differences between cities were observed. The level of lead in the milk of healthy women appears to be very similar to that of cow's milk.

For most individuals the major source of lead is the food and beverages consumed (6, 104, 134, 145a). Most dietary items contain less than 1 ppm Pb. Patterson (145a) gives a typical overall dietary concentration for "contaminated" man of 0.2 ppm, with an estimated intake from this source of 0.4 mg Pb/day. Additional daily intakes of 0.01 mg from the water supply, 0.026 mg from urban air, 0.01 mg from rural air, and 0.024 mg from cigarette smoke for individuals smoking 30 cigarettes per day, were also estimated. Monier-Williams (134) estimates that "normal, healthy individuals" in England would obtain approximately 0.22 mg from the food, 0.10 mg from the water, and 0.08 mg from inhaled dust, giving a total of 0.4 mg Pb/day. Schroeder and Tipton (168) present an average estimated total intake of 0.3 mg/day, of which 0.26 mg comes from the food, 0.02 mg from the water, and 0.02 mg from the air.

The possibility that lead in low concentrations performs some essential biochemical functions in living organisms cannot be excluded, although no such functions have yet been demonstrated in plants, animals, or microorganisms. Acute lead intoxication is manifested by colic, anemia, and neuropathy or encephalopathy and is accompanied by excretion of porphyrins and their precursors, including ALA, in the urine. The anemia is probably caused or enhanced by a lead inhibition of heme synthesis. Certain adenosine triphosphatases are sensitive to small concentrations of lead (82) and lead has been shown by Ulmer and Vallee (200) to inhibit strongly lipoamide dehydrogenase, an enzyme crucial to cellular oxidation. These workers found, further, that lead inhibits the incorporation of $^{14}$C-leucine into *Escherichia coli* transfer ribonucleic acid (t+RNA), which suggests that the metal may affect protein synthesis. Growth in *Rhodopseudomonas spheroides* is inhibited by lead, and tetrapyrrole synthesis appears to be altered in this organism through a complex metal-ion antagonism involving coproporphyrinogen oxidase or coprogenase (200). The significance of these effects and of the binding of lead to the specific subcellular constituents, metallothionein and liver RNA, to the lead exposure and tissue lead concentrations characteristic of industrialized, contaminated man, cannot yet be evaluated.

## VIII. Mercury

On present evidence, mercury must be considered a nonessential element for living organisms. It occurs widely in low concentrations in the biosphere and has long been known as a toxic element presenting occupational hazards associated with both ingestion and inhalation. At room temperature, mercury has a significant vapor pressure (0.001 mm at 18°C), which rises markedly at higher temperatures (0.27 mm at 100°C). It can, therefore, constitute a health hazard to laboratory workers and others handling the element for long periods, unless proper precautions are taken. Mercury poisoning has been prominent at times among goldsmiths and mirror makers and the expression "mad as a hatter" derives from the symptoms shown by workers engaged in the treatment of furs with mercuric nitrate. Industrial exposure to mercury is now relatively well controlled. Interest centers more upon the biological consequences of the injection of mercury into the environment of modern man, through the burning of coal and the use of mercury-containing dental amalgams, ointments, and other medicinal agents and fingerprint powders, as well as from the application of organic mercury fumigants and fungicidal agents in industry and agriculture. The use of these latter compounds in industry can result in sufficient mercury in the effluents to produce increased mercury concentrations in the marine life of rivers, lakes, and estuaries, and in the plant and animal products of the local environment (207). The magnitude of these sources of mercury has recently evoked concern in parts of Sweden (207).

The first systematic examination of the normal distribution of mercury in soils, plants, and animal tissues was undertaken over 30 years ago by Stock and his collaborators (190-193). These workers developed an ingenious microanalytical method for estimating the low concentrations present in these materials, in which the final step involved the microscopic measurement of the diameter of a drop of mercury. By such means soils were shown to contain 0.1-0.3 ppm Hg, soot 3-30 ppm, and fruits, vegetables, and cereal grains 0.005-0.035 ppm. Similar low concentrations were reported in foods of animal origin, with somewhat higher levels in animal fats (0.07-0.28 ppm) and fish (0.02-0.18 ppm). On the basis of these analyses the average daily intake of mercury in the food by adult man was estimated to be 0.005 mg. Subsequently, Gibbs *et al.* (61) gave a higher estimate of 0.02 mg Hg/day, which appears to be a more realistic figure. Little is known of normal mercury intakes by farm animals. Lunde (120) reported a mean mercury concentration of 0.18 ppm (range 0.03-0.40) for twelve commercial fishmeals from different sources.

The determination of normal levels of mercury in biological materials has been raised to a new level of accuracy and sensitivity by the advent of neutron activation techniques capable of estimating mercury in quantities as small as $10^{-3}$ $\mu$g, with an accuracy of $\pm 5\%$ (105, 183). Using such a method, Howie and Smith (91) and Lenihan and Smith (115) studied the mercury levels in the tissues of accident victims in the Glasgow area, with no known abnormal exposure to mercury, other than in dental repair. Mercury was detected in all the tissues examined. The mean concentrations mostly fell between 0.5 and 2.5 ppm Hg (dry basis), with generally higher values in tissues, such as skin, nail, and hair, which are exposed to atmospheric and other contamination. Higher values were also found for liver, kidney, and lungs. Molokhia and Smith (133) showed that the apex of lung tends to be higher in mercury than the base and healthy lung tissue to be higher than cancerous. These findings suggest that the relatively high-mercury levels in lung tissue found in some studies are affected by airborne contamination. Several other groups of workers, employing a variety of analytical methods, have reported variable concentrations of mercury in normal human tissues (53, 70, 101). All agree that the kidneys carry consistently higher levels than other organs. Thus Joselow *et al.* (101) obtained a mean of 2.75 ppm wet weight (range 0-26.3) for kidney, compared with means ranging from 0.05 to 0.30 ppm for eleven other tissues, including the liver and lungs. These workers obtained no evidence of increasing mercury concentration with age, but the numbers in the different age groups were extremely small.

Kellershohn *et al.* (105) found the average concentration in the whole blood of a small series of subjects to be 0.005 $\mu$g/g, with extremes ranging from 0.002 to 0.009 $\mu$g/g. These values compare well with those obtained earlier by Stock (190). Goldwater (66) reported that 74% of a normal population contained less than 0.005 μg Hg/ml and 98% had less than 0.05 μg/ml. The red cells contain approximately twice the mercury concentration of the plasma (105).

Individual variability in the reported mercury content of human urine is high. Howie and Smith (91) observed a range of 0.001 to 0.133 μg Hg/ml in 46 normal human subjects, with a mean of 0.023. In industrially exposed individuals, urinary mercury levels can increase beyond these levels by several orders of magnitude (100). Monier-Williams (134) suggested that the critical urinary concentration above which mercury poisoning can be suspected is of the order of 0.1 to 0.2 μg/ml.

Studies of hair, nails, and teeth have highlighted the readiness with which these structures acquire mercury from the environment. Nixon and co-workers (140) obtained a mean value of 2.8 ppm for the outer enamel and 2.3 ppm for the inner enamel of human teeth. In teeth which had not erupted into the oral cavity the mercury content of the enamel was only of the order of 0.1 ppm. The enamel of teeth in contact with silver amalgam fillings had mercury contents ranging from 153 to 1600 ppm. The opportunities for contamination are even more evident from comparisons of the mercury content of head hair and of pubic or axillary hair and of fingernails and toenails. Thus Rodger and Smith (155) give the following normal mean concentrations for subjects with no known abnormal exposure to mercury: head hair, 5.5; pubic hair, 1.6; fingernails, 7.3; and toenails, 2.4 ppm. Markedly higher levels occur in these tissues in individuals with dental, accidental, or industrial exposure. For instance Howie and Smith (91) obtained mean mercury concentrations for the head hair and fingernails of a group of 20 dental assistants of 32.3 and 68.8 ppm, respectively, compared with 8.8 and 5.1 ppm for these tissues in 26 control subjects. Higher levels, up to 98.6 ppm in head hair and 1068 ppm in the fingernail of one individual, were observed in a small group of industrial workers with mercury contamination of their laboratory. A further example of such mercury contamination of historic interest comes from an examination of a sample of hair believed to have come from the head of King Charles II of England, who died in 1685. King Charles had a private laboratory, where he worked on a process involving the distillation of large amounts of mercury and is believed to have died from chronic mercury poisoning, terminated by renal failure (212). His hair was found to contain 54.6 ppm Hg, or about 10 times the mean concentration of normal subjects (115). It is apparent that the level of mercury in hair can provide a useful indication of the mercury status of individuals and of their environment, although the normal range is extremely wide (9, 155).

Mercury balance studies at normal intakes do not appear to have been carried out with man or animals using modern methods of analysis. Orally ingested mercury is excreted mainly in the feces and only to a small extent in the urine. Absorption is followed by a rapid rise in the mercury content of the kidneys, from which the metal is slowly eliminated (134). Remarkable strain differences in the ability of chickens to retain and excrete mercury have been demonstrated by Miller and co-workers (126, 127). The preferential retention of mercury in the livers and kidneys of one strain was not affected by the mode of administration or by the dosage. It was suggested that the strain with the smaller retention has a lower renal threshold for mercury or has less mercury-binding sites. The influence of mercury on the absorption and transport of copper, zinc, and cadmium is discussed in the chapters dealing with those elements. The protective effect of selenium against the toxic effects of injected mercury, as well as of cadmium, and the influence of selenium upon plasma and liver levels of mercury in the rat is also considered in Chapter 10. In addition, it has been shown that mercury resembles cadmium in its ability to decrease markedly the respiratory excretion of volatile selenium compounds in rats given a small amount of selenite (145).

Little is known of the chemical forms or combinations of mercury present in body tissues and food substances. The abnormally large amounts of mercury present in fish from parts of Sweden were initially believed to be present either as inorganic mercury or as phenyl mercury released as industrial wastes (207). Later, Westoo (208) showed that this mercury occurred almost entirely as methyl mercury ($CH_3Hg^+$), and Jensen and Jernelov (99) demonstrated the ability of living organisms to methylate mercury compounds present in pollution. Both mono- and dimethyl mercury ($CH_3 \cdot Hg \cdot CH_3$) were shown to be produced in bottom sediments and in rotten fish. It was suggested that the formation of the volatile $CH_3 \cdot Hg \cdot CH_3$ may be a factor in the redistribution of mercury from industrial wastes into the general environment.*

## IX. Rubidium

Biological interest in rubidium has been stimulated by its close physicochemical relationship to potassium and its presence in living tissues in higher concentrations, relative to those of potassium, than in the terrestrial environment. Many years ago Ringer (152) observed that rubidium affected the contractions of the isolated frog's heart in a manner similar to potassium. A general relationship between these two elements,

*Author's Note: Methylmercury compounds are also retained longer in the tissues and appear to exert a more specific toxic effect on the brain than inorganic forms of the element.

and also between cesium and potassium, has since been demonstrated for a variety of physiological processes. These include such diverse actions as the ability of these ions to neutralize the toxic action of lithium on fish larvae, to affect the motility of spermatozoa, the fermentative capacity of yeast, and the utilization of Krebs cycle intermediates by isolated mitochondria. Their extracellular ionic concentrations also influence the resting potential in nerve and muscle preparations and the configuration of the electrocardiogram (150).

The metabolic interchangeability evident in these processes suggests that rubidium might act to some extent as a nutritional substitute for potassium. Rubidium, and to a lesser extent cesium, can replace potassium as an essential nutrient for the growth of yeast (113) and of sea urchin eggs (116). This nutritional replaceability can now be extended to bacteria (122).

The higher animals appear to be more discriminating. Additions of rubidium, or cesium, to potassium-deficient diets prevent the lesions in the kidneys and muscles characteristic of potassium depletion in rats (48) and, for a very short period, permit almost normal growth until death inevitably supervenes (81). Using purified diets with varying supplements of rubidium, sodium, and potassium, Glendenning *et al.* (63) obtained no evidence that rubidium is essential for rats, although there were slight indications that this metal possibly substitutes partially for potassium and is more toxic on low than on high-potassium diets. Purified diets containing up to 200 ppm Rb were nontoxic, whereas levels of 1000 ppm or more resulted in decreased growth, reproductive performance, and survival time.

Rubidium resembles potassium in its pattern of distribution and excretion in the animal body (14, 63, 81, 107, 150, 178, 189, 214). All plant and animal cells are freely permeable to rubidium and cesium ions, at rates comparable with those of potassium (150). All the soft tissues of the body normally contain rubidium in concentrations which are high compared with most trace elements. The element does not accumulate in any particular organ or tissue and is normally very low in the bones (63, 178, 189). The plasma also contains significantly lower concentrations than the red cells (14, 54, 63). The soft tissues of man and other vertebrates have been reported to contain 20-60 ppm Rb (dry basis) (14, 178). Higher average values for these tissues have been observed in rats (63) and in adult man, with lower levels common in infants and children (189). Concentrations ranging from 100 to 200 ppm of dry tissues were found in the muscles, liver, lungs, kidneys, heart, and brain of rats fed a standard diet, and very much higher concentrations, up to 8000 and 12,000 ppm, when the animals were fed toxic levels of rubidium (63).

The rubidium retained in the tissues as a result of such high dietary intakes, or from injections, is slowly lost from the body, mainly in the urine.

Human and animal foods were examined by Glendenning *et al.* (62) using a spectrographic method sensitive to 1 ppm Rb. Soybeans were found to contain 160-225 ppm Rb and bromegrass and sorghum silage, 130 ppm (dry basis). Human foods of plant origin averaged about 35 ppm, with the cereal grains below this average and white flour the lowest in rubidium (< 1 ppm) of those analyzed. Meat foods were relatively rich in rubidium, with beef muscle containing no less than 140 ppm of dry tissue. In a recent study, employing atomic absorption spectroscopy, Murthy *et al.* (136) found that market milk collected from various cities in the United States averaged between 0.57 and 3.39 ppm Rb (fresh basis), with significant differences between the different cities. On the basis of the limited data available, it seems that human diets high in refined cereals would supply much less rubidium than diets rich in animal foods and fruits and vegetables. There is no evidence that this, of itself, has any significance in human health and nutrition.

## X. Strontium

As soon as it became obvious that $^{90}Sr$ is an abundant and potentially hazardous radioactive by-product of nuclear fission, the biological movements of this element became an active field of study. This topic lies outside the scope of the present text but consideration of the normal distribution of stable strontium and its normal intakes and metabolism by man and animals is appropriate, especially as satisfactory methods for the estimation of stable strontium in biological materials have been developed (77, 102).

The strontium content of plants growing on normal soils in England was found, by neutron activation, to range from 1 to 169 ppm, with a mean of 36 ppm (dry basis) (19). Certain plant species growing on Sr-rich soils carried very much higher concentrations, up to 26,000 ppm Sr. These are regarded as true strontium accumulator plants, although other plant species will take up large amounts of strontium from culture solutions specially enriched with this element (32, 204). Mitchell (128) found 15 samples of red clover, growing on different soils, to range from 53-115 (mean 74) ppm Sr, and 15 comparable samples of ryegrass to range from 5-18 (mean 10) ppm Sr (dry basis). These levels are very much higher than those of animal tissues and fluids, including bone. Gerlach and Muller (60) found the strontium concentration of a wide variety of animal tissues to range from 0.01 to 0.10 ppm, with no evidence of accumulation in any particular species, organ, or tissue. Tipton and Cook (197) obtained the following mean values for adult human organs: adrenals, 0.06; muscle, 0.07; liver, 0.10; brain, 0.12; heart, 0.15; kidney, 0.35; and lung, 0.50 ppm of dry tissue. In a critical study of the strontium content of cow's milk, Jury *et al.* (102) reported levels ranging from 31 to 65 ppm of the ash or approximately 0.2-0.4 ppm of the whole milk.

The strontium content of bone has attracted particular attention because of the affinity of this tissue for this element and, therefore, its relevance to the problem of $^{90}Sr$ retention. The levels reported by different investigators are highly variable, although there is agreement that different bones in the same individual do not differ significantly in strontium concentration and that these concentrations tend to increase with age. Thus Hodges and co-workers (84) found the bone ash of human fetuses and human adults to average 0.016 and 0.024% Sr, respectively. This is substantially higher than the mean of 100 ppm Sr in the ash of bones examined by Jury *et al.* (102) by four different methods.

Strontium is poorly absorbed and retained from ordinary human diets. Harrison and associates (78) found a single individual to ingest 1.99 mg Sr daily over an 8-day period, of which 1.58 mg appeared in the feces and 0.39 mg in the urine. Tipton and co-workers (198) measured the 30-day mean daily intakes and excretion of 2 adults. The intake of one individual was 1.37 mg/day, of which 0.81 mg appeared in the feces and 0.24 mg in the urine. The corresponding figures for the other adult were 1.24, 0.97, and 0.42 mg/day. The mean daily strontium intakes of children of different ages and, hence, of different levels of food consumption, ranged from 0.67 to 3.57 mg. A large proportion of this strontium appeared in the feces (11). Little information is available on the normal intakes of strontium by farm animals. With grazing stock, these will be greatly influenced by the strontium status of the soil and by the proportion of legumes to grasses in the herbage consumed.

Strontium has been claimed to act as a plant growth stimulant (160) but has not been shown to be essential for either plants or animals. In the unconfirmed experiments of Rygh (159), the omission of strontium from the mineral supplement fed to rats and guinea pigs consuming a purified diet resulted in a growth depression, an impairment of the calcification of the bones and teeth, and a higher incidence of carious teeth. There appears to be no other evidence implicating strontium in bone calcification or caries prevention.

## XI. Tin

Tin is widely but irregularly distributed in the biosphere. There is no evidence that it serves any essential function in plants, or microorganisms, despite rather unconvincing reports of a stimulatory effect of the element upon plant growth (30, 125). Quite recently Schwarz *et al.* (172a) have demonstrated a significant growth effect in rats maintained on purified amino acid diets in trace element-controlled isolators when the diets were supplemented with tin, in various forms and in amounts similar to those normally present in foods, feeds and tissues. These findings suggest that tin is an essential trace element for mammals.

In the early spectrographic studies, tin was demonstrated in some but not all human and animal tissues examined, in concentrations ranging mostly from 0.5-4.0 ppm of dry tissues (20). These studies included human teeth (118) and spinal fluid (173), and human and cow's milk (42). Kehoe *et al.* (104) carried out a thorough spectrochemical study of human tissues and fluids. They found tin in 80% of the samples, with a fairly even distribution among 10 of the 11 tissues examined. No tin was detected in the brain, a finding substantially confirmed in a much later study (169). The actual mean values reported by Kehoe for other human tissues ranged from 0.1 ppm in muscle and 0.2 ppm in kidney, heart, spleen, and intestines, to 0.8 ppm Sn in fresh bone. The lungs showed no appreciable accumulation of this element. More recently Schroeder *et al.* (169) demonstrated a wide but highly variable distribution of tin in human tissues, with significant differences related to age and geographical location and with a predilection for intestines and lung. Little or no tin was found in the tissues of stillborn infants. This suggests that this element does not readily cross the placental barrier, but "arrives in tissues shortly after birth, *de novo*." Tin tended to accumulate in the lungs with advancing age, but not in the liver, kidney, aorta, or intestine. Variations in tissue tin among geographical regions were quite marked but were much less noticeable between 8 cities of the United States. Mean concentrations in 24-27 samples of kidney from each of these locations ranged from 0.23-0.76; of liver 0.35-1.0; of lung 0.44-1.20; and of aorta 0.22-0.94 ppm Sn wet weight. By contrast, all groups from Europe, Africa, and Asia had lower mean and median values for kidney, liver, and lung than Americans, the differences being less marked with lungs. Tin in the tissues of Africans was notably low. It was concluded that Africans are exposed to less tin than are North Americans, while people of other areas accumulate the metal in their lungs but accumulate little in their kidneys and livers.

Little information is available on particular sources of tin in normal human and animal diets. In 1940 Kehoe *et al.* (104) reported the mean daily intake of tin by a normal adult American to be 17 mg. This is many times higher than the 1.5 and 2.5 mg/day obtained 26 years later by Tipton *et al.* (198) for the 30-day mean intakes of two adults in that country and the 3.5 mg/day calculated by Schroeder *et al.* (169) to be supplied by an institutional diet, containing regular amounts of canned fruits and juices. Analytical differences may account for some of this disparity, particularly in view of possible losses of tin during the processes of drying and ashing of samples due to the relatively low boiling points of many tin compounds (172a). Variations in the amounts and proportions of different food and beverage items comprising the total diet, especially of canned foods, can result in wide differences in the amounts of tin ingested daily by man. Schroeder *et al.* (169) calculated from their data on the level of tin in foods that a 2400 -kcal diet composed largely of fresh meats, grain products, and vegetables, which usually contain less than 1 ppm Sn, would supply 1 mg/day. Such a diet which included a substantial proportion of canned vegetables and fish could supply as much as 38 mg/day.

Large amounts of tin can accumulate in foods in contact with tinplate, particularly when nonlacquered (134). The introduction of lacquered cans and the crimping of the tops, so allowing little direct contact of the food with the solder, represent two developments resulting in a reduction of this source of tin to modern civilized man. Storage of canned food, even in lacquered cans, can nevertheless result in accumulation of tin, presumably due to corrosion starting from defects in the lacquer (134). Schroeder *et al.* (169) mostly found relatively high concentrations in oily fish in uncoated cans. The higher consumption of canned foods and juices by North Americans, compared with Africans, could, therefore, account, at least in part, for their higher body burden of tin. On the other hand, this source of tin should be decreasing, since the introduction of lacquering and crimped tops, compared with a few decades earlier when Kehoe and co-workers (104) obtained their high estimates of daily tin intakes.

Data on the metabolism of tin at physiological intakes are meager but the available evidence indicates that this metal is poorly absorbed and retained by man and is excreted mainly in the feces. Thus, Perry and Perry (146) found 24 adult subjects to excrete only $16.6 \pm 1.85$ $\mu g$ Sn/liter in the urine or 23.4 $\mu g$/day. Kehoe *et al.* (104), in their earlier study, obtained a mean of 11 $\mu g$/ liter which represents a urinary excretion of approximately 16 $\mu g$/day. An even lower urinary excretion of 11 and 8 μg/day was reported for 2 adults by Tipton *et al.* (198). The fecal excretion of tin in these investigations, by contrast, was very much higher and approximated the total amount ingested with the food. Schroeder *et al.* (169) have calculated a typical U. S. daily adult human balance for tin as shown in the table:

| Intake (mg) | | Output (mg) | |
|---|---|---|---|
| Food | 4.0(range 1-40) | Feces | 3.98(range 1-40) |
| Water | 0.0(range 0-0.03) | Urine | 0.023(range 0-0.04) |
| Air | 0.003(range 0-0.007) | | |
| Total | 4.003 | | 4.003 |

Little interest has been shown in the tin content of forage and pasture plants or in the intakes of this element by farm animals. Tin has not been consistently demonstrated in all soils and plants. This may merely mean that the concentrations are below the limits of detection of the methods employed. Thus Lounamaa (117) detected tin only sporadically in the higher plants in Finland but the lower limit of his spectrographic method was 10 ppm of ash. Lichens were found to concentrate this element, so that those growing on silicic rocks contained the remarkably high mean level of 72±4.7 ppm and those on ultrabasic rocks, 37±8.0 ppm. Pasture herbage growing in Scotland was reported to contain only 0.3-0.4 ppm Sn (dry basis) (129).

Ingested tin has a low toxicity, no doubt due in part to its poor absorption and retention in the tissues. Mice and rats given 5 ppm Sn, as stannous ions, in the drinking water from weaning to natural death, grew normally, and the life-span of mice of both sexes and of male rats was unaffected. Female rats showed a reduced longevity. An increased incidence of fatty degeneration of the liver and of vacuolar changes in the renal tubules was apparent in both sexes (165, 170).

## XII. Titanium

Titanium resembles aluminum in being extremely abundant in the lithosphere and in soils and in being poorly absorbed and retained by plants and animals, so that the levels in the tissues are generally much lower than those in the environment to which the organisms are exposed. The problem of contamination from soil and dust is also similar with these elements. In fact, the titanium levels in herbage samples can be used as an index of soil contamination (8, 129), because the titanium concentration of most soils is some 10,000 times greater than it is in uncontaminated herbage. Over 40 years ago, Bertrand (15) examined a wide variety of plants for titanium and recorded levels ranging from 0.1 to 5 ppm, with a high proportion of the values lying close to 1 ppm. Very similar concentrations were reported by Mitchell (128) in his later study of the mineral composition of red clover and ryegrass grown on different soils. A mean of 1.8 ppm (dry basis; range 0.7-3.8) was obtained for the former species and of 2.0 ppm (range 0.9-4.6) for the latter. Very little is known of the titanium content of human foods. Tipton *et al.* (198) reported the 30-day mean total dietary titanium intakes of 2 individuals to be 0.37 and 0.41 mg/day. A surprising feature of this investigation was the high urinary excretion, suggesting either considerable absorption from the diet or loss from previously retained tissue titanium. Both individuals were in substantial negative titanium balance, with approximately equal excretion via the feces and the urine. No evidence that absorbed titanium performs any vital functions in animals nor that it is a dietary essential for any living organisms has yet been produced.

Bertrand and Varonia-Spirt (15) determined the titanium content of the tissues of a number of animals and marine organisms with highly variable results. In the domestic animals the highest concentrations (2 ppm of dry tissue) were found in the liver. Tipton and Cook (197) also found great variability in the levels of titanium in all adult human organs examined, with some samples of tissue below the limit of detection of the method, namely 5 ppm of the ash or approximately 0.2 ppm of dry tissue. Most of the soft tissues contained 0.3-0.6 ppm, but the lungs averaged over 20 ppm, with some samples containing over 250 ppm. Evidence was obtained that titanium accumulates with age in the lungs, but not in the liver and kidneys. A high proportion of the titanium in the lungs, like the aluminum and the silicon, presumably comes from the inhalation of atmospheric dust.

## XIII. Zirconium

Biological interest in zirconium was virtually nonexistent until Schroeder and his collaborators (164, 171) included this element in their series of studies of "abnormal" trace metals in man. This is surprising because zirconium is much more abundant in the lithosphere than a large number of other trace elements which have been extensively studied. The mean concentration in the earth's crustal rocks has been reported as 170 ppm, in soils as 300 ppm, in seawater as 4 ppb, and in marine sediments as 132 ppm (69). The data of Lounamaa (117) indicate that zirconium is present in Finnish soils and wild plants in relatively high but very variable concentrations. The mean levels reported for herbs and grasses growing on three different types of soil ranged from 7 to 22 ppm of ash. Vlamis and Pearson (202) had previously shown that radioactive zirconium is readily absorbed by plants from soil.

Schroeder and Balassa (164) determined the zirconium content of a wide range of biomaterials, by means of a spectrophotometric procedure stated to be sensitive to 0.5 to 0.8 $\mu$g/g ash, or about 0.01 to 0.02 ppm wet weight of most samples. The method is highly specific for zirconium and measures only traces of hafnium, an element which it resembles closely in physical and chemical characteristics. Using this method, zirconium was found to be widely distributed in soils and in plant and animal tissues. It was present in mammalian tissues in concentrations larger than those of copper and in human tissues in concentrations exceeded, among the trace elements, only by iron, zinc, silicon, and rubidium. Zirconium was observed in all the organs examined from 4 male accident victims, including the brain, and was especially high in fat (mean 18.7 $\mu$g/g wet weight), liver (6.3 $\mu$g/g wet weight), and red blood cells (6.2 $\mu$g/g wet weight). Similar high values were found for single samples of the aorta and gall bladder, with levels in the other tissues and blood serum mostly lying between 1 and 3 $\mu$g/g. It was calculated that the total zirconium content of the body of adult man is approximately 250 mg or some 3 times that of copper.

From analysis of an institutional diet and from calculations based on the examination of a range of common human foods, Schroeder and Balassa (164) estimated a daily oral intake of zirconium by man of about 3.5 mg. This quantity could be increased by high intakes of vegetable oils (3-6 $\mu$g/g) or tea (11.7 $\mu$g/g). These substances were the highest in zirconium of the items examined. Meat, dairy products, vegetables, cereal grains, and nuts generally contained 1-3 $\mu$g/g (fresh basis), with appreciably lower levels in most fruits and seafoods.

The metabolic movements of zirconium in the animal body do not appear to have been directly studied. Its presence in relatively high concentrations in the blood and tissues indicates that it is absorbed from ordinary diets, while its virtual absence from the urine (164) suggests that it is excreted by the intestine. Hepatic excretion may be suspected from the high concentrations in liver and gall bladder. The high concentrations in fat and aorta suggest that the element is stored in fats (164). More detailed and direct evidence of the sites of absorption, retention, and excretion of zirconium is clearly desirable.

Zirconium compounds have a low order of toxicity for rats and mice, whether injected (172) or orally ingested (171, 172). Schroeder *et al.*

(171) studied the long-term effects on mice of adding 5 ppm Zr, as zirconium sulfate, to the drinking water. No effects on growth were observed, but there was a small reduction in survival time. The element was neither carcinogenic nor tumorigenic, with little evidence of accumulation in the tissues.

## REFERENCES


1. G. V. Alexander, R. E. Nusbaum, and N. S. McDonald, *J. Biol. Chem.* **192**, 489 (1951).
2. R. Allcroft, *J. Comp. Pathol. Ther.* **60**, 190 (1950).
3. R. Allcroft, *Vet. Rec.* **63**, 583 (1951).
4. R. Allcroft and K. L. Blaxter, *J. Comp. Pathol. Ther.* **60**, 209 (1950).
5. J. G. Archibald, *J. Dairy Sci.* **38**, 159 (1955).
6. K. N. Bagchi, H. D. Ganguly, and J. N. Sirdar, *Indian J. Med. Res.* **28**, 441 (1940).
7. F. Bamford, "Poisons: Their Identification and Isolation". McGraw-Hill (Blakiston), New York, 1951.
8. R. M. Barlow, D. Purves, E. J. Butler, and I. J. McIntyre, *J. Comp. Pathol. Ther.* **70**, 396 (1960).
9. L. C. Bate and F. F. Dyer, *Nucleonics* **23**, 74 (1965).
10. E. J. Baumann, D. B. Sprinson, and D. Marine, *Endocrinology* **28**, 793 (1941).
11. J. Bedford, G. E. Harrison, W. H. A. Raymond, and A. Sutton, *Brit. Med. J.* **1**, 589 (1960).
12. K. C. Beeson, *U.S.Dep. Agr., Misc. Publ.* **369** (1941).
13. F. Bernheim and M. L. C. Bernheim, *J. Biol. Chem.* **127**, 353 (1939); **128**, 79 (1939).
14. G. Bertrand and D. Bertrand, *Ann. Inst. Pasteur, Paris* **72**, 805 (1946); **80**, 339 (1951).
15. G. Bertrand and C. Varonea-Spirt, *C. R. Acad. Sci.* **188**, 119; **189**, 73 and 122 (1929).
16. K. L. Blaxter, *J. Comp. Pathol. Ther.* **60**, 140 (1950).
17. D. K. Bosshardt, J. W. Huff, and R. H. Barnes, *Proc. Soc. Exp. Biol. Med.* **92**, 219 (1956).
18. H. J. M. Bowen, *Biochem. J.* **73**, 381 (1959).
19. H. J. M. Bowen and J. A. Dymond, *Proc. Roy. Soc., Ser. B* **144**, 355 (1955).
20. T. C. Boyd and N. K. De, *Indian J. Med. Res.* **20**, 789 (1933).
21. R. F. Brewer, *in* "Diagnostic Criteria for Plants and Soils" (H. G. Chapman, ed.), p. 213, Div. Agr. Sci., Univ. of California Press, Riverside, California, 1966.
22. R. Brock, *Nature (London)* **195**, 841 (1962).
23. D. Brune, K. Samsahl, and P. O. Wester, *Clin. Chim. Acta.* **13**, 285 (1966).
24. H. L. Cannon and J. M. Bowles, *Science* **137**, 765 (1962).
25. A. Cantarrow and M. Trumper, "Lead Poisoning". Williams & Wilkins, Baltimore, Maryland, 1944.
26. A. Casini, *Ann. Chim. Appl.* **36**, 219 (1946).
27. A. P. Cawadias, *Amer. Heart J.* **65**, 277 (1963).
28. A. C. Chapman, *Analyst* **51**, 548 (1926).
29. W. Clode, J. M. Sobral, and A. M. Baptista, *Advan. Thyroid Res. Trans. Int. Goitre Conf., 4th, 1960* p. 65 (1961).
30. B. B. Cohen, *Plant Physiol.* **15**, 755 (1940).
31. B. T. Cole and H. Patrick, *Arch. Biochem. Biophys.* **74**, 357 (1958).

32. R. Collander, *Plant Physiol.* **16**, 691 (1941).
33. E. J. Conway and J. C. Flood, *Biochem. J.* **30**, 716 (1936).
34. E. J. Coulson, R. E. Remington, and K. M. Lynch, *J. Nutr.* **10**, 255 (1935).
35. H. E. Cox, *Analyst* **50**, 3 (1925); **51**, 132 (1926).
36. K. Craemer, S. Selander, and M. K. Williams, *Lancet* **1**, 544 (1966).
37. G. Curli and D. Coppini, *Lait* **38**, 497 (1938).
38. M. A. Damiens and S. von Blaignan, *C. R. Acad. Sci.* **193**, 1460 (1931); **194**, 2077 (1932).
39. J. R. Davis and S. L. Andelman, *Arch. Environ. Health* **15**, 53 (1967).
40. H. J. Deobald and C. A. Elvehjem, *Amer. J. Physiol.* **111**, 118 (1958).
41. W. A. Dewar and J. M. A. Lenihan, *Scot. Med. J.* **1**, 236 (1956).
42. H. Dingle and J. H. Sheldon, *Biochem. J.* **32**, 1078 (1938).
43. T. F. Dixon, *Biochem. J.* **28**, 86 (1935).
44. P. P. Donovan, D. T. Feeley, and P. P. Canavan, *J. Sci. Food Agr.* **20**, 43 (1969).
45. H. S. Ducoff, W. B. Neal, F. L. Straube, L. O. Jacobson, and A. M. Brues, *Proc. Soc. Exp. Biol. Med.* **69**, 549 (1948).
46. R. E. Duggan and J. R. Weatherwax, *Science* **157**, 1006 (1967).
47. M. Fay, M. A. Andersch, and V. G. Behrmann, *J. Biol. Chem.* **144**, 383 (1942).
48. R. H. Follis, Jr., *Amer. J. Physiol.* **138**, 246 (1943).
49. R. H. Follis, Jr., *Amer. J. Physiol.* **143**, 385 (1945).
50. R. M. Forbes, A. R. Cooper, and H. H. Mitchell, *J. Biol. Chem.* **209**, 857 (1954).
51. W. P. Ford, D. W. Kent-Jones, A. M. Maiden, and R. C. Spalding, *J. Soc. Chem. Ind., London* **59**, 177 (1940).
52. J. A. Fordyce, I. Rosen, and C. N. Myers, *Amer. J. Syph.* **8**, 65 (1924).
53. R. B. Forney, and R. N. Harger, *Fed. Proc., Fed. Amer. Soc. Exp. Biol.* **8**, 292 (1949).
54. A. S. Freedberg, H. B. Pinto, and A. Zipser, *Fed. Proc., Fed. Amer. Soc. Exp. Biol.* **11**, 49 (1952).
55. D. V. Frost, *Poultry Sci.* **32**, 217 (1953).
56. D. V. Frost, *Fed. Proc., Fed. Amer. Soc. Exp. Biol.* **26**, 194 (1967).
57. D. V. Frost, and H. C. Spruth, "Symposium on Medicated Feeds." Med. Encycl. Inc., New York, 1956.
58. D. V. Frost, L. R. Overby, and H. C. Spruth, *J. Agr. Food Chem.* **3**, 235 (1955).
59. H. C. Gauch and W. M. Dugger, *Md. Agr. Exp. Sta., Tech. Bull.* **A-80** (1954).
60. W. Gerlach and R. Muller, *Virchows Arch. Pathol. Anat. Physiol.* **294**, 210 (1934).
61. O. S. Gibbs, H. Pond, and G. A. Hansmann, *J. Pharmacol. Exp. Ther.* **72**, 16 (1941).
62. B. L. Glendenning, D. B. Parrish, and W. G. Schrenk, *Anal. Chem.* **27**, 1554 (1954).
63. B. L. Glendenning, W. G. Schrenk, and D. B. Parrish, *J. Nutr.* **60**, 563, (1956).
64. J. Godonniche and G. Dastugue, *Bull. Soc. Chim. Biol.* **16**, 253 (1964).
65. J. R. Goldsmith and A. C. Hexter, *Science* **158**, 132 (1967).
66. L. J. Goldwater, *J. Roy. Inst. Pub. Health* **27**, 279 (1964).
67. L. J. Goldwater and A. W. Hoover, *Arch. Environ. Health* **15**, 60 (1967).
68. J. E. Greaves, *Soil Sci.* **38**, 355 (1934).
69. J. Green, *Geol. Soc. Amer. Bull.* **70**, 1127 (1959).
70. G. C. Griffith, E. M. Butt, and J. Walker, *Ann. Intern. Med.* **41**, 501 (1954).
71. R. E. R. Grimmett, *N. Z. J. Agr.* **58**, 383 (1939).
72. F. Guffroy, *Ann. Fals. Fraudes* **47**, 423 (1947); cited by J. G. Archibald, *Dairy Sci. Abstr.* **20**, 712 (1958).
73. C. Hackett, *Nature (London)* **195**, 471 (1962).
74. B. Haeger-Aronsen, *Scand. J. Clin. Lab. Invest.* **12**, Suppl. 47, 1 (1960).

75. P. B. Hammond, H. N. Wright, and M. H. Roepke, *Minn. Agr. Exp. Sta., Sta. Bull.* **221** (1956).
76. L. E. Hanson, L. E. Carpenter, W. J. Aunan, and E. F. Ferrin, *J. Anim. Sci.* **14**, 513 (1955).
77. G. E. Harrison and W. H. A. Raymond, *J. Nucl. Energy* **1**, 290 (1955).
78. G. E. Harrison, W. H. A. Raymond, and H. C. Tretheway, *Clin. Sci.* **14**, 681 (1955).
79. W. B. Healy, L. C. Bate, and T. G. Ludwig, *N. Z. J. Agr. Res.* **7**, 603 (1964).
80. S. Hellerstein, C. Kaiser, D. D. Darrow, and D. C. Barrow, *J. Clin. Invest.* **39**, 282 (1960).
81. L. A. Heppel and C. L. A. Schmidt, *Univ. Calif., Berkeley, Publ. Physiol.* **8**, 189 (1938).
82. S. Hernberg, V. Vihdo, and J. Hasan, *Arch. Environ. Health* **14**, 319 (1967).
83. B. J. Heywood, *Science* **152**, 1408 (1966).
84. R. M. Hodges, N. S. Mcdonald, R. E. Nusbaum, R. Stearns, F. Ezmirlian, P. Spain, and S. Macarthur, *J. Biol. Chem.* **185**, 519 (1950).
85. W. S. Hodgkiss and B. J. Errington, *Trans. K. Acad. Sci.* **9**, 17 (1940).
86. H. Hopkins and J. Eisen, *J. Agr. Food Chem.* **7**, 633 (1959).
87. B. L. Horecker, E. Stotz, and T. R. Hogness, *J. Biol. Chem.* **128**, 251 (1939).
88. E. Hove, C. A. Elvehjem, and E. B. Hart, *Amer. J. Physiol.* **123**, 640 (1938).
89. E. Hove, C. A. Elvehjem, and E. B. Hart, *Amer. J. Physiol.* **124**, 205 (1938).
90. E. Hove, C. A. Elvehjem, and E. B. Hart, *Amer. J. Physiol.* **127**, 689 (1939).
91. R. A. Howie and H. Smith, *Forensic Sci. Soc., J.* **7**, 90 (1967).
92. W. C. Hueper, *Amer. J. Med. Sci.* **181**, 820 (1931).
93. J. W. Huff, D. K. Bosshardt, O. P. Miller, and R. H. Barnes, *Proc. Soc. Exp. Biol. Med.* **92**, 216 (1956).
94. F. T. Hunter, A. F. Kip, and J. W. Irvine, *J. Pharmacol. Exp. Ther.* **76**, 207 (1942); O. H. Lowry, F. T. Hunter, A. F. Kip, and J. W. Irvine, *ibid.* p. 221.
95. G. E. Hutchinson, *Quart. Rev. Biol.* **18**, 1, 128, 242, and 331 (1943).
96. G. E. Hutchinson, *Soil Sci.* **60**, 29 (1945).
97. I. Ishikawa, *Osaka City Med. J.* **5**, 99, 109, and 117 (1959).
98. N. Iwataki and K. Horiuchi, *Osaka City Med. J.* **5**, 209 (1959).
99. S. Jensen and A. Jernelov, *Nature* (*London*) **223**, 753 (1969).
100. M. M. Joselow and L. J. Goldwater, *Arch. Environ. Health* **15**, 155 (1967).
101. M. M. Joselow, L. J. Goldwater, and S. B. Weinberg, *Arch. Environ. Health* **15**, 64 (1967).
102. R. V. Jury, M. S. Webb, and R. J. Webb, *Anal. Chim. Acta* **22**, 145 (1960).
103. R. A. Kehoe, *J. Roy. Inst. Pub. Health* **24**, 81, 101, 129, and 177 (1961).
104. R. A. Kehoe, J. Cholak, and R. V. Storey, *J. Nutr.* **19**, 579 (1940).
105. C. Kellershohn, D. Comar, and C. Lopoec, *J. Lab. Clin. Med.* **66**, 168 (1965).
106. N. L. Kent and R. A. McCance, *Biochem. J.* **35**, 837 and 877 (1941).
107. R. Kilpatrick, H. E. Renschler, D. S. Munro, and G. M. Wilson, *J. Physiol.* (*London*) **133**, 194 (1956).
108. M. Kirchgessner, *Z. Tierphysiol., Tierernaehr. Futtermittelk.* **14**, 270 and 278 (1959).
109. M. Kirchgessner, G. Merz, and W. Oelschläger, *Arch. Tierernaehr.* **10**, 414 (1960).
110. H. J. Koch, E. R. Smith, N. F. Shimp, and J. Connor, *Cancer* **9**, 499 (1956).
111. L. Kopito, R. K. Byers, and H. Shwachman, *N. Engl. J. Med.* **276**, 949 (1967).
112. W. Langham and E. C. Anderson, *U.S. At. Energy Comm., Health Safety Lab.* **42**, 282 (1958); cited by Patterson (145a).
113. A. Laznitski and E. Szorenyi, *Biochem. J.* **28**, 1678 (1934).
114. R. M. Leach, Jr. and M. C. Nesheim, *J. Nutr.* **81**, 193 (1963).

115. J. M. A. Lenihan and H. Smith, *in* "Nuclear Activation Techniques in the Life Sciences," I.A.E.A., Vienna, 1967.
116. R. F. Loeb, *J. Gen. Physiol.* **3**, 229 (1920).
117. J. Lounamaa, *Ann. Bot. Soc. Zool. Bot. Fenn. "Vanamo"* **29**, 4 (1956).
118. F. Lowater and M. M. Murray, *Biochem. J.* **31**, 837 (1937).
119. C. L. Luke and M. D. Campbell, *Anal. Chem.* **28**, 1273 (1956).
120. G. Lunde, *J. Sci. Food Agr.* **19**, 432 (1968).
121. G. E. Lynn, S. A. Shrader, O. H. Hammer, and C. A. Lassiter, *J. Agr. Food Chem.* **11**, 87 (1963).
122. R. A. MacLeod and E. E. Snell, *J. Biol. Chem.* **176**, 39 (1948); *J. Bacteriol.* **59**, 783 (1950).
123. M. J. Mason, *J. Biol. Chem.* **113**, 61 (1936).
124. J. McClachlan and J. S. Craigie, *Nature (London)* **214**, 604 (1967).
125. H. Micheels, *Rev. Sci.* [5] **5**, 427 (1906); cited by Schroeder *et al.* (169).
126. V. L. Miller, G. E. Bearse, and K. E. Hammermeister, *Poultry Sci.* **38**, 1037 (1959).
127. V. L. Miller, D. V. Larkin, G. E. Bearse, and C. M. Hamilton, *Poultry Sci.* **46**, 142 (1967).
128. R. L. Mitchell, *Research (London)* **10**, 357 (1957).
129. R. L. Mitchell, *Commonw. Bur. Soil Sci. Tech. Commun.* No. 44 (1948).
130. R. L. Mitchell and J. W. S. Reith, *J. Sci. Food Agr.* **17**, 437 (1967).
131. A. J. Mitteldorf and D. O. London, *Anal. Chem.* **22**, 828 (1952).
132. R. Mole and C. Pesaresi, *Folia Med.* **47**, 73 (1964).
133. M. M. Molokhia and H. Smith, *Arch. Environ. Health* **15**, 745 (1967).
134. G. W. Monier-Williams, "Trace Elements in Food." Chapman & Hall, London, 1949.
135. K. Morgareidge, *J. Agr. Food Chem.* **11**, 377 (1963).
136. G. K. Murthy, U. Rhea, and J. T. Peeler, *J. Dairy Sci.* **50**, 651 (1967).
137. V. C. Myers and J. W. Mull, *J. Biol. Chem.* **78**, 605 and 625 (1928).
138. M. Nakao, *Osaka Shiritsu Daigaku, Igaku Zasshi* **9**, 541 (1960); cited by Schroeder and Balassa, (163).
139. A. H. Neufeld, *Can. J. Res., Sect. B* **14**, 160 (1936).
140. G. S. Nixon, H. Smith, and H. D. Livingston, *in* "Symposium on Nuclear Activation Techniques in the Life Sciences," I.A.E.A., Vienna, 1967.
141. E. Orent-Keiles, *Proc. Soc. Exp. Biol. Med.* **44**, 199 (1941).
142. L. R. Overby and R. L. Frederickson, *J. Agr. Food Chem.* **11**, 378 (1963).
143. L. R. Overby and D. V. Frost, *J. Anim. Sci.* **19**, 140 (1960).
144. E. C. Owen, *J. Dairy Res.* **13**, 243 (1944).
145. J. Parizek, I. Benes, A. Babicky, J. Benes, V. Prochazkova, and J. Lener, *Physiol. Bohemoslov.* **18**, 105 (1969).
145a. C. C. Patterson, *Arch. Environ. Health* **11**, 344 (1965).
146. H. M. Perry, Jr. and E. F. Perry, *J. Clin. Invest.* **38**, 1452 (1959).
147. C. C. Pfeiffer, L. F. Hallman, and I. Gersh, *J. Amer. Med. Assoc.* **128**, 266 (1945).
148. K. I. Plotnikov, *Veterinariya (Moscow)* **37**, 35 (1960); *Nutr. Abstr. Rev.* **30**, 1138 (1960).
149. D. Purves, *Plant Soil* **26**, 380 (1967).
150. A. S. Relman, *Yale J. Biol. Med.* **29**, 248 (1956–1957).
151. C. E. Richards, R. O. Brady, and D. S. Riggs, *J. Clin. Endocrinol.* **9**, 1107 (1949).
152. S. Ringer, *J. Physiol. (London)* **4**, 370 (1882).
153. W. O. Robinson and G. Edginton, *Soil Sci.* **60**, 15 (1945).
154. R. O. Robinson, R. R. Whetstone, and G. Edginton, *U.S., Dep. Agr., Tech. Bull.* **1013** (1950).

155. W. J. Rodger and H. Smith, *Forensic Sci. Soc., J.* **7**, 86 (1967).
156. G. Rosenfeld, *Arch. Biochem. Biophys.* **48**, 54 (1954).
157. G. Rosenfeld and E. D. Wallace, *Arch. Ind. Hyg.* **8**, 466 (1953).
158. E. J. Rubins and G. R. Hagstrom, *J. Agr. Food Chem.* **7**, 722 (1959).
159. O. Rygh, *Bull. Soc. Chim. Biol.* **31**, 1052, 1403, and 1408 (1949); **33**, 133 (1953); *Research* (*London*) **2**, 340 (1949).
160. K. Scharrer and W. Schropp, *Bodenk. Planzenernaehr.* **3**, 369 (1937).
161. E. F. Schittmann, Hohenheim University Inaug. Dissertation, 1955; quoted by J. G. Archibald, *Dairy Sci. Abstr.* **20**, 712 (1958).
162. H. A. Schroeder, *Advan. Intern. Med.* **8**, 259 (1956).
163. H. A. Schroeder and J. J. Balassa, *J. Chronic Dis.* **19**, 85 (1966).
164. H. A. Schroeder and J. J. Balassa, *J. Chronic Dis.* **19**, 573 (1966).
165. H. A. Schroeder and J. J. Balassa, *J. Nutr.* **92**, 245 (1967).
166. H. A. Schroeder and J. J. Balassa, *J. Chronic Dis.* **20**, 211 (1967).
167. H. A. Schroeder and A. P. Nason, *J. Invest. Dermatol.* **53**, 71 (1969).
168. H. A. Schroeder and I. H. Tipton, *Arch. Environ. Health* **17**, 965 (1968).
169. H. A. Schroeder, J. J. Balassa, and I. H. Tipton, *J. Chronic Dis.* **17**, 483 (1964).
170. H. A. Schroeder, M. Kanisawa, D. V. Frost, and M. Mitchener, *J. Nutr.* **96**, 37 (1968).
171. H. A. Schroeder, M. Mitchener, J. J. Balassa, M. Kanisawa, and A. P. Nason, *J. Nutr.* **95**, 95 (1968).
172. J. Schubert, *Science* **105**, 389 (1947).
172a. K. Schwarz, D. B. Milne & E. Vinyard, *Biochem. Biophys. Res. Comm.* **40**, 22 (1970).
173. G. H. Scott and J. H. McMillen, *Proc. Soc. Exp. Biol. Med.* **35**, 287 (1936).
174. W. Seaber, *Analyst* **58**, 575 (1930).
175. M. Seibold, *Klin. Wochenschr.* **38**, 117 (1960).
176. T. L. Shaw, *in* "Physiology and Biochemistry of Algae" (R. A. Lewin, ed.) Academic Press, New York, 1962.
177. G. D. Shearer, J. R M. Innes, and E. I. McDougall, *Vet. J.* **96**, 309 (1940).
178. J. H. Sheldon and H. Ramage, *Biochem. J.* **25**, 1608 (1931).
178a. F. B. Shorland, *Proc. Roy. Soc. N. Z.* **64**, 35 (1934).
179. J. T. Skinner and J. S. McHargue, *Amer. J. Physiol.* **143**, 385 (1945); **145**, 500 (1946).
180. J. Skok, *in* "Trace Elements" (C. A. Lamb, O. G. Bentley, and J. M. Beattie, eds.), p. 227, Academic Press, New York, 1958.
181. H. Smith, *Anal. Chem.* **31**, 1361 (1959).
182. H. Smith, *Forensic Sci. Soc., J.* **4**, 192 (1964); **7**, 97 (1967).
183. H. Smith and J. M. A. Lenihan, *in* "Methods in Forensic Science" (A. S. Curry, ed.), Wiley (Interscience), New York, 1964.
184. H. Smith, A. Forshufvud, and A. Wassen, *Nature* (*London*) **194**, 725 (1962).
185. J. C. Smith and K. Schwarz, *J. Nutr.* **93**, 182 (1967).
186. T. Sollmann, "Manual of Pharmacology". Saunders, Philadelphia, Pennsylvania, 1953.
187. A. L. Somner and C. B. Lipman, *Plant Physiol.* **1**, 231 (1926).
188. K. Sonjo, *Folia Pharmacol. Jap.* **151** (1934).
188a. E. M. Sowden & S. R. Stitch, *Biochem. J.* **67**, 104 (1957).
189. S. R. Stitch, *Biochem. J.* **67**, 97 (1957).
190. A. Stock, *Biochem. Z.* **304**, 73 (1940).
191. A. Stock, F. Cucuel, and H. Köhle, *Z. Angew. Chem.* **46**, 187 (1933).
192. A. Stock, H. Lux, F. Cucuel, and H. Köhle, *Z. Angew. Chem.* **46**, 62 (1933).
193. A. Stock and N. Neuenschwander-Lemmer, *Ber. Deut. Chem. Ges.* **72**, 1844 (1939).

194. E. Szep, *Hoppe-Seyler's Z. Physiol. Chem.* **267**, 29 (1940).
195. J. D. Teresi, E. Hove, C. A. Elvehjem, and E. B. Hart, *Amer. J. Physiol.* **140**, 513 (1940).
196. H. V. Thomas, B. K. Gilmore, G. A. Heidbreder, and B. A. Kogan, *Arch. Environ. Health* **15**, 695 (1967).
197. I. H. Tipton and M. J. Cook, *Health Phys.* **9**, 103 (1963).
198. I. H. Tipton, P. L. Stewart, and P. G. Martin, *Health Phys.* **12**, 1683 (1966).
199. S. L. Tompsett and A. B. Anderson, *Biochem. J.* **29**, 1851 (1935).
200. D. D. Ulmer and B. L. Vallee, *Proc. 2nd Missouri Conf. Trace Substances Environ. Health 1968* p. 7 (1969).
201. B. L. Vallee, D. D. Ulmer, and W. E. C. Wacker, *AMA Arch. Ind. Health* **21**, 132 (1960).
202. J. Vlamis and G. A. Pearson, *Science* **111**, 112 (1950).
203. N. Wada, *Osaka City Med. J.* **4**, 113 (1957).
204. T. Walsh, *Proc. Roy. Irish Acad. Sect. B* **50**, 287 (1945).
205. K. Warrington, *Ann. Bot.* (*London*) **37**, 629 (1923); **40**, 27 (1926).
206. L. J. Webb, *Aust. J. Bot.* **2**, 176 (1954).
207. T. Westermark, *Kvicksilverfragen i. Sverige, 1964,* Stockholm, 1965.
208. G. Westoo, *Acta Chem. Scand.* **20**, 2131 (1966).
209. G. H. Whipple and F. S. Robscheit-Robbins, *Amer. J. Physiol.* **72**, 419 (1925).
210. W. B. White, P. A. Clifford, and H. O. Calvery, *J. Amer. Vet. Med. Assoc.* **102**, 292 (1943).
211. P. S. Winnek and A. H. Smith, *J. Biol. Chem.* **119**, 93 (1937); **121**, 345 (1937).
212. M. L. Wolbarsht and D. S. Sax, *Notes Rec. Roy. Soc. London* **16**, 154 (1961).
213. J. Wurker, *Biochem. Z.* **265**, 169 (1933).
214. A. Zipser and A. S. Freedberg, *Cancer Res.* **12**, 867 (1952).
215. E. G. Zook and J. Lehmann, *J. Amer. Diet. Assoc.* **52**, 225 (1968).
216. E. G. Zook and J. Lehmann, *J. Assoc. Off. Agr. Chem.* **48**, 850 (1965).

# 17

# SOIL-PLANT-ANIMAL INTERRELATIONS

## I. Introduction

All plants and all animals, including man, depend ultimately upon the soil for their supply of mineral nutrients. With plants, this relationship is direct and is simplified by the fact that the plant is stationary. On the other hand, grazing animals may derive their minerals from a variety of soil types and plant species, so long as they are free to range freely over wide areas. In this way any disabilities associated with particular soils would tend to be minimized or even eliminated. Intensification of settlement imposes restrictions on movements until some animals become dependent upon a single soil type, or a narrower range of soil types, certain of which are incapable, without appropriate treatment, of sustaining the health, fertility, and productivity of stock. The fact that animals did not thrive or suffered various disorders when restricted to some areas, and remained quite healthy in other areas, was recognized in several parts of the world as early as the 18th century. These observations focused attention upon the environmental, particularly the soil factors involved because (*a*) areas considered satisfactory for stock and those classed as unthrifty or unhealthy were often adjacent, which would tend to rule out climatic differences as causal factors, and (*b*) animals transferred from unhealthy to healthy areas usually recovered, which suggested the presence of nutritional rather than infectious disorders.

Investigations carried out during the last half century have shown that many of the nutritional maladies of the type just mentioned result from the inability of the soils of the affected areas to supply, through the plants that grow upon them, the mineral needs of man and his domestic animals in adequate, safe, or nontoxic amounts and in proper proportions. Nutritional abnormalities involving the trace elements may arise as simple, gross deficiencies or excesses of single elements. More usually they occur as deficiencies or excesses conditioned by the extent to which other mineral elements, nutrients, or organic factors are present in the environment and are capable of modifying the ability of the animal to utilize the deficient or toxic element. The conditioning factors may themselves be a reflection of the soils on which the herbage grows, since the soil conditions affect the chemical composition of plants in a variety of ways, additional to their effect on the primary element itself. The presence of particular plant species, such as selenium accumulator or goitrogenic plants, may also influence the incidence or severity of various trace element deficiencies or excesses, because of their influence upon the amounts or the availability of particular elements consumed by the animal. Soil-plant-animal interrelations are thus quite complex. They go far beyond the simple concept that deficiency or toxicity conditions in the animal are merely a reflection of a deficiency or excess of a particular mineral in the soil and, therefore, in the herbage that this soil supports.

The incidence of the disease, phalaris staggers, in sheep and cattle provides a further example of the complexity of soil-plant-animal-trace element interrelationships. This condition is unknown in most areas where the grass, *Phalaris tuberosa,* is grown. On Co-deficient, or marginally Co-deficient soils, the neurotoxic substance present in this plant species induces demyelination and a "staggers" syndrome in animals consuming the plant, unless they are treated with cobalt salts or cobalt pellets, or the soils are fertilized with cobalt salts or ores. It seems that normal soils produce herbage with cobalt levels adequate to meet the normal needs of ruminants, plus sufficient to enable them to detoxicate the neurotoxic substance in *Phalaris*. Marginally Co-deficient soils are capable of fulfilling the former needs but not the latter. Severely Co-deficient soils result in the growth of plants carrying insufficient cobalt to fulfill either of these needs satisfactorily. The incidence of phalaris staggers thus depends upon the interaction between a soil factor and a plant factor, each of which can vary independently in a particular environment.

The incidence of chronic copper poisoning in sheep and cattle, and its relation to the presence or absence of the hepatotoxic alkaloid-containing plant, *Heliotropium europaeum*, provide a different example of complex trace element interactions involving the soil, the plant, and the animal. The actual field occurrence and severity of deficiency or toxicity conditions in animals, involving the 4 trace elements iodine, cobalt, copper, and selenium, are, therefore, known to be influenced by complex interrelations between the soil, the plant, and the animal. It seems likely that similar interrelations, involving other elements and other plant species, will emerge as research proceeds.

With most trace elements the food, and not the water or the atmosphere, supplies an overwhelming proportion of total daily intakes by animals and by man. This is not true of most endemic fluorosis regions where the water supplies constitute the principal source of the high-fluoride intakes. Even in these circumstances the toxic quantities of fluoride present in the water seldom bear any relationship to the fluoride status of the soils and herbage of the affected area, because the water usually comes from deep wells or bores drawing from other soil or rock formations. A high inverse correlation between the iodine content of the drinking water and the incidence of goiter has long been known, but only some 10% of the total intakes of this element by man come from the water supply, in goitrous and nongoitrous regions alike. In a recent study carried out by Hadjimarkos (27) of the concentration of 17 trace elements in the public water supplies located in forty-four states of the United States, it was calculated that 0.3 to 10.1% of the total daily intakes came from the drinking water. Similar conclusions were reached by Schroeder (54) for individuals living in certain large cities of the United States.

Insignificant quantities of trace elements, compared with the amounts in foods, are also normally contributed by the atmosphere, except in areas adjacent to mines and factories, where substantial atmospheric pollution can occur. With the rise of modern industrial technology and with the increasing urbanization and motorization of large sections of the population, this source of trace elements, together with contamination of the water supplies, may constitute an increasingly significant source of a number of elements with possible long-term dangers to human health. Already some evidence has been produced that the level of cadmium in the air in U.S. cities is correlated with the incidence of cardiovascular diseases in those cities (18). The deposition of lead from automobile exhausts on soil and plants along highways and in urban areas is also well established (17,51).

## II. Soil Relations in Human Health

Trace element deficiencies, toxicities, and imbalances are more difficult to relate to the soil in man than they are in farm stock. Copper deficiency has never been reported in human populations, although extensive areas exist in which a deficiency of this element occurs naturally in crops and stock. The position is similar in selenium deficiency areas. Even in the severely seleniferous areas of North America and Ireland, the existence of disease conditions in man definitely attributable to selenium poisoning has not been convincingly demonstrated. In fact the only evidence of this nature available is the apparent association between high, or relatively high, dietary intakes of selenium and the prevalence of dental caries in children, presented by Hadjimarkos (28) and by Ludwig and Bibby (45).

Many factors conspire to minimize the effects on man of soil-induced variations in the trace element contents of human foods. The geographic sources of human foods and beverages are everwidening in most modern communities, so that the overall diet comprises materials grown on a range of soil types. Modern human dietaries also contain a considerable variety of types of foods. Trace element abnormalities which may be present in particular plants, parts of plants, or animal tissues and fluids can, therefore, be offset by the consumption of other food items not so affected.

Industrial treatment of an increasing proportion of materials destined for human consumption provides the opportunity for both gain and loss of trace elements during storage, transport, preservation, and processing. The impact of such processes upon dietary intakes varies markedly with different elements. They all tend to reduce the directness of the relationship between the soils of a given area and the actual intakes of mineral elements by man. For instance, the pasteurization and drying of milk, where these involve contact with metal containers, can substantially increase the levels of iron and copper, two elements in which this food is naturally deficient. The milling of grain into white flour results in large but variable losses of most of the minerals present in the original whole grain. This applies to elements that are initially high, as well as to those that are present in low concentrations. For instance, the level of selenium in white flour made from seleniferous wheat is very much lower than it is in the whole grain. This may be one of the factors contributing to the absence of signs of selenosis in man in seleniferous areas. Finally, the domestic treatment of human foods can appreciably affect the position with several elements. Significant losses of fluorine and selenium, which may be present in toxic quantities in raw vegetables, can occur in the cooking water. Rising standards of cleanliness and hygiene at the retail and domestic level, and decreasing use of iron cooking vessels and containers may significantly reduce the opportunities for contamination of foods with iron and other trace elements.

Where the choice of foods is poor, because of poverty, ignorance, or prejudice, so that the general quality of the diet is low and where the dependence on locally grown foods is high, local soil deficiencies or excesses are likely to accentuate any dietary disabilities and to affect adversely human health and nutrition. Such a combination of circumstances can almost certainly be incriminated in the iron deficiency anemia in children in Florida, mentioned in Chapter 2. Poor choice of foods, involving diets high in white sugar and refined cereals low in chromium, rather than the source of these foods, may be the causative factor in the incidence of the chromium-responsive impaired glucose tolerance which exists in a proportion of old people. The conditioned zinc deficiency which occurs in Middle East male dwarfs also appears to be unrelated to the soil conditions under which their foods are grown. Total dietary zinc intakes by these individuals compare well with those from other parts of the world where no such disorders arise in man. The zinc deficiency in these cases appears to result from a combination of circumstances, including excessive dependence on a high phytate, whole wheat or corn bread, and beans diet, plus intestinal parasitism. Nevertheless, there is ample evidence that the content of zinc and other trace elements in plants used as foods by man and animals is influenced by the soil type and fertilizer treatment applied. Regional differences in plasma zinc levels in human adults have also been demonstrated in the United States (39), which presumably reflect differences in dietary zinc intakes. These are more likely to arise from local differences in the zinc content of the foods and beverages consumed than from variations in dietary habits.

The relationship between the incidence of goiter and soil and plant iodine levels provides the most convincing evidence of a link between the composition of the soil and human disease. Subnormal levels of iodine in the soils from which the food is derived have been correlated with the incidence of goiter in man and animals in many areas. In New Zealand, Hercus and co-workers (31) found that "variation in the average amount of iodine in soils containing more than 10 ppm has little effect on the small incidence of goiter; but as the amount of soil iodine decreases, so the incidence of goiter rises."

Normal, mature soils contain about 10 times the iodine concentration of the rocks from which they are formed. Since the solubility of most iodine compounds is high and concentration during the process of soil formation would, therefore, be difficult, an extraneous source of iodine is implied. This is believed to be the iodine borne in on ocean winds. Long periods of time would be necessary for this source of iodine to effect substantial increases in soil iodine. Goldschmidt (25) contends that the low-iodine status of the soils of goitrous areas is associated with the removal of iodine-enriched surface soils by recent glaciation, coupled with insufficient time for replenishment with postglacial airborne oceanic iodine during the subsequent soil-forming period. The principal factors determining a regional iodine deficiency in soils would, therefore, be recent glaciation, distance from the sea, and low annual precipitation. With respect to the last two factors, it has been calculated that from 22 to 50 mg of iodine per acre will fall annually in the rain on the Atlantic coastal plain, compared with 0.7 mg/acre in the goitrous Great Lakes region of North America (32). However, as Beeson (9) has pointed out, all areas subject to Pleistocene glaciation are not iodine-deficient, and large areas which have not been so subjected exist in which goiter is endemic.

The relationship between soil iodine levels and the incidence of goiter is subject to several modifying influences. The nature of the plant species or strains present can be important because of their variable uptake of soil iodine. The amounts and types of goitrogens present in the plants consumed as food can also be significant. Dietary habits, involving a greater diversity of foods and wider sources of supply, also reduce the dependence of human populations upon foods and drinking water derived from local soils. The opportunities for substantially increasing iodine intakes by natural means are limited, except where they include importation of foods of marine origin.

Numerous links between human health and the soil of a more tenuous nature have been proposed. These associations rest heavily upon correlation rather than causation. Thus differential mortality from cancer of the stomach in different parts of England (42, 56) and Holland (58) have been correlated with soil type. In New Zealand, Saunders (53) has similarly related the highest prevalence of total cancer, total digestive tract cancer, and stomach cancer with particular types of soil. More recently, Shamberger and Frost (55) have demonstrated a high inverse correlation between the selenium status of several states of the United States, as indicated by the average selenium content of different forage crops, and the age and sex-adjusted death rates of the populations of those states.

Regional variations in the prevalence of human dental caries have repeatedly been demonstrated (45). In many of these studies the caries differences follow the distribution of fluoride levels in the drinking waters. They cannot, therefore, be related with any certainty to the soils of the areas because the water supplies are frequently unrelated in origin to the local soils. A similar reasoning applies to the geographic variations in the incidence of mottled enamel. However, broad relationships exist between variations in caries incidence and the nature of soils that occur in particular geographic areas in the United States (45, 46, 50) and in New Zealand (16, 33, 47, 53). In a study of 1876 twelve- to fourteen-year-old lifelong residents of nineteen communities of 3000 to 15000 population in eastern United States, situated on four different soil types and all using water containing 0.3 ppm F, or less, caries prevalence was highest on the podzol soils of New England. It was of descending prevalence on gray-brown podzols and red-yellow podzols and of lowest prevalence on the subhumic gley soils of the south Atlantic States (45). An examination of 553 children in eight similar communities on high- and low-selenium soils in the northwestern United States showed that caries was higher in the towns in the higher-selenium areas. These findings support those of Hadjimarkos (28), although the distribution of caries prevalence, in this case, suggested that other factors in addition to selenium are involved.

Comparable differences in caries incidence on different soil types have been observed in New Zealand (16, 33, 47). The adjacent cities of Napier and Hastings have significantly different caries prevalence rates, despite similar socioeconomic conditions, dietary habits, and fluoride content of the drinking water (47). In an attempt to identify the cause of the difference between the two towns, the composition of the vegetables grown on each of the soils was examined. Those grown on the Napier soil were found to be generally richer in molybdenum, aluminum, and titanium and poorer in copper, manganese, barium, and strontium than those from the Hastings soil. The higher molybdenum content of the Napier vegetables was suggested as a likely factor in the lower prevalence of caries among the children of that city (45). Direct evidence in support of this suggestion is not available. In fact, a direct cause and effect relationship between soil type and caries prevalence has yet to be established.

## III. Factors Affecting Trace Element Levels in Plants

Plant materials provide the main source of minerals to animals and to most members of the human race. The factors influencing the trace element content of plants are, therefore, major determinants of dietary intakes of these elements. The concentration of all minerals in plants depends upon four basic, interdependent factors: (*1*) the genus, species, or strain of plant, (*2*) the type of soil on which the plant has grown, (*3*) the climatic or seasonal conditions during growth, and (*4*) the stage of maturity of the plant. The relative impact of these variables depends upon the element in question and can be greatly modified by man through the use of appropriate fertilizers and soil amendments, weedicides, and pesticides, and by irrigation and different husbandry practices. In addition, the inherent capacity of particular plant species to absorb and retain trace elements from the soil can be changed by cross-breeding and selection.

### 1. *Genetic Differences*

Certain plant species have the ability to accumulate uniquely high concentrations of particular elements. The selenium accumulator plants provide an outstanding example of this genetic effect. Strontium, aluminum, and cobalt accumulator plant species are also known. Several species growing on strontium-rich soils in England were shown to contain strontium concentrations as high as 26,000 ppm, compared with 100-200 ppm in other species from similar soils (14). Black gum (*Nyssa sylvatica*) similarly takes up about 100 times as much cobalt as broom sedge growing on the same soils of the coastal plain in the eastern United States (12).

When growing together on the same soils and sampled at similar stages of maturity, leguminous plants commonly contain significantly higher concentrations than grasses or cereals, of cobalt, nickel, iron, copper, and zinc (5, 6, 23, 41, 44, 48, 57), although the differences in favor of legumes are rarely as large or consistent as they are with calcium and strontium. Grasses and cereals are commonly higher than legumes in manganese and molybdenum, and especially in silicon (6-8, 19, 44, 48, 57). Most of the evidence available indicates that this is also true for selenium, although the difference is less obvious when the soil selenium status is low (29, 60). The more limited data for iodine, chromium, titanium, and vanadium do not indicate consistent differences in concentration in favor of either grasses or legumes.

Within both the grasses and legumes, substantial species differences have been observed. Thus Beeson *et al.* (11), in an investigation of seventeen grass species grown together on a sandy loam soil and sampled at the same time, found the cobalt concentration to range from 0.05 to 0.14, the copper from 4.5 to 21.1, and the manganese from 96 to 815 ppm (dry basis). Intraspecies differences have been highlighted by New Zealand investigations of the iodine content of pasture plants. The total iodine levels of two strains of white clover were highly significantly different, and tenfold differences were observed among strains of ryegrass growing on similar soils (34). Furthermore, differences in the iodine content of single plants within strains of ryegrass were found to be large. When diallele crosses were made, analysis of the progenies revealed that herbage iodine content is a strongly inherited character (15).

The inherent genetic differences in the trace element levels in plants and in their capacity to respond to soil applications can have important consequences for the grazing animal. Any agronomic practice that results in changes in the botanical composition of the herbage must clearly be taken into consideration in assessing the adequacy or the safety of any environment in respect to trace element levels. Plant breeding and selection, resulting in the development of higher-yielding or otherwise superior strains, can also lead to significant changes in the concentrations of trace elements present in their tissues. Plant breeders and agronomists need, therefore, to give consideration to this aspect of their efforts, especially in areas deficient or marginal in a particular element.

The genetic differences in the trace element composition of the vegetative parts of plants are not necessarily paralleled by comparable differences in the seeds of those plants. The seeds of leguminous plants and the oil seeds are almost invariably higher in most trace elements than the seeds of grasses or cereals. With cobalt, this difference is usually substantial. Significant species differences within leguminous and gramineous seeds also exist. This is more striking with manganese than with other trace elements. Thus wheat and oat grains are normally 5 times higher in manganese than corn or maize grain and some 3 times higher than barley (59) or sorghum grain (21). Species differences in the manganese concentrations of the seeds of various species of lupins are even greater. Thus Gladstones and Drover (22) found the seeds of *Lupinus albus* to contain manganese concentrations ranging from 817 to 3397 ppm or 10–15 times those of other lupin species growing on the same sites. The seeds of some samples of *Lathyrus sativus* have similarly been reported to contain high levels of manganese, up to 500 ppm (52). Apart from manganese, the cereal grains do not reveal significant species differences in copper, cobalt, molybdenum, zinc, or selenium concentrations when grown under comparable conditions.

### 2. *Soil and Fertilizer Effects*

Plants normally react to a lack of an available element in the soil, either by limiting their growth, by reducing the concentration of the element in their tissues, or more usually, by both at the same time. Conversely, plants respond to soil applications of the deficient element or of soil amendments that increase the plant availability of this element, either by increasing their growth, by raising their tissue concentrations of the element, or by both together. The extent to which one or more of these responses actually takes place varies markedly among different trace elements and different plant species. Soil composition is, nevertheless, the basic factor determining the level of trace elements in plants and, therefore, their capacity to supply adequate, or nontoxic, amounts of these elements to animals and to man.

The composition of the soil is influenced primarily by the nature of the rocks from which the soil is derived. Different parent rocks not only contain the various trace elements in different amounts and proportions, they contain them in differing degrees of stability and are subject to varying influences over variable periods of time during the soil-forming processes. In this way trace elements initially present in the parent rock can be lost, concentrated, or changed in chemical form. Highly leached (podzolized) soils have usually lost an appreciable proportion of their original trace elements. Where such leaching has taken place in soils derived from granitic rocks, low in the trace element-bearing minerals, deficiencies affecting plants and animals can be predicted with some confidence. On the other hand, the amounts of certain plant-extractable trace elements are often much greater on poorly drained than on well-drained soils on the same soil association. The marked influence of this soil factor on the uptake of cobalt and nickel by pasture plants appears to have been first demonstrated by Mitchell (48). A wet soil condition similarly favors the uptake of cobalt and molybdenum by alsike clover on some soils but has no such effect upon copper (38). Adams and Honeysett (1) showed that the cobalt concentrations and to a much lesser extent the copper concentrations, of subterranean clover and ryegrass, are exceedingly sensitive to periods of water-logging in some soils. In several cases the cobalt content of the plants on water-logged soils was 10–20 times greater than that of the controls. Climatic effects upon soil-water status in different years could, therefore, provide one possible explanation of the seasonal differences in the incidence of cobalt deficiency in grazing sheep and cattle observed in some areas. The high levels of selenium in plants in the seleniferous areas of Ireland have also been claimed to result from concentration of this element in the soil due to poor drainage over long periods of time (20).

The uptake of trace elements by plants is influenced further by the acidity of the soil. Cobalt and nickel, and to a smaller extent copper and manganese, are poorly absorbed from calcareous soils, whereas molybdenum uptake is greater from such soils than from those which are acid or neutral in reaction. Teart soils, which carry herbage exceptionally high in molybdenum, are mostly derived from clays and limestones, calcareous and alkaline in character. These differences provide the opportunity for modifying the levels of individual trace elements in pasture and forage plants by the use of soil amendments, such as sulfur, which can lower soil pH, or lime, which can raise soil pH. The depressing effects of incremental dressings of calcium carbonate on the levels of cobalt, nickel, and manganese in red clover and ryegrass and the enhancing effects of such dressings upon the molybdenum levels are illustrated in Table 51.

Although the use of soil amendments or treatments, such as drainage or aeration, which alter the availability of particular trace elements is of practical value in some circumstances, the application of trace element-containing fertilizers is widely practiced as a means of raising the trace element concentration in herbage from deficient to satisfactory levels for livestock. With some soils and with some elements these applications also increase yields of herbage. The effects may be wholly or largely upon yield, or wholly or largely upon plant composition, depending upon the element. Thus, in iodine-deficient and selenium-deficient soils, treatment with iodine or selenium salts can markedly increase the concentrations of these elements in the herbage to levels well beyond those required by animals, without increasing or decreasing plant growth (3, 26). With cobalt, the effect is mainly upon the cobalt level in the plants, although growth responses to cobalt-containing fertilizers have been reported on certain soils. On copper-deficient soils, applications of copper usually raise both herbage yields and herbage copper concentrations, but the ability of most plant species to respond to such applications with

TABLE 51

*Effects of Liming on Trace Element Content of Plants Grown on a Granitic Soil*[a]

| Soil treatment | Element content (ppm; dry basis) | | | | Soil pH |
|---|---|---|---|---|---|
| | Co | Ni | Mo | Mn | |
| | | Red clover | | | |
| Unlimed | 0.22 | 1.98 | 0.28 | 58 | 5.4 |
| 115 cwt $CaCO_3$/acre | 0.18 | 1.40 | 1.48 | 41 | 6.1 |
| 216 cwt $CaCO_3$/acre | 0.12 | 1.10 | 1.53 | 40 | 6.4 |
| | | Ryegrass | | | |
| Unlimed | 0.35 | 1.95 | 0.52 | 140 | 5.4 |
| 115 cwt $CaCO_3$/acre | 0.20 | 1.16 | 1.44 | 120 | 6.1 |
| 216 cwt $CaCO_3$/acre | 0.12 | 0.92 | 1.53 | 133 | 6.4 |

[a] From Mitchell (48).

high-copper concentrations is much less with this element than with most other trace elements. Thus Gladstones and Loneragan (24), in a study of a range of crop and pasture species, found substantial increases in zinc and manganese concentrations, following soil treatment with those elements, whereas copper increases were very small following treatment with comparable amounts of that element. On molybdenum-deficient soils, spectacular increases in the yield of pasture legumes can be achieved by small applications of molybdenum, accompanied usually by small increases in the levels of molybdenum in the plant tissues. The latter are of no direct significance to the grazing animal because of its low-molybdenum requirement, but they can be of great importance in areas of high-copper status because of their ability to reduce copper retention in animals to safe levels. Conversely the increased herbage molybdenum concentrations could accentuate or precipitate copper deficiency in animals in areas of low-copper status.

### 3. *Influence of Season and Stage of Maturity*

Plants mature partly in response to internal factors inherent in their genetic constitution and partly in response to external factors, among which climatic and seasonal effects are of major importance. The effects of the latter can, of course, be modified greatly by irrigation and grazing management practices.

The marked decline in whole plant concentration with advancing maturity which occurs with phosphorus and potassium is not paralleled by comparable declines in the trace elements. Whole plant concentrations of these elements may increase, decrease, or show no consistent change with stage of growth, depending upon the element, the plant species, and the soil or seasonal conditions. Most investigators have observed a rise, with advancing maturity of the plant, in the concentrations of silicon, aluminum, and chromium and a fall in copper, zinc, cobalt, nickel, and molybdenum, together with fluctuations in iron, iodine, and manganese not clearly related to the stage of growth (5, 6, 10, 23, 35, 44, 48, 51, 61). Significant species differences in manganese concentrations with stage of maturity were observed by Loneragan and Gladstones (43). Many of the annual crop and pasture species studied, declined in manganese concentration with age, while some remained unchanged or even showed an increase. Lupin species differed from all others studied, both in their outstandingly high-manganese concentrations and in the marked net loss from the tops after flowering. Bisbjerg and Gissel-Nielsen (13) found the selenium concentration of the grain and straw of barley to be lower than in the green plant, but with mustard this decline varied with the oxidation state of the added selenium, i.e., whether selenate or selenite. Remarkable increases in the lead content of mixed pasture herbage at senescence, i.e., when active growth ceases, have been reported by Mitchell and Reith (49) for a range of Scottish soils where no problem of external contamination exists.

Changes in the trace element concentrations of forages or pastures related to the stage of growth of the plants are much more likely to be of significance to the animal in areas of marginal status of particular elements than elsewhere. In these circumstances, management can have important consequences. Seasonal differences in the severity of both deficiency and toxicity states in animals undoubtedly exist. These differences could arise from seasonal changes in (*1*) the moisture status of the soil, affecting trace element availability to plants, as mentioned earlier, (*2*) the botanical composition of the herbage, (*3*) the morphological characteristics of the plants, such as proportions of leaf to stem and seed, (*4*) the palatability and, hence, level of consumption by the animal, (*5*) the relative amounts and proportions of other elements or compounds in the plant that affect the utilization by the animal of the element in question, or (*6*) the chemical forms of the elements in the plants. Little is yet known of the chemical forms and availability to animals of trace elements in plants at different growth stages and at different total concentrations. Observations of the copper status of grazing and housed cattle in the Netherlands indicate that differences exist in the availability of copper from herbage at different growth stages. Hartmans and Bosman (30) found that feeding grass in an older growth stage, despite its lower copper content, resulted in higher liver copper levels in cattle than did young grass.

## IV. Detection and Correction of Deficiencies and Toxicities in Animals and Man

### 1. *Detection or Diagnosis*

Mild trace element deficiencies and toxicities are difficult to diagnose because their effects on the animal are often indistinguishable from those arising from a primary dietary energy deficit and because they are seldom accompanied by specific clinical signs. Loss of appetite and subnormal growth are common manifestations of most trace element deficiencies and excesses. The extent to which these take place and take precedence over other expressions of the dietary abnormality varies greatly with different trace elements. For instance, inappetence and growth failure are not conspicuous features of copper deficiency, as they are in zinc and cobalt (vitamin $B_{12}$) deficiencies. Furthermore, reduced food consumption can arise either as a direct physiological response to some metabolic defect or it can be secondary to a structural abnormality of teeth and joints which limit the animal's ability and willingness to masticate and graze.

The appearance of particular lesions or abnormalities in animals and man has long been used to define the limiting or precipitating factors in diseases of nutritional origin. Clinical and pathological studies have therefore become essential diagnostic tools in the investigation of all trace element deficiencies, imbalances, and toxicities. It is important, nevertheless, to recognize their limitations. Various functional and structural disorders apparent to the clinician and pathologist may be merely the final expression of a defect arising early or late in a chain of metabolic events. Trace element $x$ can be vital at one point in this chain, and trace element $y$ at another. A simple or conditioned deficiency of either element would, therefore, lead to the same end-result in the animal, although the cause would obviously be different. For instance, anemia can be a manifestation of iron, copper, or cobalt deficiency or of selenium, zinc or molybdenum toxicity. Similarly, abnormalities in the size, shape, strength, and composition of bones, amounting in some cases to gross deformities, can occur in copper, manganese, and zinc deficiencies and in fluorine and molybdenum toxicities. Furthermore, the nature as well as the severity of the signs of deficiency or toxicity can vary greatly with the age and sex of the animal.

The diagnostic limitations of clinical and pathological observations, particularly in mild deficiency and toxicity states, can be largely overcome by concurrent biochemical studies of appropriate body tissues and fluids, made possible by the remarkable advances in analytical techniques in recent years. The concentrations of the trace elements in the tissues or of their functional forms such as thyroxine or vitamin $B_{12}$, must be maintained within fairly narrow limits, if the growth, health, and productivity of the animal is to be sustained. Departures from these normal limits, which are now well-defined for most trace elements, therefore, constitute exceedingly useful diagnostic aids, especially as they frequently arise prior to the appearance of clinical or pathological signs of deficiency or toxicity. The organ, tissue, or fluid chosen for analysis varies with the element, but estimations of whole blood or plasma trace element concentrations have exceedingly wide applicability and represent by far the most valuable diagnostic criteria. The normal range of concentration in the blood of man and his domestic animals and the levels which can be regarded as indicative of marginal, mild, or severe deficiencies or toxicities of particular elements are presented in the individual chapters dealing with those elements. The levels of certain trace elements in hair are also of great value in the detection of deficiencies and toxicities, despite considerable individual variability. Hair analyses have proved useful indicators of zinc deficiency and of selenium and arsenic toxicities.

Chemical determination of the trace element levels in human and animal diets and their components provide the best indication of levels of intake in relation to minimum needs and toxic potential. Reasonably satisfactory standards of adequacy and safety, assessed against such criteria as growth, health, performance, and tissue concentrations, have been developed for most of the trace elements. The duration of intake, as well as the magnitude, and the criteria of adequacy employed can be particularly important in any such assessment. In human dietaries the impact of food preparation and cooking upon actual and available daily trace element intakes also requires evaluation. In stall-fed animals the actual feed intakes are usually known, and there is little opportunity for feed selection or discrimination. Total dietary trace element determinations are, therefore, particularly meaningful, so long as these are not confined to a single element and the significance of dietary balance is not neglected. With grazing stock, diagnoses of deficiency or excess based on herbage analyses cannot always be made with the same confidence because the samples analyzed may not represent the material actually consumed and because variations in total concentrations may not correspond with variations in the availability of the element. Soil contamination of herbage can be a further factor of some significance, as with cobalt where soil concentrations are usually very much higher than those of the plants which the soils support.

Attempts to correlate soil type and soil trace element content with the incidence of nutritional disabilities in animals involving these elements have met with varied success. This is hardly surprising in view of the many factors which affect mineral uptake by plants and the levels in mixed herbage. Nevertheless, the correlation with particular soil types can be quite high, as has been shown for cobalt by Mitchell (48) in Scotland, by Kubota (36) in the United States, and by Andrews in New Zealand (5). The geographic patterns of distribution of soils that produce crops having deficient amounts of selenium and excessive concentrations of molybdenum have also been delineated by Kubota and coworkers (37, 40). Correlations of this kind are required for other elements, not only for mapping the location and extent of known problem areas but for predicting their likely occurrence elsewhere.

None of the criteria mentioned in earlier paragraphs is completely satisfactory when considered in isolation. When these are used together and their combined evidence is assessed, deficiency and toxicity states can be securely recognized and confidently predicted, even when these are mild. The ultimate criterion is, of course, the improvement in health or productivity which occurs in response to changes in the intake or utilization by the animal of the element or elements in question.

### 2. *Prevention and Control*

Various direct and indirect means exist for the prevention and control of trace element deficiencies and toxicities in man and animals. The method of choice varies with different elements and with different animal species and their normal feeding practices. For instance, with iodine deficiency in man, direct supplementation through iodination of the domestic salt supplies has proved the most convenient and effective procedure. With iron deficiency in man, the direct approach is also accepted, either through iron fortification of a staple food such as flour and bread or through the prescription of iron tablets to infants and pregnant women during their periods of special iron need. Direct supplementation of the diet of animals normally housed and hand-fed for extended periods is also the cheapest and most convenient means of preventing trace element deficiencies in farm livestock. The element can be incorporated into the whole diet or, more usually, it can be included in prescribed proportions in the mineral mixtures required for other reasons. This is now common practice for zinc and manganese in pig and poultry rations and in some countries for the prevention of selenium deficiency in these species. Direct supplementation may be achieved also by oral dosing, by injections of slowly absorbed organic forms of the element, or by the provision of trace-mineralized salt "licks," available for voluntary consumption. With ruminants, the administration of heavy cobalt or selenium pellets into the reticulorumen constitutes an additional direct means of preventing deficiencies of those elements that are of particular value for grazing animals not normally subject to frequent handling.

Indirect means of controlling trace element intakes by animals, i.e., by raising or lowering the concentrations in the plant materials as grown and consumed are successfully practiced in many environments. The agronomic practices which offer alternative opportunities toward achieving such control have been categorized by Allaway (2) as follows: (*a*) soil selection; (*b*) trace element fertilization; (*c*) soil management, or practices, directed toward increasing or decreasing the availability to plants of the trace elements in the soil, including the use of competitive elements; (*d*) crop selection; and (*e*) crop management and utilization.

The applicability of one or more of these procedures varies with different elements and in different environments.

Trace element fertilization of the soil is widely practiced as a means of raising herbage concentrations of cobalt and copper to satisfactory levels for animals. Difficulties due to dangerously high initial concentrations have so far limited the use of this procedure for controlling selenium deficiency. Furthermore, under sparse grazing or range conditions, trace element fertilization is usually uneconomic and unreliable because of low herbage productivity per unit area, variable uptake of the element, and high application and transport costs. In these circumstances direct supplementation through the use of trace element-mineralized salt licks, periodic oral dosing or injections, or, with cobalt and selenium, the use of heavy pellets are the preferred methods of control. Mineral licks are generally the easiest and cheapest form of treatment but they suffer from two disadvantages, namely irregular consumption and loss of the element due to leaching or volatilization.

Trace element toxicities in animals are usually more difficult to control than deficiencies, especially under grazing or range conditions, although various procedures have been successfully adopted. Soil selection, involving the actual elimination of particularly toxic soils from use for grazing, then becomes an important control procedure. An example of soil selection to avoid excessive concentrations of selenium in crops is described by Anderson (4). In endemic fluorosis areas the only practical form of protection is periodic removal of the animals from dependence upon the fluoridated waters. The control of molybdenosis in animals can be achieved by regular massive oral doses of copper sulfate or periodic injections of this salt. Where the herbage molybdenum levels are lower but still potentially toxic, either moderate dosage with copper or treatment of the soil with copper-containing fertilizers to raise the copper concentrations in the herbage provides a satisfactory means of control. In areas where chronic copper poisoning in sheep occurs as a consequence of normal to high copper, accompanied by very low levels of molybdenum in the pastures, the provision of molybdate-containing salt licks, to achieve a better Cu-Mo dietary ratio, can reduce or eliminate the incidence of the disease. Finally, the "dilution" technique offers some possibilities for control. This involves the importation of feeds known to be low in the toxic element which can be used in conjunction with the local toxic feeds, so reducing overall intakes to safe levels.

## REFERENCES

1. S. N. Adams and J. L. Honeysett, *Aust. J. Agr. Res.* **15**, 357 (1964).
2. W. H. Allaway, *Advan. Agron.* **20**, 235 (1968).
3. W. H. Allaway, D. P. Moore, J. E. Oldfield, and O. H. Muth, *J. Nutr.* **88**, 411 (1966).
4. M. S. Anderson, *U.S., Dep. Agr., Agr. Handb.* **200**, 53 (1961); cited by Allaway (2).
5. E. D. Andrews, *N. Z. J. Agr.* **92**, 239 (1956).
6. M. Anke, *Z. Acker Pflanzenbau* **112**, 113 (1961).
7. G. Baker, L. H. P. Jones, and I. D. Wardrop, *Aust. J. Agr. Res.* **12**, 462 (1961).
8. A. B. Beck, *Aust. J. Exp. Agr. Anim. Husb.* **2**, 40 (1962).
9. K. C. Beeson, *in* "Trace Elements" (C. A. Lamb, O. G. Bentley, and J. M. Beattie, eds.), p. 67. Academic Press, New York, 1958.
10. K. C. Beeson and A. H. McDonald, *Agron. J.* **43**, 589 (1951).
11. K. C. Beeson, L. C. Gray, and M. G. Adams, *J. Amer. Soc. Agron.* **39**, 356 (1947).
12. K. C. Beeson, V. A. Lazar, and S. G. Boyce, *Ecology* **36**, 155 (1955).
13. B. Bisbjerg and G. Gissel-Nielsen, *Plant Soil* **31**, 287 (1969).
14. H. J. M. Bowen and J. A. Dymond, *Proc. Roy. Soc., Ser. B* **144**, 355 (1955).
15. G. W. Butler and A. T. Johns, *J. Aust. Inst. Agr. Sci.* **27**, 123 (1961).
16. P. D. Cadell, *Aust. Dent. J.* **9**, 32 (1964).
17. H. L. Cannon and J. M. Bowles, *Science* **137**, 765 (1962).
18. R. L. Carroll, *J. Amer. Med. Assoc.* **198**, 267 (1966).
19. A. T. Dick, C. W. E. Moore, and J. B. Bingley, *Aust. J. Agr. Res.* **4**, 44 (1953).
20. G. A. Fleming and T. Walsh, *Proc. Roy. Irish Acad., Sect. B* **58**, 151 (1957).
21. R. J. W. Gartner and J. O. Twist, *Aust. J. Exp. Agr. Anim. Husb.* **8**, 210 (1968).
22. J. S. Gladstones and D. P. Drover, *Aust. J. Exp. Agr. Anim. Husb.* **2**, 46 (1962).
23. J. S. Gladstones and J. F. Loneragan, *Aust. J. Agr. Res.* **12**, 427 (1967).
24. J. S. Gladstones and J. F. Loneragan, *Proc. 11th Int. Grassland Congr., 1969* Sect. iv/18 (1970).
25. V. M. Goldschmidt, "Geochemistry," Oxford Univ. Press (Clarendon), London and New York, 1954.
26. G. P. Gurevich, *Izv. Akad. Nauk SSSR., Ser. Biol.* No. 5, p. 791 (1962); *Fed. Proc., Fed. Amer. Soc. Exp. Biol.* **23**, Trans. Suppl., T511 (1964).
27. D. M. Hadjimarkos, *J. Pediat.* **70**, 967 (1967).
28. D. M. Hadjimarkos, *Caries Res.* **3**, 14 (1969).
29. W. J. Hartley, A. B. Grant, and C. Drake, *N. Z. J. Agr.* **101**, 343 (1960).
30. J. Hartmans and M. S. N. Bosman, *Proc. 1st Int. Symp. Trace Element Metab. Anim., 1969* p. 362 (C. F. Mills, ed.) Livingstone, Edinburgh, 1970.
31. C. E. Hercus, H. H. Aitken, H. M. Thompson, and G. H. Cox, *J. Hyg.* **31**, 493 (1931).
32. C. E. Hercus, W. N. Benson, and C. L. Carter, *J. Hyg.* **24**, 321 (1925).
33. R. E. T. Hewat and D. F. Eastcott, Rep. Med. Res. Council, New Zealand, 1955.
34. J. M. Johnson and G. W. Butler, *Physiol. Plant.* **10**, 100 (1957).
35. M. Kirchgessner, G. Merz, and W. Oelschläger, *Arch. Tierernaehr.* **10**, 414 (1960).
36. J. Kubota, *Soil Sci. Soc. Amer., Proc.* **28**, 246 (1964); *Soil Sci.* **106**, 122 (1968).
37. J. Kubota, W. H. Allaway, D. L. Carter, E. E. Cary, and V. A. Lazar, *J. Agr. Food Chem.* **15**, 448 (1967).
38. J. Kubota, E. R. Lemon, and W. H. Allaway, *Soil Sci. Soc. Amer., Proc.* **27**, 679 (1963).

39. J. Kubota, V. A. Lazar, and F. L. Losee, *Arch. Environ. Health* **16,** 788 (1966).
40. J. Kubota, V. A. Lazar, G. H. Simonsen, and W. W. Hill, *Soil Sci. Soc. Amer., Proc.* **31**, 667 (1967).
41. H. J. Lee, *Proc. Spec. Conf. Agr., 1949* p. 262 (1951).
42. C. D. Legon, *Brit. Med. J.* **2,** 700 (1952).
43. J. F. Loneragan and J. S. Gladstones, unpublished data (1970).
44. G. M. Loper and D. Smith, Res. Rep. No. 8. Dept. Agron., University of Wisconsin, Madison, Wisconsin, 1961.
45. T. G. Ludwig and B. G. Bibby, *Caries Res.* **3**, 32 (1969).
46. T. G. Ludwig, P. D. Cadell, and R. S. Malthus, *Int. Dent. J.* **14,** 443 (1964).
47. T. G. Ludwig, W. B. Healy, and F. L. Losee, *Nature (London)* **186**, 695 (1960); F. L. Losee, W. B. Healy, and T. G. Ludwig, *U.S. Nav. Med. Res. Inst. Res. Rep.* No. 9 (1960).
48. R. L. Mitchell, *Research (London)* **10**, 357 (1957).
49. R. L. Mitchell and J. W. S. Reith, *J. Sci. Food Agr.* **17**, 437 (1966).
50. A. E. Nizel and B. G. Bibby, *J. Amer. Dent. Assoc.* **31**, 1619 (1944).
51. D. Purves, *Plant Soil* **26**, 380 (1967).
52. T. S. Sadavisan, C. B. Gulochana, V. T. John, M. R. Subbarain, and C. Gopalan, *Curr. Sci.* **29**, 86 (1960).
53. J. L. Saunders, *Trans. Roy. Soc. N. Z.* **75**, 57 (1945).
54. H. A. Schroeder, *J. Amer. Med. Assoc.* **195**, 125 (1966).
55. R. J. Shamberger and D. V. Frost, *Can. Med. Assoc. J.* **100**, 682 (1969).
56. G. W. Smith, "Soil and Cancer," Medical Press, London, 1960.
57. B. Thomas, A. Thompson, V. A. Oyenuga, and R. H. Armstrong, *Emp. J. Exp. Agr.* **20**, 10 (1952).
58. S. W. Tromp and J. C. Diehl, *Brit. J. Cancer* **9**, 349 (1955).
59. E. J. Underwood, T. J. Robinson, and D. H. Curnow, *J. Dep. Agr., West. Aust.* **24**, 259 (1947).
60. J. H. Watkinson and E. B. Davies, *N. Z. J. Agr. Res.* **10**, 122 (1967).
61. N. Wells, *N. Z. J. Sci. Technol., Sect. B* **37**, 473 (1956).

# AUTHOR INDEX

Numbers in parentheses are reference numbers and indicate that an author's work is referred to although his name is not cited in the text.

Numbers in italics show the page on which the complete reference is listed.

## A

## B

## C

## D

## G

## L

## M

## Q



## R

## S

## U

## Y



## Z

# SUBJECT INDEX

## A

## D



## E

# I